Lecture Notes in Computer Science 4781

Commenced Publication in 1973
Founding and Former Series Editors:
Gerhard Goos, Juris Hartmanis, and Jan van Leeuwen

Editorial Board

David Hutchison
  Lancaster University, UK
Takeo Kanade
  Carnegie Mellon University, Pittsburgh, PA, USA
Josef Kittler
  University of Surrey, Guildford, UK
Jon M. Kleinberg
  Cornell University, Ithaca, NY, USA
Friedemann Mattern
  ETH Zurich, Switzerland
John C. Mitchell
  Stanford University, CA, USA
Moni Naor
  Weizmann Institute of Science, Rehovot, Israel
Oscar Nierstrasz
  University of Bern, Switzerland
C. Pandu Rangan
  Indian Institute of Technology, Madras, India
Bernhard Steffen
  University of Dortmund, Germany
Madhu Sudan
  Massachusetts Institute of Technology, MA, USA
Demetri Terzopoulos
  University of California, Los Angeles, CA, USA
Doug Tygar
  University of California, Berkeley, CA, USA
Moshe Y. Vardi
  Rice University, Houston, TX, USA
Gerhard Weikum
  Max-Planck Institute of Computer Science, Saarbruecken, Germany

Lecture Notes in Computer Science

Guoping Qiu Clement Leung
Xiangyang Xue Robert Laurini (Eds.)

Advances in
Visual Information Systems

9th International Conference, VISUAL 2007
Shanghai, China, June 28-29, 2007
Revised Selected Papers

Springer

Volume Editors

Guoping Qiu
University of Nottingham
School of Computer Science
Jubilee Campus, Nottingham NG8 1BB, UK
E-mail: qiu@cs.nott.ac.uk

Clement Leung
Hong Kong Baptist University
Department of Computer Science
Kowloon Tong, Hong Kong, China
E-mail: clement@comp.hkbu.edu.hk

Xiangyang Xue
Fudan University
Department of Computer Science & Engineering
Shanghai Key Laboratory of Intelligent Information Processing
Shanghai 200433, China
E-mail: xyxue@fudan.edu.cn

Robert Laurini
Institut National des Sciences Appliquées (INSA) de Lyon
Laboratoire d'InfoRmatique en Image et Systèmes d'information (LIRIS)
Bât. Blaise Pascal, 7 av. Jean Capelle, 69621 Villeurbanne, France
E-mail: Robert.Laurini@insa-lyon.fr

Library of Congress Control Number: 2007940861

CR Subject Classification (1998): I.4, I.5, I.2.6-10, I.3, H.3, H.5, H.2

LNCS Sublibrary: SL 6 – Image Processing, Computer Vision, Pattern Recognition, and Graphics

ISSN   0302-9743
ISBN-10 3-540-76413-5 Springer Berlin Heidelberg New York
ISBN-13 978-3-540-76413-7 Springer Berlin Heidelberg New York

This work is subject to copyright. All rights are reserved, whether the whole or part of the material is concerned, specifically the rights of translation, reprinting, re-use of illustrations, recitation, broadcasting, reproduction on microfilms or in any other way, and storage in data banks. Duplication of this publication or parts thereof is permitted only under the provisions of the German Copyright Law of September 9, 1965, in its current version, and permission for use must always be obtained from Springer. Violations are liable to prosecution under the German Copyright Law.

Springer is a part of Springer Science+Business Media

springer.com

© Springer-Verlag Berlin Heidelberg 2007
Printed in Germany

Typesetting: Camera-ready by author, data conversion by Scientific Publishing Services, Chennai, India
Printed on acid-free paper   SPIN: 12177349   06/3180   5 4 3 2 1 0

Preface

The Visual Information Systems International Conference series is designed to provide a forum for researchers and practitioners from diverse areas of computing including computer vision, databases, human–computer interaction, information security, image processing, information visualization and mining, as well as knowledge and information management to exchange ideas, discuss challenges, present their latest results and to advance research and development in the construction and application of visual information systems. Following previous conferences held in Melbourne (1996), San Diego (1997), Amsterdam (1999), Lyon (2000), Taiwan (2002), Miami (2003), San Francisco (2004) and Amsterdam (2005), the Ninth International Conference on Visual Information Systems, VISUAL2007, was held in Shanghai, China, June 28–29, 2007.

Over the years, the visual information systems paradigm continues to evolve, and the unrelenting exponential growth in the amount of digital visual data underlines the escalating importance of how such data are effectively managed and deployed. VISUAL2007 received 117 submissions from 15 countries and regions. Submitted full papers were reviewed by more than 60 international experts in the field. This volume collects 54 selected papers presented at VISUAL2007. Topics covered in these papers include image and video retrieval, visual biometrics, intelligent visual information processing, visual data mining, ubiquitous and mobile visual information systems, visual semantics, 2D/3D graphical visual data retrieval and applications of visual information systems.

Two distinguished researchers delivered keynote talks at VISUAL2007. Wei-Ying Ma from Microsoft Research Asia gave a talk on "The Challenges and Opportunities of Mining Billions of Web Images for Search and Advertising." Michael Lew from Linden University, The Netherlands, gave a talk on "Visual Information Retrieval: Grand Challenges and Future Directions."

We would like to thank Hong Lu, Yue-Fei Guo and their team for the significant organization effort that they put in. We are grateful to the Department of Computer Science and Engineering, Fudan University for hosting the conference. In particular, we would like to express our gratitude to members of the Program Committee for their part in reviewing the papers to ensure a high-quality and successful conference.

September 2007

Guoping Qiu
Clement Leung
Xiang-Yang Xue
Robert Laurini

Organization

General Chairs

Robert Laurini
INSA of Lyon, France

Xiang-Yang Xue
Fudan University, China

Technical Program Chairs

Clement Leung
Hong Kong Baptist University, Hong Kong

Guoping Qiu
University of Nottingham, UK

Program Committee

Ching-chih Chen, Simmons College, USA
Yixin Chen, University of Mississippi, USA
Mingmin Chi, Fudan University, China
Zheru Chi, Hong Kong Polytechnic University, Hong Kong
Arjen P. de Vries, CWI, The Netherlands
Martin Dzbor, Knowledge Media Institute, The Open University, UK
Peter Enser, University of Brighton, UK
Jianping Fan, University of North Carolina at Charlotte, USA
Graham Finlayson, University of East Anglia, UK
Xiaodong Gu, Thomson R&D Centre, Beijing, China
Alan Hanjalic, Delft University of Technology, The Netherlands
Xian-Sheng Hua, Microsoft Research Asia, Beijing, China
Jesse Jin, Newcastle University, Australia
Joemon Jose, University of Glasgow, UK
Irwin King, The Chinese University of Hong Kong, Hong Kong
Markus Koskela, Dublin City University, Ireland
Igor V. Kozintsev, Intel Microprocessor Research Lab, USA
Jorma Laaksonen, Helsinki University of Technology, Finland
Kenneth Lam, Hong Kong Polytechnic University, Hong Kong
Robert Laurini, INSA, Lyon, France
Bongshin Lee, Microsoft Research, Redmond, USA
Wee-Kheng Leow, National University of Singapore, Singapore

Clement Leung, Victoria University, Melbourne, Australia
Michael Lew, University of Leiden, The Netherlands
Ze-Nian Li, Simon Fraser University, Canada
Rainer W. Lienhart, University of Mannheim, Germany
Tie-Yan Liu, Microsoft Research Asia, China
Hong Lu, Fudan University, China
Stephane Marchand-Maillet, University of Geneva, Switzerland
Graham Martin, Warwick University, UK
Jean Martinet, National Institute of Informatics, Japan
Dalibor Mitrovic, Vienna University of Technology, Austria
Keith Nesbitt, Charles Stuart University, Australia
Chong-Wah Ngo, City University of Hong Kong, Hong Kong
Fernando Pereira, Institute of Telecommunications, Portugal
Tony Pridmore, University of Nottingham, UK
Guoping Qiu, University of Nottingham, UK
Mark Sanderson, University of Sheffield, UK
Bertrand Le Saux, Ecole Normale Supérieure de Cachan, France
Gerald Schaefer, Aston University, UK
Raimondo Schettini, DISCO, University of Milano-Bicocca, Italy
Linlin Shen, Shenzhen University, China
Timothy Shih, Tamkang University, Taiwan
Yap-Peng Tan, Nanyang Technological University, Singapore
Qi Tian, University of Texas at San Antonio, USA
Martin Varley, University of Central Lancashire, UK
Giuliana Vitiello, University of Salerno, Italy
James Z. Wang, The Pennsylvania State University, USA
Roland Wilson, Warwick University, UK
Raymond Wong, National ICT Australia, Australia
Fei Wu, Zhejiang University, China
Jian Kang Wu, Institute for Infocomm Research, Singapore
Yihong Wu, Institute of Automation, Chinese Academy of Science, China
Xiang-Yang Xue, Fudan University, China
Pong Chi Yuen, Hong Kong Baptist University, Hong Kong
Matthias Zeppelzauer, Vienna University of Technology, Austria
Zhi-Hua Zhou, Nanjing University, China

Table of Contents

Keynote Paper

Visual Information Retrieval – Future Directions and Grand
Challenges 1
   Michael Lew

Image and Video Retrieval

Approximation-Based Keypoints in Colour Images – A Tool for
Building and Searching Visual Databases 5
   Andrzej Sluzek

A Knowledge Synthesizing Approach for Classification of Visual
Information 17
   Le Dong and Ebroul Izquierdo

Image Similarity – From Fuzzy Sets to Color Image Applications 26
   *Mike Nachtegael, Stefan Schulte, Valerie De Witte,
Tom Mélange, and Etienne E. Kerre*

A Semi-automatic Feature Selecting Method for Sports Video Highlight
Annotation.. 38
   Yanran Shen, Hong Lu, and Xiangyang Xue

Face Image Retrieval System Using TFV and Combination of
Subimages 49
   Daidi Zhong and Irek Defée

Near-Duplicate Detection Using a New Framework of Constructing
Accurate Affine Invariant Regions 61
   Li Tian and Sei-ichiro Kamata

Where Are Focused Places of a Photo? 73
   Zhijun Dai and Yihong Wu

Region Based Image Retrieval Incorporated with Camera Metadata 84
   Jie Ma, Hong Lu, and Yue-Fei Guo

Empirical Investigations on Benchmark Tasks for Automatic Image
Annotation 93
   Ville Viitaniemi and Jorma Laaksonen

Automatic Detection and Recognition of Players in Soccer Videos 105
   *Lamberto Ballan, Marco Bertini, Alberto Del Bimbo, and
Walter Nunziati*

A Temporal and Visual Analysis-Based Approach to Commercial
Detection in News Video 117
  Shijin Li, Yue-Fei Guo, and Hao Li

Salient Region Filtering for Background Subtraction 126
  Wasara Rodhetbhai and Paul H. Lewis

A Novel SVM-Based Method for Moving Video Objects Recognition 136
  Xiaodong Kong, Qingshan Luo, and Guihua Zeng

Image Classification and Indexing by EM Based Multiple-Instance
Learning 146
  Hsiao T. Pao, Yeong Y. Xu, Shun C. Chuang, and Hsin C. Fu

Visual Biometrics

Palm Vein Extraction and Matching for Personal Authentication 154
  Yi-Bo Zhang, Qin Li, Jane You, and Prabir Bhattacharya

A SVM Face Recognition Method Based on Optimized Gabor
Features 165
  Linlin Shen, Li Bai, and Zhen Ji

Palmprint Identification Using Pairwise Relative Angle and EMD 175
  Fang Li, Maylor K.H. Leung, and Shirley Z.W. Tan

Finding Lips in Unconstrained Imagery for Improved Automatic Speech
Recognition 185
  Xiaozheng Jane Zhang, Higinio Ariel Montoya, and Brandon Crow

Intelligent Visual Information Processing

Feature Selection for Identifying Critical Variables of Principal
Components Based on K-Nearest Neighbor Rule 193
  Yun Li and Bao-Liang Lu

Denoising Saliency Map for Region of Interest Extraction 205
  Yandong Guo, Xiaodong Gu, Zhibo Chen, Quqing Chen, and
  Charles Wang

Cumulative Global Distance for Dimension Reduction in Handwritten
Digits Database 216
  Mahdi Yektaii and Prabir Bhattacharya

A New Video Compression Algorithm for Very Low Bandwidth Using
Curve Fitting Method 223
  Xianping Fu, Dequn Liang, and Dongsheng Wang

The Influence of Perceived Quality by Adjusting Frames Per Second
and Bits Per Frame Under the Limited Bandwidth 230
   Huey-Min Sun, Yung-Chuan Lin, and LihChyun Shu

An Evolutionary Approach to Inverse Gray Level Quantization 242
   *Ivan Gerace, Marcello Mastroleo, Alfredo Milani, and
Simona Moraglia*

Visual Data Mining

Mining Large-Scale News Video Database Via Knowledge
Visualization 254
   Hangzai Luo, Jianping Fan, Shin'ichi Satoh, and Xiangyang Xue

Visualization of the Critical Patterns of Missing Values in Classification
Data 267
   Hai Wang and Shouhong Wang

Visualizing Unstructured Text Sequences Using Iterative Visual
Clustering 275
   Qian You, Shiaofen Fang, and Patricia Ebright

Enhanced Visual Separation of Clusters by M-Mapping to Facilitate
Cluster Analysis 285
   Ke-Bing Zhang, Mehmet A. Orgun, and Kang Zhang

Multimedia Data Mining and Searching Through Dynamic Index
Evolution 298
   Clement Leung and Jiming Liu

Ubiquitous and Mobile Visual Information Systems

Clustering and Visualizing Audiovisual Dataset on Mobile Devices in a
Topic-Oriented Manner 310
   Lei Wang, Dian Tjondrongoro, and Yuee Liu

Adaptive Video Presentation for Small Display While Maximize Visual
Information 322
   *Yandong Guo, Xiaodong Gu, Zhibo Chen, Quqing Chen, and
Charles Wang*

An Efficient Compression Technique for a Multi-dimensional Index in
Main Memory 333
   *Joung-Joon Kim, Hong-Koo Kang, Dong-Suk Hong, and
Ki-Joon Han*

RELT – Visualizing Trees on Mobile Devices 344
   Jie Hao, Kang Zhang, and Mao Lin Huang

Auto-generation of Geographic Cognitive Maps for Browsing Personal Multimedia 358
Hyungeun Jo, Jung-hee Ryu, and Chang-young Lim

Semantics

Automatic Image Annotation for Semantic Image Retrieval............ 369
Wenbin Shao, Golshah Naghdy, and Son Lam Phung

Collaterally Cued Labelling Framework Underpinning Semantic-Level Visual Content Descriptor... 379
Meng Zhu and Atta Badii

Investigating Automatic Semantic Processing Effects in Selective Attention for Just-in-Time Information Retrieval Systems 391
John Meade and Fintan Costello

News Video Retrieval by Learning Multimodal Semantic Information ... 403
Hui Yu, Bolan Su, Hong Lu, and Xiangyang Xue

2D/3D Graphical Visual Data Retrieval

Visualization of Relational Structure Among Scientific Articles 415
Quang Vinh Nguyen, Mao Lin Huang, and Simeon Simoff

3D Model Retrieval Based on Multi-Shell Extended Gaussian Image 426
Dingwen Wang, Jiqi Zhang, Hau-San Wong, and Yuanxiang Li

Neurovision with Resilient Neural Networks 438
Erkan Beşdok

Applications of Visual Information Systems

Visual Information for Firearm Identification by Digital Holography 445
Dongguang Li

GIS-Based Lunar Exploration Information System in China 453
Sheng-Bo Chen and Shu-Xin Bao

Semantic 3D CAD and Its Applications in Construction Industry – An Outlook of Construction Data Visualization 461
Zhigang Shen, Raja R.A. Issa, and Linxia Gu

A Fast Algorithm for License Plate Detection........................ 468
Vahid Abolghasemi and Alireza Ahmadyfard

Applying Local Cooccurring Patterns for Object Detection from Aerial Images 478
  Wenjing Jia, David Tien, Xiangjian He, Brian A. Hope, and Qiang Wu

Enticing Sociability in an Intelligent Coffee Corner 490
  Khairun Fachry, Ingrid Mulder, Henk Eertink, and Maddy Janse

Geometric and Haptic Modelling of Textile Artefacts 502
  Fazel Naghdy, Diana Wood Conroy, and Hugh Armitage

A Toolkit to Support Dynamic Social Network Visualization........... 512
  Yiwei Cao, Ralf Klamma, Marc Spaniol, and Yan Leng

The Predicate Tree – A Metaphor for Visually Describing Complex Boolean Queries 524
  Luca Paolino, Monica Sebillo, Genoveffa Tortora, and Giuliana Vitiello

Potentialities of Chorems as Visual Summaries of Geographic Databases Contents... 537
  Vincenzo Del Fatto, Robert Laurini, Karla Lopez, Rosalva Loreto, Françoise Milleret-Raffort, Monica Sebillo, David Sol-Martinez, and Giuliana Vitiello

Compound Geospatial Object Detection in an Aerial Image 549
  Yi Xiao, Brian A. Hope, and David Tien

Texture Representation and Retrieval Using the Causal Autoregressive Model 559
  Noureddine Abbadeni

An Approach Based on Multiple Representations and Multiple Queries for Invariant Image Retrieval 570
  Noureddine Abbadeni

Author Index 581

Visual Information Retrieval – Future Directions and Grand Challenges

Michael Lew

LIACS Media Lab, Leiden University
mlew@liacs.nl

Abstract. We are at the beginning of the digital Age of Information, a digital Renaissance allowing us to communicate, share, and learn in novel ways and resulting in the creation of new paradigms. However, having access to all of the knowledge in the world is pointless without a means to search for it. Visual information retrieval is poised to give access to the myriad forms of images and video, comprising knowledge from individuals and cultures to scientific fields and artistic communities. In this paper I summarize the current state of the field and discuss promising future directions and grand challenges.

Keywords: Visual Information Retrieval, Grand Challenges.

1 Visual Information Retrieval

Millennia ago, the Egyptians created the Library of Alexandria, an attempt to collect all of the knowledge in the world and store it in one vast library. The people who were responsible for collecting, indexing, and storing the knowledge were the earliest members of our field. They had the challenge of preserving the knowledge of their culture for future generations. This paper is a summary of the recent work in the field of Visual Information Retrieval (VIR) as described in the survey found in [1].

Today, we live in a world flooded with limitless data from every corner of the globe. The goal of the field of Visual Information Retrieval (VIR) is to develop new paradigms and theories for how to collect, store, analyze, search, and summarize visual information [2,3]. Raw bits are not enough. We must convert the bits into semantic concepts, meaningful translations of the data, and thereby bridge the semantic gap between the computer and humans.

Even though VIR encompasses many areas such as new features, selection algorithms, and similarity measures [11,12,13] and aspects such as high performance data structures and algorithms for very large data repositories, the focus in this article is on human centered aspects which are crucial to the future success of VIR. Next we discuss visual concept detection.

Early systems for face detection came from the field of Computer Vision and were limited to common assumptions such as (1) simple background such as a single color; (2) only one face per image; (3) no facial expressions; (4) no glasses; and (5) no occlusions over any part of the face. While the early face detection systems [16] had intriguing theory underlying the methods, the systems were not useful in VIR because the assumptions were too strict. In the past ten years major leaps forward have occurred as our field has addressed the difficult challenge of detecting objects in complex backgrounds. By the mid 90s, VIR researchers had succeeded in creating robust face detection systems founded on new paradigms such as information theory [7,8] which had eliminated the previous limitations and assumptions.

The next step was to generalize the face detector system to different kinds of visual concepts. Instead of only detecting faces, one would want to detect trees, sky, etc. Using similar methods as in the detection of faces, by the early 21st century, researchers also created robust systems for detection of simple visual concepts [4-6] such as sky, water, trees, and rocks.

The importance of the detection of visual concepts can not be understated. By automatically detecting visual concepts in imagery, we are directly bridging the semantic gap, bringing meaning to raw or senseless data. The visual concepts are the words upon which we can then build languages to describe and query for knowledge.

While research was indeed progressing forward, our field was still in the early stages of formation. In the late 90s conferences such as the *International Conference on Visual Information Systems* (VISUAL) were primarily focused on scientific researchers sharing their work with other researchers. To make our field stronger it was felt that we had to address at least two new challenges. First, it was clear that we needed researchers and practitioners to share their collective knowledge. Second, although we had succeeded in creating several promising systems, it was not clear how to perform quantitative evaluation. How could we scientifically say that one system outperformed another system or compare systems? These two challenges sparked the founding of several important meetings, notably, the International Conference on Image and Video Retrieval (CIVR) [3,14] and NIST TRECVID [15]. The mission of CIVR was *"to illuminate the state of the art in image and video retrieval between researchers and practitioners throughout the world."* The goal of TRECVID was *"to encourage research in information retrieval by providing a large test collection, uniform scoring procedures, and a forum for organizations interested in comparing their results."*

2 Current Research

Thus, in the early 21st century we were at a stage where Visual Information Retrieval had made substantial progress [1], but there were still major problems ahead. Beyond detecting faces, it is important to consider the temporal aspect such as understanding facial expressions such as emotional states like *happy* or *sad* in temporal sequences [9,10], or more generally, detecting visual concepts in time dependent imagery. The early visual concept detectors also had accuracy deterioration as the number of concepts grew. Reasonable results could be found at ten to twenty concepts, but not higher. If we could develop a visual concept detector which could grasp a thousand concepts, that might be sufficient for the development of a general language for describing the world.

In the area of interactive search [1], considerable research has addressed the algorithmic and learning issues, which have included varying feedback ratings, different architectures for well known learning algorithms such as Support Vector

Machines and Neural Networks, integration of multiple modalities, etc. However, few have addressed the problem of finding subimages which can occur when a user is interested in one part of an image, but not the rest. Furthermore, human users do not want to manually classify hundreds to thousands of examples for every search session, which has been called the small training set problem - How can we optimize the interactive search process for a small amount of interaction feedback?

Overall, recent research has been quite promising in terms of both incremental advances and in proposing new theories and paradigms. One of the next anticipated steps is for contributions in content based visual information retrieval to be used in a widely used application such as the next Google. Toward that end, we as a community would need to continue our work toward robust systems which work effectively under a wide range of real world imagery.

3 Grand Challenges

Current visual concept detection algorithms are only effective on small numbers of visual concepts, which can be said to be promising but is certainly insufficient for a generally useful vocabulary. Therefore, the first grand challenge is to create visual concept detectors which work robustly on hundreds of visual concepts instead of twenty.

The second grand challenge is to perform multi-modal analysis which exploits the synergy between the various media and uses knowledge sources in a deeper way. An example would be Wikipedia as an extensive knowledge source which we could tap into for fundamental knowledge of the world.

Beyond browsing and searching, how do we develop systems to allow users to gain insight and education? This is the third challenge which is to develop systems which focus on human-centered interaction not just toward searching but also toward gaining insight and knowledge.

The fourth grand challenge stems from the core interactive dialog between a librarian and a user. In some way the librarian asks context dependent, relevant, and intelligent questions to determine what the user really wants. How can we achieve this deeper level of relevance feedback between computers and humans?

In the early 21st century, TRECVID was an excellent example of combining researchers and users toward scientific benchmarking and evaluation. It is frequently forgotten that test sets can be used not just for benchmarking, but also for improving a system, revealing insights into strengths and weaknesses. The fifth grand challenge is to develop test databases and situations with emphasis on truly representative test sets, usage patterns, and aiding the researcher in improving his algorithm. For example, how can we make a good test set for relevance feedback?

4 Final Remarks

Regarding the future, the most important strength of our community is the ongoing sharing of knowledge between researchers and practitioners. As long as researchers keep the systems centered on humans, VIR will continue to bring significant advances to the world.

Acknowledgments

I would like to thank Leiden University, the Dutch National Science Foundation (NWO), and the BSIK/BRICKS research funding programs for their support of my work.

References

[1] Lew, M.S., Sebe, N., Djeraba, C., Jain, R.: Content-based Multimedia Information Retrieval: State-of-the-art and Challenges. ACM Transactions on Multimedia Computing, Communication, and Applications 2(1), 1–19 (2006)
[2] Lew, M.S.: Principles of Visual Information Retrieval. Springer, London (2001)
[3] Sebe, N., Lew, M.S., Zhou, X., Huang, T.S., Bakker, E.: The State of the Art in Image and Video Retrieval. In: Bakker, E.M., Lew, M.S., Huang, T.S., Sebe, N., Zhou, X.S. (eds.) CIVR 2003. LNCS, vol. 2728, pp. 1–8. Springer, Heidelberg (2003)
[4] Lew, M.S.: Next Generation Web Searches for Visual Content, pp. 46–53. IEEE Computer, Los Alamitos (2000)
[5] Buijs, J.M., Lew, M.S.: Learning Visual Concepts. In: ACM-MM. Proceedings of the Seventh ACM International Conference on Multimedia, vol. 2, pp. 5–7 (1999)
[6] Lew, M.S., Sebe, N.: Visual Websearching Using Iconic Queries. In: CVPR. Proceedings of the IEEE Conference on Computer Vision and Pattern Recognition, vol. 2, pp. 2788–2789 (2000)
[7] Lew, M.S., Huijsmans, N.: Information Theory and Face Detection. In: ICPR. Proceedings of the International Conference on Pattern Recognition, Vienna, Austria, August 25-30, pp. 601–605 (1996)
[8] Lew, M.S.: Information theoretic view-based and modular face detection. In: Proceedings of the IEEE Face and Gesture Recognition Conference, Killington, VT, pp. 198–203 (1996)
[9] Sebe, N., Lew, M.S., Cohen, I., Sun, Y., Gevers, T., Huang, T.S.: Authentic Facial Expression Analysis. In: Proceedings of the International Conference on Automatic Face and Gesture Recognition, Seoul, Korea, pp. 517–522 (May 2004)
[10] Cohen, I., Sebe, N., Garg, A., Lew, M.S., Huang, T.S.: Facial expression recognition from video sequences. In: ICME. Proceedings of the IEEE International Conference Multimedia and Expo, Lausanne, Switzerland, vol. I, pp. 641–644 (2002)
[11] Sebe, N., Lew, M.S., Huijsmans, N.: Toward Improved Ranking Metrics, IEEE Transactions on Pattern Analysis and Machine Intelligence, 1132–1143 (October 2000)
[12] Lew, M.S., Huang, T.S., Wong, K.W.: Learning and Feature Selection in Stereo Matching, IEEE Transactions on Pattern Analysis and Machine Intelligence, pp. 869–881 (September 1994)
[13] Sebe, N., Lew, M.S.: Wavelet Based Texture Classification. In: Proceedings of the 15th International Conference on Pattern Recognition, Barcelona, Spain, vol III, pp. 959–962 (2000)
[14] Webpage: http://www.civr.org
[15] Webpage: http://en.wikipedia.org/wiki/TRECVID
[16] Yang, M., Kriegman, D.J., Ahuja, N.: Detecting Faces in Images: A Survey. IEEE Transactions on Pattern Analysis and Machine Intelligence 24, 34–58 (2002)

Approximation-Based Keypoints in Colour Images – A Tool for Building and Searching Visual Databases

Andrzej Sluzek[1,2]

[1] Nanyang Technological University, School of Computer Engineering,
N4 Nanyang Avenue, Singapore 639798
[2] SWPS, Warszawa, Poland
assluzek@ntu.edu.sg

Abstract. The paper presents a framework for information retrieval in visual databases containing colour images. The concept of *approximation-based keypoints* is adapted to colour images; building and detection of such keypoints are explained in details. The issues of matching images are only briefly highlighted. Finally, the idea of higher-level keypoints is proposed.

Keywords: Colour images, keypoints, visual information retrieval, moment methods, shape matching.

1 Introduction

Visual exploration of unknown environments (a typical task for robotic vision) and search operations in visual databases have similar foundations. In both cases the objective is to extract visually relevant fragments from a wider (or very wide) visual content. Such systems unavoidably emulate human vision which is considered the ultimate benchmark. Although systems with visual capabilities comparable to human skills are apparently unachievable in the near future, but in realistically limited scenarios vision systems may significantly replace/supplement humans. Browsing visual databases seams to be one of such scenarios. First, visual databases are generally too large for human inspection. Secondly, search results are usually further verified by humans so that false returns are not as critical as in other applications of machine vision. Actually, visual information retrieval becomes one of the fundamental issues in information systems and numerous methods and techniques have been recently developed in this area (e.g. [1], [2], [3]). This paper is another attempt in the same direction. The proposed framework has been originally developed for vision-based robotic navigation. However, the applicability of the method to visual information retrieval has been quickly identified.

The proposed method is applicable to certain types of visual search, namely for retrieval of images/frames containing the same/similar objects as the query image, or for retrieval of images/frames containing objects already known by the system. Our works are motivated by three facts:

  (a) Humans generally recognise known objects as collections of local visual clues (see Fig. 1) which are subsequently, if enough of them are perceived, "interpolated" into the object (sometimes optical illusions may happen, however). Although several theories exist (e.g. [4]) explaining the human perception of objects, the conclusion is that systems emulating human vision should also use local features as the fundamental tool for matching the scene content to the models of known objects/scenes. Such a mechanism allows visual detection of known objects under various degrading conditions (occlusions, cluttered scenes, partial visibility due to poor illumination, etc.).

Fig. 1. A collection of local features that can be easily recognized by human vision as a star

(b) Local features (also referred to as corner points, keypoints, interest points, characteristic points, local visual saliencies, etc.) are used in computer vision since the 80's (e.g. [5]) and gradually their application areas have expanded (from stereovision and motion tracking to image matching and detection of known objects under distorting conditions, e.g. [6]). Once the issues of scale changes (the weakest point of the early keypoint detectors) and perspective distortions had been satisfactorily handled (e.g. [7], [8]) the method was found useful for many applications of machine vision.

(c) There is a conceptual gap between "human-vision keypoints" (e.g. geons – see [9]) and machine-vision keypoints that are usually defined for gray-level images only and based on differential properties of intensity functions. Such keypoints are highly uniform (no structural information is used in defining, detecting or matching such keypoints) and there are usually too many of then in typical images. Thus, search for visual contents based on those keypoints is ineffective and tedious.

We propose another category of keypoints for colour images (actually, similar keypoints have been proposed for grey-level images in our previous papers, e.g. [10], [11]). The keypoints, originally intended for robotic vision, are defined by approximating fragments of analysed images by selected patterns (defined over circular windows for rotational invariance and for computational efficiency). The concept of such keypoint is explained in detail in Section 2 (which is the core part of this paper). Section 3 briefly explains and illustrates the method of using such keypoints to supplement visual data in visual databases and for searching visual databases. Section 4 concludes the paper and proposes directions for future researches.

2 Approximation-Based Keypoints

In our previous papers (e.g. [10], [11]) a method has been proposed for approximating gray-level circular images with predefined patterns. Although corners and corner-like patterns (e.g. junctions) are particularly important patterns, the method is applicable to any parameter-defined patterns. For colour images, similar patterns can be proposed and Fig. 2 presents exemplary circular patterns of radius R (*Corner, Round_tip, Round_corner_with_hole, T-junction*). Each pattern can be uniquely characterised by several configuration parameters (indexed β characters in Fig. 2) and several colours. More complex patterns can be proposed as well, but from the practical perspective patterns with 2-3 configuration parameters and a similar number of colour parameters are the most useful ones (low complexity combined with sufficiently useful structural information).

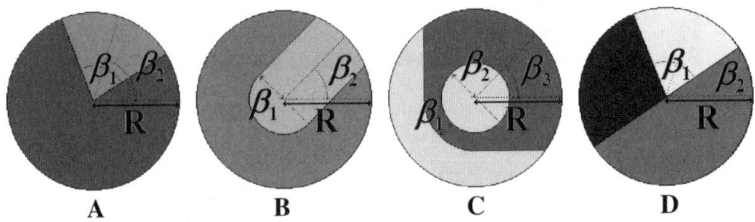

Fig. 2. Exemplary circular colour patterns

The configuration parameters and colours of circular patterns can be expressed as solutions of certain equations derived from the colour functions over the circular image. Moment-based equations for gray-level images have been proposed in [10]. For example, the angular width β_1 of a corner (Fig. 2A) can be expressed for gray-level corners as

$$\beta_1 = 2\arcsin\sqrt{1 - \frac{16[(m_{20} - m_{02})^2 + 4m_{11}^2]}{9R^2\left(m_{10}^2 + m_{01}^2\right)}} \quad (1)$$

while β_2 orientation angle of a T-junction (Fig. 2D) for a gray-level pattern can be computed from:

$$m_{01}\cos\beta_2 - m_{10}\sin\beta_2 = \pm\frac{4}{3R}\sqrt{(m_{20} - m_{02})^2 + 4m_{11}^2} \quad (2)$$

where m_{10}, m_{20}, etc. are moments of the corresponding orders computed in the coordinate system attached to the centre of a circular window. More details are available in [10] (including the equations for determining intensities of the approximations).

The same equations can be applied to any circular image (not necessarily containing the pattern of interest). Then, the solution (if it exists) of the equations defines the optimum approximation of the circular image by the given pattern. Thus, circular patterns can be used as templates that are matched to circular windows of a larger image in order to determine how well that image can be locally approximated by given patterns. Exemplary approximations of gray-level image fragments by selected patterns are given in Fig. 3.

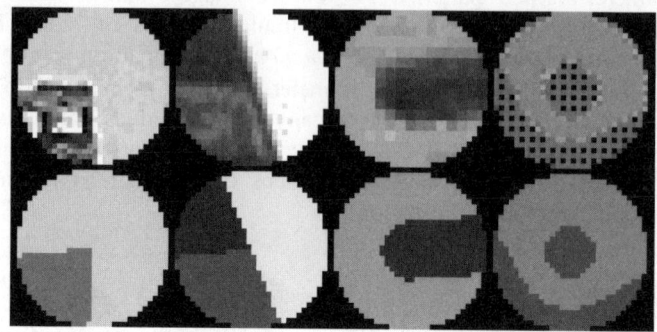

Fig. 3. Exemplary circular fragments (the radius is 15 pixels) approximated by various patterns

2.1 Approximation by Colour Patterns

The approximation method based on moment equations can be easily adapted to colour images. Moment of colour images are 3D vectors consisting of moments of RGB (or other similar colour representation) components of the intensity functions. Thus, instead of m_{20} moment the vector $\vec{m_{20}} = [m_{20}(R), m_{20}(G), m_{20}(B)]$, etc. would be used. Then, the equations for approximation parameters with colour patterns would be changed correspondingly. If scalar moments can be directly replaced by moment vectors, the gray-level equations remain basically unchanged. For example, to compute the angular width β_1 of a corner (see Fig. 2A) Eq.1 is converted into:

$$\beta_1 = 2\arcsin\sqrt{1 - \frac{16\left(\|\vec{m_{20}} - \vec{m_{02}}\|^2 + 4\|m_{11}\|^2\right)}{9R^2\left(\|m_{10}\|^2 + \|m_{01}\|^2\right)}} \qquad (3)$$

However, if the direct replacement of scalars by vectors of moment is not possible (e.g. Eq.2) the scalar moments would be replaced by the dominant components of vector-moments (i.e. the components with the largest magnitudes). For example, Eq.2 would be replaced by

$$m_{01}(X)\cos\beta_2 - m_{10}(X)\sin\beta_2 = \pm\frac{4}{3R}\sqrt{\|\vec{m_{20}} - \vec{m_{02}}\|^2 + 4\|m_{11}\|^2} \qquad (4)$$

where X is R, G or B, depending for which colour the value of $|m_{10}(X)|+|m_{01}(X)|$ is the largest one.

Equations for computing colours of the approximations are generally the same as for gray-level images (see [10]). The equations are just separately applied to each component of the colour space (e.g. RGB).

Exemplary approximations of circular image fragments by colour patterns are shown in Fig. 4.

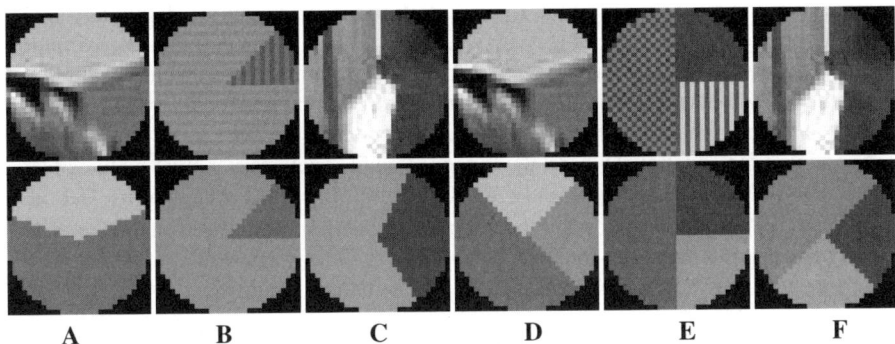

Fig. 4. Exemplary circular fragments (the radius is 15 pixels) approximated by two colour patterns (*corner* and *90°T-junction*)

Pattern approximations of circular images may exist even if the content of an image is (visually) distant from the geometry of the pattern (e.g. Figs 4C and 4F). Thus, a certain quantitative method is needed to estimate how accurate the approximation is, i.e. how closely the pattern of interest is depicted in the tested circular image. The initially proposed (for gray-level images) method described in [10] was found computationally too expensive, and later equally satisfactory results have been achieved by comparing moments of the circular images to the moments of approximations (these moments can be immediately computed using configuration parameters and colours of approximations) as presented in [12]. For colour images, the similarity between a circular image I and its pattern approximation A would be expressed as:

$$sim(I, A) = 1 - \frac{\left\|\overrightarrow{mi_{20}} - \overrightarrow{ma_{20}}\right\| + \left\|\overrightarrow{mi_{02}} - \overrightarrow{ma_{02}}\right\| + \left\|\overrightarrow{mi_{11}} - \overrightarrow{ma_{11}}\right\|}{\left\|\overrightarrow{mi_{20}}\right\| + \left\|\overrightarrow{mi_{02}}\right\| + \left\|\overrightarrow{mi_{11}}\right\|} \quad (5)$$

where $\overrightarrow{mi_{pq}}$ and $\overrightarrow{ma_{pq}}$ are vectors of moments for I and A (respectively). Other moment-based similarity measured can be used alternatively (another similarity function for gray-level images is proposed in [12]).

Numerous experiments have confirmed that the visual similarity between window contents and pattern approximation very well correlates with the value of the above *sim* function. However, there are some cases when high *sim* value can be obtained although the visual similarity does not exist. This effect is caused by mathematical properties of the equations used determine parameters of the approximations. The practical solution of this problem is presented in Subsection 2.3.

2.2 Single-Scale Keypoint Candidates

For an image scanned by a circular window of radius R, there might be many locations where pattern approximations exist (i.e. the corresponding equations have solutions). However, very few of such locations should be considered local features corresponding to the patterns of interest. First, the approximation may exist even the content of a window does not look like the pattern (some examples shown in Fig. 4). Thus, a sufficient level of similarity between a window and its pattern approximation should be postulated, i.e. the value of *sim* function should be sufficiently high (an experimentally determined threshold, depending on the expected quality of processed images and the application objectives, is recommended).

Secondly, for locations where the image contains patterns, a high value of *sim* function would be found not only at the actual locations of patterns but also within certain neighbourhoods around such locations. Fig. 5 shows a sequence of *90° T-junction* approximations obtained when the scanning window moves across a junction actually existing in the analysed image of natural quality.

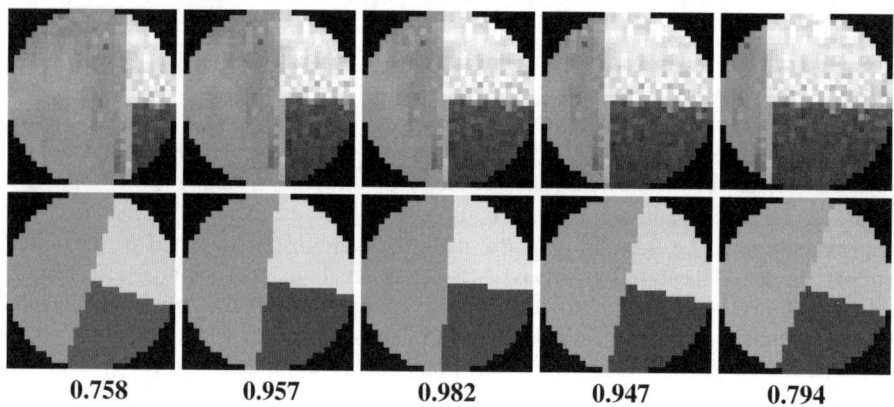

Fig. 5. A sequence of *90° T-junction* approximations near the actual junction in the images. The values of similarity function *sim(I,A)* defined by Eq.5 are given.

We can, therefore, conclude that the pattern of interest P presumably exists in an image at (x,y) location if:

1. For the content I of the scanning circular window of radius R located at (x,y), the approximation A by the pattern P can be found (i.e. the solution of the corresponding equations exists).
2. The approximation A should be accurate enough, i.e. the value of similarity function $sim(I, A)$ exceeds a predefined threshold.
3. The similarity function *sim* reaches a local maximum at (x,y) location.

If the above condition are satisfied, (x,y) pixel is considered a *single-scale keypoint candidate* for pattern P. The scale is obviously defined by the radius R of the scanning window.

In the proposed approach, the search for single-scale keypoint candidates (for a given selection of patterns) is the fundamental operation performed on analysed images, i.e.

either on database or query images. The selected radius R of the scanning window defines the *reference scale* of the analysis. This scale should correspond to the typical size of the objects of interest potentially present in the images. In other words, the radius R should be small enough to avoid other elements of the image to be captured within the scanning window together with a pattern, and should be large enough to ignore minor visual artifacts that may accidentally look like patterns of interest.

2.3 Multi-scale Keypoints

Using the definition of single-scale keypoint candidates presented above, window contents at some locations can be wrongly recognized as candidates (because of mathematical properties of moment-based equations and/or discretization effects). Examples of such incorrectly identified keypoint candidates are shown (for synthesized images) in Fig. 6.

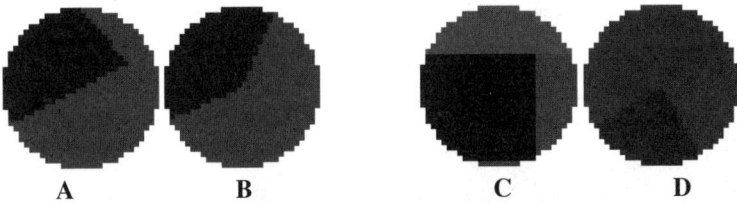

Fig. 6. An incorrect *corner* keypoint candidate (A) and an incorrect *90° T-junction* keypoint candidate (C). The approximations are shown in (B) and (D), respectively. The scanning window radius is 15 pixels.

To verify whether a keypoint actually exists, we use windows of other sizes. In most cases patterns should be consistently detected irrespective to the radius of window (at least for a certain range of radii). This straightforward fact is used to validate the keypoint candidate.

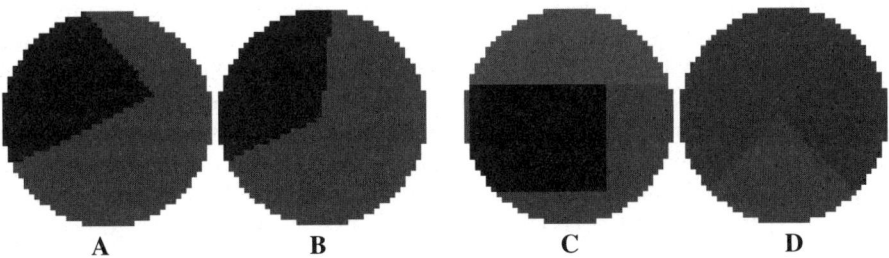

Fig. 7. Approximations obtained for Fig. 6 keypoint candidates by using windows of 20-pixel radius

To a single-scale keypoint candidate (for a certain pattern P) detected at (x,y) location, windows of gradually incremented and decremented radius (starting from the original radius R) are applied. If the approximations exist and are consistently similar over a sufficiently wide range of radii <R_{min}, R_{max}>, the presence of the actual pattern at (*x,y*) location is considered verified. Such locations are referred to as *multi-scale keypoints* for the pattern ***P***.

Fig. 7 shows results obtained by applying the window of 20-pixel radius to the keypoint candidates given in Figs 6A and 6C. In both cases, the approximations still exist but they have significantly different configurations (both cases) and significantly different colours (only the T-junction approximation). Therefore, both keypoint candidates are eventually rejected.

An example as a confirmed multi-scale keypoint (detected in a real image) is illustrated in Fig. 8.

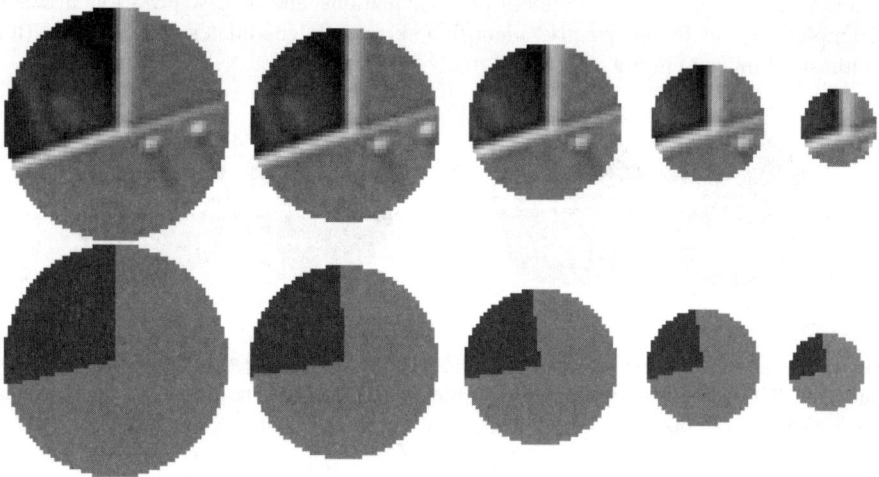

Fig. 8. Consistently similar *corner* approximations obtained for the window radius ranging from 10 to 30 pixels

Multi-scale keypoints obtained using various colour patterns are the local features proposed for warehousing images/frames in databases and for visual information retrieval. Such keypoints are stable, prominent features that are likely to be preserved in any other image that contains the same objects even if the viewing conditions are changed. Secondly, the number of such high-quality keypoints is usually limited (for a single pattern) even in complex scenes. However, if several different patterns are used, an image can still contain enough keypoints for a reliable detection of similar contents under partial occlusions and other conditions reducing visibility.

3 Using Keypoints for Warehousing and Retrieval of Visual Data

Multi-scale keypoints are proposed as the fundamental tool for detecting images of similar visual contents. If two images contain a sufficient number of similar keypoints (that additionally are consistently located in both images) we can assume that their visual contents are similar.

Generally, two pattern-based keypoints match if their approximations are similar (both configuration parameters and colours). Usually, however, the similarity of orientations is not required so that rotation-invariant matching is possible. Orientation plays an important role at the later stage when consistency within a set of matched keypoints is verified. A set of query keypoints can be considered a match for a database set of keypoints only if orientation changes between keypoints of each matched pair are the same (at least approximately). More details about matching individual keypoints are in Subsection 3.1.

To apply the proposed mechanism for searching visual databases, it is proposed that each database image would be memorized with the following supplementary data:

(a) Types of patterns used to find approximations in the image.
(b) Characteristics of each multi-scale keypoint detected in the images. The following entries are used:

- type of pattern approximating the keypoint;
- coordinates of the keypoint;
- configuration parameters;
- color parameters;
- the range of scales (i.e. the range of the scanning window radii).

Query images (to be matched with the database images) are either images from another database or new images, e.g. images captured by a camera (possibly working within a real-time system). In the first-case scenario, query images are presumably similarly warehoused so that the effectiveness of search procedures is determined by the efficiency of algorithms matching keypoints of query and database images.

Generally, a keypoint-based "search for visual similarity" between a query image and a database image is a voting scheme where a match is considered found if a sufficient number of query keypoints are consistently corresponding to database keypoints (see e.g. [13]). However, matching images where contents are shown in arbitrary scales has been a problem for which the proposed solutions (e.g. [14]) are not fully satisfactory yet. The proposed concept of multi-scale keypoints seems to be a straightforward method to match corresponding image fragments even the images are in different scales (within a certain range of scales).

The necessary condition to find a match between a query keypoint and a database one is that their ranges of window radii intersect (taking into account the relative scale between images) i.e.

$$\sigma R_1(q) \leq R_2(d) \quad \text{and} \quad \sigma R_2(q) \geq R_1(d) \tag{6}$$

where $<R_1(q), R_2(q)>$ is the range of radii for the query keypoint, $<R_1(d), R_2(d)>$ is the range of radii for the query keypoint, and σ is the *relative scale* (more in [14]) between the database image and the query image.

However, if query images are captured in real time, the complexity of the multi-scale keypoint detection may be too high the existing time constraints. Then, single-scale keypoint candidates (which can be detected at much lower costs and hardware accelerators can be used for moment computations) would be used as substitutes of query multi-scale keypoints. In such cases the number of matches will be obviously larger (i.e. more potentially relevant images can be found in the database) but otherwise the principles remain identical.

If a sufficient number of matches between query and database keypoints are found (more about the matching principles in the next sub-section) a hypothesis about the content similarity between the query image and the database image is proposed. All such hypotheses are subsequently verified and eventually some of them can be accepted.

Hypotheses are verified using the concept of *shape graphs* (see [14]). *Shape graphs* are built for both query and database images using the sets of matching keypoint. Nodes of the graph represent keypoints and each edge of the graph is labeled by the distance between the adjoined nodes (keypoints). An iterative algorithm is used to find the maximum sub-graphs of both *shape graphs* for which all corresponding pairs of edges have approximately proportional labels (i.e. the geometric distribution of keypoints is similar). This algorithm converges very fast and in most cases only a few iterations are needed. The generated sub-graphs (if they contain enough nodes) specify the final set of query and database keypoints confirming the validity of the hypothesis. Fig. 9 shows a b/w example (from [12]) of two images matched using *corner* and *90°T-junction* patterns.

Fig. 9. Matched keypoints in a database image (A) and a query image (B). Crossed circles indicate *corner* keypoints, and squares are *90°T-junction* keypoints.

3.1 More About Keypoint Matching

The number of keypoints detected in images can be controlled. More keypoints can be obtained if less accurate and/or less contrasted pattern-based approximations are accepted keypoints. Similarly, the keypoint matching process can be controlled by using more or less relaxed rules

The configuration parameters of keypoint approximations generally should have a higher priority as they specify geometry of the local structures in the compared images. It should be remembered that the orientation angles, though usually ignored when pairs of keypoints are matched, play an important role when hypotheses about visual similarities are created and verified. For example, in Fig. 9 images, orientations of query keypoints (Fig. 9B) are rotated by 25° to 36° with respect to their database matches (Fig. 9A). Since in this example we consider images of deformable objects, this range of orientation changes can be assumed sufficiently consistent.

The color parameters of keypoint approximations can be used for keypoint matching in more diversified ways. For example, a match can exist if colours are similar in terms of their normalized RGB coefficients (but not necessarily in terms of absolute coefficients). Alternatively, only the dominant components should match (e.g. the colour should be "predominantly green" in both kepoints). In the extreme scenarios the colour parameters might be totally ignored (i.e. only geometry of the local structures in the matched images would be considered important).

4 Summary

The method has been derived from previous algorithms originally developed for robotic vision, in particular for robotic vision-based navigation (e.g. [14]) using b/w cameras. Its effectiveness has been already tested for natural images and its ability to handle images of high complexity and with distortions typically present in natural scenes has been verified. By applying the method to colour images, we just further improve the reliability/robustness. Therefore, we believe that the approach is a promising solution for information search and retrieval in visual databases.

Moreover, the presented technique strongly refers to principles of human visual perception. Humans generally perceive known objects as a collection of local visual clues subsequently "interpolated" into a known entity. It can be concluded, therefore, that the proposed method can be seen as a balanced combination of biologically inspired approach and practical limitations.

Fig. 10. A higher-level pattern deducted and subsequently approximated using low-level approximations

4.1 Future Works

The major direction of our future works is the idea of higher-level features built from multi-scale keypoints. If a certain geometric configuration of consistent keypoints is detected in an image, this may be an indicator that a more complex pattern exists in the image. The presence of such a pattern may be confirmed by either placing a larger window (with a set of equations to approximate its content by the pattern, etc.) or it may be left as a hypothesis added to the image data (and possibly exploited when this image is matched to another image). The idea is briefly illustrated in Fig. 10.

There are also other research and commercial initiatives undertaken using fundamentals of the proposed methodology.

References

1. Eidenberger, H.: A new perspective on visual information retrieval. Proc. of SPIE - Int. Society for Optical Enginering 5307, 133–144 (2004)
2. Sethi, I.K., Coman, I.: Image retrieval using hierarchical self-organizing feature maps. Pattern Recognition Letters 20, 1337–1345 (1999)
3. Prasad, B.G., Biswas, K.K, Gupta, S.K.: Region-based image retrieval using integrated color, shape, and location index. Computer Vision & Image Understanding 94, 193–233 (2004)
4. Edelman, S.: Computational theories of object recognition. Trends in Cognitive Sciences 1, 298–309 (1997)
5. Harris, C., Stephens, M.: A combined corner and edge detector. In: Proc. of 4th Alvey Vision Conference, Manchester, pp. 147–151 (1988)
6. Schmid, C., Mohr, R.: Local grayvalue invariants for image retrieval. IEEE Transactions on Pattern Analysis and Machine Intelligence 24, 530–535 (1997)
7. Lowe, D.: Distinctive image features from scale-invariant keypoints. Int. Journal of Computer Vision 60, 91–110 (2004)
8. Mikolajczyk, K., Schmid, C.: Scale & affine invariant interest point detectors. Int. Journal of Computer Vision 60, 63–86 (2004)
9. Biederman, I.: Recognition-by-components: A theory of human image understanding. Psychological Review 94, 115–147 (1987)
10. Sluzek, A.: On moment-based local operators for detecting image patterns. Image and Vision Computing 23, 287–298 (2005)
11. Sluzek, A.: A new local-feature framework for scale-invariant detection of partially occluded objects. In: Chang, L.-W., Lie, W.-N. (eds.) PSIVT 2006. LNCS, vol. 4319, pp. 248–257. Springer, Heidelberg (2006)
12. Sluzek, A., Islam, M.S.: New types of keypoints for detecting known objects is visual search tasks. In: Obinata, G., Dutta, A. (eds) Vision Systems, Application. ARS, Vienna, pp. 423–442 (2007)
13. Wolfson, H.J., Rigoutsos, I.: Geometric hashing: an overview. IEEE Computational Science & Engineering 4, 10–21 (1997)
14. Islam, M.S.: Recognition and localization of objects in relative scale for robotic applications. PhD Thesis, School of Comp. Engineering, Nanyang Technological University, Singapore (2006)

A Knowledge Synthesizing Approach for Classification of Visual Information

Le Dong and Ebroul Izquierdo

Department of Electronic Engineering, Queen Mary, University of London,
London E1 4NS, U.K.
{le.dong, ebroul.izquierdo}@elec.qmul.ac.uk

Abstract. An approach for visual information analysis and classification is presented. It is based on a knowledge synthesizing technique to automatically create a relevance map from essential areas in natural images. It also derives a set of well-structured representations from low-level description to drive the final classification. The backbone of this approach is a distribution mapping strategy involving a knowledge synthesizing module based on an intelligent growing when required network. Classification is achieved by simulating the high-level top-down visual information perception in primates followed by incremental Bayesian parameter estimation. The proposed modular system architecture offers straightforward expansion to include user relevance feedback, contextual input, and multimodal information if available.

Keywords: classification, essence map, knowledge synthesizing.

1 Introduction

Classification and retrieval of visual information is a critical task for high-level computer based understanding of visual information. Current systems for classification of visual information are mostly based on the analysis of low-level image primitives [1], [2]. Relying on low-level features only, it is possible to automatically extract important relationships between images. However, such approaches lack potential to achieve accurate classification for generic automatic retrieval. A significant number of semantic-based approaches address this fundamental problem by utilizing automatic generation of links between low- and high-level features. For instance, Dorado *et al.* introduced in [3] a system that exploits the ability of support vector classifiers to learn from relatively small number of patterns. Based on a better understanding of visual information elements and their role in synthesis and manipulation of their content, an approach called "computational media aesthetics" studies the dynamic nature of the narrative via analysis of the integration and sequencing of audio and video [4]. Semantic extraction using fuzzy inference rules has been used in [5]. These approaches are based on the premise that the rules needed to infer a set of high-level concepts from low-level descriptors can not be defined a priori. Rather, knowledge embedded in the database and interaction with an expert user is exploited to enable learning.

Closer to the models described in this paper, knowledge and feature based classification as well as topology preservation are important aspects that can be used to improve classification performance. The proposed system uses a knowledge synthesizing approach to approximate human-like inference. The system consists of two main parts: knowledge synthesizing and classification. In this paper a knowledge synthesizing approach is exploited to build a system for visual information analysis following human perception and interpretation of natural images. The proposed approach aims at, to some extent, mimicking the human knowledge synthesizing system and to use it to achieve higher accuracy in classification of visual information. A method to generate a knowledge synthesizing based on the structured low-level features is developed. Using this method, the preservation of new objects from a previously perceived ontology in conjunction with the colour and texture perceptions can be processed autonomously and incrementally. The knowledge synthesizing network consists of the posterior probability and the prior frequency distribution map of each visual information cluster conveying a given semantic concept.

Contrasting related works from the conventional literature, the proposed system exploits known fundamental properties of a suitable knowledge synthesizing model to achieve classification of natural images. An important contribution of the presented work is the dynamic preservation of high-level representation of natural scenes. As a result, continually changing associations for each class is achieved. This novel feature of the system together with an open and modular system architecture, enable important system extensions to include user relevance feedback, contextual input, and multimodal information if available. These important features are the scope of ongoing implementations and system extensions targeting enhanced robustness and classification accuracy. The essence map model for feature extraction is described in Section 2. The knowledge synthesizing approach is given in Section 3. A detailed description of the high-level classification is given in Section 4. The selected result and a comparative analysis of the proposed approach with other existing methods are given in Section 5. The paper closes with conclusions and an outline of ongoing extensions in section 6.

2 Essence Map Model

Five features of intensity (I), edge (E), colour (C), orientation (O), and symmetry (S) are used to model the human-like bottom-up visual attention mechanism [6], as shown in Fig. 1. The roles of retina cells and LGN are reflected in previously proposed attention models [7]. The feature maps are constructed by centre-surround difference and normalization (CSD & N) of the five bases. This mimics the on-centre and off-surround mechanism in the human brain. Subsequently, they are integrated using a conventional independent component analysis (ICA) algorithm [8]. The symmetry information is used as a joint basis to consider shape primitives in objects [9], which is obtained by the noise tolerant general symmetry transform ($NTGST$) method [7]. The ICA can be used for modelling the role of the primary visual cortex for the redundancy reduction according to Barlow's hypothesis and Sejnowski's results [8].

Barlow's hypothesis is that human visual cortical feature detectors might be the end result of a redundancy reduction process [10]. Sejnowski's result states that the ICA is the best way to reduce redundancy [8].

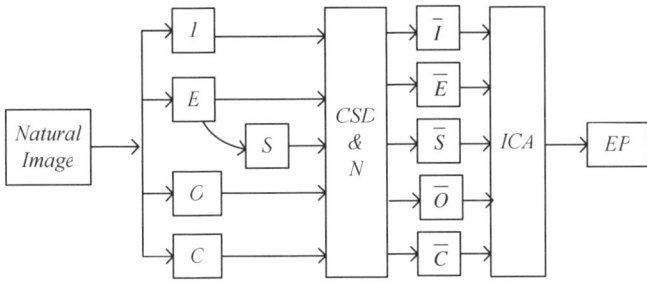

Fig. 1. The architecture of essence map model. $\bar{I}$: normalized intensity feature map, $\bar{E}$: normalized edge feature map, $\bar{S}$: normalized symmetry feature map, $\bar{O}$: normalized orientation feature map, $\bar{C}$: normalized colour feature map, *EP*: essence point.

Using a similar notation to that used in [11], and after the convolution between the channel of feature maps and filters obtained by ICA, the essence map is computed by the summation of all feature maps for every location [12]. In the course of preprocessing, a Gaussian pyramid with different scales from 0 to n level is used [7]. Each level is obtained by subsampling of 2^n, thus constructing five feature maps. Subsequently, the centre-surround mechanism is implemented in the model as the difference between the fine and coarse scales of Gaussian pyramid images [7]. Consequently, five feature maps are obtained by the following equations.

$$I(c,s) = |I(c) \bullet I(s)|, E(c,s) = |E(c) \bullet E(s)|, S(c,s) = |S(c) \bullet S(s)| \cdot \tag{1}$$

$$O(c,s) = |O(c) \bullet O(s)|, C(c,s) = |C(c) \bullet C(s)| \cdot \tag{2}$$

Here, $\bullet$ represents interpolation to the finer scale and point-by-point subtraction, N stands for the normalization operation, c and s are indexes of the finer scale and the coarse scale, respectively. Feature maps are combined into five characteristic maps.

$$\bar{I} = \oplus_{c,s} N(I(c,s)), \bar{E} = \oplus_{c,s} N(E(c,s)), \bar{S} = \oplus_{c,s} N(S(c,s)), \bar{O} = \oplus_{c,s} N(O(c,s)), \bar{C} = \oplus_{c,s} N(C(c,s)) \cdot \tag{3}$$

Here, $\bar{I}, \bar{E}, \bar{S}, \bar{O}$, and $\bar{C}$ are obtained through across-scale addition "$\oplus$". To obtain ICA filters, the five feature maps are used for input patches of the ICA. The basis functions are determined using the extended infomax algorithm [13]. Each row of the basis functions represents an independent filter and that is ordered according to the length of the filter vector. The resulting ICA filters are then applied to the five feature maps to obtain the essence map according to [7]:

$$E_{qi} = FM_q * ICs_{qi} \text{ for } i = 1, \cdots, M, \; q = 1, \cdots, 5, \; EM(x,y) = \sum E_{qi}(x,y) \text{ for all } i. \tag{4}$$

Here, M denotes the number of filters; FM_q denotes feature maps, ICs_{qi} denotes each independent component accounting for the number of filters and feature maps, $EM(x, y)$ denotes the essence map. The convolution result E_{qi} represents the influences of the five feature maps on each independent component and the most essential point is computed by maximum operator, then an appropriate essential area centred by the most essential location is masked off and the next essential location in the input visual information is calculated using the essence map model.

3 Knowledge Synthesizing

In this section the knowledge synthesizing approach is described. The proposed knowledge synthesizing approach automatically creates a relevance map from the essential areas detected by our proposed essence map model. It also derives a set of well-structured representations from low-level description to drive the high-level classification. The backbone of this technique is a distribution mapping strategy involving knowledge synthesizing based on growing when required network (GWR).

The precise steps of the GWR algorithm will now be detailed as follows [14].

Let A be the set of map nodes, and $C \subset A \times A$ be the set of connections between nodes in the map field. Let the input distribution be $p(\xi)$ for inputs ξ. Define w_n as the weight vector of node n.

Initialisation. Create two nodes for the set A, $A = \{n_1, n_2\}$, with n_1, n_2 initialised randomly from $p(\xi)$.

Define C, the connection set, to be the empty set $C = \emptyset$. Then, each iteration of the algorithm looks like this:

1. Generate a data sample ξ for input to the network.
2. For each node i in the network, calculate the distance from the input $\| \xi - w_i \|$.
3. Select the best matching node, and the second best, that is the nodes $s, t \in A$ such that $s = \arg\min_{n \in A} \| \xi - w_n \|$ and $t = \arg\min_{n \in A / \{s\}} \| \xi - w_n \|$, where w_n is the weight vector of node n.
4. If there is not a connection between s and t, create it $C = C \cup \{(s,t)\}$, otherwise, set the age of the connection to 0.
5. Calculate the activity of the best matching unit $a = \exp(-\| \xi - w_s \|)$.
6. If the activity $a <$ activity threshold a_T and firing counter $<$ firing threshold h_T then a new node should be added between the two best matching nodes (s and t)
    - Add the new node r, $A = A \cup \{r\}$.
    - Create the new weight vector, setting the weights to be the average of the weights for the best matching node and the input vector $w_r = (w_s + \xi)/2$.
    - Insert edges between r and s and between r and t, $C = C \cup \{(r,s), (r,t)\}$.
    - Remove the link between s and t, $C = C / \{(s,t)\}$.

7. If a new node is not added, adapt the positions of the winning node and its neighbours, i, that is the nodes to which it is connected, $\Delta w_s = \varepsilon_b \times h_s \times (\xi - w_s)$, $\Delta w_i = \varepsilon_n \times h_i \times (\xi - w_i)$, where $0 < \varepsilon_n < \varepsilon_b < 1$ and h_s is the value of the firing counter for node s.
8. Age edges with an end at s, $age_{(s,i)} = age_{(s,i)} + 1$.
9. Reduce the counter of how frequently the winning node s has fired according to $h_s(t) = h_0 - \frac{S(t)}{\alpha_b}(1 - e^{(-\alpha_b t / \tau_b)})$ and the counters of its neighbours, (i), $h_i(t) = h_0 - \frac{S(t)}{\alpha_n}(1 - e^{(-\alpha_n t / \tau_n)})$, where $h_i(t)$ is the size of the firing variable for node i, h_0 the initial strength, and $S(t)$ is the stimulus strength, usually 1. α_n, α_b and τ_n, τ_b are constants controlling the behaviour of the curve. The firing counter of the winner reduces faster than those of its neighbours.
10. Check if there are any nodes or edges to delete, i.e. if there are any nodes that no longer have any neighbours, or edges that are older than the greatest allowed age, in which case, delete them.
11. If further inputs are available, return to step (1) unless some stopping criterion has been reached.

The input of the algorithm is a set of extracted low-level features generated by essence map model. Various topology maps of the network subtly reflect the characteristics of distinct visual information groups which are closely related to the order of the forthcoming visual information. Furthermore, the extracted information from perceptions in colour and texture domains can also be used to represent objects.

4 Classification

Using the output generated by the knowledge synthesizing approach, high-level classification is achieved. The proposed high-level classification approach follows a high-level perception and classification model that mimics the top-down attention mechanism in primates' brain. A proposed high-level perception and classification model uses a generative approach based on an incremental Bayesian parameter estimation method. The input features of this generative object representation are the low-level information generated by knowledge synthesizing module. A new class can be added incrementally by learning its class-conditional density independently of all the previous classes. In this paper n training data samples from a class ω are considered. Each class is represented by f ($f < n$) codebook vectors. Learning is conducted incrementally by updating these codebook vectors whenever a new data vector u is entered. The used generative approach learns the class prior probabilities $p(\omega)$ and the class-conditional densities $p(u|\omega)$ separately. The required posterior probabilities area then obtained using the Bayes' theorem:

$$p(\omega|u) = \frac{p(u|\omega)p(\omega)}{p(u)} = \frac{p(u|\omega)p(\omega)}{\sum_j p(u|j)p(j)}. \quad (5)$$

In order to estimate the class-conditional density of the feature vector u given the class ω, a vector quantizer is used to extract codebook vectors from training samples. Following Vailaya et al. in [15], the class-conditional densities are approximated using a mixture of Gaussians (with identity covariance matrices), each centred at a codebook vector. Then, the class-conditional densities can be represented as,

$$p_U(u|\omega) \propto \sum_{j=1}^{f} m_j * \exp\left(-\|u - v_j\|^2 / 2\right). \quad (6)$$

where $v_j (1 \le j \le f)$ denotes the codebook vectors, m_j is the proportion of training samples assigned to v_j. When human beings focus its attention in a given area, the prefrontal cortex gives a competition bias related to the target object in the inferior temporal area [16]. Subsequently, the inferior temporal area generates specific information and transmits it to the high-level attention generator which conducts a biased competition [16]. Therefore, the high-level perception and classification model can assign a specific class to a target area, which gives the maximum likelihood. If the prior density is assumed essentially uniform, the posterior probability can be estimated as follows [15],

$$\arg\max_{\omega \in \Omega}\{p(\omega|u)\} = \arg\max_{\omega \in \Omega}\{p_U(u|\omega)p(\omega)\}. \quad (7)$$

where Ω is the set of pattern classes. In addition, the high-level perception and classification model can generate a specific attention based on the class detection ability. Moreover, it may provide informative control signals to the internal effectors [16]. This in turn can be seen as an incremental framework for knowledge synthesizing with human interaction.

5 Experimental Evaluation

Given a collection of completely unlabelled images, the goal is to automatically discover the visual categories present in the data and localize them in the topology preservation of the network. To this end, a set of quantitative experiments with progressively increasing level of topology representation complexity was conducted. The Corel database containing 700 images was used, which was labelled manually with eight predefined concepts. The concepts are "building", "car", "autumn", "rural scenery", "cloud", "elephant", "lion", and "tiger". In order to assess the accuracy of the classification, a performance evaluation based on the amount of missed detections (*MD*) and false alarms (*FA*) for each class from the large dataset of the Corel database was conducted. In this evaluation recall (*R*) and precision (*P*) values were estimated and used: $recall = \frac{D}{D+MD}$, $precision = \frac{D}{D+FA}$, where D is a sum of true memberships for the corresponding recognized class, MD is a sum of the complement of the full true memberships and FA is a sum of false memberships. The obtained results are given in Table 1.

Table 1. Recall/Precision Results of Classification and Retrieval

Class	D	MD	FA	R (%)	P (%)
Building	84	16	12	84	88
Autumn	42	14	10	75	81
Car	89	11	8	89	92
Cloud	90	10	10	90	90
Tiger	87	13	10	87	90
Rural scenery	36	8	9	82	80
Elephant	93	7	14	93	87
Lion	88	12	10	88	90

The proposed technique was compared with an approach based on multi-objective optimization [17] and another using Bayesian networks for concept propagation [18]. Table 2 shows a summary of results on some subsets of the categories coming out from this comparative evaluation. It can be observed that the proposed technique outperforms the other two approaches. Even though multi-objective optimization can be optimized for a given concept, the result of the proposed technique performs better in general. Except for the class "lion", in which the Bayesian belief approach delivers the highest accuracy, the proposed technique performs substantially better in other cases. This summary of results truly represents the observed outcomes with other classes and datasets used in the experimental evaluation and evidences our claim that the proposed technique has good discriminative power and it is suitable for retrieving natural images in large datasets.

We also compared the performance of the proposed approach with two binary classifiers: one based on ant colony optimization and constraints of points with K-Means approach (ACO/COP-K-Means) [19], and the other using particle swarm optimization and self organizing feature maps (PSO/SOFM) [20]. A summary of results on some subsets of the categories is given in Table 3.

According to these results, it can be concluded that the proposed technique also outperforms other classical approaches and works well in the case of multi-mode classification.

Table 2. Precision Results of the Proposed Technique Compared with Two Other Approaches

(%)	Proposed Technique	Bayesian Belief	Multi-Objective Optimization
Building	88	72	70
Cloud	90	84	79
Lion	90	92	88
Tiger	90	60	60

Table 3. Results of the Proposed Technique Compared with Two Other Binary Classifiers

(%)	ACO/ COP-K-Means		PSO/ SOFM		Proposed technique	
	P	R	P	R	P	R
Lion	55	62	48	69	90	88
Elephant	71	71	74	65	87	93
Tiger	63	58	68	64	90	87
Cloud	62	57	69	63	90	90
Car	65	56	70	64	92	89
Building	65	62	51	74	88	84

6 Conclusion

A knowledge synthesizing approach for classification of visual information is presented. By utilizing biologically inspired theory and knowledge synthesizing, this system simulates the human-like classification and inference. Since the knowledge synthesizing base creation depends on information provided by expert users, the system can be easily extended to support intelligent retrieval wit enabled user relevance feedback. The whole system can automatically generate relevance maps from the visual information and classifying the visual information using learned information. Additional expansion capabilities include learning from semantics and annotation-based approach and the use of multimodal information.

References

1. Smeulders, A., Worring, M., Santini, S., Gupta, A., Jain, R.: Content Based Image Retrieval at the End of the Early Years. IEEE Trans. Patt. Anal. Mach. Intell. 22(12), 1349–1380 (2000)
2. Manjunath, B.S., Salembier, P., Sikora, T.: Introduction to MPEG-7. In: Multimeida Content Description Interface, John Wiley & Sons, West Sussex (2003)
3. Dorado, A., Djordjevic, D., Pedrycz, W., Izquierdo, E.: Efficient Image Selection for Concept Learning. IEE Proc. on Vision, Image and Signal Processing 153(3), 263–273 (2006)
4. Dorai, C., Venkatesh, S.: Bridging the Semantic Gap with Computational Media Aesthetics. IEEE Multimedia 10(2), 15–17 (2003)
5. Dorado, A., Calic, J., Izquierdo, E.: A Rule-based Video Annotation System. IEEE Trans. on Circuits and Systems for Video Technology 14(5), 622–633 (2004)
6. Goldstein, E.B.: Sensation and Perception, 4th edn. An international Thomson Publishing Company, USA (1996)
7. Park, S.J., An, K.H., Lee, M.: Saliency Map Model with Adaptive Masking Based on Independent Component Analysis. Neurocomputing 49, 417–422 (2002)
8. Bell, A.J., Sejnowski, T.J.: The Independent Components of Natural Scenes Are Edge Filters. Vision Research 37, 3327–3338 (1997)
9. Vetter, T., Poggio, T., Bülthoff, H.: The Importance of Symmetry and Virtual Views in Three-Dimensional Object Recognition. Current Biology 4, 18–23 (1994)
10. Barlow, H.B., Tolhust, D.J.: Why Do You Have Edge Detectors? Optical Society of America Technical Digest 23(172) (1992)

11. Itti, L., Koch, C., Niebur, E.: A Model of Saliency-Based Visual Attention for Rapid Scene Analysis. IEEE Trans. Patt. Anal. Mach. Intell. 20(11), 1254–1259 (1998)
12. Ratnaparkhi, A.: Maximum Entropy Models for Natural Language Ambiguity Resolution. Ph.D. Dissertation. Comp.and Inf. Science, Univ. of Pennsylvania, USA (1998)
13. Jasinschi, R.S., Dimitrova, N., McGee, T., Agnihotri, L., Zimmerman, J., Li, D., Louie, J.: A Probabilistic Layered Framework for Integrating Multimedia Content and Context Information. In: Proc. IEEE Int. Conf. Acoustics, Speech, and Signal Processing, vol. 2, pp. 2057–2060. IEEE Computer Society Press, Los Alamitos (2002)
14. Marsland, S., Shapiro, J., Nehmzow, U.: A Self-organising Network That Grows When Required. Neural Networks 15, 1041–1058 (2002)
15. Vailaya, A., Figueiredo, M.A.T., Jain, A.K., Zhang, H.J.: Image Classification for Content-based Indexing. IEEE Trans. on Image Processing 10(1), 117–130 (2001)
16. Lanyon, L.J., Denham, S.L.: A Model of Active Visual Search with Object-based Attention Guiding Scan Paths. Neural Networks Special Issue: Vision & Brain 17(5-6), 873–897 (2004)
17. Zhang, Q., Izquierdo, E.: A Multi-feature Optimization Approach to Object-based Image Classification. In: Sundaram, H., Naphade, M., Smith, J.R., Rui, Y. (eds.) CIVR 2006. LNCS, vol. 4071, pp. 310–319. Springer, Heidelberg (2006)
18. Li, F.F., Fergus, R., Perona, P.: A Bayesian Approach to Unsupervised One-shot Learning of Object Categories. In: Proc. IEEE Int. Conf. on Computer Vision, vol. 2, pp. 1134–1141. IEEE Computer Society Press, Los Alamitos (2003)
19. Saatchi, S., Hung, Ch.: Hybridization of the Ant Colony Optimization with the K-Means Algorithm for Clustering. In: Kalviainen, H., Parkkinen, J., Kaarna, A. (eds.) SCIA 2005. LNCS, vol. 3540, pp. 511–520. Springer, Heidelberg (2005)
20. Chandramouli, K., Izquierdo, E.: Image Classification Using Chaotic Particle Swarm Optimization. In: Proc. IEEE Int. Conf. on Image Processing, Atlanta, USA, pp. 3001–3004. IEEE Computer Society Press, Los Alamitos (2006)

Image Similarity – From Fuzzy Sets to Color Image Applications

M. Nachtegael, S. Schulte, V. De Witte, T. Mélange, and E.E. Kerre*

Ghent University, Dept. of Applied Mathematics and Computer Science
Fuzziness and Uncertainty Modelling Research Unit
Krijgslaan 281 - S9, 9000 Gent, Belgium
Mike.Nachtegael@UGent.be

Abstract. Image similarity is an important topic in the field of image processing. The goal is to obtain objective measures that express the similarity between two images in a way that matches human evaluation. Such measures have both theoretical and practical applications. In this paper, we show how similarity measures for fuzzy sets have been modified in order to be applied in image processing. We also discuss a new application of these measures in the context of color image retrieval, indicating the potential of this class of similarity measures.

Keywords: Image processing, similarity, retrieval, color images.

1 Introduction

Objective quality measures or measures of comparison are of great importance in the field of image processing: they enable us to automatically evaluate image quality, they serve as a tool to evaluate and to compare different algorithms designed to solve particular problems (e.g. noise reduction, image enhancement, decompression), and they can be applied for example in image retrieval and noise reduction ([16,17]). The application in the context of image retrieval, which we will illustrate in this paper, is quite promising: if one has a reference image and a database of images, similarity measures can be used to retrieve those images from the database that are most similar to the reference image. The advantage of this approach is that it is not required to index all images, which is often partially done manually.

The classical quality measures, such as the $RMSE$ (Root Mean Square Error) or the $PSNR$ (Peak Signal to Noise Ratio) have the disadvantage that they do not always correspond to visual observations. Therefore, we cannot rely on them for automated quality evaluation or image retrieval, and need to find new tools. In this paper we give an overview of the evolution of similarity measures for fuzzy sets: orginally introduced to express the similarity between fuzzy sets, they have been modified for application in image processing (Section 2). In Section 3 we discuss a recent application in color image processing, which illustrates the potential of this class of similarity measures. Conclusions are made in Section 4.

---
* Corresponding author.

2 From Fuzzy Set Similarity to Image Similarity

An n-dimensional image can be represented as a mapping from a universe $\mathcal{X}$ (a finite subset of $\mathbb{R}^n$, usually a $M \times N$-grid of points which we call pixels) to a set of values. The set of possible pixel values depends on the fact whether the image is binary, grayscale or color. Binary images take values in $\{0, 1\}$ (black = 0; white = 1), and grayscale images in $[0, 1]$. The representation of color images depends on the specific color model, e.g. RGB, HSV, La*b* [12].

Fuzzy set theory is widely recognized as a tool to model imprecision and uncertainty. The basic idea behind fuzzy set theory is that an element can belong to a set, i.e. an element can satisfy a property, to a certain degree [22]. A fuzzy set A in a universe $\mathcal{X}$ is characterized by a membership function, also denoted by A, that associates a degree of membership $A(x) \in [0, 1]$ with each element x of $\mathcal{X}$. In summary, a fuzzy set A can be represented as a $\mathcal{X} - [0, 1]$ mapping; the class of all fuzzy sets in $\mathcal{X}$ is denoted as $\mathcal{F}(\mathcal{X})$. From a mathematical point of view, it follows that grayscale images and fuzzy sets are represented in the same way. This explains the possible exchange of techniques between fuzzy set theory and image processing.

Similarity measures for fuzzy sets were first directly applied to grayscale images. The measures were then modified, taking into account homogeneity properties and histogram characteristics, finally resulting in so-called combined similarity measures. Experiments show a very good performance of these combined measures.

2.1 The Classical Approach: MSE, RMSE, PSNR

The mean square error $MSE(A, B)$ of two images A and B is defined as:

$$MSE(A, B) = \frac{1}{MN} \sum_{(i,j) \in \mathcal{X}} |A(i,j) - B(i,j)|^2 \ . \tag{1}$$

The Root Mean Square Error ($RMSE$) is the square root of MSE. The $PSNR$ is obtained by dividing the square of the luminance range R of the display device by the MSE and expressing the result in decibels:

$$PSNR(A, B) = 10 \log_{10} \frac{R^2}{MSE(A, B)} \ . \tag{2}$$

2.2 Direct Application of Similarity Measures for Fuzzy Sets

Expressing similarity is expressing a relation between objects. In the literature a lot of measures are proposed to express the similarity between two fuzzy sets [5,23]. There is no unique definition, so we will consider a similarity measure purely as a measure to compare two fuzzy sets or images, imposing properties that are considered relevant in image processing, namely: reflexivity (identical images are similar), symmetry (similarity is independent of the order of the images), low reaction and decreasing w.r.t. noise, and low reaction and decreasing w.r.t. enlightening or darkening.

From a total of more than 40 different similarity measures, 14 similarity measures satisfy the above mentioned properties [18,19,21]. Here, we will restrict ourselves to some examples.

Measures based on the Minkowski distance:

$$M_1(A,B) = 1 - \left(\frac{1}{MN}\sum_{(i,j)\in\mathcal{X}}|A(i,j)-B(i,j)|^r\right)^{\frac{1}{r}}, \text{ with } r \in \mathbb{N}\backslash\{0\} \quad (3)$$

$$M_2(A,B) = 1 - \max_{(i,j)\in\mathcal{X}}|A(i,j)-B(i,j)|. \quad (4)$$

A measure based on the notion of cardinality ($|A| = \sum_{x\in\mathcal{X}} A(x)$):

$$M_6(A,B) = \frac{|A\cap B|}{|A\cup B|} = \frac{\sum_{(i,j)\in\mathcal{X}}\min(A(i,j),B(i,j))}{\sum_{(i,j)\in\mathcal{X}}\max(A(i,j),B(i,j))}. \quad (5)$$

Another example is the measure M_{20}:

$$M_{20}(A,B) = \frac{1}{MN}\cdot\sum_{(i,j)\in\mathcal{X}}\frac{\min(A(i,j),B(i,j))}{\max(A(i,j),B(i,j))}. \quad (6)$$

2.3 Neighbourhood-Based Similarity Measures

The direct application of similarity measures to images is pixel-based, i.e., the result depends on the comparison between individual pixels. In the context of image processing one should also take homogeneity properties into account. This is achieved by neighbourhood-based similarity measures. The construction is as follows: we divide the images A and B in disjoint 8×8 image parts A_i, B_i ($i = 1,\ldots,P$, with P the number of parts), we calculate the similarity $M(A_i, B_i)$ between each of corresponding parts (using pixel-based similarity measures), we define the similarity $M^h(A,B)$ between A and B as the weighted average of these similarities:

$$M^h(A,B) = \frac{1}{P}\sum_{i=1}^{P} w_i \cdot M(A_i, B_i), \quad (7)$$

where the weight w_i is the similarity between the homogeneity h_{A_i} of image part A_i and the homogeneity h_{B_i} of image part B_i. The homogeneity is computed as the similarity between the gray value of the pixel in the image part with maximum intensity and the gray value of the pixel in the image part with minimum intensity, using a resemblance relation s:

$$h_{A_i} = s(\max_{(i,j)\in A_i} A(i,j), \min_{(i,j)\in A_i} A(i,j)). \quad (8)$$

So the weight w_i is given by $w_i = s(h_{A_i}, h_{B_i})$. Resemblance relations [6] were introduced as an alternative model for the modelling of approximate equality. A resemblance relation on $\mathcal{X}$ is a fuzzy relation E in a pseudo-metric space $(\mathcal{X}, d)$ satisfying the following properties: $(\forall x \in \mathcal{X})(E(x,x) = 1)$ and $(\forall (x, y, z, u) \in \mathcal{X}^4)(d(x,y) \leq d(z,u) \Rightarrow E(x,y) \geq E(z,u))$.

Using the similarity measures from section 2.2 we obtain 14 neighbourhood-based similarity measures which also satisfy the relevant properties and are denoted as M_i^h.

2.4 Histogram-Based Similarity Measures

Similarity between images can be calculated at the pixel or neighbourhood level, but also at the histogram level. Indeed, similar images will have similar histograms. This explains why we have investigated histogram-based similarity measures. Similarity measures can be applied to two different types of histograms: normalized histograms and ordered normalized histograms.

Normalized histograms. The histogram of an image can be transformed to a fuzzy set in the universe of gray levels by dividing the values of the histogram by the maximum number of pixels with the same gray value. In this way we obtain the membership degree of the gray value g in de fuzzy set Fh_A associated with the histogram h_A of the image A: $Fh_A(g) = \frac{h_A(g)}{h_A(g_M)}$, with $h_A(g_M) = \max_{g \in G} h_A(g)$.
As histograms can be identified with fuzzy sets (which are in fact normalized histograms), similarity measures can be applied.

A profound experimental study [20] resulted in 15 similarity measures which are appropriate for histogram comparison, i.e., they satisfy the list of relevant properties. Some examples:

$$H_1(A,B) = 1 - \left(\frac{1}{L} \sum_{g \in G} |Fh_A(g) - Fh_B(g)|^r \right)^{\frac{1}{r}}, \text{ with } r \in \mathbb{N}\setminus\{0\} \quad (9)$$

$$H_9(A,B) = \frac{\min(|Fh_A|, |Fh_B|)}{|Fh_A \cup Fh_B|}, \quad (10)$$

with L the total number of possible gray values and G the universe of gray levels.

Ordered normalized histograms. The values of a histogram can be ordered in ascending order, i.e., in such a way that the least occurring gray values are placed in the first positions of the histogram. This ordered histogram can be normalized as before, such that similarity measures can be applied. In this case the frequency of the most occurring gray value in the image A is compared with the frequency of the most occurring gray value in the image B, and so on. If the ordered histogram of an image A is denoted as o_A, we obtain the following expression for the fuzzy set associated with the ordered histogram: $Oh_A(i) = \frac{o_A(i)}{o_A(L)}$, for $i = 1, \ldots, L$ (L is the number of occurring gray levels) and with $o_A(L) = \max_{g \in G} h_A(g)$.

If the similarity measures are applied to ordered histograms we obtain the same 15 measures as for normalized histograms, and 7 additional measures which again satisfy the list of relevant properties. For example:

$$OH_6(A,B) = \frac{|Oh_A \cap Oh_B|}{|Oh_A \cup Oh_B|} = \frac{\sum\limits_{g \in G} \min(Oh_A(g), Oh_B(g))}{\sum\limits_{g \in G} \max(Oh_A(g), Oh_B(g))}. \quad (11)$$

2.5 Combined Similarity Measures

In order to optimally incorporate the image characteristics we propose combined similarity measures which consist of a combination of neighbourhood-based similarity measures and similarity measures which are applied to ordered histograms:

$$Q_{v,w}(A,B) = M_v^h(A,B) \cdot OH_w(A,B). \quad (12)$$

The obtained measures take into account information on pixel, neighbourhood and histogram level, resulting in a better perceptual behaviour. The latter is illustrated in the next subsection.

2.6 Experimental Results

In this section we will illustrate the different approaches with some examples. Obviously, we will not illustrate all the different similarity measures, but we will restrict ourselves to a limited amount of similarity measures [18,19,20,21]:

- the similarity measures M_1 ($r=2$), M_6 and M_{20};
- the neigbhourhood-based similarity measures M_1^h, M_6^h and M_{20}^h;
- the histogram similarity measures H_1, H_6 and H_9;
- the ordered histogram similarity measures OH_1, OH_6 and OH_9;
- the combined similarity measures $Q_{2,1}$, $Q_{6,6}$, $Q_{9,6}$ and $Q_{11,1}$.

The performance of the different similarity measures has been tested extensively with several experiments; here we show and discuss the results of the "brick" image. We add a variety of corruptions to the image (salt & pepper noise, Gaussian noise, enlightening, blur and JPEG compression), in such a way that the corruptions lead to the same $RMSE$ relative to the original image (see Figure 1). The results of the different similarity measures are shown in Table 1. From these numerical results we can conclude the following:

- the similarity measures that are applied directly, i.e., on the pixel level, do not succeed in expressing the differences in visual quality;
- the similarity measures that also incorporate homogeneity succeed partially in uncovering the visual differences, however the results for the compressed version of the "brick" image (image f) are still too large;
- the similarity measures that are applied directly to normalized histograms yield relatively good results, however the results for the enlightened version are too small to be in correspondence with visual quality evaluation;

- the similarity measures that are applied to ordered normalized histogram give rise to a good ordering of the different results, but the results of the "brick" image corrupted with salt & pepper noise are too large;
- the combined similarity measures succeed, in contrast to the other similarity measures, in uncovering the differences in visual quality. For every combined similarity measure we obtain the highest value for the enlightened version of the "brick" image, and the lowest value for the compressed version of the "brick" image.

If we order the different images according to image quality, we obtain, based on the numerical results in Table 1, the following order: (1) enlightening, (2) salt & pepper noise, (3) Gaussian noise, (4) blur, (5) JPEG-compression. This order is in accordance with the visual quality of the different images (see Figure 1).

Table 1. Results of the different similarity measures applied to the different versions of the "brick" image

measure	(a)-(b)	(a)-(c)	(a)-(d)	(a)-(e)	(a)-(f)
M_1	0.94495	0.94504	0.9451	0.94587	0.9457
M_6	0.98896	0.91868	0.90391	0.9272	0.9221
M_{20}	0.99162	0.91501	0.89878	0.92344	0.9183
M_1^h	0.61754	0.68617	0.9451	0.58405	0.79757
M_6^h	0.62547	0.66582	0.90288	0.5727	0.77551
M_{20}^h	0.62631	0.66334	0.89878	0.57209	0.77361
H_1	0.96833	0.93806	0.83253	0.91756	0.62283
H_6	0.9801	0.85611	0.64874	0.77094	0.08465
H_9	0.9885	0.87454	0.82437	0.80749	0.08911
OH_1	0.99249	0.94884	1	0.94006	0.63134
OH_6	0.98356	0.88266	1	0.82837	0.08951
OH_9	0.99023	0.88705	1	0.83368	0.08951
$Q_{2,1}$	0.5978	0.44652	0.9451	0.37209	0.23278
$Q_{6,6}$	0.62213	0.44755	0.90288	0.31666	0.16855
$Q_{9,6}$	0.62213	0.44755	0.90288	0.31666	0.16855
$Q_{11,1}$	0.61637	0.46939	0.9451	0.28749	0.12217
$RMSE$	14.03	14.01	14	13.80	13.84
$PSNR$	12.59	12.59	12.60	12.66	12.65

3 Extension to and Application in Color Image Processing

The increasing availability of images and the corresponding growth of image databases and users, make it a challenge to create automated and reliable image retrieval systems [2,3,10,11,13,15]. We consider the situation in which a reference image is available, and that similar images from a database have to be retrieved. A main drawback of most existing systems is that the images are characterized by textual descriptors (describing features like color, texture,

Fig. 1. The different versions of the "brick" image

morphological properties, and so on), which usually have to be made by a person [4,7,14]. In other words, the retrieval is not fully automated. In this section we discuss the results of a new color image retrieval system that does succeed in extracting images from a database of images automatically, without relapsing into the characterization of images by assigning some fundamental properties.

Suppose that our image database contains n digital color images, and that we have a source image A_0. In order to obtain the most similar image w.r.t. A_0 from the database, we calculate the similarity between the source image A_0 and every image A_j. The images are then ranked with respect to decreasing similarity, and only the most similar images (e.g. the first 10) are retrieved from the database. So in this case, the calculation of the similarity between two color images is the key to a successful image retrieval system. Therefore, we introduce the fuzzy HSI color space.

3.1 The Tools: A Fuzzy Partition of the HSI Color Space

The RGB color space (Red, Green, Blue) is widely used to represent colors, e.g. on computer screens. However, for color image retrieval purposes it is more convenient to use a color model that characterizes color with one dimension instead of three. Therefore, we prefer the HSI color space (Hue, Saturation and Intensity). This model attempts to produce an intuitive representation of color [12]. Hue is the color as described by wavelength, saturation is the amount of the color that is present, intensity is the amount of light.

The hue component is enough to recognize the color, except when the color is very pale or very somber. In order to perform an extraction based on dominant colors, we limit ourselves to 8 fundamental colors, that are modelled with trapezoidal fuzzy numbers [3]; a trapezoidal fuzzy number A is a fuzzy set with a trapezoidal-shaped membership function [9]. In that way we obtain a fuzzy partition in the sense of Ruspini [8] of the hue component. In those cases where there is nearly no color present in the image we will use the intensity component to identify the dominant "color". Also for this component we use a fuzzy partition to model the intensity component (see Figure 2).

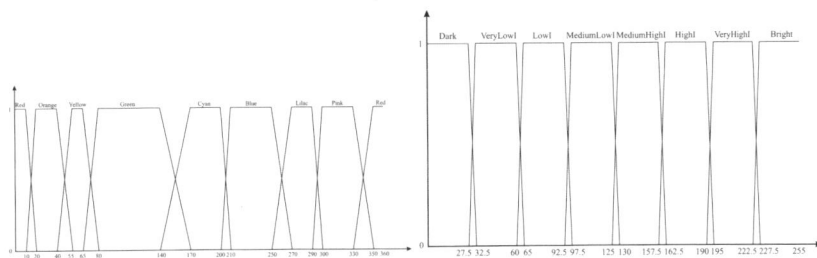

Fig. 2. Fuzzy partition of the hue component (left) and intensity component (right)

3.2 Image Retrieval Using Fuzzy Similarity Measures

First, we calculate the membership degree of all the pixels in every image with respect to the fundamental colors modelled by the trapezoidal fuzzy numbers. In that way we obtain 8 new "images".

Secondly, we consider the histogram of each of these 8 images, and normalize this by dividing all the values by the maximum value. In that way we obtain for each image 8 fuzzy sets, representing the frequency distribution of the membership degrees with respect to the 8 fundamental colors. For an image A_j and a color c, these histograms will be denoted as $h^c_{A_j}$.

To calculate the similarity $S(h^c_{A_0}, h^c_{A_j})$ between two histograms $h^c_{A_0}$ and $h^c_{A_j}$, we use the fuzzy similarity measure H_6 that turned out to be useful for histogram comparison [20]:

$$H_6(h^c_{A_0}, h^c_{A_j}) = \frac{|h^c_{A_0} \cap h^c_{A_j}|}{|h^c_{A_0} \cup h^c_{A_j}|} = \frac{\sum_{x \in s_c} \min(h^c_{A_0}(x), h^c_{A_j}(x))}{\sum_{x \in s_c} \max(h^c_{A_0}(x), h^c_{A_j}(x))}. \quad (13)$$

This value can be considered as the degree of similarity between A_0 and A_j w.r.t. color c; note that s_c is the support of the fuzzy number representing color c.

Finally, the similarities between A_0 and A_j with respect to the 8 fundamental colors are merged into one single overall similarity value $S^h(A_0, A_j)$ for the hue component, using the standard average as aggregation operator:

$$S^h(A_0, A_j) = \frac{\sum_c H_6(h^c_{A_0}, h^c_{A_j})}{8}. \quad (14)$$

In those cases where both histograms $h^c_{A_0}$ and $h^c_{A_j}$ only contain values equal to zero, the value $S(h^c_{A_0}, h^c_{A_j})$ will not be taken into account to calculate the average (this means that this color is not present in both images, thus a comparison with respect to this color is not relevant).

It is necessary to also consider the intensity component because in extreme cases, where there is hardly no color present in the images, black and white will be considered as highly similar. This is of course not satisfactory, and the intensity component will make a distinction between black (intensity equals zero) and white (intensity equals one). The procedure to calculate the overall similarity is the same as for the hue component. For an image A_j and an intensity degree d, the histograms will be denoted as $h^d_{A_j}$. The overall similarity value $S^i(A_0, A_j)$ for the intensity component is then given by:

$$S^i(A_0, A_j) = \frac{\sum_c H_6(h^d_{A_0}, h^d_{A_j})}{8}. \qquad (15)$$

The overall similarity between the images A_0 and A_j is defined as:

$$S(A_0, A_j) = \frac{S^h(A_0, A_j) + S^i(A_0, A_j)}{2}. \qquad (16)$$

Calculating membership degrees of all the pixels with respect to the 8 fundamental colors and 8 degrees of intensity is a rather time-consuming process (e.g. for 256 × 256 pixels). We can increase the calculation speed by first considering the standard histogram, followed by calculating the membership degree of every bin in the histogram (e.g. for 256 bins) with respect to the 8 fundamental colors and 8 degrees of intensity.

3.3 Experimental Results

Let us note first that it is not necessary that all the images in the database have the same dimension, because we consider histograms to calculate the similarity. We have performed several experiments to investigate the performance of the described color image retrieval system. One of the experiments concerned the retrieval of flag images from a database of 130 flags. This experiment showed that the retrieved (most similar) images only contained the colors of the source image.

Here we discuss another experiment with more complex natural images. It concerns a database of over 500 natural images of animals, flowers, buildings, cars, texture images, ... The results using a flower image as query image are shown in Figure 3, where the 10 most similar images are displayed together with the numerical result.

The results are quite good: the three most similar retrieved images are flowers in the same color as the one in the query image. The other retrieved images do not contain flowers but have a very similar layout, i.e., they all show an object with a natural background. This illustrates that the proposed approach has potential w.r.t. color image retrieval.

Fig. 3. Retrieval result for the natural image experiment

4 Conclusion and Future Research

Modified similarity measures, originally introduced to model similarity between fuzzy sets, can be applied successfully in image processing. Psycho-visual experiments confirm their good relation with human evaluation, which is not the case for the classical measures such as $RMSE$ and $PSNR$. A new application in color image retrieval also shows the potential of these measures: they overcome the drawback of indexing all images in a database before a query can be done. This leads to the conclusion that the discussed similarity measures deserve further research. This research should focus on the optimization of similarity measurs for color images (including psycho-visual experiments), the optimization of image retrieval applications (including comparative studies with other existing retrieval systems, study of the use of more advanced aggregation operators [1]), and the extension to video (where not only color but also time has to be taken into account).

References

1. Beliakov, G.: Definition of general aggregation operators through similarity relations. Fuzzy Sets and Systems 114, 437–453 (2000)
2. Brunelli, R., Mich, O.: Histograms analysis for image retrieval. Pattern Recognition 34, 1625–1637 (2001)
3. Chamorro-Martínez, J., Medina, J.M., Barranco, C., Galán-Perales, E., Soto-Hidalgo, J.M.: An approach to image retrieval on fuzzy object-relational database using dominant color descriptors. In: Proceedings of the 4th Conference of the European Society for Fuzzy Logic and Technology, EUSFLAT, pp. 676–684 (2005)
4. Chang, S.: Content-based indexing and retrieval of visual information. IEEE Signal Processing Magazine 14/4, 45–48 (1997)
5. Chen, S.M.: Measures of similarity between vague sets. Fuzzy Sets and Systems 74, 217–223 (1995)
6. De Cock, M., Kerre, E.E.: On (un)suitable fuzzy relations to model approximate equality. Fuzzy Sets and Systems 133/2, 137–153 (2003)
7. Del Bimbo, A.: Visual Information Retrieval. Morgan Kaufmann Publishers, San Francisco (2001)
8. Dubois, D., Prade, H. (eds.): Fundamentals of Fuzzy Sets. Kluwer Academic Publishers, Dordrecht (2000)
9. Fuller, R.: On product-sum of triangular fuzzy numbers. Fuzzy Sets and Systems 41/1, 83–87 (1991)
10. Lu, G., Phillips, J.: Using perceptually weighted histograms for colour-based image retrieval. In: Proceedings of the 4th International Conference on Signal Processing, pp. 1150–1153 (1998)
11. Omhover, J.F., Detyniecki, M., Rifqi, M., Bouchon-Meunier, B.: Ranking invariance between fuzzy similarity measures applied to image retrieval. In: Proceedings of the 2004 IEEE International Conference on Fuzzy Systems, pp. 1367–1372. IEEE, Los Alamitos (2004)
12. Sharma, G.: Digital Color Imaging Handbook. CRC Press, Boca Raton (2003)
13. Smeulders, A.W.M., Worring, M., Santini, S., Gupta, A., Jain, R.: Content-based image retrieval at the end of the early years. IEEE Transactions on Pattern Analysis and Machine Intelligence 22/12, 1349–1379 (2000)
14. Stanchev, P.: Using image mining for image retrieval. In: Proceedings of the IASTED International Conference on Computer Science and Technology, pp. 214–218 (2003)
15. Stanchev, P., Green, D., Dimitrov, B.: High level color similarity retrieval. International Journal of Information Theories & Applications 10/3, 283–287 (2003)

16. Tolt, G., Kalaykov, I.: Fuzzy-similarity-based noise cancellation for real-time image processing. In: Proceedings of the 10th FUZZ-IEEE Conference, pp. 15–18. IEEE Computer Society Press, Los Alamitos (2001)
17. Tolt, G., Kalaykov, I.: Real-time image noise cancellation based on fuzzy similarity. In: Nachtegael, M., Van der Weken, D., Van De Ville, D., Kerre, E.E. (eds.) Fuzzy Filters for Image Processing, pp. 54–71. Springer, Heidelberg (2002)
18. Van der Weken, D., Nachtegael, M., Kerre, E.E.: The applicability of similarity measures in image processing. Intellectual Systems 6, 231–248 (2001) (in Russian)
19. Van der Weken, D., Nachtegael, M., Kerre, E.E.: An overview of similarity measures for images. In: Proceedings of the IEEE International Conference on Acoustics, Speech and Signal Processing, pp. 3317–3320. IEEE, Los Alamitos (2002)
20. Van der Weken, D., Nachtegael, M., Kerre, E.E.: Using Similarity Measures for Histogram Comparison. In: De Baets, B., Kaynak, O., Bilgiç, T. (eds.) IFSA 2003. LNCS, vol. 2715, pp. 396–403. Springer, Heidelberg (2003)
21. Van der Weken, D., Nachtegael, M., Kerre, E.E.: Using Similarity Measures and Homogeneity for the Comparison of Images. Image and Vision Computing 22/9, 695–702 (2004)
22. Zadeh, L.: Fuzzy Sets. Information Control 8, 338–353 (1965)
23. Zadeh, L.: Similarity Relations and Fuzzy Orderings. Information Sciences 3, 177–200 (1971)

A Semi-automatic Feature Selecting Method for Sports Video Highlight Annotation

Yanran Shen, Hong Lu, and Xiangyang Xue

Dept. of Computer Science and Engineering, Fudan University
220 Handan Road, 200433 Shanghai, China
{yanran,honglu,xyxue}@fudan.edu.cn

Abstract. When accessing contents in ever-increasing multimedia chunks, indexing and analysis of video data are key steps. Among different types of videos, sports video is an important type of video and it is under research focus now. Due to the increasing demands from audience, highlights extraction become meaningful. This paper proposed a mean shift clustering based semi-automatic sports video highlight annotation method. Specifically, given small pieces of annotated highlights, by adopting Mean Shift clustering and earth mover's distance (EMD), mid-level features of highlight shots are extracted and utilized to annotate other highlights automatically. There are 3 steps in the proposed method: First, extract signature of different features – Camera Motion Signature (CMS) for motion and Pivot Frame Signature (PFS) for color. Second, Camera motion's co-occurrence value is defined as Camera Motion Devotion Value (CMDV) and calculated as EMD distance between signatures. Decisive motion feature for highlights' occurrences is thus semi-automatically detected. Finally highlights are annotated based on these motion parameters and refined by color-based results. Another innovation of this paper is to combine semantic information with low-level feature aiding highlight annotation. Based on Highlight shot feature (HSF), we performed hierarchical highlight annotation and got promising results. Our method is tested on four video sequences comprising of different types of sports games including diving, swimming, and basketball, over 50,000 frames and experimental results demonstrate the effectiveness of our method.

Keywords: Sports video, highlight annotation, Mean Shift, EMD, CMS, PFS, CMDV, HSF, hierarchical annotation.

1 Introduction

As digital videos become more pervasive, indexing and searching methods become more important research topic these days due to lack of human resource to perform annotation and indexing. Sports video, as a special type of video data possessing good structuring property, can be analyzed and annotated adopting different methods. Previous works include both high-level rooted and low-level rooted methods. For high-level based methods, semantic structures were adopted for basketball games [1], but purely empiristic semantic knowledge is neither robust nor flexible for different sports videos. More heuristic methods include those to use low-level features for parsing video into shots, [2,3,4,5] or to use low-level features for clustering video frames into similar classes.[6,7] Recently, interesting patterns in videos such as "dialog" or "action like" patterns are being researched for video indexing[8,9,10]. To take use of both semantic information and low-level feature, our method utilizes mean shift[11] for automatic mode identification, and uses EMD distance[12] for similarity measurement, thus constituting a scheme that can find low-level features semi-automatically based on semantic knowledge for highlight annotation.

While different sports games may vary in forms, each of them holds clear-structured intrinsic order due to predefined sports rules. In this case, sports games video can be considered as sequences of repeated structured events, and it's helpful to project human's knowledge of these rules to the process of annotating sports highlights. To construct mid-level concept bridging high-level and low-level features, we take consecutive shots of one highlight as a whole and develop highlight shot feature (HSF). As will be shown in part 4.1, an HSF is comprised of structured color signatures, highlight candidate choosing criterion (CCC) and corresponding thresholds, etc. In this way, high-level semantic information is associated with low-level features and a novel framework for semi-automatic sports video highlight feature selection is constructed.

The process of this method is demonstrated in Fig.1. First we train our method scheme on annotated training data, get camera motion vectors for each frame, and extract highlight frames' color histograms as color feature. Then camera motion devotion value (CMDV) for each motion parameter is calculated, different motion parameter is assigned different weight, considered as weighted classifier. These classifiers vote for each frame and judge whether the frame is highlight or not. A hierarchical framework is adopted here: after motion based annotation, further annotation is performed using boundary information extracted using color feature and annotation results are thus refined. Considering the whole highlight part's features as a whole, a mid-level concept (HSF) is defined aiding further extraction and annotation of certain structured parts of sports video in our case.

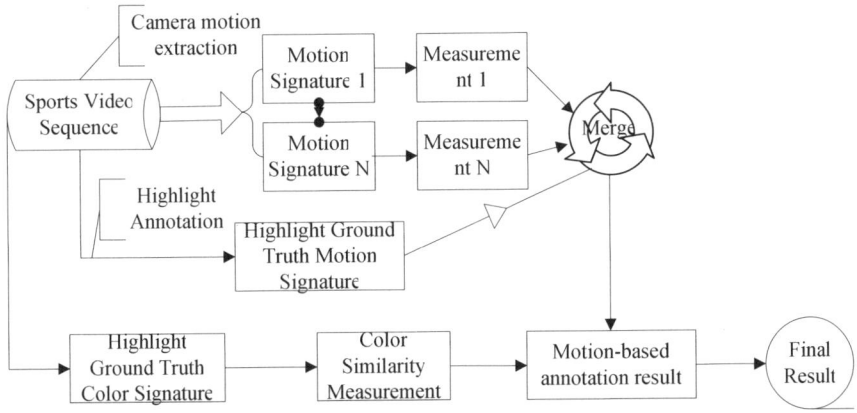

Fig. 1. Procedure of video highlight extraction and re-influence on highlight annotation

This paper is organized as follows: section 2 analyzes motion feature extraction and selection of video highlights; section 3 demonstrates color feature selection for video highlights. Section 4 combines these low-level features with high-level semantic structure definition, proposes the mid-level highlight shot feature (HSF) and analyzes its function for highlight annotation and indexing. Experiment results are discussed in section 5. Experiments were made on three sports video sequences including diving, swimming and basketball possessing different camera motion features and color features, proving robustness and effectiveness of our method.

2 Motion Analysis of Sports Video

Low-level feature extraction is the first step for semantic concept modeling. Low-level feature representation is fundamental for building concept. To construct low-level model for highlight, we adopted two kinds of features: camera motion parameters and color information of frames.

2.1 Camera Motion Analysis of Highlights

Motion is an important feature in video analysis and can be derived from available motion vector information in MPEG bit stream [14, 15, 16, 17]. Video motion vectors are usually classified into two types - local motion and global motion. Local motion refers to motion of objects in a video. Global motion indicates camera motion of a video.

Cameras usually track athletes or concerned scenes during highlights with a particular contrail, so we focus on camera motion parameters to track highlight motion modes for video sequences. The linearly independent three camera motion parameters are picked: panning, tilting, and zooming. The motion feature extraction procedure is demonstrated in Fig.2. Take diving games as example, during highlight segments, camera motion parameter "tilt" possesses extremely exquisite fluctuation. Assuming that $<f_1, f_2, ..., f_n>$ is the video frame sequence and f_n stands for the n_{th} frame, we set parsing window width w_d according to empiristic knowledge of highlight duration and calculate camera motion variance over the window through the whole video sequence, as indicated in Fig. 2. Because camera motion variance value curve during highlight periods possesses particular shape, camera motion variance reflect certain camera motion's co-occurrence property with highlights to some extent.

2.2 Camera Motion Devotion Value (CMDV)

In this paper, a new method is proposed to extract camera motion feature, calculating camera motion devotion value (CMDV) to evaluate devotion of a camera motion for determining sports highlights. The reason to calculate CMDV is to aid results merging – for sports videos possessing particular motion features, highlight parts have certain camera motion modes and they are represented by camera motion features. We adopted panning, tilting and zooming variance values separately for extract highlight modes over the video sequence and classify each frame as highlight or non-highlight. Three motion parameters lead to three result sets. In this paper we simply choose the camera motion possessing highest CMDV to be the decisive motion feature. After that, results are refined using color features. Classifiers' results merging method is a popular research area and we adopted standard HR merging method to compare with our method. The procedure of CMDV production is shown in Fig.2.

Camera Motion Signature (CMS) production. Getting camera motion parameters for each frame, we calculate motion variance for each frame using it as window center with preset parsing window width. By normalizing motion variance values to [0,1], we applied mean shift clustering and get local maximum points standing for local modes. Local maximum points stand for locations where motion vectors vary most intensively, possessing largest probability to be highlight regions. Other clustering methods would also be useful, like k-means, but mean shift is proved practically best performer for this case. Broadening each local mode point to a region with each point having motion vector variance above threshold t set by training, we got candidate regions as Fig.2 shows. Then we get a 0-1 sequence as <s1, s2, ..., sn>. n is the index of frame over the whole video sequence and si is 1 while the frame is in the highlight candidate region and 0 otherwise. Thus camera motion signature (CMS) is created for each camera motion parameter. (pan, tilt and zoom in this paper, can also be extended to 7 camera motions [1]).

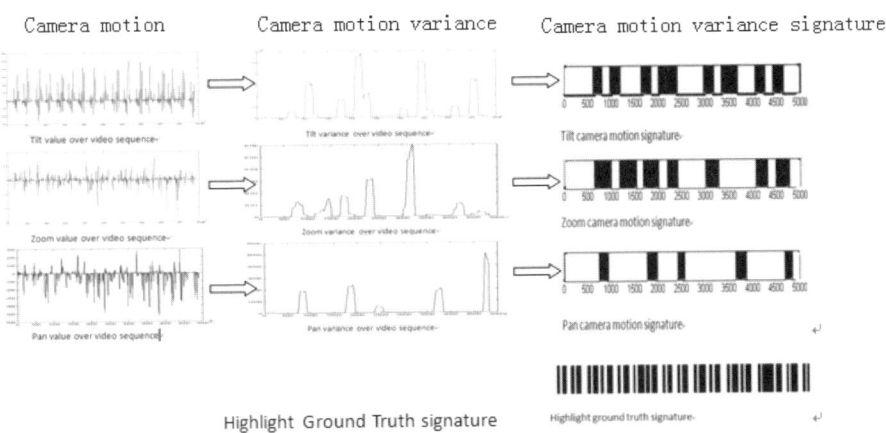

Fig. 2. Taking diving video as example, pre-annotated diving highlights are trained and parsing window width is got. Then parsing window parses video sequence and camera variance values are produced. Mean shift is adopted on variance values and CMS associated with cluster center as well as cluster weights are produced. Ground truth CMS is also shown in Fig.2, which is annotated by human in training.

Camera Motion Devotion Value (CMDV) production. After CMS extraction, next step is to measure camera motion's co-occurrence with highlight ground truth regions. We adopt earth mover's distance (EMD) method here to calculate co-occurrence of certain camera motion (CM) parameter and ground truth highlight regions for two reasons. First, highlights and CM variance distribution both possess periodicities. After normalization, they can be considered probability distribution over the video sequence. To measure probability distribution's distance, EMD is a powerful measurement with tolerance for different distribution bin numbers. Ground distance is L1 distance in our case, L1 distance is a reasonable choice because co-occurrence associates with overlapping of distributions, which is represented as L1 distance in this case. Another reason to choose L1 as ground distance is that the metric space is even and we don't have to worry about proportional spacing property of metric space. As indicated by Table 1, we got tilting as highest ranked CM for diving, panning for swimming and zooming for basketball games. The second reason to choose EMD distance is that it performs co-occurrence counting well. Given a highlight ground truth sequence H with length l_h and a video sequence C with length l_c, getting their separate CMS h=<h_1, h_2, ..., h_{lh}> and c=<c_1, c_2, ..., c_{lc}>, CMDV of sequence C is defined as EMD(h, c), intuitively standing for a camera feature's co-occurrence with highlights. CMDV for different CM features are used as weights for classifier merging in more sophisticated conditions.

Table 1. Different CMDV values and mean shifted results for camera motions

Sports Type	CM Feature	CMDV	Meanshifted cluster number
Diving	Tilt	0.918	8
	Pan	0.770	5
	Zoom	0.856	8
Swimming	Tilt	0.359	7
	Pan	0.626	4
	Zoom	0.077	3

To ensure that EMD distance between these signatures is true metric, CMDV is calculated after transferring CMS into normalized signatures with sum equaling to 1 over the video sequence.

2.3 Merging Method of Camera Motion Based Classifiers

After we have got CMDV for different CM, we adopt them and merge classifier results. In this paper, we adopt the simplest case, just use the result set of the feature possessing highest CMDV. For comparison we also give the results for HR method. HR method is popular and widely used method for measurement results merging.[17]

To better utilize semantic information of video sequence for annotation, based on the results of camera motion highlight extraction, we constructed a novel hierarchical refining framework for highlight extraction, as will be demonstrated in section 3.

3 Color Analysis of Sports Video

3.1 Color Analysis of Highlights

Color is another important low-level feature in video sequence analysis. Some sports highlights possess particular color feature modes as indicated in Fig.3. And Fig.4 shows color hue value distributions as histograms over frames that possess apparent features. These frames can be considered pivots for highlight indexing. For example, diving highlights begin with close take of the athlete's frontal, which comprises mainly of skin colors and end with blue scenes casting water surface. Similarly, swimming highlights comprise of scenes casting swimming pool presenting blue color in the main part, while tennis games can be segmented due to identification of color of the ground. However, not all kind of sports videos possess this property. Gymnastics, for example, does not hold any color regularity and thus color is not a robust feature for highlight extraction. So color can be only used as supplementary information for highlight annotation.

For those video highlights possessing apparent color features, we construct pivot color frames (PCF) with particular color feature representational of corresponding highlight. There are several cases listed in Table 2 indicating different PCF values for highlights.

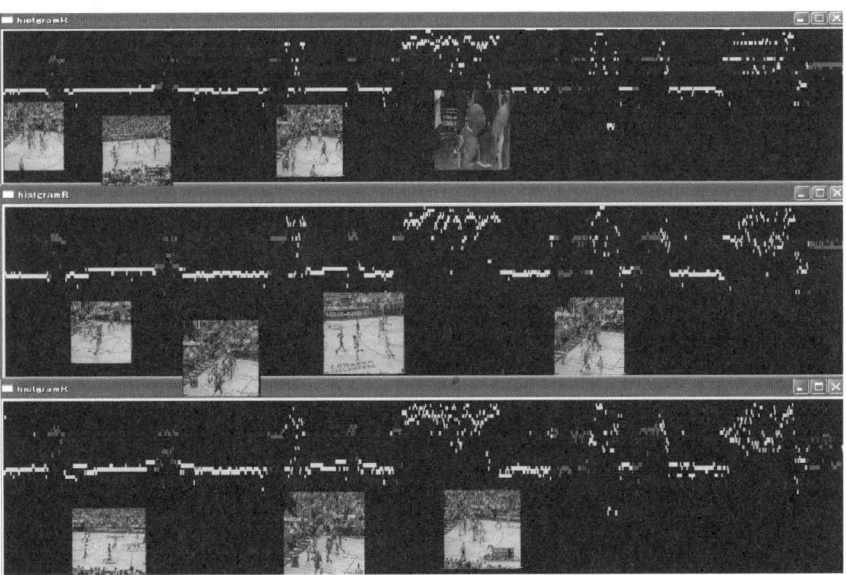

Fig. 3. Highlight hue value distribution over basket ball game video sequence. This shows a figure consisting of highest ranked three hue values alone y-axis, which is divided into 50 bins in all, ranging over the whole hue value region. X-axis stands for frame index over time line.

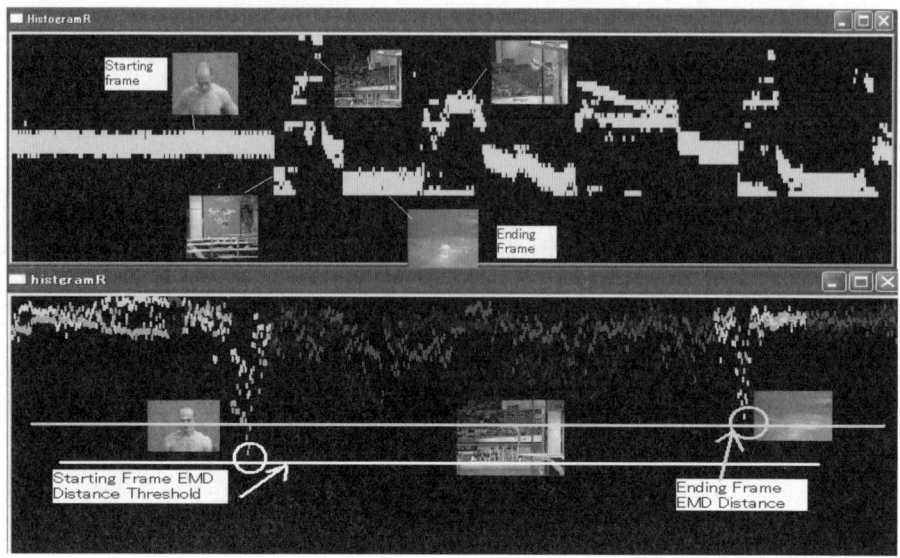

Fig. 4. The upper image is highlight hue value distribution for diving video. The upper one shows top five hue bin values over y-axis indicating complete hue range. The below one is the curve of EMD distance between each frame and two prototypes (frame 357 for starting and 753 for ending), starting frame hue color signature and ending frame hue color signature. Green and blue curves are EMD distance between starting color prototype and frames, while yellow and red curves are EMD distance of frames and ending color prototype. For one curve, two different colors indicates highlight region and non-highlight region, in this case we set green and yellow parts as non-highlight parts, while blue and red parts indicate highlight regions.

We can easily tell that around starting and ending of sports highlights, EMD distance of frames are especially close to prototype's values, thus leading to smaller EMD distances than that of other frames. (5.86 for starting frame and 6.46 for ending frame in our case)

3.2 Color Signature Extraction and Similarity Measurement

Extraction of color signatures [12] is based on mean shift clustering [11]. To compute similarity between a frame's hue and the PCF's hue values, we applied mean shift clustering to each frame; extract its color mode in CIE-lab space and label clusters with mean shift values, thus get signatures for similarity comparison. Taking video frame size and color difference for main scenes into consideration, we set mean shift thresholds with sp=10 and 10 iterations for mean shift clustering here. Top five highest ranked hue values are remained in the signatures for computing efficiency. Then earth mover's distances (EMD) are calculated between color signatures of each frame and the pivot frames (starting pivot frames and ending pivot frames), frames possessing smallest EMD are most likely to be pivot frames. Using this combined method, aided by semantic information, highlights can be annotated with better performance.

The reason to apply mean shift clustering is that it efficiently and adaptively produces color signatures. Using the form of signature to represent color information, we only refer to those existing colors, omitting non-existing colors, bringing efficiency of similarity computation. Earth mover's distance is a similarity measurement that is in accordance with perceptual difference (color difference in our experiments), and it can also measure difference-length signatures, which brings extensibility for our signatures and is suitable for our case.

4 Mid-Level Highlight Shot Feature(HSF)

4.1 Color Analysis of Highlights

Based on both camera motion and color similarity, we developed mid-level feature of sports highlights in video sequences to represent both low-level features and semantic information about the video. Sports games usually possess good structuring properties. While different sports games may vary in forms, each of them holds clear-structured intrinsic order due to predefined sports rules. In this case, sports games video can be considered as sequences of repeated structured events, and it's very helpful to project human's knowledge of these rules to the process of annotating sports highlights.

To construct the mid-level concept to bridge high-level and low-level features, we take consecutive shots of one highlight as a whole and developed highlight shot feature (HSF). As is shown below, a HSF is comprised of structured color signatures, highlight candidate choosing criterion (CCC) and corresponding thresholds and decisive camera motion parameter's CMDV values as weights, etc, as (1) shows:

<CMDV$_1$, CMDV$_2$, ..., CMDV$_m$, PFS1, PFS2, ..., PFSn, CCC, Threshold Sets>   (1)

Taking diving games for example, different diving games may possess different structures, but the common ground is they all comprise of the following components - Preparation Action (PA) -> Run-up and Jump(RJ) -> Disembark(D)-> Jump Replay (JR)-> Scoring (S). Each part has its pivot color signature (PCF) and decisive camera motion (panning, tilting or zooming). HSF of this diving video is like:

<Tilt, PA signature, D signature, variance over trained parsing window, mean   (2)
shift threshold, variance truncation thresholds>

Applying HSF, highlights in sports videos can be annotated semi-automatically. A very good point is that once we know CCC of a HSF, CM set of it is extracted automatically according to the criterion we set. For example, for diving video we can decide automatically which camera motion is more important, and this is usually set by experience in all the previous works [1].

Without loss of generality, we applied the proposed annotation algorithm on different sports videos and formed the semantic structure of each game. For each sports game, we construct HSV based on training and universal knowledge of the game, but not too specific structure dependent on certain sports sequences. This makes our method heuristic. Another point is that since we are applying motion and color features aiding our annotation, choices of sports games possess different motion and color properties.

4.2 Hierarchical Highlight Annotation Method

We tested different methods including: our method, use tilt variance only, use pan variance only, use zoom variance only and use merged result combining these three motion vector's results altogether. Experiments were made on different types of sports videos including diving videos, swimming videos and basketball videos. We calculated precision and recall for highlight annotation results, and compared these methods. Experiment results indicate that our method is comparatively more effective and more robust than other methods. Part of the reason is because that the sports videos we tested have good semantic structures, when projected to low-level features, lead to apparent color features as well as dominant motion modes for highlights, thus produce clear boundary between highlights and non-highlights.

Table 2. Highlight annotation results for diving video 1

Sports Type	CM Feature	Frame pr	Frame rc
Diving1	Tilt variance	0.024	0.227
	Pan variance	0.020	0.185
	Zoom variance	0.046	0.435
	HR Method	0.090	0.868
	Our Method	0.096	0.925

Table 3. Highlight annotation results for diving video 2

Sports Type	CM Feature	Frame pr	Frame rc
Diving2	Tilt variance	0.052	0.434
	Pan variance	0.034	0.278
	Zoom variance	0.045	0.370
	HR Method	0.074	0.615
	Our Method	0.082	0.681

Table 4. Highlight annotation results for swimming video

Sports Type	CM Feature	Frame pr	Frame rc
Swimming	Tilt variance	0.020	0.134
	Pan variance	0.068	0.467
	Zoom variance	0.050	0.342
	HR Method	0.103	0.709
	Our Method	0.119	0.819

Table 5. Highlight annotation results for basketball video

Sports Type	CM Feature	Frame pr	Frame rc
Basketball	Tilt variance	0.022	0.114
	Pan variance	0.054	0.281
	Zoom variance	0.081	0.423
	HR Method	0.111	0.575
	Our Method	0.141	0.732

5 Discussion and Future Work

There are many issues that may lead to future work in the area of video annotation. For example, to use semantic knowledge combined with low-level features and further improve our method may lead to some automatic video feature recognizing method and thus greatly improve video indexing and annotation works. In our method, still, different algorithm to draw camera motion feature and annotated highlight ground truth and some other factors should be given. How to develop an automatic annotation scheme is also an interesting topic. And there are still sports videos with highlights possessing no apparent motion or color features, how to efficiently annotate these segments using semantic knowledge is also a promising and interesting research direction.

References

1. Tan, Y.-P., Saur, D.D., Kulkami, S.R., Ramadge, P.J.: Rapid Estimation of Camera Motion from Compressed Video with Application to Video Annotation. Circuits and Systems for Video Technology, IEEE Transactions 10(1), 133–146 (2000)
2. Aigrain, P., Joly, P.: The Automatic and Real-time Analysis of Film Editing and Transition Effects and Its Applications. Computer and Graphics 1, 93–103 (1994)
3. Zhang, H.J., Low, C.Y., Smoliar, S.W.: Video Parsing and Browsing Using Compressed Data. Multimedia Tools and Applications 1(1), 89–111 (1995)
4. Meng, J., Yujue, J., Chang, S.-F.: Scene Change Detection in a MPEG Compressed Video Sequence. In: ISLT/SPIE Symposium Proceedings, vol. SPIE-2419, pp. 14–25 (1995)
5. Yeo, B.-L., Liu, B.: Rapid Scene Analysis on Compressed Videos. IEEE Transactions on Circuits and Systems for Video Technology 5, 533–544 (1995)
6. Zhong, D., Zhang, H.-J., Chang, S.-F.: Clustering Methods for Video Browsing and Annotation. Storage and Retrieval for Still Image and Video Database IV SPIE-2670, 239–246 (1996)
7. Yeung, M.M., Yeo, B.-L.: Time –Constrained Clustering for Segmentation of Video into Story Units. In: Int. Conf. on Pattern Recog., vol. C, pp. 375–380 (August 1996)
8. Aigrain, P., Joly, P., Longueville, V.: Medium Knowledge-Based Macro-Segmentation of Video into Sequences, Intell, Multimedia Inf, Retrieval (1996)
9. Hisashi, A., Shimotsuji, S., Hori, O.: A Shot Classification Method of Selecting Effective Key-Frames for Video Browsing, ACM Multimedia, pp. 1–10 (1996)

10. Minerva, M.Y., Yeo, B.-L.: Video Content Characterization and Compaction for Digital Library Applications. Storage and Retrieval for Still Image and Video Database IV SPIE-3022, 45–58 (1997)
11. Comaniciu, D., Meer, P.: Mean Shift, a Robust Approach Toward Feature Space Analysis. Pattern Analysis and Machine Intelligence, IEEE Transaction 24, 603–619 (2002)
12. Rubner, Y., Tomasi, C., Guibas, L.J.: The Earth Mover's Distance as a Metric for Image Retrieval. International Journal of Computer Vision 40, 1405–1573 (2000)
13. Yuh-Lin, C., Wenjung, Z., Ibrahim, K., Rafael, A.: Integrated Image and Speech Analysis for Content-based Video Indexing. In: Proceedings of IEEE Multimedia, vol. 2996, pp. 306–313
14. Chang, S.F., Messerschmitt, D.G.: Manipulation and Composition of MC-DCT Compressed Video, IEEE Journal of Selected Areas in Communications, Special Issue on Intelligent Signal Processing, 1–11 (January 1995)
15. ISO/IEC 11172-1: Information Technology – Coding of moving pictures and associated audio for digital storage media at up to 1.5 MBit/s – Part 1: Systems (1993)
16. Vasudev, B., Konstantinos, K.: Image and Video Compression Standards – Algorithms and Architectures. Kluwer Academic Publishers, Dordrecht (1995)
17. Lu, Y., Meng, W., Shu, L., Yu, C., Liu, K.-L.: Evaluation of Result Merging Strategies for Metasearch Engines. In: Ngu, A.H.H., Kitsuregawa, M., Neuhold, E.J., Chung, J.-Y., Sheng, Q.Z. (eds.) WISE 2005. LNCS, vol. 3806, pp. 53–66. Springer, Heidelberg (2005)

Face Image Retrieval System Using TFV and Combination of Subimages

Daidi Zhong and Irek Defée

Department of Information Technology, Tampere University of Technology,
P.O. Box 553, FIN-33101 Tampere, Finland
{daidi.zhong,irek.defee}@tut.fi

Abstract. Face image can be seen as a complex visual object, which combines a set of characterizing facial features. These facial features are crucial hints for machine to distinguish different face images. However, the face image also contains certain amount of redundant information which can not contribute to the face image retrieval task. Therefore, in this paper we propose a retrieval system which is aim to eliminate such effect at three different levels. The Ternary Feature Vector (TFV) is generated from quantized block transform coefficients. Histograms based on TFV are formed from certain subimages. Through this way, irrelevant information is gradually removed, and the structural and statistical information are combined. We testified our ideas over the public face database FERET with the Cumulative Match Score evaluation. We show that proper selection of subimage and feature vectors can significantly improve the performance with minimized complexity. Despite of the simplicity, the proposed measures provide results which are on par with best results using other methods.

Keywords: Face image retrieval, FERET, TFV, subimage.

1 Introduction

Face image retrieval is a highly complex pattern recognition problem due to the enormous variability of data. The variability is due to a combination of both statistical and structural factors. Despite this variability, images are very efficiently processed for pattern extraction by biological systems. This means that those systems are robust in extracting both the statistics and structure from the data. The details of how their processing is done are not deciphered yet, and in fact even the formulation what efficiency and robustness means in the biological context is not clear. However, it seems that combination of structure and statistics-based processing is used in this process. In this sense, the face images are mixtures of structure and statistics which makes the description problem hard because its complexity looks like unbounded. In addition, the image quality often suffers from the noise and different light conditions, which make the retrieval tasks more difficult.

Some previous works focused on extracting and processing global statistical information by using the whole image [1], while some other researchers start from some key pixels [2] to represent the structural information. Based on their achievement, a reasonable way to further improve the retrieval performance is to extract the visual information in a way like a mixture of statistical and structural information.

In this paper, we illustrate our idea by proposing a retrieval system which is based on subimages and combinations of feature histograms. This can also be seen as a pathway to from local information (pixels) to the global information (whole image). The experimental results disclose that the usage of subimage and local feature vectors can lead to the combination of statistical and structural information, as well as minimized impact of noise, which finally improve the performance of the approach.

In order to achieve a comparable result, we tested our method over a public benchmark of face image database. The evaluation method of this database has been standardized, which allow us to see the change of performance clearly. However, using face images as an example here does not mean our method is limited to the application of face image retrieval; it also has the potentiality to be applied to other image retrieval tasks.

2 Transform and Quantization

Visual images are usually in the format of 2-D matrix of pixels. Such representation contains a large amount of redundant information. Several image and video compression standards are intend to minimize such redundancy by utilizing lossy compression methods. Transform and quantization are two common steps of these methods. From the view of compression, they can reduce the redundancy; from the view of retrieval, they help us to find out the most distinguishable features.

Some transforms have been found useful in extracting local visual information from images. Popular transforms include: Gabor Wavelet, Discrete Wavelet Transform, Discrete Cosine Transform (DCT), and Local Steerable Phase. Specially, DCT and Wavelets have already been adopted to the image and video compression standards [3],[4]. The specific block transform we used in this paper was introduced in the H.264 standard [5] as particularly effective and simple. The transform matrix of the transform is denoted as **B**$_f$ and the inverse transform matrix is denoted as **B**$_i$. They are defined as

$$B_f = \begin{bmatrix} 1 & 1 & 1 & 1 \\ 2 & 1 & -1 & -2 \\ 1 & -1 & -1 & 1 \\ 1 & -2 & 2 & -1 \end{bmatrix} \quad B_i = \begin{bmatrix} 1 & 1 & 1 & 1 \\ 1 & 0.5 & -0.5 & -1 \\ 1 & -1 & -1 & 1 \\ 0.5 & -1 & 1 & -0.5 \end{bmatrix} \quad (1)$$

The 4x4 pixel block **P** is forward transformed to block **H** using (2), and the block **R** is subsequently reconstructed from **H** using (3). The 'T' means linear algebraic transpose here.

$$H = B_f \times P \times B_f^T \quad (2)$$

$$R = B_i^T \times H \times B_i \tag{3}$$

A 4x4 block contains 16 coefficients after applying the transform. The main energy is distributed around the DC coefficient, namely, the upper-left coefficient. The rest are called AC coefficients, which are usually quite small. The small coefficients contain small energies, which is generally regarded as irrelevant information or noise. Therefore, quantization can be applied immediately after the transform to remove the redundancy. The reduction of the amount of information greatly facilitates our information retrieval tasks.

Furthermore, it also has the effect of limiting the dynamic range of coefficients. Larger quantization level tends to reduce the range of possible coefficients. In the ultimate case, when the quantization level is large enough, all the coefficients will be thresholded to zero; thus, no distinguishing ability can be achieved from them, which is not the case we wish to see. Therefore, we use certain training process to find out the optimal range of quantization level. The objective is a good compromise between removing non-important visual information and preserving the main perceptually critical visual information.

3 Feature Vectors

Block transform and quantization arranged the local information in a suitable way for retrieval. Based on this merit, we utilize the specific feature vector defined below to further group the local information in the neighboring blocks. The grouping process can be applied separately or jointly over DC and AC coefficients for all transform blocks of an image.

Considering a 3x3 block matrix containing nine neighboring blocks, the DC coefficients from them can form a 3x3 coefficient matrix. The eight DC coefficients surrounding the center one can be thresholded to form a ternary vector with length eight. This vector is called DC Ternary Feature Vectors (DC-TFV), which encode the local information based on those quantized transform coefficients.

The threshold is defined as a flexible value related to the mean value of all the nine DC coefficients.

$$Threshold_1 = M + (X - N) \times f$$
$$Threshold_2 = M - (X - N) \times f \tag{4}$$

where f is real number from the interval (0, 0.5), X and N are maximum and minimum values in the 3x3 coefficient matrix, and M is the mean value of the coefficients. Our initial experiments have shown that performance with changing f has broad plateau for f in the range of 0.2~0.4. From this reason, we use f = 0.3 in this paper. The thresholded values can be either 0, 1 or 2.

If the coefficient value ≤ $Threshold_2$ put 0
If the coefficient value ≥ $Threshold_1$ put 2
   otherwise   put 1

The resulting thresholded vectors of length eight are subsequently converted to decimal numbers in the range of [0, 6560], where 6560=3^8. However, not all of these 6561 TFV are often present in the face images. In fact, only a small part of them often appear. Therefore, we only utilize the most common TFVs for retrieval tasks. The proper selection is done based on training process.

The same process as above can be applied to AC coefficients. The resulted feature vectors are called AC-TFV. Due to the quantization and intrinsic property, not all the AC coefficients are suitable for the retrieval task. Especially, those coefficients representing high frequency energy tends to have little dynamic range of values. Their distinguishing ability is too poor to achieve good performance. Therefore, we only select those coefficients which can show high distinguishing ability. Such selection is done based on training process.

On the other hand, using more coefficients will certainly increase the complexity. The proper selection can be conducted with training set. However, we only present the results with one very capable AC coefficient, which can already show fairly good retrieval accuracy. The information obtained from it will be further combined with the information obtained from DC coefficient to enhance the performance. Since they are representing different information, the combination of them is expected to show more aspects of the visual object.

4 Representation Based on Subimages

Certain facial areas, such as eye, nose and mouth, are generally believed to contain more distinguishing ability than other areas. Here we refer to such areas as subimage. It is not sagacious enough to treat all the subimages of face image equally. Some researchers have already noticed this point. For example, Ahonen tried to manually assign different weights to the features generated from a set of predefined subimages [6]. Alternatively, in [7], the features obtained from subimages are combined with the features obtained from the whole image. Proper decision is made by certain data fusion strategy. In [8], Chunghoon manually selected several subimages and generate PCA features based on them. In this paper, we do it in a different way: we randomly defined some rectangular subimages over the original image. TFV is extracted from each subimage separately. These TFV vectors are represented by special histograms, which may further be combined to serve the retrieval tasks.

Rather than manually select the eye or noise areas, we randomly selected 512 subimages, which can be overlapped to cover the whole image. There is a wide dynamic range of the sizes of these subimages: the smallest one is about 1/150 of the size of whole image, while the largest one is about 1/5. Both the large and small subimages are selected from every region of the face image. Some examples of subimage are shown in Fig.1.

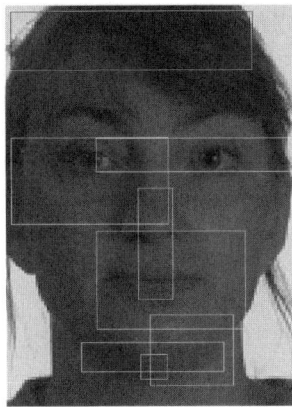

Fig. 1. Examples of subimage (each rectangle is a subimage)

5 Histograms of Feature Vectors and Similarity Measure

The aforementioned block transform, TFV and subimage can be regarded as three levels of collecting local information. Such system gradually moves from local to global aspects of the image. They are linked by using a histogram of TFV features. The generation of histograms is done in the following way:

1. The 4x4 H.264 AC Block Transform is applied to a subimage.
2. Quantization is applied separately to all the AC and DC coefficients.
3. TFV is generated from certain coefficient.
4. Histogram is generated from this subimage by simply counting the number of each occurring TFV.
5. Histogram is normalized according to the size of subimage.

Specifically for AC-TFV histogram, there is one bin which is too dominant comparing to other bins. This is caused by the smooth area in image and quantization. Such areas will generate many all-one vectors, like [1 1 1 1 1 1 1 1]. Our retrieval does not use this bin, since it decreases the discriminate ability.

Histogram based on DC-TFV and AC-TFV can be used separately or collectively. Since they represent different information, the combination of them can leads to better performance, which will be shown in the following experiment. The combination is done by simply concatenating each histogram one by one. For example, two subimages (sub1 and sub2) are used to generate the DC-TFV and AC-TFV histograms, the combined histogram can be shown like:

[Combined Histogram] =

[DC-TFV-sub1 AC-TFV-sub1 DC-TFV-sub2   AC-TFV-sub2]   (5)

There is no specific weight for each histogram. In another word, the information coming from different subimages are treated equally.

The retrieval task is completed by classifying any input image to a known person, according to the distance between their corresponding histograms. There are several well-known similarity measures which are widely used in pattern recognition society, e.g., L1 norm, L2 norm, Cosine distance and Mahalanobis distance. The *pdist* function in MATLAB [9] has implemented several of them, which are used by us to test over some training data. Finally, we choose to use the L1-norm. It is simple to be calculated and suitable for the TFV histograms. Thus, the distance between two histograms $H_i(b)$ and $H_j(b)$, b= 1, 2, ... B are calculated as:

$$\text{Distance}(i,j) = \sum_{b=1}^{B} \left| H_i(b) - H_j(b) \right| \qquad (6)$$

6 Experiments with FERET Database

6.1 FERET Database

For testing the performance of the proposed method we use the FERET face mage database [10]. The advantage of using this database is its standardized evaluation method of based on performance statistics reported as Cumulative Match Scores (CMS) which are plotted on a graph [11]. Horizontal axis of the graph is retrieval rank and the vertical axis is the probability of identification (PI) (or percentage of correct matches). On the CMS plot higher curve reflects better performance. This lets one to know how many images have to be examined to get a desired level of performance since the question is not always "is the top match correct?", but "is the correct answer in the top n matches?"

National Institute of Standard and Technology (NIST) have published several releases of FERET database. The release which we are using is the one published at October 2003, called Color FERET Database. It contains overall more than 10,000 images from more than 1000 individuals taken in largely varying circumstances. Among them, the standardized FA and FB sets are used here. FA set contains 994 images from 994 different objects, FB contains 992 images. FA serves as the gallery set, while FB serves as the probe set.

Before the experiments, all the source images are cropped to contain face and a little background. They are normalized to have the same size. The eyes are located in the similar position according to the given information from FERET. Simple histogram normalization is taken to the entire image to tackle the luminance change. However, we did not apply any mask for the face images, which are used by some researchers to eliminate the background.

6.2 Retrieval System with the Training Process

We made experiments over certain number of subimages. During the training and retrieval process, the same set of subimages is applied to all the images. TFV histograms are generated from them, and are subsequently used for retrieval. The training process is described in Fig.2. To ensure the independence of training set, five different groups of images are randomly selected to be the training sets. Each group contains 50 image pairs (from 992 pairs). Five parameter sets are obtained from them, which will be applied for evaluation of performance of the rest 942 images of the whole database. The resulting five CMS curves are averaged, which is the final performance result.

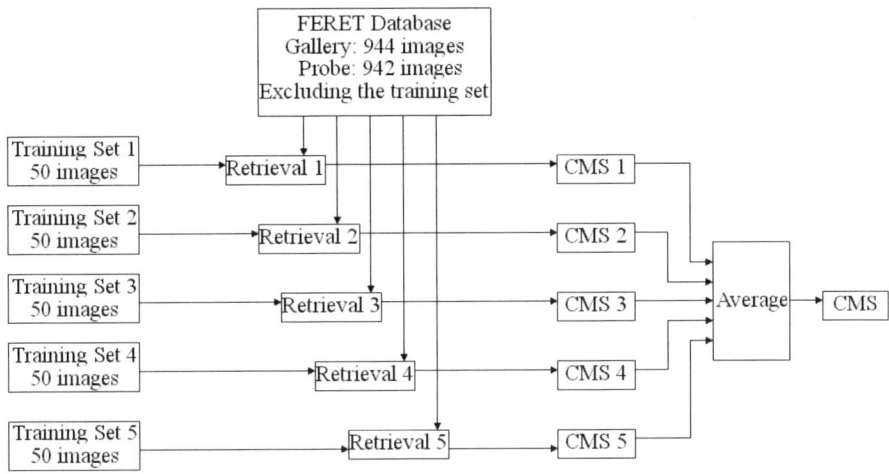

Fig. 2. Training and Retrieval process. The optimal parameter set from five training sets are utilized separately, which give five CMS scores. The overall performance of given subimage will be evaluated as the average of above five CMS scores.

6.3 Performance of TFV Histograms Using the Single Subimage

The first experiment is conducted over the aforementioned randomly selected 512 subimages. Since we have five training sets, the final result is actually a matrix of 5x512 CMS scores. They are further averaged to be a 1x512 CMS vector. The maximum, minimum and average of these 512 CMS scores are shown in Table 1. One can see from it that there is very wide performance variation for different subimages. The selection of subimage is thus critical for the performance which can be achieved. However, this is also reasonable since there are certain subimages which are too small to provide enough information. Fig.3 shows the distribution of rank-1 CMS scores using DC-TFV from single subimage.

When interpreting the results in Table 1, one may also conclude that: the DC-TFV subimage histograms always perform markedly better than DC-TFV histograms but their combination performs still better in the critical high performance range. When comparing them, one may use the average CMS scores as the criterion. It is representing the overall performance, rather than the performance of any specific subimage.

Table 1. The Rank-1 CMS results of using single subimage

Rank-1 CMS (%)	DC-TFV	AC-TFV	DC-TFV + AC-TFV
Maximum	93.77	60.77	95.30
Minimum	9.01	1.69	12.94
Average	56.59	20.99	62.11

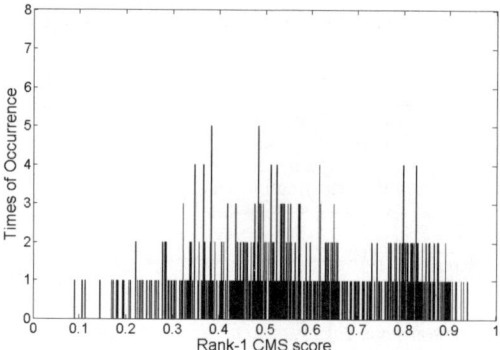

Fig. 3. The distribution of rank-1 CMS scores using DC-TFV generated from single subimage

As we mentioned before, the training process is conducted over five training sets. Therefore, we have five set of CMS scores which are obtained from the same database. It is thus interesting to study their differences. Table 2 shows the maximum, minimum and average difference between five training sets. As one can see, the average difference between different training sets is only about one percent. This is quite good considering the size of training set is about only 5.2% of the probe set.

Table 2. The difference between five training sets, using DC-TFV from single subimage, represented by Rank-1 CMS scores

	Maximum	Minimum	Average
Rank-1 CMS (%)	3.73	0.1	1.26

6.4 Performance of TFV Histograms Using the Single Subimage

Based on above results, a reasonable way to improve the performance is to combine multiple subimages. In the following experiment, we utilized 216 pairs of subimage for retrieval. Most of these selected subimages can provide relatively better performance in the previous experiment. In addition, when any two subimages are making a pair, they must be coming from different region of the face image. For example: one is from eye region, and the other is from mouth region. The retrieval task is performed over them and the best pairs are identified.

Furthermore, one additional subimage from different region is added to above 216 pairs. Therefore, we have another experiment which is using three subimages. Totally, we have 432 combinations of 3-subimage. All of these three subimages are coming from different regions of the face image.

The maximum, minimum and average of the resulted scores for both the 2-subimage and 3-subimage cases are shown in Table 3. Clearly we can see the improvement from 1-subimage, 2-subimage to 3-subimage. In addition, we show in Table 4 the average difference between five training sets when doing these two sets of experiments. The difference is less than one percent, which can be safely regarded that the results are not sensitive to the selection of training sets.

Table 3. The Rank-1 CMS results of using two/three subimage

Rank-1 CMS (%)	DC-TFV		AC-TFV		DC-TFV + AC-TFV	
Subimage	Two	Three	Two	Three	Two	Three
Maximum	97.76	97.50	81.94	89.19	97.70	98.23
Minimum	47.54	76.21	13.47	24.60	52.50	78.99
Average	79.06	92.03	43.89	63.15	82.56	93.59

Again, similar to the Fig.2, we also show the distribution of rank-1 CMS scores in Fig.4 and Fig.5. They are obtained by using DC-TFV histograms generated from 2-subimage and 3-sbuimage respectively.

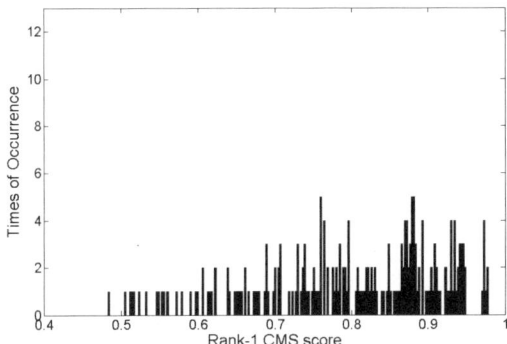

Fig. 4. The distribution of rank-1 CMS scores using DC-TFV generated from two subimages

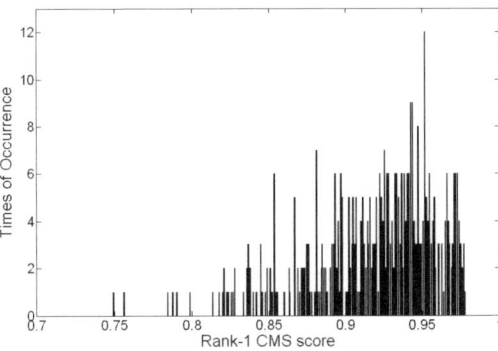

Fig. 5. The distribution of rank-1 CMS scores using DC-TFV generated from three subimages

Table 4. The average difference between five training sets, using DC-TFV from one/two/three subimage, represented by Rank-1 CMS scores

	1-subimage	2-subimage	3-subimage
Average difference of Rank-1 CMS (%)	1.26	0.94	0.62

To give a better understanding of the improvement from 1-subimage to 3-subimage, we would like to compare the result of any combination and its corresponding component subimages. The CMS score of each 3-subimage combination is compared to the corresponding CMS score of 2-subimage combination; and the CMS score of each 2-subimage combination is compared to the corresponding CMS scores of two component 1-subimage. The compared results are shown in Table 5. The notation I_1, I_2 and I_3 means the subimages which constitute the combinations.

Table 5. The average difference between five training sets, using DC-TFV from one/two/three subimage, represented by Rank-1 CMS scores

	$I_1+I_2 <$ min (I_1, I_2)	otherwise	$I_1+I_2 >$ max(I_1, I_2)
2-sub VS. 1-sub (among 216 combinations)	0	37/216	179/216
	$I_1+I_2+I_3 <$ min(I_1+I_2, I_3)	otherwise	$I_1+I_2+I_3 >$ max(I_1+I_2, I_3)
3-sub VS. 2-sub (among 432 combinations)	0	220/432	212/432

We can notice that the possibility to get improvement from 1-subimage to 2-subimage is higher than the possibility from 2-subimage to 3-subimage. In fact, the performance of 2-subimage is already in a saturation range, which is quite difficult to get improvement. Furthermore, using more subimages is not necessarily to achieve better performance, especially when the previous subimages have already represented quite distinguishing contents, adding the more subimages usually do not result in any improvement. This is possible when the added subimage has too small size, or it contains too much noise or useless texture, and so on. However, we can also conclude that adding new subimage at least does not reduce the performance. Based on above results, we also believe that using four or five subimages may further improve the performance, but such improvement will be relatively small. Proper decision has to be made by also taking the complexity into consideration.

6.5 Comparison with Other Methods

In order to compare the performance of our system with other methods, we list below some reference results from other researches for the FERET database. These results are all obtained by using the FA and FB set of the same release of FERET database. In [12], the eigenvalue weighted bidimensional regression method is proposed and applied to biologically meaningful landmarks extracted from face images. Complex principal component analysis is used for computing eigenvalues and removing correlation among landmarks. An extensive work of this method is conducted in [2], which comparatively analyzed the effectiveness of four similarity measures including the typical L1 norm, L2 norm, Mahalanobis distance and eigenvalue-weighted cosine (EWC) distance. The author of [13] employs a simple template matching method to complete a verification task. The input and model faces are expressed as feature vectors and compared using a distance measure between them. Different color channels are utilized either separately or jointly. A combined subspace method is proposed in [8], using the global and local features obtained by applying the LDA-based method to either the whole or part of a face image respectively. The combined subspace is constructed with the projection vectors corresponding to large eigenvalues of the between-class scatter matrix in each subspace. The combined sub-space is evaluated in view of the Bayes error, which shows how well samples can be classified. Table 6 lists the result of above papers, as well as the result of 3-subimaghe case of our method. The results are expressed by the way of Rank-1 CMS score.

Table 6. Referenced results based on release 2003 of FERET

References	[12]	[2]	[13]	[8]	Proposed Method
Rank-1 CMS (%)	79.4	60.2	73.08	97.9	98.23

7 Conclusions

We proposed a hierarchical retrieval system based on block transform, TFV and subimage for visual image retrieval. Such histogram gradually integrates the structural and statistical information in the face images. The performance is illustrated using a public face image database. This system achieves good retrieval results due to the fact it efficiently combines the statistical and structural information. Future research will try to find out any simpler way to select the optimal subimages.

Acknowledgments. The first author would like to thank for the financial grant from Tampere Graduate School in Information Science and Engineering (TISE).

References

1. Ekenel, H.K., Sankur, B.: Feature selection in the independent component subspace for face recognition. Pattern Recognition Letter 25, 1377–1388 (2004)
2. Shi, J., Samal, A., Marx, D.: How effective are Landmarks and Their Geometry for Face Recognition. Computer Vision and Image Understanding 102(2), 117–133 (2006)
3. Pennebaker, W.B., Mitchell, J.L.: JPEG still image compression standard. Van Nostrand Reinhold, New York (1993)
4. ISO/IEC 14496-2: Information Technology - Coding of Audio-Visual Objects - Part 2: Visual (1999)
5. ITU-T Rec. H.264 | ISO/IEC 14496-10 AVC: Draft ITU-T Recommendation and Final Draft International Standard of Joint Video Specification (2003)
6. Ahonen, T., Hadid, A., Pietikainen, M.: Face recognition with local binary patterns. In: Pajdla, T., Matas, J(G.) (eds.) ECCV 2004. LNCS, vol. 3021, pp. 469–481. Springer, Heidelberg (2004)
7. Rajagopalan, A.N., Srinivasa, R.K., Anoop, K.Y.: Face recognition using multiple facial features. Pattern Recognition Letters 28, 335–341 (2007)
8. Chung, H.K., Jiyong, O., Chong-Ho, C.: Combined Subspace Method Using Global and Local Features for Face Recognition. In: IJCNN 2005 (2005)

9. MathWorks, Inc.: Documentation for MathWorks Products (2007), Available at http://www.mathworks.com
10. FERET Face Database: (2003), Available at http://www.itl.nist.gov/iad/humanid/feret/
11. Phillips, P.J., Moon, H., Rauss, P.J., Rizvi, S.: The FERET evaluation methodology for face recognition algorithms. IEEE Pattern Analysis and Machine Intelligence 22, 10 (2000)
12. Shi, J., Samal, A., Marx, D.: Face Recognition Using Landmark-Based Bidimensional Regression. In: proceeding of ICDM 2005 (2005)
13. Roure, J., Faundez, Z.M.: Face recognition with small and large size databases. In: proceeding of ICCST 2005 (2005)

Near-Duplicate Detection Using a New Framework of Constructing Accurate Affine Invariant Regions

Li Tian and Sei-ichiro Kamata

Graduate School of Info., Pro. & Sys., Waseda University,
2-7, Hibikino, Wakamatsu-ku, Kitakyushu, 808-0135, Japan
tianli@ruri.waseda.jp,
kam@waseda.jp

Abstract. In this study, we propose a simple, yet general and powerful framework for constructing accurate affine invariant regions and use it for near-duplicate detection problem. In our framework, a method for extracting reliable seed points is first proposed. Then, regions which are invariant to most common affine transformations are extracted from seed points by a new method named the Thresholding Seeded Growing Region (TSGR). After that, an improved ellipse fitting method based on the Direct Least Square Fitting (DLSF) is used to fit the irregularly-shaped contours of TSGRs to obtain ellipse regions as the final invariant regions. At last, SIFT-PCA descriptors are computed on the obtained regions. In the experiment, our framework is evaluated by retrieving near-duplicate in an image database containing 1000 images. It gives a satisfying result of 96.8% precision at 100% recall.

Keywords: Invariant region, Thresholding Seeded Growing Regions, ellipse fitting, image matching, near-duplicate detection.

1 Introduction

Affine Region Detection is an essential problem in image processing and computer vision fields. Recent years, different types of detectors including the Harris-Affine, the Hessian-Affine [1], the Maximally Stable Extremal Region (MSER) [2], the Intensity extrema-Based Region (IBR), the Edge Based Region (EBR) [3] and the Salient [4] detectors have been proposed and widely used in applications such as object recognition [5,6], image matching [7,3], image retrievals [8,9,10], and video data mining [11]. Generally, these detectors can be divided into two categories:

1. structure-based detectors such as the Harris-Affine, the Hessian-Affine and the EBR,
2. and region-based detectors including the MSER, the IBR and the Salient.

There does not exist one detector which outperforms the other detectors for all image types and all types of transformation. Points of interest at regions

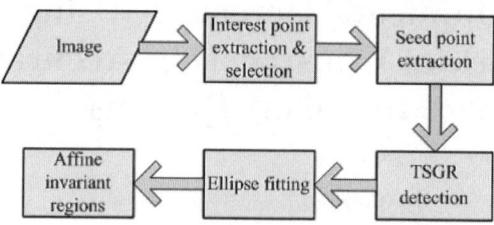

Fig. 1. The scheme of the proposed framework

containing a considerable amount of image structure are located in structure-based detectors, but they tend to fail in uniform regions and regions with smooth transitions. Region-based detectors deliver blob like structures of uniform regions but highly structured regions are not detected. Consequently, detectors which can cope with both the structured and uniform regions are highly desired. In addition, the ellipse fitting methods for fitting the irregularly-shaped contours of detected regions in the above detectors using moment such as the MSER and the IBR, often generate some ellipses out of the bound of the images. In this study, we propose an simple, yet general and powerful framework for constructing accurate affine regions to solve the existing problems.

We concentrate on detectors computed on gray-value images in this study. Our framework involves three main factors: the seed extraction from interest points, the Thresholding Seeded Growing Region (TSGR) detection and the improved Direct Least Square Fitting (DLSF). In our framework, a method for extracting reliable seed points from selected interest points according to the types of images is first proposed. Then, regions are generated from seed points by the TSGR which integrates the Seeded Region Growing (SRG) algorithm and the MSER detection algorithm. An important characteristic of the contour of the TSGR is that it is invariant to most common affine transformations. Finally, an improved ellipse fitting method based on the Direct Least Square Fitting (DLSF) is used to fit the irregularly-shaped contour of TSGRs to obtain ellipse regions. A simple scheme of our framework is shown in Fig. 1. In the experiment, our framework is evaluated on the near-duplicate detection problem [10]. It is shown our framework gives a near perfect result in the application to near-duplicate detection problem for image retrieving.

Our work makes four contributions to the affine invariant detection problem. First, the seed extraction method has flexibilities under different requirements and different types of images; Second, we first introduce the concept of TSGR based on the MSER and the SRG in our framework; Third, the regions detected by the TSGR are invariant to most general transformations; Finally, we improve an ellipse fitting method based on the DLSF which is invariant to most transformations to enhance the invariance of the whole framework.

The rest of the paper is organized as follows. First, we give a review of the related work to our framework in Section 2. Then, we describe the whole framework for constructing accurate affine invariant regions in detail in Section 3.

Section 4 is about the experimental results. We conclude this study and suggest the future work in the last section.

2 Related Work

Mikolajczyk *et al.* gave a performance evaluation framework for comparing different affine region detectors including the Harris-Affine, the Hessian-Affine [1], the MSER [2], the IBR, the EBR [3] and the Salient [4] against changes in viewpoint, scale, illumination, blurring and image compression in [12]. They indicated that there does not exist one detector which outperforms the other detectors for all image types and all types of transformation. Hence, detectors combining superiorities of different detectors are desired.

Both the Harris-Affine and the Hessian-Affine detectors detect interest points in the scale-space. The Harris-Affine region is based on the Harris corner detector [13]. The second moment matrix is used to extract corner point in it. This matrix describes the gradient distribution in a local neighborhood of a point:

$$M = \mu(\mathbf{x}, \sigma_I, \sigma_D) = \begin{bmatrix} \mu_{11} & \mu_{12} \\ \mu_{21} & \mu_{22} \end{bmatrix}$$
$$= \sigma_D^2 g(\sigma_I) * \begin{bmatrix} I_x^2(\mathbf{x}, \sigma_D) & I_x I_y(\mathbf{x}, \sigma_D) \\ I_x I_y(\mathbf{x}, \sigma_D) & I_y^2(\mathbf{x}, \sigma_D) \end{bmatrix} \quad (1)$$

where the derivatives are first computed with Gaussian kernels of scale σ_D, and then averaged by smoothing with a Gaussian window of scale σ_I. The eigenvalues of this matrix represent two principal signal changes in a neighborhood of the point. A similar idea is also used in Hessian-Affine detector where, however, Hessian matrix is used:

$$H = H(\mathbf{x}, \sigma_D) = \begin{bmatrix} h_{11} & h_{12} \\ h_{21} & h_{22} \end{bmatrix}$$
$$= \begin{bmatrix} I_{xx}(\mathbf{x}, \sigma_D) & I_{xy}(\mathbf{x}, \sigma_D) \\ I_{xy}(\mathbf{x}, \sigma_D) & I_{yy}(\mathbf{x}, \sigma_D) \end{bmatrix}. \quad (2)$$

The MSER is a connected component of an appropriately thresholded image. All pixels inside the MSER have either higher (bright extremal regions) or lower (dark extremal regions) intensity than all the pixels on its outer boundary. MSER detection is related to *shresholding*. The IBR detects affine covariant regions that starts from intensity extrema and explores the image around them in a radial way by using an evaluation function along each emanated ray:

$$f_I(t) = \frac{abs(I(t) - I_0)}{max(\frac{\int_0^t abs(I(t)-I_0)dt}{t}, \sigma)} \quad (3)$$

where t is an arbitrary parameter along the ray, I(t) is the intensity at t, I_0 is the intensity value at starting point and σ is a small number to prevent a division by zero. The EBR is based on geometry method and two different cases are considered: one is developed for curved edges while the other one is applied in case of straight edges. Salient region detector is based on the probability distribution function (pdf) of intensity values computed over an elliptical region. The detection process includes two steps: entropy evaluation for candidate salient regions and selection of regions. Because the obtained regions by the MSER, the IBR and EBR are in arbitrary shapes, they are always replaced by ellipses having the same first and second moments as the originally detected regions. In addition, the DLSF [14] is a fitting method for fitting ellipses to scattered data. It is robust, efficient and easy to implement. The SRG [15] is used for segmentation of intensity images. It is controlled by choosing a number of pixels, known as seeds and choose regions as homogeneous as possible.

Our framework is motivated and is built on the above works. The idea for choosing seeds and using thresholding in the TSGR is inspired by both the MSER and the SRG; The criterions for choosing seeds are from the Harris-Affine, the Hessian-Affine and the MSER; The fitting method is improved from the DLSF. However, our framework differs from them in many ways:

- A TSGR is obtained from a seed point, but an MSER is obtained without seed point.
- The SRG grows regions by an evaluation of intensity function of neighborhoods while the TSGR generates regions relating to shresholding process.
- Seed extraction is based on multiple types of candidates in our framework for different types of images whereas a single type of point is used in the SGR.
- Fitting methods for the obtained irregularly-shaped regions are also different between our framework and those in the IBR, the EBR and the MSER.
- The normal DLSF fits all points in a point set while the improved DLSF only fits a part of points satisfying some constraints.

3 The Proposed Framework

In this section, we first describe how to extract seed points from images. Then, the TSGR will be discussed in detail. After that, the method of fitting ellipse for the irregular-shaped contour of the TSGR will be introduced. Finally, we analyze the invariant characteristic of the detected regions by giving detailed examples.

3.1 Seed Extraction

This process involves three steps: *interest point extraction, interest point selection* and *seed point extraction*. Notice that an interest point is not exactly a seed point here. As mentioned previously, corner points are usually used as interest points in structure-based detectors such as in the Harris-Affine, the Hessian-Affine and the EBR, while local extremum points in intensity are used as interest points in region-based detectors such as in the IBR. Thus, a method integrating the advantages of both the structure-based and region-based detectors is required.

Actually, any point in the image can be used as an interest point depending on the purposes of applications and image characteristic. Since the Harris-Affine, the Hessian-Affine and the MSER are the most three effective detectors related to certain types of images as reported in [1], three types of interest points from them are used in this study. They are: corner points in the Harris-Affine, Hessian-Affine detectors and central points of the MSER. Because the MSER is an irregularly-shaped region, the central point uses the one of the fitted ellipse with the same first and the second moment of the region.

Why interest point selection is necessary? This is according to the purposes of image matching or recognition problems. In some cases, exclusively correct region matches are required to identify images, however, the number or rate of correct matches is not important. For example, in applications that finding the affine parameters between two images, only three pairs of correct matches (not in a line) are enough to compute the affine parameters. In other cases, more correct matches are desired, hence as many feature points as possible should be selected. For example, as in the experiment for near-duplicate detection, a large number of correct matches is required to identify the categories of the images. Therefore, we propose this selection step for different requirements. In this step, we introduce two constraints for selecting interest points. The two constraints are:

- The importance of the interest point.
- The minimum distance between any two interest points.

The importance is presented by the values of Eqn. 1 and Eqn. 2 for the corner points in the Harris-Affine and the Hessian-Affine, and by a neighborhood gradient estimation value for the central point (x, y) in the MSER in a local $(2m+1) \times (2n+1)$ window defined as

$$P(x,y) = \frac{\sum_{i=-m}^{m} \sum_{j=-n}^{n} g(x+i, y+j)}{(2m+1)(2n+1)} \qquad (4)$$

where g is the local gradient. Notice that the larger the values are, the more important the points are in Eqn. 1 and Eqn. 2. While the smaller the value is, the more important the point is in Eqn. 4. The importance can be used for rejecting unimportant interest points. The more interest points are selected, the more seed point will be extracted and more invariant regions will be detected as well as the correct matches. The first constraint makes a tradeoff between processing speed and the number of detected regions. The second constraint removes the points that are too close and all selected points can be located in a well distribution. Fig. 2 gives an illustration of selecting interest points by using our method from corner points in the Harris-Affine. Fig. 2(a) shows the sparsely selected interest points where cornerness in Eqn. 1 larger than 0.1, and the distance between any two corner points larger than 5 are specified. Fig. 2(b) shows the densely selected interest points where no minimum cornerness and no minimum distance are specified.

After interest points were extracted and selected, seed points can be generated from the selected interest points. In our study, because three types of interest

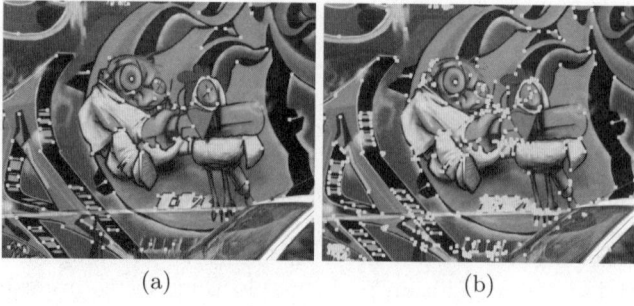

Fig. 2. Selected interest points. (a) Sparsely selected interest points. (b) Densely selected interest points.

points are used, three types of seed points also can be extracted. For corner points from the Harris-Affine and the Hessian-Affine, seed points are extracted by computing the second moment descriptor with automatically *integration* and *differentiation* by defining a scalar $\Sigma_D = s\Sigma_I$. Then, the *shape adaptation matrix* is estimated and is used to normalize the point neighborhood. For more details, please refer to [1]. The central point of the neighborhood of an interest point is used as a seed point. For interest points from the MSER, seed points are the interest points themselves.

3.2 Thresholding Seeded Growing Region

In this section, we introduce the TSGR based on the MSER and the SGR in detail. The TSGR includes two types: the bright TSGR and the dark TSGR. The typical bright TSGR is a region having bright homogeneous pixels with dark boundaries and the typical dark one is a region having dark homogeneous pixels with bright boundaries when a seed point is a local extremum point in intensity. If a seed point is a corner point, the TSGR will be a structured region with homogeneous pixels as many as possible in its local neighborhoods.

We first discuss the bright TSGR. For a given seed point with intensity I_0, we first set a range of thresholds **t** from I_0 decreasing to I_0'. Then, we change the threshold t_i and start to threshold the image as

$$I_{t_i} = \begin{cases} 255 & \text{if } I \geq t_i \\ 0 & \text{else} \end{cases} \quad (5)$$

which refers to the pixels below the threshold t_i as 'black' and to those above or equal as 'white'. As a result, we can obtain several thresholded images corresponding to all the thresholds. Then, for the given seed point, we find its connected 'white' components **R** corresponding to all thresholds. Subsequently the connected 'white' region will appear and grow. Two 'white' regions will merge

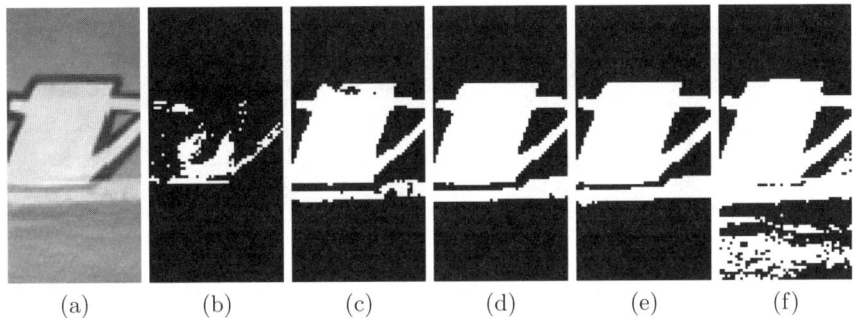

Fig. 3. Detecting a bright TSGR region. (a) The original image (a seed point with intensity 230 is inside the region of inverse number 4). (b) Thresholded image by $t = 230$. (c) Thresholded image by $t = 210$. (d) Thresholded image by $t = 190$. (e) Thresholded image by $t = 170$. (f) Thresholded image by $t = 150$.

at some point. After growing through all thresholds, for each connected 'white' region, a stability value Ψ_t is calculated. This value Ψ_t is defined as

$$\Psi_t = \frac{|R_{t-\Delta}| - |R_{t+\Delta}|}{|R_t|}, \qquad (6)$$

where Δ is a stability range parameter and $|\,.\,|$ denotes the cardinality, in other words, the number of pixels in the connected regions. Finally, if the Ψ_t is smaller than a threshold Ψ_0, the corresponding connected 'white' region will be treated as a bright TSGR region. The dark TSGR also can be detected by the same way but inverting the intensities.

There are four items should be clarified here.

- First, multiple TSGRs can be detected from one seed point if we use a larger Ψ_0 or no TSGR will be detected from one seed points if the Ψ_0 is too small. To prevent these, we can use some alternatives such as finding the smallest value among all Ψ values.
- Second, if a seed point is a corner point, the detected region may be an atypical one—the seed point itself. It is recommended that a small area around a seed point be used instead of the single seed point when using a corner point.
- Third, constraints such as the minimum and the maximum size also can be specified to the detected regions.
- Finally, the detected region by the proposed TSGR will be very similar to the one by the MSER when two conditions are satisfied: the seed point in TSGR happens to locate inside a MSER region and the range of thresholds is the same to that in the MSER.

Fig. 3 shows an example of the bright TSGR detection. Fig. 3(a) shows the original image cropped from Fig. 2 and a seed point with intensity 230 locates inside the inverse number 4 of the image. Figs. 3(b)-(e) show the thresholded

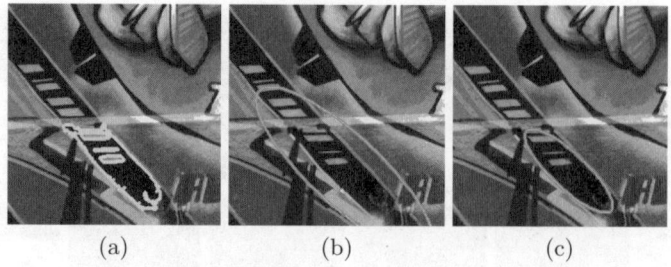

(a)   (b)   (c)

Fig. 4. Fitting irregularly-shaped region. (a) An Irregularly-shaped region region. (b) Fitted ellipse by traditional method. (c) Fitted ellipse by our method.

images by using five different thresholds from 230 to 150 decreasing 20 at each time. All connected 'white' regions of the seed point are detected and the stabilities are computed using the Eqn. 6. By comparing their stability values, the connected regions of the seed point in Fig. 3(d) is detected as a bright TSGR since it is the most stable one among all the connected regions.

3.3 Ellipse Fitting

Now, we are arriving at how to fit ellipse areas for the obtained TSGR regions. The obtained regions are always irregularly-shaped regions. In the IBR, the EBR and the MSER, the irregularly-shaped region is fitted by ellipses having the same first and second moments as the originally detected regions. This often results in an ellipse which is out of the bound of the image. Fig. 4(a) shows an irregularly-shaped region detected by the MSER and Fig. 4(b) shows the fitted ellipse by the moment method. We can observe that a part of the fitted ellipse is out of the bound of the image and this phenomenon is not favored in many applications such as image retrieves since it introduces the uncertainty.

In our framework, we improve an ellipse fitting method based on a simple and straightforward one—the DLSF [14]. A general quadratic curve can be presented as an implicit second order polynomial:

$$F(\mathbf{a}, \mathbf{x}) = ax^2 + bxy + cy^2 + dx + fy + g = 0 \qquad (7)$$

where $\mathbf{a} = [a\ b\ c\ d\ f\ g]^T$ and $\mathbf{x} = [x^2\ xy\ y^2\ x\ y\ 1]^T$. In the DLSF, $F(\mathbf{a}, \mathbf{x}_i)$ is called the *algebraic distance* of a point (x, y) to the quadratic curve $F(\mathbf{a}, \mathbf{x}) = 0$. The DLSF uniquely yields elliptical solutions that, under the normalization $4ac - b^2 = 1$, minimize the sum of squared algebraic distances from the points to the ellipse. It has the following advantages:

- it can provide useful results under all noise and occlusion conditions,
- it is invariant to affine transformation,
- it is robust to noise,
- it is computational efficient.

In our case, before an ellipse fitted, we improve an elimination process to remove unreliable contour points using a rule as follow: for all the points on this

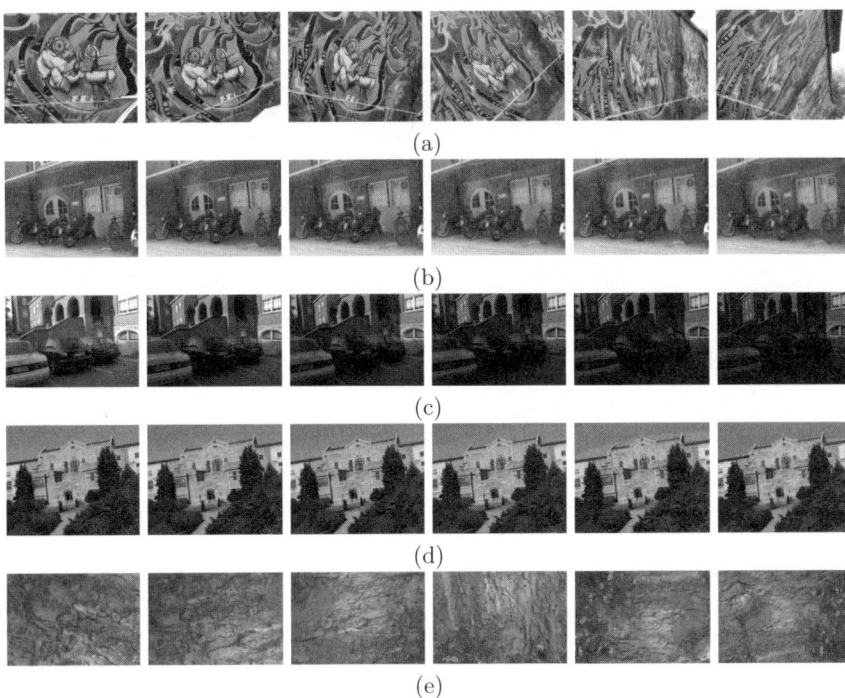

Fig. 5. Near-duplicate images used in the experiments. (a) Viewpoint change. (b) Blur. (c) Illumination change. (d) Compression. (e) Rotation and scaling.

contour $P = \{\mathbf{p}_i = (x_i, y_i) | i = 1, 2, \ldots, n\}$, we first compute all distances d_i from each contour point to their seed point; and then compute the mean value $\overline{d}$ of all d_i. Finally, if the distance d_i from an end point to its seed point does not satisfy $\alpha_1 \overline{d} < d_i < \alpha_2 \overline{d}$, where α_1 and α_2 are two thresholds, this contour point will be eliminated. Fig. 4(c) shows an example of the fitted region from the same irregularly-shaped region in Fig. 4(a) using our method. From the figure, we can see that the fitted ellipse is more distinguishable to describe the irregularly-shaped region than using the moment method in traditional methods, and the fitted ellipse locates exactly inside the bound of the image. Note that the resulting elliptical regions may not locate centered around the original seed point.

4 Experiments

Our framework is supposed to be related with near duplicate problem [10,16,17,18]. In this experiment we will study the performance of our framework for retrieving near duplicate images. The experiment setup is as follows:

Test images. There are 1000 images in the database: 30 near-duplicates of the query images in Fig. 5 are from the Oxford Affine Invariant Feature

Fig. 6. Some images in confusing data

data set [19] including the general transformations such as viewpoint change, image blur, illumination change, compression, and scaling+rotation, in which the images in the most left column are the five query images; Rest 970 images serving as confusing data. The confusing data including similar objects or scenes as in query images are randomly selected from Caltech101 image database [20] and some images are shown in Fig. 6.

Descriptor. Since the speed of retrieving depends on the dimension of the descriptor, the PCA-SIFT [21] with 36 dimension is used in this experiment to save the computational time.

Evaluation criterions. *Recall* and *precision* are used to evaluate the retrieving results:

$$recall = \frac{\sharp correct\ positives}{\sharp positives} \qquad (8)$$

and

$$precision = \frac{\sharp correct\ positives}{\sharp matches(correct\ and\ false)}. \qquad (9)$$

A correct positive is a correct match between a query image and one of its near-duplicates in the image database.

For more details about the experiment setup, please refer to [10].

Because the types of image always are not known in retrieving problem, the interest points are all extracted from the MSER in this experiment. For two seed points p and q, if the χ^2 distance between their PCA-SIFT descriptors H_p, and if H_q defined as

$$\chi^2(p,q) = \frac{1}{2} \sum_{n=1}^{36} \frac{(H_p(n) - H_q(n))^2}{|H_p(n)| + |H_q(n)|} \qquad (10)$$

is no larger than d, the two descriptor will be treated as matching. And if more than θ matches are found between two images, the one in the database is defined

Table 1. Recall-precision results by varying different θ and d

recall-precision(%)	d=1000	d=3000	d=5000
θ=1	90.0-100.0	100-71.4	100-6.2
θ=3	76.7-100.0	**100-96.8**	100-12.8
θ=5	73.3-100.0	**100-96.8**	100-19.6

as a near-duplicate. The recall-precision results by varying different θ and d are shown in Table 1. From the table, we see that when $\theta = 3$ or 5 and $d = 3000$, we get a very satisfying result of 96.8% precision at 100% recall for retrieving near-duplicate. We also make a comparison between our framework and the one in [10]. Because affine parameters are not estimated in our framework, we also skip this step in [10] for fair. Thus, Differences of Gaussian (Dog) is used as interest point detector, and PCA-SIFT descriptors are calculated on fixed size patches. Since Dog produces much more duplicate interest points, $\theta = 300$ and $d = 3000$ are the best parameter settings and only 100% precision at 83.3% recall for retrieving near-duplicate are achieved.

5 Conclusions and Future Work

In this paper, we propose a new framework integrating the advantages of both the structured-based and region-based detectors for constructing accurate affine invariant regions for intensity images. The detected regions are invariant to general transformations especially the viewpoint change. In our work, we proposed a method for selecting interest points and extracting seed points. Then, we introduce the Thresholding Seeded Growing Region (TSGR) to obtain invariant regions. Fitting ellipse of the irregularly-shaped regions using an improved method from the the Direct Least Square Fitting (DLSF) is also different from traditional methods, and it is more efficient but still straightforward. A retrieving experiment are tested to evaluate our framework in the experiment and we get near perfect accuracy of 96.8% precision at 100% recall on a image database including 1000 images.

However, it is seemed that our framework sometimes is sensitive to some types of transformed images such as rotation and scaling. Using RANSAC liking in [10] may solve this problem. In addition, it is a little computational expensive for retrieving problem. Future work will aim at improving its performance on all types of transformed images and an efficient implementation of it.

Acknowledgement

We would like to thank all people providing their free code and test images for this study. This work was supported at part by the JSPS Research Fellowships for Young Scientists.

References

1. Mikolajczyk, K., Schmid, C.: Scale & affine invariant interest point detectors. IJCV 60, 63–86 (2004)
2. Matas, J., Chum, O., Urban, M., Pajdla, T.: Robust wide-baseline stereo from maximally stable extremal regions. Image and Vision Computing 22(10), 761–767 (2004)
3. Tuytelaars, T., Gool, L.V.: Matching widely separated views based on affine invariant regions. IJCV 59(1), 61–85 (2004)
4. Kadir, T., Zisserman, A., Brady, M.: An affine invariant salient region detector. In: Pajdla, T., Matas, J(G.) (eds.) ECCV 2004. LNCS, vol. 3021, pp. 228–241. Springer, Heidelberg (2004)
5. Ferrari, V., Tuytelaars, T., Gool, L.V.: A simultaneous object recognition and segmentation by image exploration. In: Pajdla, T., Matas, J(G.) (eds.) ECCV 2004. LNCS, vol. 3021, pp. 40–54. Springer, Heidelberg (2004)
6. Lowe, D.: Distinctive image features from scale-invariant keypoints. IJCV 60(2), 91–110 (2004)
7. Schaffalitzky, F., Zisserman, A.: Multi-view matching for unordered image sets. In: Heyden, A., Sparr, G., Nielsen, M., Johansen, P. (eds.) ECCV 2002. LNCS, vol. 2350, pp. 414–431. Springer, Heidelberg (2002)
8. Mikolajczyk, K., Schmid, C.: Indexing based on scale invariant interest points. In: Proc. 8th ICCV, pp. 525–531 (2001)
9. Schmid, C., Mohr, R.: Local grayvalue invariants for image retrieval. IEEE Trans. on PAMI 19(5), 530–534 (1997)
10. Ke, Y., Sukthankar, R., Huston, L.: Efficient near-duplicate detection and sub-image retrieval. In: Proc. 12th ACM Intl. Conf. on Multimedia, pp. 869–876 (2004)
11. Sivic, J., Zisserman, A.: Video google: A text retrieval approach to object matching in videos. In: Proc. 9th ICCV, pp. 1470–1477 (2003)
12. Mikolajczyk, K., Tuytelaars, T., Schmid, C., Zisserman, A., Matas, J., Schaffalitzky, F., Kadir, T., Gool, L.V.: A comparison of affine region detectors. IJCV 65, 43–72 (2005)
13. Harris, C., Stephens, M.: A combined corner and edge detector. In: Harris, C., Stephens, M. (eds.) Proc. of the 4th Alvey Vision Conference, pp. 147–151 (1988)
14. Fitzgibbon, A., Pilu, M., Fisher, R.: Direct least square fitting of ellipses. IEEE Trans. on PAMI 21, 476–480 (1999)
15. Adams, R., Bischof, L.: Edge region growing. IEEE Trans. on PAMI 16, 641–647 (1994)
16. Zhang, D.Q., Chang, S.F.: Detecting image near-duplicate by stochastic attributed relational graph matching with learning. In: Proc. 12th ACM Intl. Conf. on Multimedia, pp. 877–884. ACM Press, New York (2004)
17. Chen, C.Y., Kurozumi, T., Yamato, J.: Poster image matching by color scheme and layout information. In: Proc. ICME2006, pp. 345–348 (2006)
18. Meng, Y., Chang, E.Y., Li, B.: Enhancing dpf for near-replica image recognition. In: Proc. CVPR2003, pp. 416–423 (2003)
19. http://www.robots.ox.ac.uk/~vgg/research/affine/index.html
20. http://www.vision.caltech.edu/Image_Datasets/Caltech101/Caltech101.html
21. Ke, Y., Sukthankar, R.: PCA-SIFT: A more distinctive representation for local image descriptors. In: Proc. of the CVPR, vol. 2, pp. 506–513 (2004)

Where Are Focused Places of a Photo?

Zhijun Dai and Yihong Wu

National Laboratory of Pattern Recognition, Institute of Automation,
Chinese Academy of Sciences, P.O. Box 2728, Beijng 100080, P.R. China
{zjdai, yhwu}@nlpr.ia.ac.cn

Abstract. Focused places of a photo act as a significant cue for image concept discovery and quality assessment. Therefore, to find them is an important issue. In this paper, we design a focusing degree detector by which a focusing degree map is generated for a photograph. The results could be used to obtain focused places of photographs. As a concrete example of their applications, image retrieval and image quality assessment are investigated in this work. The experimental results show that the retrieval algorithm based on this detector and map can get more accurate retrieval results and the proposed assessment algorithm has a high ability to discriminate photos from low quality to high quality.

1 Introduction

With the popularization of digital cameras, many image retrieval systems have collected more and more digital photos, for example, the Google image search engine: http://images.google.com. When you input the "dog" keyword to search, the system will return about 2.9 million results. Due to the amount is great, it is needed to rank these images. Image quality can be an important factor for the ranking and recently has been considered in the works of [1] [2]. Besides keyword search, Content-Based Image Retrieval (CBIR) is another way for image search. In the case of query by example [3], the main concept of an image has a strong relation with the retrieval results. There are tremendous previous works on main concept discovery from images [4] [5] [6]. However, the semantic gap between high level concepts and low level features is still an open problem.

Photos are taken by people with cameras. When a professional photographer takes a photo, the main concept of the photo will be focused by his or her camera, so the focused places are usually the main concept places of the photo. A photo usually contains both places out of focus and places in focus, especially when the difference is large between depth information of main concept objects and other objects. Moreover sometimes, in order to enhance impressiveness of the main concept, background of a photo is intentionally unfocused by photographers.

The most distinctive effect between places of in focus and out of focus is blurring effect. We present a blurring detection method to discover the focused places of a photo. Then based on the method, two applications have been explored from the resulting focusing degree measures. One application is CBIR. When a user does the photo query by example, he or she not only wants the results to be relevant with the query, but also wants the relevant places to have high quality and to be the main concept of the photo, which needs the relevant places in focus. Thus the focused places discovery is necessary for the similarity computation between photos. Another application is photo quality assessment. We will show the focusing degree measure is useful for photo quality assessment.

2 Related Work

There are a lot of meaningful previous works on CBIR and photo quality assessment. In the recent research of CBIR, the sematic gap has become a hot spot and huge amounts of works have been done to reduce this gap. The interested reader is referred to a survey paper [7] for more information. One of the promising methods for this problem is the saliency based method [8] [9] [10] or attention-based method [11], where the salient or attentive place of an image takes more weight than other places. Photos come from photographers and the photographers always make the salient or attentive place of a photo focused. Otherwise, the photo quality will be low. Therefore, it is reasonable to assume the salient or attentive place of a high quality photo is in focus.

The research on focusing degree measure is through computing the blur degree of a whole photo. Based on the explicit or implicit assumption that a given photo is blurred, many methods [12] [13] [14] can be used to determine the blur parameters. Edge sharpness and edge types were considered for blur estimation in the works of Li [15] and Tong et al. [16]. Ke et al. [2] used the normalized number of frequencies of Fourier Transform to denote the inverse blur degree. Blur feature also acts an important role on image quality assessment [15] [17] [18]. As far as we know, few research efforts have been made to judge which part of a photo is in focus up till now. Background of some professional photos may be intentionally blurred to make main concepts appear sharper. Thus the focusing degree of a photo should be determined from the focused places.

3 Focusing Degree Extraction and Applications

In this section, we are to introduce the focusing degree detector and focusing degree map. Then their applications to image concept discovery, CBIR, and photo quality assessment are reported.

3.1 Focusing Degree Detector and Map

In a camera imaging model, one of the biggest differences between places of in focus and out of focus is the blurring effect. The focused places appear sharper than unfocused places. There are three main sources for image blurring: focus out, motion, and compression. As the blur from motion [19] [20], if the motion object is out of focus and background is in focus, the focused background will appear sharper than the motion object. In this case, the focused place is background which can easily be detected by the blur contrast. If a photo's motion blur is significant and the motion object is in focus, the photo will have poor quality, so these motion blur photos can be filtered out by computing the overall blur degree. Consider blurring by compression [17], if the blur degree is small, the place in focus will also appear sharper than the place out of focus. If the blur degree is large, the photo has poor quality and can be filtered out by the overall blur degree thresholding. So we can assume that the place in focus of a photo locates at the sharpest area.

Let I_b be an image taken by a camera, and the corresponding ideal image be I_s. The relation between the two images can be modelled as

$$I_b = PSF * I_s, \qquad (1)$$

where PSF denotes point spread function of the image. In many cases, an image contains both region out of focus and region in focus, so the effects of PSF to I_s would be changed greatly from region in focus to region out of focus. Therefore, we divide images into different square blocks to extract the corresponding focusing degrees. Each block has a width n. We use B_s to denote a block of the image I_s, and B_b to the corresponding block of the image I_b, the PSF is modelled as a Gaussian smoothing filter G_σ in the block's position. The two blocks are related as

$$B_b = G_\sigma * B_s. \qquad (2)$$

We would like to recover the parameter σ only from the given blurred block B_b. This parameter would be proportional to the blur degree of the block. We use the two dimensional Fourier transform FFT to the (2) and get

$$FFT(B_b) = FFT(G_\sigma * B_s) = FFT(G_\sigma) \cdot FFT(B_s). \qquad (3)$$

The frequency response of G_σ function is also a Gaussian function. The $FFT(G_\sigma)$ function likes a low frequency pass filter. The less the σ value, the more the high frequency will pass. Let F be $FFT(B_b)$ and $|F|$ be the frequency power of $FFT(B_b)$. Then, we obtain the sum for high frequency component in B_b as:

$$C = \sum_{(u,v) \in B_b} |F(u,v)| W(u,v), \qquad (4)$$

where $W(u,v)$ is the weight of frequency $F(u,v)$ defined by:

$$W(u,v) = e^{-((n-u)^2 + (n-v)^2))/\sigma_w} \qquad (5)$$

and the value of σ_w is chosen as $0.25n^2$. The bigger the u, v value, the bigger the weight $W(u,v)$ should be. Therefore the weight defined in (5) is reasonable.

Based on (4), the focusing degree for the block B_b is given by:

$$f = (\sum_{(u,v) \in B_b} |F(u,v)| W(u,v)) / (\sum_{(u,v) \in B_b} |F(u,v)|) \qquad (6)$$

f can be a focusing degree detector for a photo. Different blocks have different focusing degree values. All the values form a focusing degree map for this photo. Figure 1 illustrates the focusing degree map of two photos. The top one is the standard Lena photo and we can see the focusing degree is significant around the Lena's hair. Another is a

Fig. 1. The focusing degree maps of 2 photos. The chosen block size is 8×8 pixels. On the left side are source photos, where the top one is standard Lena image and the bottom one is "Smiling Right At You" by Joshuamli, 2001. The focusing degree values are illustrated by the box sizes in the middle. The focusing degree maps are shown on the right side.

professional high quality photo which comes from http://www.dpchallenge.com/image.php?IMAGE_ID=150. Undoubtedly, the main concept is the insect face and we can get the most significant focusing degree value around the face area.

A focusing degree map for a photo could display the focusing degree distribution intuitively. All the blocks with focusing degree values larger than a threshold constitute the focused places of a photo.

3.2 Focusing Weight for Main Concept Discovery and CBIR

Now, we consider the established focusing degree map on a region and then provide a focusing weight for this whole region. The obtained result is further applied to CBIR, focused regions or main concepts discovery.

An image is segmented into different regions $\{r_1, r_2, ..., r_n\}$ and then each region r_i is divided into different blocks $\{b_{i,1}, b_{i,2}, ..., b_{i,l_i}\}$. We use $f_{i,j}$ to denote the focusing degree of $b_{i,j}$ and then compute a sum as

$$f_i = \sum_{j=1}^{j=l_i} f_{i,j}. \tag{7}$$

The sum is further normalized by:

$$w_i = f_i / \sum_{j=1}^{j=n} f_j. \qquad (8)$$

And the result is called the focusing weight of r_i.

The weight w_i can be used to find focused regions of a photo as follows. To judge whether a region is focused or not should base on the photo context and should compare with other regions' focusing degree. Therefore, we define Region Focusing Degree (RFD) for r_i by

$$RFD_i = w_i / p_i, \qquad (9)$$

where p_i is the area percentage of r_i. After having RFD_i, we can judge whether a region r_i is focused or not by the following condition:

$$\begin{cases} r_i \text{ is focused} & RFD_i > 1 \\ r_i \text{ is unfocused} & RFD_i \leq 1 \end{cases} \qquad (10)$$

The threshold is taken as 1 because we think the focusing degree should be paid on by more consideration than the area percentage for a prominent region. This is a reasonable criterion also since RFD of a region more than 1 indicates this region is more focused than other regions with smaller RFD, and vice versa. The discovered focused regions by this criterion can just be the main concepts of a photo. The corresponding experiment is shown in Subsection 4.1.

Besides the above application to find main concepts of a photo, w_i in (8) can also be used for region weight assignment in a CBIR system. In order to test the effectiveness of the focusing weight w_i, we implemented Integrated Region Match (IRM) algorithm [21] and Unified Feature Match (UFM) [22] algorithm for region-based image retrieval by replacing the previous region weight assignment of region area percentage by our w_i. The experiments in Subsection 4.2 show using our assignment can improve the retrieval precision.

3.3 Application of Focusing Degree Map to Photo Quality Assessment

Blur feature is useful for photo quality assessment, in the work of Ke et al. [2], they used many features for photo quality classification and the blur feature has the most discriminative ability. Focusing degree is inverse proportional to blur degree in a photo. After a photo's focusing degree map is computed, we can get some quality information from this map. For a given threshold ϵ, if a photo has blocks with focusing degrees $> \epsilon$, then this photo contains focused places. The more these blocks the photo has, the higher quality the photo is. Therefore, we compute the sum of all these blocks' focusing degrees as the overall focusing degree of the photo and then use the result as a measure for quality assessment. If there is no block with focusing degree larger than ϵ, we use the average of all focusing degree of blocks as the focusing degree of the photo to assess the photo quality.

4 Experimental Results

4.1 On Main Concept Discovery

By using the method in Subsection 3.2, the main concepts of a photo can be discovered by discovering the focused regions. Here are some examples.

Although there are at least hundreds of segmentation algorithms in the literature, image segmentation can hardly be perfect and is still an ill-posed inverse problem. In order to make the results unbiased by segmentation, we use the images from [23] for test. All these human hand-labeled images are benchmarks for image segmentation. Three examples by our method are shown in Figure 2, from which we can see that birds, stone, child, and flowers are focused in the photos. These focused regions conform to the results by human perception.

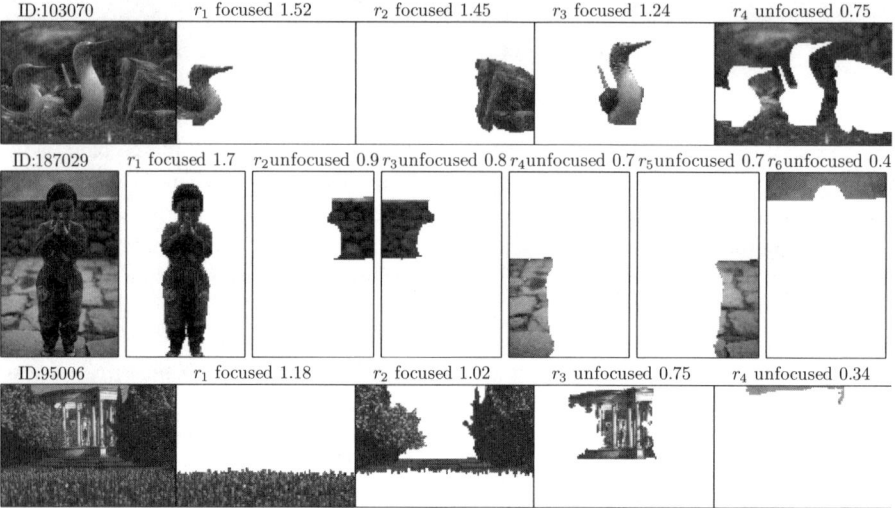

Fig. 2. Focused region discovery according to RFD in (9). The first image of each row is the source image and the followings are the human segmented regions. On the top of each region image, r_i denotes the index of the region, "focused" or "unfocused" means the region is focused or unfocused, the following number is the RFD value of the region. Small regions with area percentage less than 1% are not displayed.

4.2 On Content-Based Image Retrieval

We tested the efficiency of the defined focusing weight w_i in Subsection 3.2 for image retrieval. The details are below.

For testing the performance of CBIR systems, there seems no standard data set. The only dataset used by a large number of research groups are the Corel Photo CDs [24]. In our experiments, the image data set consists of 2,000 images of 20 categories from COREL. Each category representing a distinct concept contains 100 images. The categories are: fireworks, solider, lion, penguin, tiger, dog, hippopotamus, cloud, woman, flower, bus, car, mushroom, building, surfing, racing, plant, train, dolphin, and elephant.

Images should be segmented into regions. Since segmentation is not the main concern of this work, we used a matured k-means algorithm [25] for image segmentation. Images were divided into 4×4 blocks then the color and texture features were computed for each of these blocks. The features we used are the same as the work of Wang et al. [21].

Figure 3 are some examples of image segmentation and the computed focusing weights of regions defined in (8), where the values of f are the focusing weights and the values of p are the area percentage. In the photo of car, car is the focused place, so region r_2 is assigned a larger focusing weight than the percentage assignment. Due to the background road is smooth and out of focus, the region r_4 is assigned less focusing weight than the percentage assignment. In the photo of horse, the head of horse is focused and the background grassland is out of focus, so the focusing weight r_1 is significantly smaller than the percentage, and other regions are assigned much larger focusing weights. These assignments are consistent with human perception of photos.

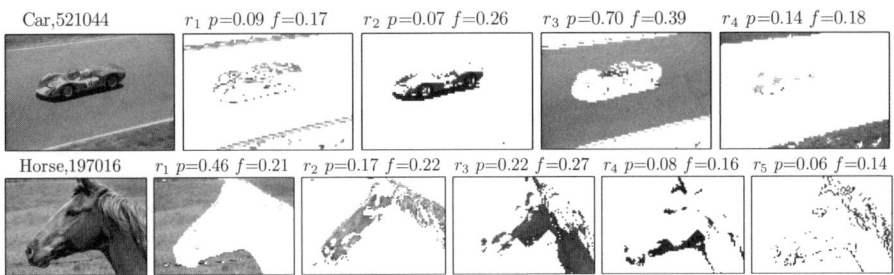

Fig. 3. Image segmentation and weight assignment. The first image of each row is the original image and the followings are the segmented regions. r_i is the index of a segmented region, p denotes region area percentage, and f denotes region focusing degree. Row 1) The place in focus is car. Row 2) The place in focus is horse.

We implemented Integrated Region Match (IRM) algorithm [21] and Unified Feature Match (UFM) algorithm [22] for region-based image retrieval by different ways. In the previous IRM and UFM algorithms, region area percentage was used as region weight assignment. We would like to replace the area percentage assignment by the defined focusing weight assignment in (8) of Subsection 3.2. We use IRM_p to denote the IRM algorithm of using area percentage assignment, UFM_p to denote UFM algorithm of using area percentage, IRM_f to denote IRM algorithm of using focusing weight assignment, and UFM_f to denote UFM algorithm of using focusing weight assignment. For each of these four retrieval methods, every image in the database was tested as a query. A retrieved image is considered a correct match if and only if it is in the same category as the query. The precisions within the first 20, 40, 70, and 100 retrieved images were computed for each query. We used the averaged precision of all 2,000 queries in each method as the final retrieval precision of this method. The results are shown in

Table 1. Results of retrieval precisions. $P(n)$ denotes the precision in the n top images from retrieval results. There are four retrieval methods for precision computation: IRM_p, IRM_f, UFM_p, and UFM_f. UFM outperform IRM, and the focusing weight assignment outperform area percentage weight assignment. UFM_f can achieve the best retrieval results.

Method	P(20)	P(40)	P(70)	P(100)
IRM_p	45.68%	37.19%	30.71%	26.72%
IRM_f	47.46%	39.07%	32.53%	28.19%
UFM_p	46.57%	38.20%	31.38%	27.19%
UFM_f	47.99%	39.54%	32.57%	28.26%

Table 1, where $P(n)$ denotes the precision in the n top images from retrieval results. As seen, the UFM_f method achieves the best retrieval results. This demonstrates the high effectiveness of the proposed focusing weight assignment.

4.3 On Photo Quality Assessment

The test of our method for photo quality assessment as well as a comparison with some previous methods are listed as follows.

For computing the overall focusing degree of a photo by our method, the threshold ϵ was taken as 0.2 and the block size as 8×8 pixels. In order to make the computation less influenced by noise, the blocks with gray value less than 15 were not considered (the used gray value is in the range of $0 \sim 255$).

Photo data were from a photo contest website, DPChallenge.com. It contains many kinds of high and low quality photos from many photographers. Furthermore, the most of the photos have been rated by its community of users, where the rated score is from 1 to 10. We used the average of these rated scores as the reference for the quality of this photo. In the previous work of Ke et al. [2], they used 60,000 photos and selected the top and bottom 10% of the photos as high quality professional photos and low quality snapshots, respectively. In our work, we used 115,226 photos. Each photo has been rated by at least 50 users. The top and bottom 5,000 photos were assigned as high quality photos and low quality ones respectively. The 5,000 high quality photos' scores range from 6.59 to 8.60, and the low quality photos' scores range from 1.81 to 3.97, so these 10,000 test photos have a clear consensus on their quality.

We applied Ke et al. method [2], Tong et al. method [16], and our method of Subsection 3.3 to this dataset. Tong et al. method uses Harr wavelet transform to discriminate different types of edges and recover sharpness according to edge types. Ke et al. method estimate the maximum frequency of an image by Fourier transform and counting the number of frequencies whose power is greater than some threshold, and the feature they used is the counting number normalized by the image size. Given the varying thresholds, the precision and recall of each method were computed by

$$recall = \frac{\text{\# high quality photos above threshold}}{\text{\# total high quality photos}} \qquad (11)$$

and

$$precision = \frac{\text{\# high quality photos above threshold}}{\text{\# photos above threshold}}. \qquad (12)$$

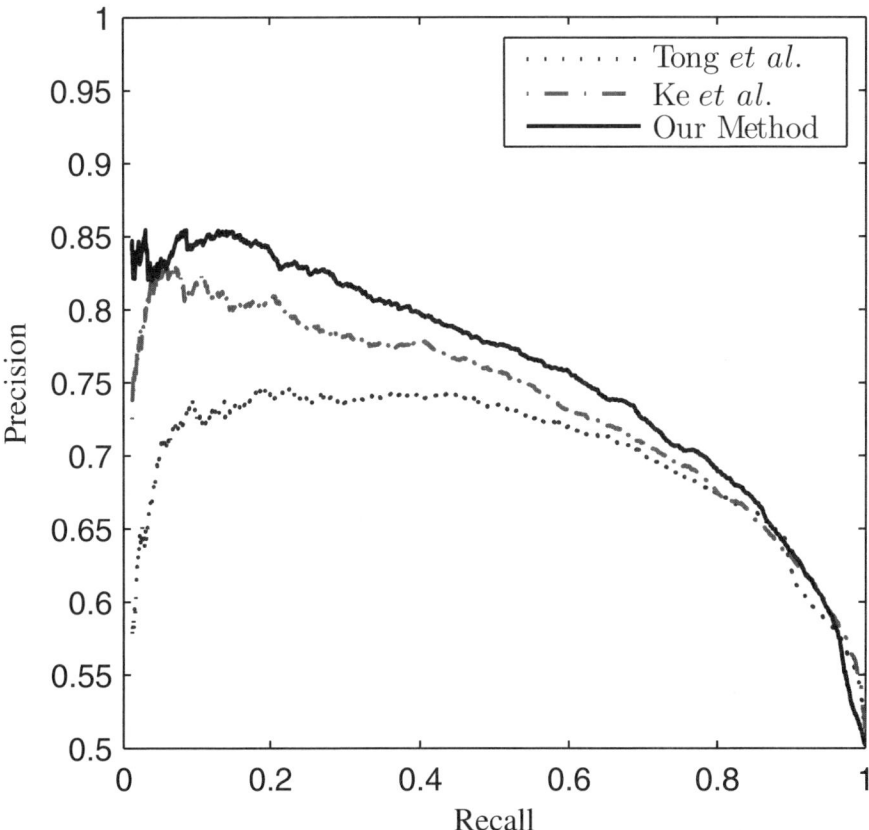

Fig. 4. Quality assessment results of Tong et al. method, Ke et al. method, and our method. Our method achieves the best results, in particular when the recall is low.

The results are shown as the precision-recall curves in Figure 4. From the figure, we see that our method achieves the best results, in particular when the recall is low. The good performance of our method is due to that the focusing degree measure of our method only depends on focused places, which is more reasonable.

5 Conclusions

Focused places play an important role in photo concept discovery and quality assessment. In this work, we have established a focusing degree detector and map to discover the focused places of a photo and applied them to image retrieval and photo quality assessment. Experiments demonstrated the high effectiveness of our method. From the successful focused region discovery in this study, we thought that the proposed focusing degree map could also be used as a cue for image segmentation. This will be our future attention point.

Acknowledgment. This work was supported by the National Natural Science Foundation of China under grant 60773039 and CASIA Innovation Fund For Young Scientists.

References

1. Datta, R., Joshi, D., Li, J., Wang, J.Z.: Studying aesthetics in photographic images using a computational approach. In: Leonardis, A., Bischof, H., Pinz, A. (eds.) ECCV 2006. LNCS, vol. 3953, pp. 288–301. Springer, Heidelberg (2006)
2. Ke, Y., Tang, X., Jing, F.: The design of high-level features for photo quality assessment. CVPR 1, 419–426 (2006)
3. Flickner, M., Sawhney, H.S., Ashley, J., Huang, Q., Dom, B., Gorkani, M., Hafner, J., Lee, D., Petkovic, D., Steele, D., Yanker, P.: Query by image and video content: The QBIC system. IEEE Computer 28(9), 23–32 (1995)
4. Chen, Y., Wang, J.Z.: Image categorization by learning and reasoning with regions. Journal of Machine Learning Research 5, 913–939 (2004)
5. Sivic, J., Russell, B.C., Efros, A.A., Zisserman, A., Freeman, W.T.: Discovering objects and their localization in images. In: ICCV, vol. 1, pp. 370–377 (2005)
6. Zhang, R., Zhang, Z.M.: Hidden semantic concept discovery in region based image retrieval. In: CVPR (2), pp. 996–1001 (2004)
7. Liu, Y., Zhang, D., Lu, G., Ma, W.Y.: A survey of content-based image retrieval with high-level semantics. Pattern Recognition 40, 262–282 (2007)
8. Heidemann, G.: The long-range saliency of edge- and corner-based salient points. IEEE Transactions on Image Processing 14(11), 1701–1706 (2005)
9. Wang, W., Song, Y., Zhang, A.: Semantics-based image retrieval by region saliency. In: International Conference on Image and Video Retrieval, pp. 29–37 (2002)
10. vande Weijer, J., Gevers, T., Bagdanov, A.D.: Boosting color saliency in image feature detection. IEEE Trans. Pattern Analysis and Machine Intelligence 28(1), 150–156 (2006)
11. Fu, H., Chi, Z., Feng, D.D.: Attention-driven image interpretation with application to image retrieval. Pattern Recognition 39(9), 1604–1621 (2006)
12. Buzzi, J., Guichard, F.: Uniqueness of blur measure. In: ICIP, pp. 2985–2988 (2004)
13. Pavlovic, G., Tekalp, A.M.: Maximum likelihood parametric blur identification based on a continuous spatial domain model. IEEE Transactions on Image Processing 1(4), 496–504 (1992)
14. Reeves, S.J., Mersereau, R.M.: Blur identification by the method of generalized cross-validation. IEEE Transactions on Image Processing 1(3), 301–311 (1992)
15. Li, X.: Blind image quality assessment. In: ICIP (1), pp. 449–452 (2002)
16. Tong, H., Li, M., Zhang, H., Zhang, C.: Blur detection for digital images using wavelet transform. In: ICME, IEEE, pp. 17–20. IEEE Computer Society Press, Los Alamitos (2004)
17. Sheikh, H.R., Wang, Z., Bovik, A.C.: No-reference perceptual quality assessment of JPEG compressed images. In: ICIP (1), pp. 477–480 (2002)
18. Tong, H., Li, M., Zhang, H., Zhang, C., He, J., Ma, W.Y.: Learning no-reference quality metric by examples. In: Chen, Y.P.P. (ed.) MMM, pp. 247–254. IEEE Computer Society Press, Los Alamitos (2005)
19. Jianchao, Y.: Motion blur identification based on phase change experienced after trial restorations. In: ICIP (1), pp. 180–184 (1999)
20. Liu, X., Li, M., Zhang, H., Wang, D.: Bayesian motion blur identification using blur priori. In: ICIP (2), pp. 957–960 (2003)
21. Wang, J.Z., Li, J., Wiederhold, G.: Simplicity: Semantics-sensitive integrated matching for picture libraries. IEEE Transactions on Pattern Analysis and Machine Intelligence 23(9), 947–963 (2001)

[22] Chen, Y., Wang, J.Z.: A region-based fuzzy feature matching approach to content-based image retrieval. IEEE Transactions on Pattern Analysis and Machine Intelligence 24(9), 1252–1267 (2002)
[23] Martin, D., Fowlkes, C., Tal, D., Malik, J.: A database of human segmented natural images and its application to evaluating segmentation algorithms and measuring ecological statistics. In: Proc. 8th Int'l Conf. Computer Vision, vol. 2, pp. 416–423 (July 2001)
[24] Muller, H., Maillet, S.M., Pun, T.: The truth about corel: Evaluation in image retrieval. In: Lew, M.S., Sebe, N., Eakins, J.P. (eds.) CIVR 2002. LNCS, vol. 2383, pp. 38–49. Springer, Heidelberg (2002)
[25] Hartigan, J.A., Wong, M.A.: Algorithm as136: A k-means clustering algorithm. Applied Statistics 28, 100–108 (1979)

Region Based Image Retrieval Incorporated with Camera Metadata

Jie Ma, Hong Lu, and Yue-Fei Guo

Shanghai Key Laboratory of Intelligent Information Processing
Department of Computer Science & Engineering
Fudan University, Shanghai, China
{052021181,honglu,yfguo}@fudan.edu.cn

Abstract. Content based image retrieval (CBIR) has been researched for decades. However, the "semantic gap" which exists between low-level features and human semantics still remains an unsolved problem. Region based image retrieval (RBIR) was proposed to bridge this gap in some extent. Beyond the pixel values in the image, what other information can also be used? The other information we use is Exif, which records the snapping condition of camera metadata. In this paper we propose an method that combines region low level features of image and camera metadata for image retrieval. Experimental results show the efficiency of our method than the traditional CBIR.

Keywords: Image retrieval, content-based image retrieval, region-based method, Exif.

1 Introduction

As the digital camera becomes more and more popular, people use it taking pictures when they visit interesting places. And there are more and more images for home use, especially on the internet. Thus, efficient and effective image analysis and retrieval are becoming a hot research area. Traditional content based image retrieval (CBIR) methods extract low-level features of images and these methods only consider the pixel values in the images. The CBIR was introduced in early 1990s to overcome the two difficulties faced to the manual image annotation. The one is the vast amount of labour needed to manual annotation image, the other is the semantics of images are rich and subject to human annotation. For example, an image in different people's viewpoints it may has different meanings. Since then, many research and commercial CBIR systems were emerged, Such as QBIC (Query by Image Content) of IBM [1], Informedia of CMU University [2], etc.

Although CBIR solve the manual annotation problem, it also has its own drawback. That is semantic gap between the low-level features and human semantics. In order to bridge this gap, many works has been done. Liu *et al.* [3] outlined three main methods to bridge this gap: region based image retrieval (RBIR), relevance feedback (RF), and modeling high-level semantic feature. In [3], the author used the high level semantic color names to improve the region based image retrieval. In it the HSV color space was mapped to the 93 semantic color names, such as (0.36, 0.99, 0.65) meaning green grass, while (0.62, 0.67, 0.92) standing for light sky blue, etc. That was very rough and imprecision because (0.36, 0.99, 0.65) can mean green grass but also other object if it had that color values.

The above methods are usually only considering the value of pixels. However, images are not taken in a vacuum. There is certain context information within an image or between images. This information can be use for image analysis. For example, Boutell *et al.* [4] used the camera metadata cues and low-level features in the semantic scene classification. And Luo *et al.* [5] gave an overview of exploiting context for semantic content understanding. Their analysis suggested that there were three kinds of context information in and between images, i.e., temporal context, spatial context, and image context from camera metadata. Furthermore, this diffident context information can be used as complement of the low-level features of image for image content analysis.

Exchangeable image file format(Exif) [6] is a standard of digital camera metadata included in the JPEG image when image is snapped. In it there are many tags relate to the image snapped conditions, such as FocalLength, ExposureTime, Aperture FNumber, FlashUsed, ISOequivalent, Distance and ExposureBias, etc. In Boutell *et al.* [4], it showed the validity of such information for image content analysis such as Flash tends to be more used in the outdoor than indoor. The larger the ApertureFNumber value, the image may be outdoor scene not a specific object. In this paper, we use the region features of images and also use the camera metadata to improve the image retrieval performance.

The following paper is organized as: section 2 describes the framework of our region based image retrieval. The experiment results and analysis are described in section 3. The paper is concluded in section 4.

2 The Framework of System

The framework of our system includes the following components: image segmentation, extracting low-level features of regions, feature similarity computing, ranking image retrieval results, and modifying the ranking by fusion Exif information.

2.1 Image Segmentation and Low Level Feature Extraction of Regions

Image segmentation is also the hot research area in the computer vision. There are many segmentation methods have being proposed [7, 8 and 9, etc.]. Using our experiment images, we find The JSEG segmentation method performing well on them. The JSEG method consists of two independent steps, i.e. colour quantization and spatial segmentation. In the first step, image colours are quantized to several representative classes which can be used for differentiating the regions in the image. In the second spatial segmentation step, using the criterion of "good" segmentation and colour-class map, we get the "J-image". After using multiscale J-image for region growing, we get the final segmentation result. Fig. 1 gives the image segmentation results of two images by using JSEG. The parameters are set according to that in [7]. And the resulted image regions in these two images are 16 and 15.

Fig. 1. Image segmentation results for two images by using JSEG

After the image segmentation, we ignore the small region that its size smaller than the 1/10 of whole image and do the low-level feature extracting on other salient regions. To represent the regions' features, we use the MPEG-7 [10] color descriptors CLD (Color Layout Descriptor) and SCD (Scalable Color Descriptor) [11].

2.2 Image Similarity Metric and Ranking

Given a query image, after segmentation and low-level feature extracting, we represent the query image as:

$$Q_i = \{(f_1, w_1), ... , (f_n, w_n)\} \quad (1)$$

Where (f_j, w_j) describe j region of query image i, f_j standing for the low-level feature of region j, w_j standing for the weight of that region defined in Eq.2.

$$w_j = size\_of(region\_j)/size\_of(image\_i) \quad (2)$$

The source images in the database are also represented in this form. Then the problem is how to evaluate the similarity of two images with different number of salient regions. I. e.

Query Image: $Q_i = \{(f_1, w_1), ... , (f_n, w_n)\}$ with n regions

Source Image: $S_t = \{(f_1, w_1), ... , (f_m, w_m)\}$ with m regions

EMD [12] method allows for partial matching satisfying our problem. Specifically, EMD method treats one distribution as the holes in the earth, the other distribution as lots of earth spread on the earth. The solution of EMD method is that use those earths to fill the holes and make the transportation distance minimal at the same time. There is a ground distance between one hole and one mass of earth. In the EMD algorithm, the ground distance can be defined as Euclidean distance, L-1 Normal, etc. In this paper we use the Euclidean distance to evaluate the low level feature distance of one region to one region because of its easy of computation and good performance for image retrieval.

$$D(f_j^Q, f_k^S) = \sqrt{(f_{j,1}^Q - f_{k,1}^S)^2 + (f_{j,2}^Q - f_{k,2}^S)^2 + ... + (f_{j,n}^Q - f_{k,n}^S)^2} \quad (3)$$

In Eq.3, n stands for the dimension of low-level feature. Then the EMD distance of two images can be formulated as:

$$EMD(Q_i, S_t) = \frac{\sum_{m \in Q_i} \sum_{n \in S_t} c_{mn} D(f_m^Q, f_n^S)}{\sum_{m \in Q_i} \sum_{n \in S_t} c_{mn}} \quad (4)$$

Where c_{mn} is optimal flow of regions m of query image i to the region block n of source image t, D function is formulated in Equ. (3). the denominator is the normalization factor. The computing of EMD algorithm is based the old transportation problem [13]. EMD algorithm minimizes Eq. 4 by finding the optimization c_{mn} of image Q_i and S_t iteratively. We rank the retrieval results by the value of EMD in the ascend order. In our experiment environment of CPU P4 3.0, 1G Memory and Windows OS, the search overall search process needs about 20 seconds for our algorithm with the image size of 320*240. We can shorten this time by resizing the image size or using more powerful computers to achieve the online operation.

2.3 Modify the Image Ranking by Exif Information

Beyond the pixel values of images, we also use Exif to modify the similarity score of two images for image retrieval. There are some relationship between the Exif value and semantics. According the [1], the focal length (FL), exposure time (ET), f-number (FN) and flash fired (FF) of the camera metadata have more relationship with semantics of image. The value of ET is different in different times of a day and the value of FN is different with the size of object in the image. Analysis our images, we find ET and FN vary different in different concepts. As shown in Table 1, statistic analysis on images in our database shows that FN tends to be smaller in the car images than the other images and the ET tends to be smaller in the Guanghua Tower Building images than in the other images. So we use this two metadata for refining the retrieval results.

Table 1. Mean of exposure time and mean of F-Number of six concepts

Concept	Mean_ET	Mean_FN
Car	0.00725466	3.31852
Garden	0.00520588	4.36471
Grass	0.00496721	5.18689
Guanghua Tower	0.00190569	5.6
Chair Mao sculpture	0.00402576	5.6
Red Building	0.00354569	5.49231

We use the Euclidean distance to measure to similarity of Exif information. For query image q and random image t_i in database, we measure the similarity in the formula 5.

$$S_E = D_{ET}(q,t_i) + D_{FN}(q,t_i) \qquad (5)$$

Then the refined similarity of two images is decided by the formula 6.

$$S(q,t_i) = w_1 S_L(q,t_i) + w_2 S_E(q,t_i) \qquad (6)$$

Where w_1 and w_2 are two weight parameters of low-level feature S_L and Exif S_E similarity. Through adjusting the weight of w_1 and w_2, we get different retrieval results. In the experimental, we set the weight of $w_1 = 0.7$ and $w_2 = 0.3$, getting the comparative good results.

3 Experiment Results

In the experiment, we use the images snapped by ourselves in order to use the Exif information. The Experimental database includes 300 images, 50 images per concept. The six concepts are: grass, garden, car, Tower building, chair Mao sculpture and red building all taken in Fudan University. We do the following four experiments: 1) traditional CBIR by using global features, 2) traditional CBIR by using global features with Exif, 3) RBIR and 4) RBIR with Exif. Following is the demo system of our experiment.

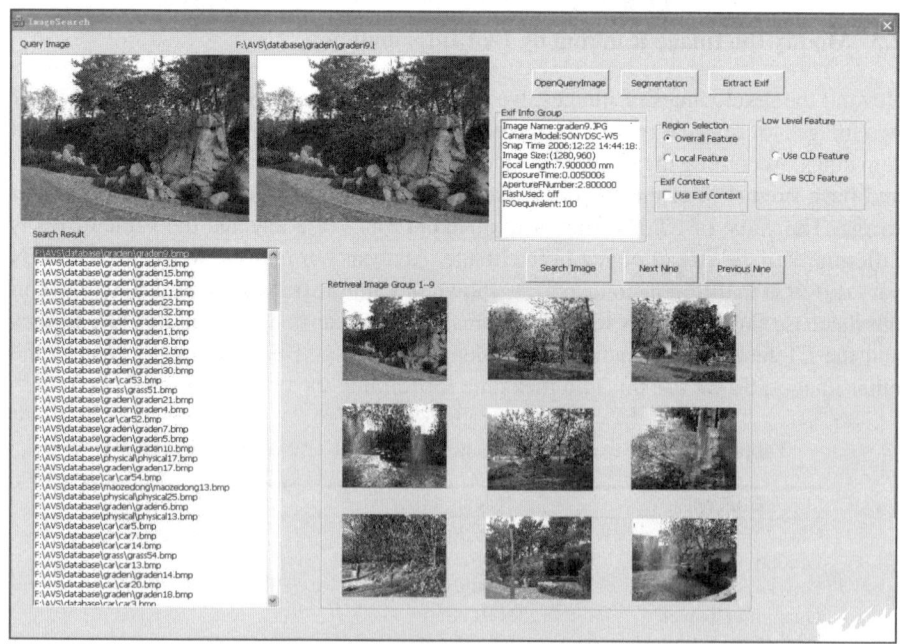

Fig. 2. The demo system of experiment

In the demo you can select the search combination, such as global feature plus SCD low level feature and without Exif information, etc.

We use the precision value to evaluate the performance of the four methods. The experiment results of our proposed method superior of the traditional content-based image retrieval. Figure 3 and figure 4 illustrate two concepts precision compared with other three concepts. In the Fig.3 and Fig.4, the X-axis means the retrieval number including the similar images and dissimilar images, while Y-axis means images which are similar to the query image. We see that the method we proposed in the experiments gets the best precision results when the retrieval number reaches to 160. To the concept of car, the improvement is obvious, although the result of Chair Mao sculpture concept is not as well as the traditional CBIR when the retrieval number is small.

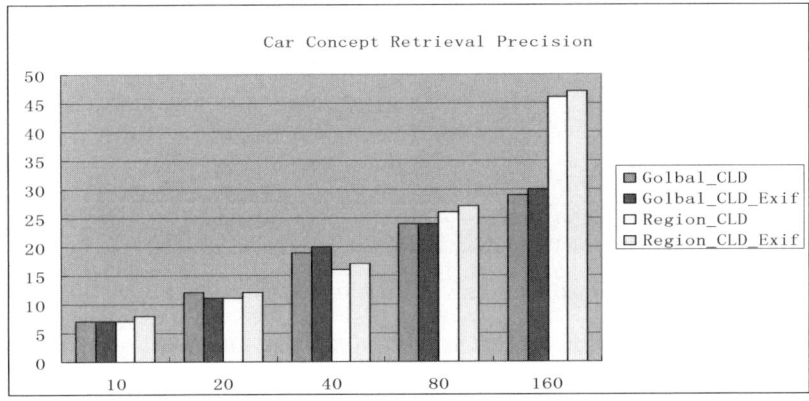

Fig. 3. Car concept image etrieval precision results comparison

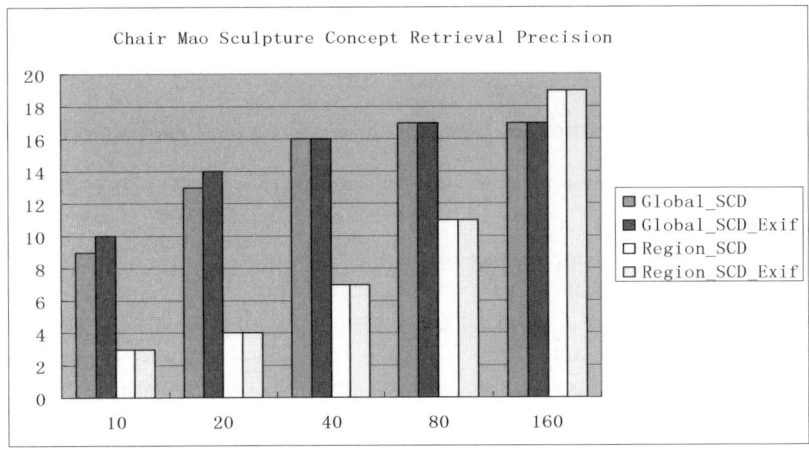

Fig. 4. Chair mao sculpture concept image retrieval precision results comparison

Figure 5 is the query images of concept car and Chair Mao sculpture and figure 6 is the top sixth retrieval results of query images in Fig.5 using our proposed method.

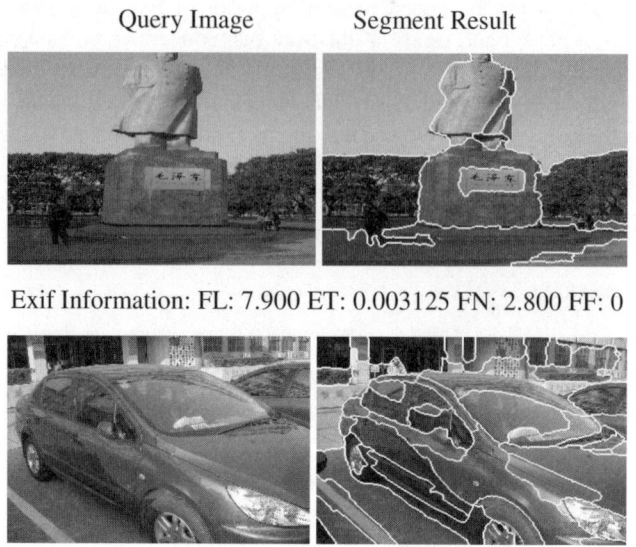

Fig. 5. Query images and Exif information of car and Chair Mao sculpture

Fig. 6. Query result of top sixth results of our algorithm (From left to right of 1st-6th)

From the Fig.3-6, we can see that the method we proposed gets the best precision results when the retrieval number is counting to 160. For the concept of car, we get 47 relevant images of Region_CLD_Exif query while the traditional content-based image retrieval only got 28 relevant images. The proposed method improves the precision of 67.8%. To the Chair Mao sculpture concept, we find the precision of our method smaller than the Golbal_SCD when the retrieval number is small. We analysis the Chair Mao image in the database, finding the image also including blue sky, tree, and grass large regions in it. When we use the region feature for the retrieval, those regions are noise and might be mistaken as regions in the garden or grass concept. At the same time, we find the color of Guanghua Tower and Chair Mao sculpture is similar, both having blue sky and architecture color displayed in the Fig.4. And we can also see from the experimental results that using Exif information improves the retrieval results compared with only using low-level feature for the car concept.

4 Conclusion and Future Work

In this paper we proposed a region based image retrieval method incorporated with camera metadata. Compared with traditional content based image retrieval, the region based image retrieval can narrow the semantics influence in some extent. Furthermore, we use the Exif information to refine the low-level feature based RBIR. Experimental results show the efficiency of our method than the traditional CBIR. Our future work will focus on the salient region determination in the image and the adaptive weight setting of regions. Specifically, user's attention selection to improve the image retrieval result. We will also deeply study the value of ET and FN with the semantics, such as snapping an object with different time in a day, or in the same time of different days.

Acknowledgement. This work was supported in part by Natural Science Foundation of China under contracts 60533100 and 60402007, and Shanghai Municipal R&D Foundation under contracts 05QMH1403, 065115017 and 06DZ15008.

References

1. QBIC System, http://www.qbic.almaden.ibm.com/
2. Informedia System, http://www.informedia.cs.cmu.edu/
3. Liu, Y., Zhang, D., Lu, G., et al.: Region-based Image Retrieval with High-Level Semantic Color Names. In: Proceedings of the 11th International Multimedia Modeling Conference, pp. 180–187. IEEE Press, Melbourne (2005)
4. Boutell, M., Luo, J.: Bayesian Fusion of Camera Metadata Cues in Semantic Scene Classification. In: Proceedings of the 2004 IEEE Computer Society Conference on Computer Vision and Pattern Recognition, pp. 623–630. IEEE Press, Washington (2004)
5. Luo, J., Boutell, M., Brown, C.: Pictures Are Not Taken in a Vacuum. IEEE Signal Processing Maganize 23(2), 101–114 (2006)
6. Exif Standard: http://www.exif.org/
7. Deng, Y., Manjunath, B.: Unsupervised Segmentation of Color-Texture Regions in Images and Video. IEEE Transaction on Pattern Analysis and Machine Intelligeance 23(8), 800–810 (2001)
8. Shi, J., Malik, J.: Normalized cuts and image segmentation. In: Proceedings of the IEEE Conference on Computer Vision and Pattern Recognition, pp. 731–737. IEEE Press, San Juan (1997)
9. Vese, L., Chan, T.: A multiphase level set framework for image segmentation using the Mumford and Shah model. UCLA Department of Mathematics CAM Report 01-25, Int'l. J. Computer Vision (2001)
10. Martinez, J.: Overview of the MPEG-7 standard. Technical Report 5.0, ISO/IEC, Singapore (2001)

11. Xue, Y., Lu, H., Yu, H., et al.: Fudan University at TRECVID 2006. NIST (2006)
12. Rubner, Y., Tomasi, C., Guibas, L.: A Metric for Distributions with Applications to Image Databases. In: Proceedings of the 1998 IEEE International Conference on Computer Vision, pp. 59–66. IEEE Press, Bombay (1998)
13. Dantzig, G.: Application of the simplex method to a transportation problem. In: Activity Analysis of Production and Allocation, pp. 359–373. John Wiley and Sons, Chichester (1951)

Empirical Investigations on Benchmark Tasks for Automatic Image Annotation[*]

Ville Viitaniemi and Jorma Laaksonen

Adaptive Informatics Research Centre, Helsinki University of Technology,
P.O. Box 5400, FIN-02015 TKK, Finland
{ville.viitaniemi, jorma.laaksonen}@tkk.fi

Abstract. Automatic image annotation aims at labeling images with keywords. In this paper we investigate three annotation benchmark tasks used in literature to evaluate annotation systems' performance. We empirically compare the first two of the tasks, the 5000 Corel images and the Corel categories tasks, by applying a family of annotation system configurations derived from our PicSOM image content analysis framework. We establish an empirical correspondence of performance levels in the tasks by studying the performance of our system configurations, along with figures presented in literature. We also consider ImageCLEF 2006 Object Annotation Task that has earlier been found difficult. By experimenting with the data, we gain insight into the reasons that make the ImageCLEF task difficult. In the course of our experiments, we demonstrate that in these three tasks the PicSOM system—based on fusion of numerous global image features—outperforms the other considered annotation methods.

1 Introduction

The problem of matching words and images has been investigated intensively during this decade. Image annotation is closely related to the more general problem of image content understanding. The capabilities of image content analysis techniques can be demonstrated by applying them to the image annotation task. A wealth of image annotation methods have been proposed and studied in literature. Comparing the relative merits of the methods is made somewhat difficult by the methods being demonstrated in a variety of different annotation tasks and evaluated using different performance measures.

In our work we have tried to experimentally clarify the matter by generating a family of annotating system configurations based on our general-purpose PicSOM image content analysis framework (e.g. [13]). By evaluating the performance of the same set of system configurations in a variety of image annotation

---

[*] Supported by the Academy of Finland in the projects *Neural methods in information retrieval based on automatic content analysis and relevance feedback* and *Finnish Centre of Excellence in Adaptive Informatics Research*. Special thanks to Kobus Barnard, Xiaojun Qi and Yutao Han for helping with the experimental setup.

tasks and with different performance measures, we can gain empirical insight how the various performance figures compare. In our earlier work [20] we have compared three different annotation performance measures in the de facto standard 5000 Corel images annotation task. In this paper we extend the set of the considered annotation tasks with two tasks: the Corel categories task and the ImageCLEF 2006 Object Annotation Task. For the Corel tasks, we experimentally correlate the performance of our annotating system configurations, thereby establishing an empirical correspondence of performance levels in the two tasks. For the ImageCLEF task, we content ourselves to a less exhaustive investigation as we feel our system to be somewhat incompatible with the problem setting of the task. However, we empirically study the aspects of the task that make it difficult both to our annotation system and to those reported in literature.

The rest of the paper is organised as follows. In Sect. 2 we define the image annotation problem and the performance measures we are going to use. Section 3 outlines the way the PicSOM framework is used for image annotation. In Sect. 4 the three annotation tasks are described. In Sect. 5 we report and discuss the results of applying our annotation system to the tasks. Finally, in Sect. 6 we present conclusions and general discussion motivated by the experiments.

2 Automatic Image Annotation

In this paper we consider the problem of automatically annotating images with keywords in a supervised learning setting: the goal of an annotation system is to predict keywords for previously unseen images, based on a set of training images and keywords annotating them. A closely related term to image annotation problem is *image categorisation*. The terms may refer to either similar or slightly different variants of image content analysis tasks. In literature the use of the terms is somewhat inconsistent. For sake of definiteness, we call these variants *one-to-many* and *one-to-one* annotation. In a one-to-one annotation task, the keywords are exclusive, i.e. each image is annotated with exactly one keyword. One-to-many annotation can be thought as a generalisation of this. In these tasks any number of keywords can be associated with an image. In literature, the term annotation is sometimes reserved for one-to-many annotation, and the term categorisation used for one-to-one annotation.

The goodness of the predicted annotations is measured by comparing them with a manually-specified ground truth. Several performance measures have been defined to quantify the difference between predicted and ground truth annotations. For the one-to-one annotation tasks, we follow the practice adopted in literature and measure performance by total classification accuracy, i.e. the fraction of correct predictions of all predictions. For many-to-one image annotation, we have discussed three performance measures—the normalised score (NS), precision/recall statistics (PR) and de-symmetrised mutual information (DTMI)—in our earlier work [20]. Due to space limitations, we select one of these—the PR-statistics—for the reason of widespread use in literature, although we claim

DTMI to have the most desirable properties of these three measures as such. For an individual keyword w the precision and recall are defined as

$$P_w = |w_c|/|w_{\text{pred}}|, \quad R_w = |w_c|/|w_{\text{gt}}|, \tag{1}$$

where $|w_{\text{gt}}|$ and $|w_{\text{pred}}|$ are the number of occurrences of keyword w in the ground truth and the predicted annotations, respectively. $|w_c|$ of the predicted annotations are correct. The precision and recall values are averaged uniformly over all keywords to obtain the PR statistics.

3 Annotating Images Using PicSOM Framework

For the experiments in this paper, we have annotated images by techniques we have devised in our earlier work. We predict the annotations as follows. For each keyword, a separate classifier is constructed using the PicSOM image content analysis framework, and trained using the training images. The test images are scored using these classifiers (Sect. 3.1). A model connecting the scores to keyword probabilities is estimated (Sect. 3.2). Given the probability estimates, annotations approximately maximising a specified annotation performance measure are then determined (Sect. 3.3).

3.1 PicSOM Framework for Image Classification

In the PicSOM image content analysis framework, the input to the image classifiers consists of three sets of images: training images annotated with a keyword (positive examples), training images not annotated with the keyword (negative examples), and test images. The task of the classifier is to associate a score to each test image that reflects simultaneously visual similarity with the positive examples and dissimilarity with the negative ones.

As a basis for the classification, a set of global visual features is extracted from the images. Some of the features are truly global, such as the global colour histogram, others encode some spatial information using a fixed grid, such as the ColorLayout feature, very similar to the corresponding feature in the MPEG-7 standard [8]. Altogether, each image is described with 10 feature extraction techniques. Each of the extracted feature vectors has several components, and the combined dimensionality of the joint feature vector is 682. The extracted feature vectors are available upon request.

The PicSOM framework classifies images emphasising features that separate well the positive and negative example images of a particular task. The components of the feature vectors are divided into several overlapping subsets, feature spaces. Each of the feature spaces is quantised using a TS-SOM [12], a tree-structured variant of the Self-Organising Map [11]. To each two-dimensional quantised SOM grid representation we place positive and negative impulses corresponding to the locations of positive and negative example images. The impulses are then normalised and low-pass filtered. This associates a partial classifier score to each part of each feature space. A total classifier score for a test image is obtained by projecting the test image to each of the quantised feature spaces and summing the partial classifier scores from the corresponding feature space locations.

3.2 Probability Model

To facilitate the selection of annotating keywords, we need to estimate the connection between the classifier score value $s_w(i)$ and the probability of the keyword w to appear in the annotation of the image i. We have observed in practice that a simple logistic sigmoid model

$$p_i(w|s_w(i)) = \frac{1}{1+e^{-\theta_1^w s_w(i) - \theta_0^w}} \qquad (2)$$

relatively well suits this purpose. The model parameters $\theta_{\{0,1\}}^w$ are estimated separately for each keyword w by maximising the likelihood of the training data numerically with the Newton-Rhapson method while regarding each of the training images as an independent observation.

3.3 Selecting the Annotations

Given the probability estimates for the keywords, this stage of the system selects annotations that maximise the expected value of a specified annotation performance measure. In many cases, optimal selection of annotations seems to be an overwhelmingly difficult task to be performed exactly. In an such event, we split the problem into two parts: 1) evaluating the expected value of the measure, given the probability estimates and predicted annotations, and 2) searching for the maximising annotations to be predicted. For the reported results, we heavily approximate both of the subproblems.

4 Benchmark Tasks

In the experiments, we applied an annotation system derived from the PicSOM image content analysis framework to three different annotation benchmark tasks that have received somewhat widespread use in literature. This way we aim at a conception of comparable performance levels in each of the tasks.

4.1 5000 Corel Images Task

The first of the considered benchmark tasks is the 5000 Corel images task, defined by Duygulu et al. [6]. The task has been subsequently used by many others, e.g. [2,9,14]. The data set consists of 5000 images from 50 commercial Corel stock photograph CDs. The images are partitioned into 4500 training and 500 test images. For the purpose of determining the free parameters of our annotating systems, we have further separated a 500 image validation set from the original training set. In this one-to-many annotation task, a 1–5 word ground truth annotation is given for each image. The vocabulary of the annotations consists of 374 keywords. 371 of the keywords appear in the training set and 263 in the test set.

4.2 Corel Categories Task

The second of the benchmark tasks—Corel categories—is a one-to-one annotation task. The task was first defined by Chen and Wang [4] for 20 image categories. The image data consists of images from 20 Corel stock photograph CDs. Each of the CDs contains 100 images from a distinct topic and forms a target category for the task. A label is chosen to describe each of the categories. Table 1 shows the labels of the 20 categories. The task was later extended to 60 categories by Qi and Han [19] by introducing 40 more Corel CDs. We use all the 60 categories in our experiments.

Table 1. Labels of the first 20 categories of the Corel categories task

African people and villages	Beach	Historical buildings	Buses
Dinosaurs	Elephants	Flowers	Horses
Mountains and glaciers	Food	Dogs	Lizards
Fashion	Sunsets	Cars	Waterfalls
Antiques	Battle ships	Skiing	Desert

4.3 ImageCLEF 2006 Object Annotation Task

As the third benchmark we used the Object Annotation Task (OAT) of the ImageCLEF 2006 evaluation campaign [5]. In this one-to-one annotation task, images of a generic photograph collection (provided by LTU Technologies [16]) are annotated according to 21 categories. In comparison to the Corel categories task, the categories correspond somewhat more consistently to man-made physical objects. The labels of categories can be read from Table 3. The images are partitioned into a training set of 13963, a development set of 100 and a test set of 1000 images. In the final testing phase, both the training and the development sets are used as a composite training set.

This task violates the standard supervised learning model where both training and test data comes from the same distribution. In this case, there are two kinds of differences between the sets. First of all, the images of the various sets are visually different. In the development and test sets the objects are embedded in cluttered backgrounds whereas the training images usually portray objects on clear backgrounds. Secondly, the proportions of the different categories are different in all the three image sets.

5 Experiments and Results

In this section we present and discuss the results obtained by applying the PicSOM annotation system to the three benchmark tasks of Sect. 4.

5.1 5000 Corel Images and Corel Categories Tasks

In the first set of experiments we evaluate the annotation performance in the 5000 Corel images and Corel categories tasks. We compare the performance of annotation systems derived using the PicSOM image content analysis framework with methods proposed in literature. For the 5000 Corel images task we report the PR-performances under the constraint of at most 5 keywords annotating an image. In the Corel categories task the performance is measured by total classification accuracy. This practice has been adopted by the earlier literature employing the Corel categories task. For both of the tasks the two first stages of the annotation up to probability estimation (Sections 3.1 and 3.2) are identical. The difference is in the final annotation selection stage (Sect. 3.3) that is specific to each task. The stage finds exactly optimal annotations in case of the total accuracy measure, for the PR-measure the solution is more approximate.

Figures 1 and 2 and compare the performance of one of the generated annotation system configurations with results reported in literature. The parameters of the system, most importantly the set of visual features used, have been chosen to maximise the PR-performance in the validation set of the 5000 Corel images task. For the Corel categories task, we have used the PR-optimising parameters of the 5000 Corel images task. In Fig. 1, the dashed operating curve corresponds to varying a precision-recall tradeoff parameter in the system. The circled operating point was chosen on basis of the validation set.

The solid lines in Fig. 2 represent the average accuracies in the Corel categories task over several random choices of the training and test sets. The dashed lines indicate the 95% confidence intervals. The relatively narrow confidence interval of the PicSOM system is explained by us performing 20 random trials, as opposed to the five trials of the other results. As expected, the accuracies generally degrade as new categories are added to the set of images, but occasionally the added categories are visually distinct enough from the other categories and the accuracies momentarily increase.

5.2 Empirical Correspondence on the Tasks in Family of Annotating System Configurations

In the second set of experiments the empirical correlation of the performance levels in the 5000 Corel images and the Corel categories tasks is extended by generating a family of intentionally inferior annotation system configurations. Their performances are evaluated in the both tasks, thereby giving a fuller picture of correspondence between the performance levels. The different system configurations differ in terms of sets of used visual features. The feature sets 1–3 optimise the PR-, NS- and DTMI-performances in the 5000 Corel images validation set. The feature sets 4–17 are rather arbitrary subsets of feature sets 1–3.

Figures 3a and 3b show the PR-performance in the 5000 Corel images task and the classification accuracy in the Corel categories task, respectively, for the generated system configurations. From the figures it can be seen that the ordering of the configurations is similar in both tasks. Within clusters of similar performance,

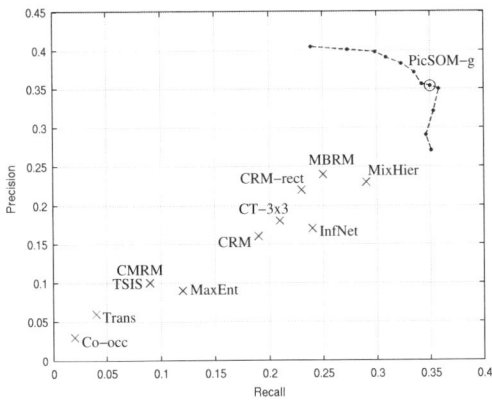

Fig. 1. PR annotation results reported for 5000 Corel images task. Key to the method abbreviations: Co-occurence Model (Co-occ)[18], Translation Model (Trans)[6], Cross-Media Relevance Model (CMRM)[9], Text Space to Images Space (TSIS)[3], Maximum Entropy model (MaxEnt)[10], Continuous Relevance Model (CRM)[14], 3x3 grid of colour and texture moments (CT-3x3)[21], CRM with rectangular grid (CRM-rect)[7], Inference Network (InfNet)[17], Multiple Bernoulli Relevance Models (MBRM)[7], Mixture Hierarchies model (MixHier)[2], and the present method with global image features (PicSOM-g).

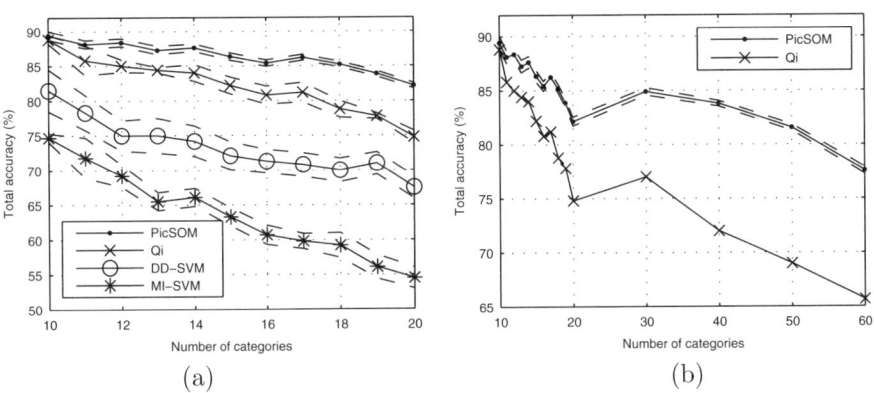

Fig. 2. Comparison of the accuracy of the PicSOM system with methods presented in literature in the Corel categories task as new categories are progressively added to the image set. Subfigure a) shows close-up of the situation with 11–20 categories. In subfigure b) the number of categories is further increased up to 60. The abbreviation Qi refers to method proposed in [19]. DD-SVM method was introduced in [4] and the MI-SVM in [1]. Experimental results were not available for DD-SVM and MI-SVM with more than 20 categories.

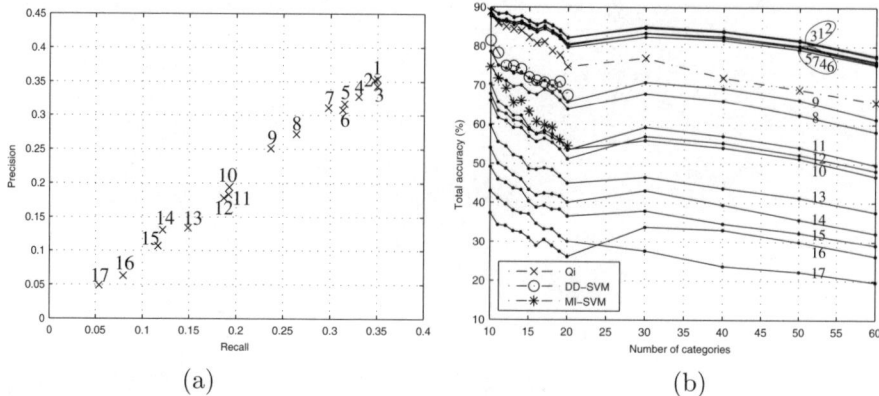

Fig. 3. The performance of family of annotation system configurations in the 5000 Corel images task (a), and in the Corel categories task (b). The numbers are used to label different PicSOM system configurations. For reference, the literature methods are shown in subfigure (b) with lines marked with symbols x,o and *. The abbreviations of the methods are the same as in Fig. 2.

however, the ordering sometimes varies between the different tasks. Based on these experiments, there in deed seems to exist a roughly monotonous correspondence between the performance levels in the two tasks. Based on this correspondence, one could predict, for instance, that the method proposed by Qi and Han in [19] could be used to obtain PR-performance between clusters {4,5,6,7} and {8,9} in the 5000 Corel images task. Yet, such predictions might be misleading since the obtained empirical correspondence is based on variations of just one annotation system. Different types of systems could, e.g., react differently to the increase in number of categories. Comparisons with more heterogeneous set of annotation systems would be required to make the correspondence firmer.

5.3 ImageCLEF 2006 Object Annotation Task

The ImageCLEF OAT benchmark violates the assumption of similar training and test set distributions, on which our PicSOM annotation system is based. Therefore, we did not consider it worthwhile to perform such comprehensive correlation of performance levels of a family of system configurations for this task as for the two Corel annotation tasks. The results would not give a justified picture of the difficulty of the ImageCLEF annotation task as our methods are not well suited for the task. Therefore we settle for results of a single annotation system configuration, the same configuration that performed best in the 5000 Corel images task validation set.

In any case, it will be interesting to compare the performance of the PicSOM annotation system to that of the ImageCLEF 2006 participants [5]. Also their systems suffer from differences between the training and test distributions. Table 2 summarises the achieved total categorisation accuracies in the test set.

Table 2. Classification accuracies in the ImageCLEF 2006 Object Annotation Task [5]

PicSOM – weighted	27.0%
PicSOM – unweighted	22.2%
RWTHi6 (RWTH Aachen, Germany)	22.7%
CINDI (Concordia University, Montreal, Canada)	16.8%
DEU CS (Dokuz Eylul University, Tinaztepe, Turkey)	11.8%
MedGIFT (University and Hospitals of Geneva, Switzerland)	9.5%

In the accuracy every test set image has equal weight. This is essentially the performance measure used also in the ImageCLEF evaluation. For the PicSOM results, we resorted to some task-specific tricks in order to be able to tackle the different distribution of training and test images at least to some degree. The sigmoidal probability estimation of (2) produced very bad results because of the very different proportion of categories in the training and test data. The sigmoids were therefore abandoned and raw classification scores used instead.

Table 2 shows two results for the PicSOM system. The row "unweighted" corresponds to simply combining the training and development images to a single training set. For the row "weighted" the images in the training and development sets were given different weights. The weights were chosen among a small set of possibilities so that the categorisation accuracy in the development set was maximised. Since the development set is very small and has different class proportions from those of the test set, it comes as no surprise that the weights selected by this criterion are quite far from optimal for the test set.

As a curiosity, it may be mentioned that leaving the small 100 image development set out from the composite training set was more detrimental to the performance than leaving away the original 13963 image training set. This testifies of the qualitative difference in the distribution of the image sets.

Compared to the Corel categories task with 20 categories, the performance levels of both the PicSOM system and the ImageCLEF participants are poor in this task. In these seemingly similar tasks, there are two evident candidates for the reason behind the performance differences: firstly, the categories of the ImageCLEF task may be visually more challenging. Secondly, the difference between the training and test distributions may make learning more difficult. To investigate the visual aspect we exclude the effect of different distributions by examining the leave-one-out cross-validation accuracy within the test set. Table 3 shows details of our results in the ImageCLEF task, including such cross-validation accuracies in the third column. From the summary rows on the top of the table we observe that also within the test set, the cross-validation accuracy (52%) is lower than the accuracies in the Corel categories task (above 80% in Fig. 2) by a wide margin. The difference is clear enough not to be explained by the somewhat different category statistics. On the other hand, the test set cross-validation accuracy is clearly larger then the 27% accuracy achieved in the whole annotation task involving different distributions. We can thus state that both the increased visual difficulty and the different distributions contribute to the ImageCLEF 2006 OAT being more difficult than the Corel categories task.

To further investigate the ImageCLEF OAT benchmark, we also evaluated the visual difficulty of the training set using the leave-one-out cross-validation accuracy as a measure. These results are included in the rightmost column of Table 3. Based on this experiment, we find that apart from being visually different, the test image set is also visually more challenging than the training image set. In the tabulated cross-validation results we have used the sigmoidal probability model. Comparing the results with and without the probability model (not shown), we get confirmation that the use of the probability model improves the accuracy when the training and test images of the annotating system have similar statistics.

Table 3. The detailed accuracies of the PicSOM annotation system for the data of the ImageCLEF 2006 Object Annotation Task, along with leave-one-out cross-validation accuracies within the test and training sets

	Unweighted	Weighted	Cross-validation, test set	Cross-validation, training set
Total accuracy	**22.2%**	**27.0%**	**52.0%**	**75.6%**
Ashtrays	0.0%	0.0%	25.0%	48.0%
Backpacks	3.6%	10.7%	39.3%	70.3%
Balls	10.0%	20.0%	40.0%	43.8%
Banknotes	15.6%	20.0%	33.3%	93.8%
Benches	15.9%	11.4%	79.5%	59.0%
Books	10.8%	29.2%	64.6%	60.4%
Bottles	5.3%	8.4%	62.1%	75.8%
Cans	5.0%	5.0%	20.0%	80.0%
Calculators	21.4%	7.1%	35.7%	52.1%
Chairs	35.6%	50.8%	66.7%	46.3%
Clocks	0.0%	0.0%	31.9%	81.5%
Coins	0.0%	0.0%	61.5%	82.9%
Computer equipment	2.5%	6.3%	43.0%	90.2%
Cups and mugs	36.1%	41.7%	52.8%	69.7%
Hifi equipment	0.0%	0.0%	37.5%	74.0%
Cutlery	73.3%	68.6%	66.3%	85.7%
Plates	25.0%	30.8%	51.9%	50.0%
Sofas	50.0%	40.9%	40.9%	60.6%
Tables	21.7%	21.7%	13.0%	56.8%
Mobile Phones	10.3%	25.6%	43.6%	71.0%
Wallets	35.3%	35.3%	41.2%	40.0%

6 Conclusions and Discussion

In this paper we have established an empirical connection of performance levels in two benchmark tasks for automatic image annotation: the 5000 Corel images task and the Corel categories task. This is achieved by evaluating the performances of several annotation system configurations derived using the PicSOM image content analysis framework. However, the obtained correspondence should not be considered definitive but suggestive. This is because the correspondence is empirical by its nature and the used annotation system configurations have been quite homogeneous.

We also took a look at the ImageCLEF 2006 Object Annotation Task, for which the results of our PicSOM system as well as others' are relatively poor, for example in comparison with the Corel categories task. With the experiments we gained some insight into the reasons that make this task difficult. Both the visual difficulty level of the categories and the training and test data coming from different distributions seem to be significant factors. In the task, the training and test images are visually different.

In the experiments we observed the performance of the PicSOM annotation system to compare very well with the other considered methods. In comparison with the other systems and models, our approach excels in the ability to automatically extract and adapt to a large set of visual descriptors of statistical nature. On the other hand, some other methods appear to use smaller collections of visual features more efficiently than PicSOM. Many of the methods partition images into individual segments or subimages. In contrast, the feature set we used for these experiments only included global image features.

Altogether, PicSOM seems to offer an alternative approach to automatic annotation with performance at least comparable to the methods in the literature. The viability of the different approaches will have to be weighed on case-by-case basis. We are not claiming that the absolute performance of an annotation method would be the only factor that matters. Comparison of ideas and models on a smaller scale is also valuable. Still, only evaluations of the maximum obtainable performance can demonstrate whether models scale up to higher performance levels when applied to more extensive data sets. By pursuing the maximum performance one can also assess the sensibility of the benchmark tasks and gain knowledge on the magnitudes of performance that can be considered reasonably good in the tasks.

In the future we will examine the possibility of improving the classification efficiency on basis of individual features while still retaining the adaptiveness and reasonable computational cost characteristics of the PicSOM architecture. We will also continue to look after still more expressive image features. In this direction, the SIFT-like interest point features [15] have shown promise in other image content analysis tasks we have performed.

References

1. Andrews, S., Tsochantaridis, I., Hoffman, T.: Support vector machines for multiple-instance learning. In: Advances in Neural Information Processing Systems 15, pp. 561–568. MIT Press, Cambridge (2003)
2. Carneiro, G., Vasconcelos, N.: Formulating semantic image annotation as supervised learning problem. In: Proc. of IEEE Conference on Computer Vision and Pattern Recognition, pp. 163–168. IEEE Computer Society Press, Los Alamitos (2005)

3. Celebi, E., Alpkocak, A.: Combining textual and visual clusters for semantic image retrieval and auto-annotation. In: EWIMT. Proc. of European Workshop on the Integration of Knowledge, Semantic and Digital Media Technologies, UK, pp. 219–225 (November 2005)
4. Chen, Y., Zwang, J.Z.: Image categorization by learning and reasoning with regions. Journal of Machine Learning Research 5, 913–939 (2004)
5. Clough, P., Grubinger, M., Deselaers, T., Hanbury, A., Müller, H.: Overview of the ImageCLEF 2006 photographic retrieval and object annotation tasks. In: CLEF working notes, Alicante, Spain (September 2006)
6. Duygulu, P., Barnard, K., de Freitas, N., Forsyth, D.: Object recognition as machine translation: Learning a lexicon for a fixed image vocabulary. In: Heyden, A., Sparr, G., Nielsen, M., Johansen, P. (eds.) ECCV 2002. LNCS, vol. 2350, pp. 97–112. Springer, Heidelberg (2002)
7. Feng, S.L., Manmatha, R., Lavrenko, V.: Multiple Bernoulli relevance models for image and video annotation. Proc. of IEEE CVPR 2, 1002–1009 (2004)
8. ISO/IEC. Information technology - Multimedia content description interface - Part 3: Visual, 15938-3:2002(E) (2002)
9. Jeon, J., Lavrenko, V., Manmatha, R.: Automatic image annotation and retrieval using cross-media relevance models. In: Proceedings of 26th Annual International ACM SIGIR Conference on Research and Development in Information Retrieval, Canada, pp. 119–126 (July-August 2003)
10. Jeon, J., Manmatha, R.: Using maximum entropy for automatic image annotation. In: Proc. of International Conference on Image and Video Retrieval, pp. 24–32 (2004)
11. Kohonen, T.: Self-Organizing Maps. Springer, Heidelberg (2001)
12. Koikkalainen, P., Oja, E.: Self-organizing hierarchical feature maps. In: Proc. IJCNN, San Diego, CA, USA, vol. II, pp. 279–284 (1990)
13. Laaksonen, J., Koskela, M., Oja, E.: PicSOM—Self-organizing image retrieval with MPEG-7 content descriptions. IEEE Transactions on Neural Networks 13(4), 841–853 (2002)
14. Lavrenko, V., Manmatha, R., Jeon, J.: A model for learning the semantics of pictures. In: Proc. NIPS, vol. 16, pp. 553–560 (2003)
15. Lowe, D.G.: Distinctive image features from scale-invariant keypoints. International Journal of Computer Vision 60(2), 91–110 (2004)
16. LTU Technologies: (Accessed 2007-5-18), http://www.LTUtech.com
17. Metzler, D., Manmatha, R.: An inference network approach to image retrieval. In: Enser, P.G.B., Kompatsiaris, Y., O'Connor, N.E., Smeaton, A.F., Smeulders, A.W.M. (eds.) CIVR 2004. LNCS, vol. 3115, pp. 42–50. Springer, Heidelberg (2004)
18. Mori, Y., Takahashi, H., Oka, R.: Image-to-word transformation based on dividing and vector quantizing images with words. In: Proc. of First International Workshop on Multimedia Intelligent Storage and Retrieval Management (1999)
19. Qi, X., Han, Y.: Incorporating multiple SVMs for automatic image annotation. Pattern Recognition 40, 728–741 (2007)
20. Viitaniemi, V., Laaksonen, J.: Evaluating performance of automatic image annotation: example case by fusing global image features. In: Proc. of International Workshop on Content-Based Multimedia Indexing, Bordeaux, France (June 2007)
21. Yavlinsky, A., Schofield, E., Rüger, S.: Automated image annotation using global features and robust nonparametric density estimation. In: Leow, W.-K., Lew, M.S., Chua, T.-S., Ma, W.-Y., Chaisorn, L., Bakker, E.M. (eds.) CIVR 2005. LNCS, vol. 3568, pp. 507–517. Springer, Heidelberg (2005)

Automatic Detection and Recognition of Players in Soccer Videos

Lamberto Ballan, Marco Bertini, Alberto Del Bimbo, and Walter Nunziati

Dipartimento di Sistemi e Informatica, University of Florence
Via di S. Marta 3, 50139 Florence, Italy
{ballan, bertini, delbimbo, nunziati}@dsi.unifi.it

Abstract. An application for content-based annotation and retrieval of videos can be found in the sport domain, where videos are annotated in order to produce short summaries for news and sports programmes, edited reusing the video clips that show important highlights and the players involved in them. The problem of detecting and recognizing faces in broadcast videos is a widely studied topic. However, in the case of sports videos in general, and soccer videos in particular, the current techniques are not suitable for the task of face detection and recognition, due to the high variations in pose, illumination, scale and occlusion that may happen in an uncontrolled environment.

In this paper we present a method for face detection and recognition, with associated metric, that copes with these problems. The face detection algorithm adds a filtering stage to the Viola and Jones Adaboost detector, while the recognition algorithm exploits *i)* local features to describe a face, without requiring a precise localization of the distinguishing parts of a face, and *ii)* the set of poses to describe a person and perform a more robust recognition.

1 Introduction and Previous Work

Semantic detection and recognition of objects and events contained in a video stream is required to provide effective content-based annotation and retrieval of videos. The goal of this annotation activity is to allow reuse of the video material at a later stage and to maintain an archive of the video assets produced. An example is that of sports videos, where live video streams are annotated in order to re-edit the video clips that show important highlights and key players to produce short summaries for news and sports programmes. Most of the recent works have dealt with the detection of sports highlights (e.g. shots on goal for soccer videos, pitching for baseball, shots for basketball, etc.). A comprehensive review of these works can be found in [1,2].

A method that could be used to select the most interesting actions is to analyze the shots that follow an highlight: in fact the shots that contain a key action are typically followed by close-ups of the players that had an important role in the action. For example in the case of soccer videos scored goals or near misses are followed by shots that show the player that carried on the action; after a foul the injured and the offending players are framed, etc. Therefore the automatic identification of these players would add considerable value to the annotation and retrieval of both the important highlights and the key players of a sport event.

The problem of detecting and recognizing faces in broadcast videos is a widely studied topic. A survey of the vast literature on face detection and recognition has been presented in [3,4]. Most of the face recognition methods are evaluated on videos whose content has been filmed in controlled environments and for relatively limited sets of faces and poses (e.g. serials or movies). However, in the case of soccer videos, and sports videos in general, the current techniques are not suitable for the task of face recognition, due to the high variations in pose, illumination, settings, scale and occlusion that may happen in an uncontrolled environment. Recently it has been shown that SIFT local features, usually used for object recognition and classification tasks [5], could be used for finding and grouping faces of same person in video shots [6,7]. In our previous work [8] players recognition was performed using textual cues obtained from superimposed captions and jerseys' numbers, while face matching was performed using SIFT descriptors centered on the eyes. In [9,10] authors investigate the problem of automatically labelling appearances of characters in TV videos and they have demonstrated that high precision can be achieved by combining visual and textual informations. In these works it has been studied the problem of detecting faces from TV or film video sequences, and of finding instances of the same person among all the detected faces. Similarity of faces is measured using distances between local descriptors based on statistical parameters (such as the χ^2 statistic).

In this paper an improved Haar-based face detection algorithm, and a new method that performs face recognition using local visual features, are presented. The local features are extracted from patches that include the most distinguishing parts of the face, without requiring a precise localization of these areas, differently from previous works. A similarity metric that is able to match faces obtained at different image scales is introduced, and experiments that assess the effectiveness and robustness of the metric over a highly varied, in terms of poses, illumination, settings, size and facial expressions test set are presented. Finally a representation of faces as sets of poses is introduced, to overcome the problem of face matching in the highly varied cases.

The structure of the paper is as follows: in Sect. 2 the improved face detection algorithm is presented; in Sect. 3 the face modeling technique, based on local features (SIFT points) is presented, along with the representation of a person with a set of poses; In Sect. 4 a new similarity metric used to compare faces is introduced. Experimental results performed on a large and highly diversified test set for both face detection and recognition are presented in Sect. 5, assessing the effectiveness of the representation of a person using a set of poses. Finally conclusions are drawn in Sect. 6.

2 Face Detection

The annotation process begins when a face is detected in the video. To reliably detect faces we have modified the Adaboost face detector of [11]; first, the algorithm has been tuned to soccer videos by using negative examples taken from actual soccer videos. Second, a filtering step that takes into account color information has been added to the output of the Adaboost detector. This is required to reduce the relatively high number of false detections obtained by the standard algorithm. In fact soccer videos contain a high number of visual elements (e.g. the playfield lines) that present a strong response to the Haar-based feature detector used in the algorithm. The result is that typically among the detected faces there is a high number of false detections. To reduce these false detections, our strategy is to discriminate actual faces using color, a feature that is not taken into account in the Adaboost step. We have collected a number (a few hundreds) of image patches that were identified as faces from the Adaboost detector. This set has been manually labeled into actual faces and false detection, and each patch has been then described with a 64-bin color correlogram ([12]), computed in the RGB color space. The color correlogram is a statistical descriptor that takes into account the spatial distribution of the colors in an image patch, estimating the probability that a pixel of some color lies within a particular distance of pixel of another color. It appears to be particularly well-suited for the task at hand, because false detection present patterns similar to faces in terms of edges, but usually very different in terms of color and color distribution.

Fig. 1. An example of face detection and tracking

The faces detected during the first step are normalized to a 64×64 matrix, and the RGB color space is uniformly quantized in 64 colors. The color autocorrelogram of each face I is calculated using:

$$\alpha_c^{(k)}(I) = Pr_{p_1, p_2 \in I_c} [|p_1 - p_2| = k] \tag{1}$$

A simple K-Nearest Neighbor classifier has been then designed to separate good and false detections. This method has proven to be effective, while maintaining the low computational requirements of the original algorithm, and avoiding a complex training stage. Preliminary experiments have shown that it is possible to use the average autocorrelogram, due to the low variance of all the autocorrelograms; this has allowed to reduce the size of the data. Since K-NN classification may be unreliable in high-dimensional spaces, we have projected the correlogram vectors onto a linear subspace obtained through principal component analysis. The PCA subspace has been designed to retain about $\sim 90\%$

of the original variance, resulting in keeping the first six principal components. Another parameter that has to be chosen is the pixel distance k used in the calculation of the color correlogram (Eq. 1). An analysis of the correlograms of the false detections of the Haar-based face detector has been performed; this has been done because the goal of the filtering step of the face detection algorithm is exactly to eliminate these false detections, and because the variability of the instances of false detections is higher than that of correct face detections. This means that while the six principal components may be enough to describe statistically the true positive faces, but may have problems with the false positives, that bring more information. Analysis of the variance of the six principal components shows that the maximum is obtained when $k = 35$. In general with $k \leq 35$ the variance is between 88% and 90%, and it can be observed a sudden drop, with values between 83% and 86%. Fig. 2 a) shows the variance associated with each principal component; Fig. 2 b) shows the residual variance related to the pixel distance k used to calculate the autocorrelogram, once that the six principal components have been kept.

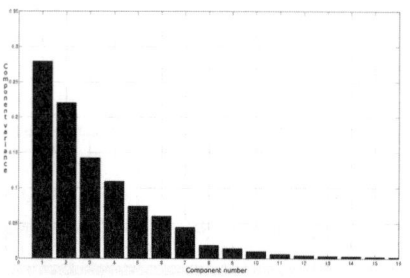 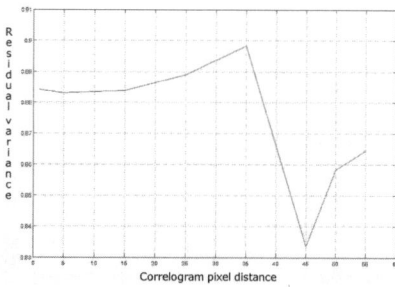

(a) Number of components vs. variance   (b) Autocorrelogram pixel distance k vs. residual variance

Fig. 2. Face detection parameters: component numbers and autocorrelogram pixel distance k

To choose the K parameter of the K-NN classifier a 5-fold cross-validation has been performed; the maximum accuracy has been obtained for $K = 3$. To speed up the K-NN classifier the points have been organized in a KD-Tree.

3   Face Model Based on SIFT Features

The annotation process begins when a frontal face is detected in a close–up shot, using the algorithm described in the previous section. To ensure reliable identification, only faces bigger than a minimum size trigger the subsequent processing. Once a face is detected, it is tracked throughout the entire shot, to collect a face sequence of the same individual, using a tracker based on particle–filtering (Fig. 1). The tracker use as observation the same color correlogram collected for classification, while the tracker state follows a first-order dynamic model (i.e., we suppose a constant image velocity for the patch to track). Details of the tracking algorithm can be found in [13]. Face tracking allows to cope with the appearance of multiple players within the same close-up sequence, letting to process each face separately, without mixing descriptors of different faces together.

Faces are modeled as a collection of local patches located in the most distinguishing part of a face. Using a part-based approach allows to cope with the frequent cases of occlusion and with all the different poses of the players. This choice is motivated also by the fact that several other elements of the image such as superimposed captions, shirt logos and numbers located near the face may be highly responsive to the SIFT detector, and lead to wrong matches. Following [4] the patches have been located in the following areas: *i)* left and right eye; *ii)* the halfway point between the eyes; *iii)* nose; *iv)* mouth.

Keypoint localization inside facial patches. The keypoint selection process of the standard SIFT technique [14], that choose the scale-space extrema, has proved to be inadequate for the task of face representation: in fact when applied to faces the number of keypoints that is obtained is not enough to guarantee a robust representation of the face. This is due partly to the low resolution of the TV videos, and partly to the fact that this process has been devised for the goal of object recognition; therefore a different selection criteria has to be devised when selecting the SIFT descriptors within the face patches.

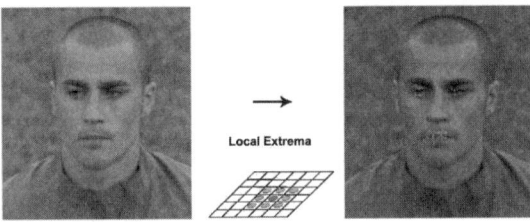

Fig. 3. Keypoint detection within local face patches. The image is processed to extract the face patches, then in each patch the local extrema are detected. In the right image the detected keypoints are shown.

During the initial experiments the naïve approach of selecting all the points of a patch to build the SIFT descriptors has proved to be unsuitable, since the large number of descriptors makes both the descriptors generation and the comparison computationally expensive. Moreover the matching of the descriptors is unstable, due to the fact that the descriptors of nearby keypoints have a distribution of the local gradients that will be largely coincident.

To solve this problem an alternative approach has been chosen: only the points of the patch that are classified as local extrema are selected as keypoints. Each pixel $p(x, y)$ of the patch is compared with the eight neighbours; $p(x, y)$ is considered as a keypoint only if it is a local min/max (Fig. 3).

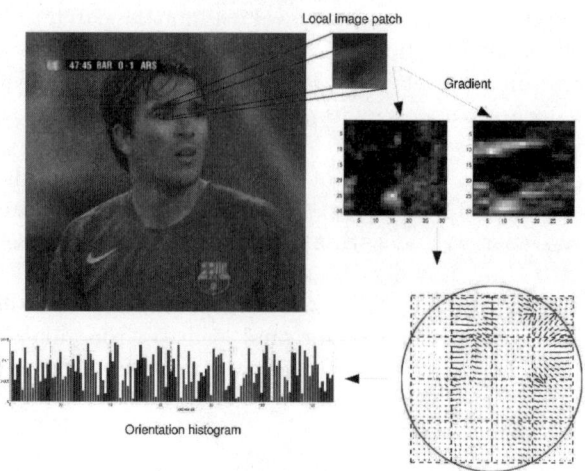

Fig. 4. The process of face patch extraction and representation through local features. Left eye patch extraction; creation of local gradients; creation of the histogram of the orientations, used to compute the SIFT descriptor.

The local extrema are used to extract the orientation histograms of the local gradient and then to compute the SIFT descriptors (Eq. 2 for gradient magnitude and Eq. 3 for orientation). Orientation histograms obtained in this way constitute the list of final patch descriptors.

$$m(x,y) = \sqrt{((L(x+1,y) - L(x-1,y))^2 + ((L(x,y+1) - L(x,y-1))^2} \quad (2)$$

$$\theta(x,y) = atan\{L(x,y+1) - L(x,y-1)\}/\{L(x+1,y) - L(x-1,y)\} \quad (3)$$

The whole process is shown in Fig. 4. The left eye patch is selected and the local gradients are computed from the pixels of the patch. Then the histogram of the orientations, used to obtain the SIFT descriptors, are computed.

Fig. 5 shows the effects of using the standard SIFT keypoint selection process and the selection of local extrema within face patches: in the first image the standard DoG detector has been used to select keypoints and perform the match, in the second image only keypoints selected within the face patches have been used. It has to be noted that none of the keypoints obtained with our selection process has been selected by the standard SIFT detection stage.

Set of poses. Initial experiments have shown that the pose of a face has a strong impact on the matching of the descriptors, and in certain cases it is possible to match two faces of different players that have a similar pose. To avoid this problem the solution that has been adopted is to describe a player P_i using a set of n video sequences of different poses, each set of poses is composed by m frames I_j:

$$P_i = \{P_{i,1}, \ldots, P_{i,n}\}, P_{i,j} = \{I_1, \ldots, I_m\}$$

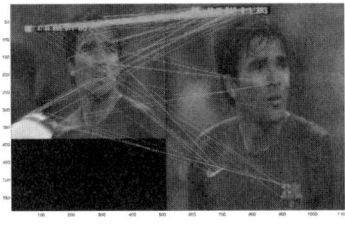 

(a) DoG keypoint selection   (b) Using Patch Model

Fig. 5. Matching SIFT descriptors extracted using the standard DoG keypoint selection (image a) and using the face patch model (image b)

instead of a single image. For each frame of the sequences of the set of poses the face descriptors are extracted, and a player is described by the list of all the keypoints of the pose set.

Experiments have shown that a set of 6 to 10 images I_j for each pose $P_{i,j}$, if sufficiently different, guarantees a performance improvement w.r.t. the one-to-one image comparison. The average size of the descriptor list of a player is about 1200 keypoints. Fig. 6 shows the effect of using set of poses to perform the face matching.

Fig. 6. Matching descriptors of different players using set of poses instead of single images

4 Face Matching

Face matching between the frames of pose sequences of players is computed comparing the face patches. Each keypoint is compared to the others using their SIFT descriptors, and searching the nearest one. To determine two corresponding SIFT descriptors the original algorithm computes the ratio of the Euclidean distances of the two nearest descriptors, w.r.t. the query descriptor. If the ratio is greater than 0.8 the match is discarded. This matching techniques has been employed also in our approach, but in order to speed up the processing the Euclidean distance has been replaced by the dot product of the SIFT descriptors.

In [7] the similarity of face descriptors take into account statistical parameters computed from the distances of the SIFT descriptors. However preliminary

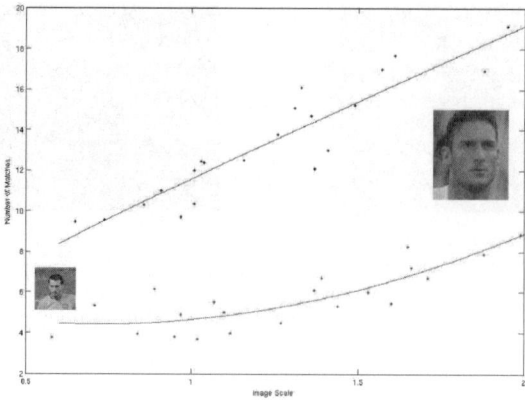

Fig. 7. Graphs that show the number of keypoints matches at different image scales. The scale has been normalized w.r.t. the most common face size of 210 × 270 pixels. The blue graph (upper curve) shows the number of correct matches, the red graph (lower curve) shows the number of wrong matches.

experiments have proved that this approach is not robust when applied to sports videos, while the figures of the number of keypoints matches suggest that this parameter may be used evaluate the face similarity. However this number alone is not significant enough since there is a direct correspondence between the resolution of the image and the number of matches, and in general we can expect that increasing the resolution of the face patches also the number of matches will increase. To evaluate the effects of face resolution on the number of correct matches experiments have been carried on, and results are reported in Fig. 7. The graph shows the number of keypoints matches depending on the image scale. The scale value S_i has been normalized w.r.t. the most common face resolution ($S_0 = 1$ corresponds to a face of 210 × 270 pixels).

The keypoints matches have been classified as "correct" if a keypoint was matched with a keypoint of a face of the same player, otherwise if it is matched to a keypoint of a different player it is classified as "wrong". Analysis of the graph shows that the number of "correct" matches (blue upper line) grows almost linearly; the number of matches is about 8 when the scale is half of the average face size, and grows to 20 for faces whose resolution is the double of the average resolution (the face almost fills the TV frame). Also the number of wrong matches (red lower curve) grows, although much more slowly, with the resolution of the faces, from a minimum of 2 to a maximum of 10 matches.

The curves of the correct and wrong matches are distant enough for each scale value, and allow to define a similarity metric for the faces: let S_q be the number of keypoints matches normalized w.r.t. the threshold value associated to the scale of the query image I_q and an image I_r of player P_i, if $S_q > 1$ then I_q is assigned to P_i. This metric can be applied to images that belong to the same shot as well to different shots, and also to sets of poses; the value of the metric can be used to rank the results of a search.

5 Experimental Results

The proposed algorithms have been tested using a dataset built from ten soccer matches of the last FIFA World Cup (Germany 2006) and UEFA Champions League (seasons 2005/06 and 2006/07). Videos have been digitally recorded at full PAL resolution (720 × 576 pixels, 25 fps). For the face detection task about 135000 frames have been selected (about 90 minutes). For the face recognition test, from the previous test set the shots that contained the most frequently appearing players have been selected (typically the most important 10–15 players of a soccer match), and for each sequence 10 to 20 frames have been selected; the resulting test set comprises about 2000 frames.

To evaluate the improvement of the proposed face detection filtering step in Table 1 a) are reported the results of the Haar-based face detector available in the OpenCV library. In Table 1 b) are reported the results of the proposed algorithm (Haar-based + filtering step); the number of missed detection due to the filtering step is extremely small. The processing performance is about 8 fps on a standard PC. Retrieval examples on the full dataset are shown in figure 8 where precision-recall curves are presented for four different players. Those experiments have been performed to obtain a recall value of 1 and are obtained as averages of different queries of the same player.

Table 1. Comparison of face detection algorithms

(a) Haar-based face detection results

	detected faces	false det.
front face	12176	1090 (9.83%)
side face	11346	1684 (17.43%)

(b) Haar-based and color-based filtering results

	detected faces	false det.	improvement	missed det.
front face	12176	36 (0.33%)	9.5%	244
side face	11346	143 (1.52%)	15.91%	239

In order to evaluate the robustness of the face model and similarity metric w.r.t. differences in pose, illumination, facial expression, shooting style, etc. experiments have been performed on two subsets of the dataset. The frames have been selected from the matches of two teams: *Barcelona* and *Chelsea*, in order to have a sufficient number of sequences of the same players; 20 players that were frequently framed in different poses and settings have been selected. In the first series of experiments the face matching has been performed on images that belong to the same shot (intra–shot); in the second series the match has been evaluated for frames that belong to different shots, and possibly to different matches (inter–shot). Furthermore the shots have been divided in two classes: one that contains shots in which the face presents a relatively small variation in terms of pose, facial aspect, scale (e.g. due to zooming), occlusions, illumination and occlusions (*class A*), and one that is characterized by larger variations due to these causes (*class B*).

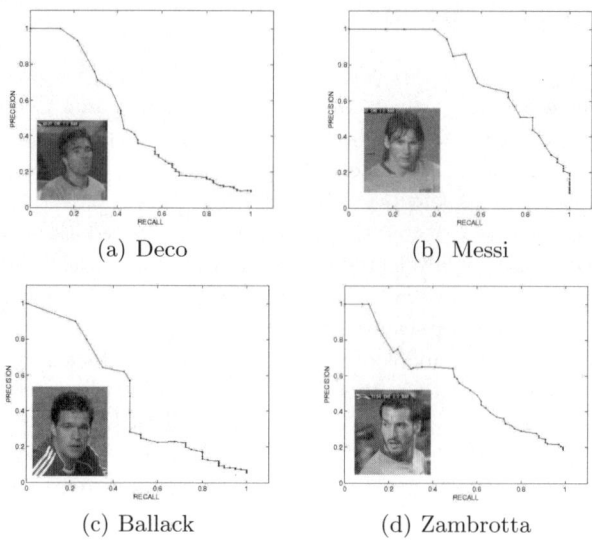

Fig. 8. Retrieval examples of four different players. The graphs show Precision (y-axis) vs. Recall (x-axis).

Table 2 a) reports results for face matching performed on frames that belong to the same shot, for class A and B shots (when available). For each player it is reported the recognition percentage; this value has been computed as the ratio of the number of matched faces vs. the number of frames of the same shot that show the player. The average values for both classes are reported, showing very good results for faces that belong to shots of class A; as expected in the case of frames that belong to class B there's a relatively small reduction of the recognition rate. Table 2 b) reports the results in the case of inter–shot face matching. Experiments have been conducted on the same 20 players of the intra–shot experiment. In this case it has been more difficult to obtain class A and B shots, so the table is more sparse than the previous. It has to be noted that in this case the class A shots are more variable than the previous case since the scene illumination and scale has changed, due to the different settings of the matches and the position of the cameras in different stadiums. This difference is even greater in the case of class B shots. As expected the recognition rate, computed as in the previous experiment, is relatively low also for shots belonging to class A, while for class B shots the average value drops to 35%.

To overcome the problems of direct face matching shown by the results of Table 2 the solution is to perform face matching using sets of poses. Each player is described by the keypoints located in the face patches of the images that belong to the poses; on average 1200 keypoints are used to describe a player. From the 20 players used in the previous experiments, 12 have been selected, based on the availability of the largest number of different shots in both A and B classes. In the experiments 8 poses have been used for each player ($P_{i,j} = \{I_1, \ldots, I_8\}$), and the face matching has been performed between a single image of a player and

Table 2. a) Face recognition results for the **intra–shot** experiment. – b) Face recognition results for the **inter–shot** experiment. The figures are computed as the ratio between the number of correctly recognized faces and the number of frames in which the face is shown. Averaged values for the test–set are reported, for both class A and B shots.

(a) Intra–shot results

Player	Class A	Class B	Player	Class A	Class B
P1	0.82	-	P11	0.81	-
P2	0.90	0,78	P12	0.71	-
P3	0.84	0.71	P13	0.80	0,70
P4	-	0.78	P14	0.80	-
P5	0.86	-	P15	0.81	0.58
P6	0.88	0.61	P16	-	0.78
P7	-	0.73	P17	0.93	-
P8	1.00	-	P18	0.78	-
P9	0.87	-	P19	-	0.65
P10	-	0.67	P20	-	0.63

Avg. (Class A)	Avg. (Class B)
0.84	0.69

(b) Inter–shot results

Player	Class A	Class B	Player	Class A	Class B
P1	-	-	P11	-	0.23
P2	0.67	0.48	P12	-	-
P3	-	-	P13	0.64	-
P4	-	-	P14	-	-
P5	0.57	-	P15	-	0.29
P6	-	-	P16	0.62	-
P7	0.50	-	P17	-	0.31
P8	-	0.37	P18	0.53	-
P9	0.68	-	P19	-	0.40
P10	-	-	P20	-	-

Avg. (Class A)	Avg. (Class B)
0.60	0.35

all the poses of the players. Results are reported in Table 3. In the "Queries" column it is reported the number of queries for each player, using images from poses that are not part of the test set, the "Correct" column reports the number of correct matches, the average percentage of correct matches is 58%. These figures show a strong improvement w.r.t. the inter–shot experiments.

Table 3. Face recognition results using **sets of poses**

Player	Queries	Correct	Player	Queries	Correct	Player	Queries	Correct
P1	18	12	P5	18	12	P9	30	12
P2	24	12	P6	42	30	P10	24	12
P3	24	12	P7	30	18	P11	36	30
P4	30	18	P8	24	12	P12	36	18

6 Conclusions and Future Work

In this paper a system that performs face detection and recognition in videos has been presented. The method is able to deal with videos that show a very high degree of variability of poses, facial expression, settings, scale and illumination as in the sports video domain; in particular experiments have been performed on soccer videos. The face detection method improves the Viola and Jones detector, adding color features to reduce the number of false detections. The face recognition method represents faces as a set of patches loosely located on distinguishing parts of the face such as eyes, nose and mouth. In each patch local visual features are extracted; the SIFT descriptors have been used to represent the keypoints within the patches. A new similarity metric that allows to match the keypoints of faces at different image scales has been introduced. Experiments of face recognition within the same shot and between different shots, that differentiate between relatively similar and highly different types of shots have been carried on, to assess the robustness of the local descriptors and the similarity metric used. Finally a solution for face recognition between extremely different types of shots, using a representation of faces as sets of poses, has been presented, showing a strong improvement over the standard face-to-face comparison.

Our future work will deal with the study of algorithms for the automatic selection of the sets of poses and the definition of a more compact representation of the sets.

Acknowledgments. This work is partially supported by the IST Program of the European Commission as part of the DELOS NoE on Digital Libraries and as part of VIDI-Video STREP Project.

References

1. Kokaram, A., Rea, N., Dahyot, R., Tekalp, A.M., Bouthemy, P., Gros, P., Sezan, I.: Browsing sports video. trends in sports-related indexing and retrieval work. IEEE Signal Processing Magazine 23(2), 47–58 (2006)
2. Yu, X., Farin, D.: Current and emerging topics in sports video processing. In: Proceedings of IEEE ICME, IEEE Computer Society Press, Los Alamitos (2005)
3. Yang, M., Kriegman, D., Ahuja, N.: Detecting faces in images: A survey. IEEE Transactions on Pattern Analysis and Machine Intelligence 24(34-58) (2002)
4. Zhao, W., Chellappa, R., Phillips, P., Rosenfeld, A.: Face recognition: A literature survey. ACM Computing Surveys 35(4), 399–458 (2003)
5. Fergus, R., Perona, P., Zisserman, A.: Object class recognition by unsupervised scale-invariant learning. Proceedings of IEEE CVPR 2, 264–271 (2003)
6. Arandjelovic, O., Zisserman, A.: Automatic face recognition for film character retrieval in feature-length films. In: Proceedings of IEEE CVPR, San Diego, IEEE Computer Society Press, Los Alamitos (2005)
7. Sivic, J., Everingham, M., Zisserman, A.: Person spotting: video shot retrieval for face sets. In: Leow, W.-K., Lew, M.S., Chua, T.-S., Ma, W.-Y., Chaisorn, L., Bakker, E.M. (eds.) CIVR 2005. LNCS, vol. 3568, Springer, Heidelberg (2005)
8. Bertini, M., Del Bimbo, A., Nunziati, W.: Automatic detection of player's identity in soccer videos using faces and text cues. In: Proc. of ACM Multimedia, ACM Press, New York (2006)
9. Ozkan, D., Duygulu, P.: A graph based approach for naming faces in news photos. In: Proceedings of IEEE CVPR, IEEE Computer Society Press, Los Alamitos (2006)
10. Everingham, M., Sivic, J., Zisserman, A.: Hello! my name is... Buffy – automatic naming of characters in tv video. In: Proceedings of BMVC (2006)
11. Viola, P., Jones, M.: Robust real-time face detection. International Journal of Computer Vision 57(2), 137–154 (2004)
12. Huang, J., Kumar, S.R., Mitra, M., Zhu, W.J., Zabih, R.: Image indexing using color correlograms. In: Proceedings of IEEE CVPR, pp. 762–768. IEEE Computer Society Press, Los Alamitos (1997)
13. Bagdanov, A., Del Bimbo, A., Dini, F., Nunziati, W.: Adaptive uncertainty estimation for particle filter-based trackers. In: Proceedings of ICIAP (2007)
14. Lowe, D.G.: Distinctive image features from scale-invariant keypoints. International Journal of Computer Vision 60(2), 91–91 (2004)

A Temporal and Visual Analysis-Based Approach to Commercial Detection in News Video

Shijin Li[1], Yue-Fei Guo[2], and Hao Li[1]

[1] School of Computer & Information Engineering, Hohai University, Nanjing, China
lishijin@hhu.edu.cn
[2] Department of Computer Science & Engineering, Fudan University, Shanghai, China
yfguo@fudan.edu.cn

Abstract. The detection of commercials in news video has been a challenging problem because of the diversity of the production styles of commercial programs. In this paper, the authors present a novel algorithm for the detection of commercials in news program. By the method suggested, firstly shot transition detection and anchorman shot recognition are conducted, then clustering analysis is employed to label commercial blocks roughly, finally the accurate boundaries of the commercials are located by analyzing the average duration of preceding and subsequent shots and the visual features of the shots, such as color, saturation and edge distribution. The experiment results show that the proposed algorithm is effective with high precision.

Keywords: Commercial detection, clustering, temporal and visual features.

1 Introduction

In news TV program, commercial blocks are often inter-mixed with news reports for commercial pursuits. However, in some applications like video on demand (VOD) and media asset management (MAM), these commercial blocks need to be removed or to be substituted for better retrieval and browsing of the news video. To take news program VOD as an example, the users want to watch their requested news program without the annoyance of commercials. So the removal of commercials will make the program hot and thus bring more profits. Thereby, an automatic and accurate commercial detection system is of highly practical value.

In recent years, researchers have proposed some algorithms to deal with the commercial detection problem.

Generally speaking, there are two main categories: One makes use of the prior-knowledge like black frames [1], TV logos [2] and so on as commercial clues, which is usually with high detection rates but only can be applied to specific news videos. The other one is becoming predominant nowadays, which takes the content-based features to distinguish commercial blocks from news reports. These features include visual [3,4], aural [5] and textual [6] information, which can be applied to all kinds of news videos. Ref. [7] proposed a learning-based commercial detection scheme based on visual, audio and derived features. Duan et al proposed a multimodal analysis based algorithm to segment commercial clips and carry out a further semantic analysis [8].

Our work aims at finding an efficient method to automatically label all commercial blocks in a full-length news program and mark their boundaries precisely based on temporal and visual features. Previous work did not pay sufficient attention to visual features, which allows much room for improvements.

Ref. [3] proposed a commercial detection algorithm which firstly analyzes "strong-cuts in a minute" in sliding windows to label candidate commercial blocks, and then used a video scene detector to mark their boundaries, sometimes still needs human intervention. Compared with that in [3], ours is based on the elaborate analysis on the structure and content of news video.

During rough labeling of commercial blocks, we firstly assume that all shots in the news video could be partitioned into several groups, thus global clustering method is then proposed to label candidate blocks which will overcome the tedious threshold tuning problem in [3].

To mark the boundaries of the candidate blocks precisely, we treat it as a pattern recognition problem: commercial shot vs. news report shot. Five features are extracted, which include duration ratio between current shot and preceding shots, duration ratio between current shot and subsequent shots, accumulative histogram similarity between current shot and preceding shots, saturation and intensity/color edge ratio. To our knowledge, it is the first time that the last two features are utilized to detect commercials in videos. Then with the help of support vector machine classifier (SVM), we mark the real boundaries of commercial blocks precisely without any human interventions.

The rest of the paper is organized as follows: Section 2 introduces the technical features like the structure of the news video and the characteristics of the commercial we utilized. In section 3 the newly proposed commercial detection algorithm is described in detail. Section 4 gives the experimental results. At last, we conclude in section 5 and discuss the future directions of research.

2 Technical Features of Commercials in News Video

2.1 The Structure of News Video

The structure of the news video considered in this paper is orderly arranged and easy to be modeled logically. Taking the news program of the City Channel from Jiangsu province of China as an example, before each news report an anchorman shot is set up to make a brief news introduction, and a commercial block which consists of several consequently commercial spots (like AD_1, AD_2...AD_n in Fig. 1) may be inserted after a certain news report and before another anchorman shot.

2.2 Characteristics of Commercials

Since we aim at detecting all commercial blocks but not to recognize them, we just need to consider the characteristics at the boundaries of commercial block. After observations on various news videos, we have found the following important characteristics of commercials:

1) There is a common difference on the time of duration between the commercial shots and news shots.

2) The visual style like dominant colors and saturation varies drastically while a commercial block starts.

Thus in this paper, we use these two kinds of features to label commercial blocks and mark their boundaries.

Fig. 1. Structure of a news program of City Channel from China Jiangsu TV station

3 The Proposed Algorithm

We propose a three-step refining detection scheme as shown in Fig. 2.

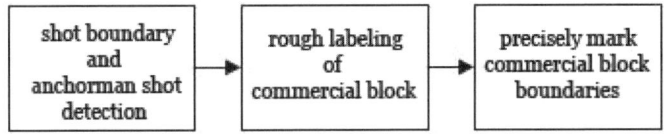

Fig. 2. The framework of our commercial detection algorithm in news video

Firstly, a complete news video is segmented into shots, while each anchorman shot is labeled at the same time. We use twin comparison method [9] to detect shot transition, and the last frame of each shot is chosen as the key-frame. The news video we process in the paper is a complete news program, which starts with an anchorman shot. So we define the first key-frame as a template frame, and match every key frame with the template frame to detect all the anchorman shots.

Then we label the commercial blocks roughly and obtain candidate commercial blocks by clustering analysis. At last, we employ SVM to recognize the temporal and visual features of shots in the candidate blocks to mark real boundaries of commercial blocks precisely.

3.1 Rough Labeling of Commercial Block

3.1.1 Global Clustering of Shots

Ref. [3] used a sliding window to calculate the parameter "strong cuts in a minute" which is the reciprocal of shot duration. And finally some successive windows are labeled as candidate if the parameter satisfies the thresholds.

In our research, shot duration is still taken into consideration since it is one of the most important features of commercial shots. But global clustering analysis is adopted to make labeling, not just based on pre-selected thresholds.

Firstly, we assume that the shots in the news video could be classified as anchorman shots, news report shots, commercial shots and other shots like weather forecasts and news preview. In another word, anchorman shots consist of group A, news report shots consist of group B, commercial shots to group C and some other shots to group D.

Those four groups of shots should be distinct from each other: Shot durations in group A are longest; Shots in group B are arranged successively but shot durations are diverse from long to short; Shots in group C are also successive to each other and shot durations are shortest; Shots in group D all last for long time and are approximate to each other in duration.

Then a sliding window of length Tw is applied to run over the whole video, the sliding speed is set to Tw / 2. In the paper Tw is set to 30 seconds. Each time we calculate the shot number, the mean shot duration and the variance of shot duration of the window. These three measures of each window make up of the feature vector for clustering.

Finally, K-means clustering algorithm is applied to partition the shots of each window into four clusters which correspond to A, B, C, D four groups we assumed.

3.1.2 Labeling Commercial Blocks

After clustering, we use an approximate method to label commercial blocks roughly:

M consecutive windows will be labeled as a commercial seed if the following conditions are satisfied:

   1) No windows of shots belong to group A in these M consecutive windows.

   2) Most windows of shots belong to group C in these M consecutive windows. We set the percentage to be 80% here.

   3) Since most commercial blocks last for 2 minutes approximately which is equal to the length of (240 / Tw - 1) windows, M should be more than 240 / Tw − 1.

Then, we intersect the set of commercial seeds with the set of anchorman shots which have been obtained previously. A commercial block is roughly labeled between the two anchorman shots which are located before-and-after a commercial seed. And the first shot of the roughly labeled commercial block is considered as the start of the candidate block, the last shot of the commercial seed is treated as the end of the candidate commercial block.

3.1.3 The By-Product of Clustering

With clustering analysis, we can recognize the commercial blocks between two anchorman shots and obtain the statistics of the duration of news report shots and commercial shots respectively, in addition to the classification of video shots into 4 classes/clusters.

In the target news programs, the inserted commercials can follow the news report, or directly between two anchorman shots. Since the duration of commercial shot is often very short, many of the sliding windows will be marked as candidate commercial blocks between two anchorman shots. So we can directly classify such kind of blocks as commercials in this kind of situation.

In addition, the statistics of the duration of news report shots and commercial shots can be computed after the clustering analysis. Let **Dn** denote the average duration of news shot, and ***Da*** the average duration of commercial shot. A threshold ***th*** can be set to determine the start point of the shot, which should be fed into the SVM classifier in the next section.

$$th = Dn \times p + Da \times (1-p) \tag{1}$$

where $p \in [0,1]$, represents time duration closeness to news shot.

The heuristics is to search from the last shot of the candidate commercial block until there is a shot whose duration is larger than ***th***. The next shot is taken as the real start position of candidate commercial block. In this way, the scope of shots to be classified can be reduced largely. Usually, about one to three shots of news report will be fed into the classifier and thus can eliminate many false alarms in the middle of news reports.

3.2 Precisely Locating of the Boundaries

As is shown in Fig. 3, the end of commercial block has been detected. So the problem is simplified as marking the end of the news report in candidate marking block as the start of commercial block.

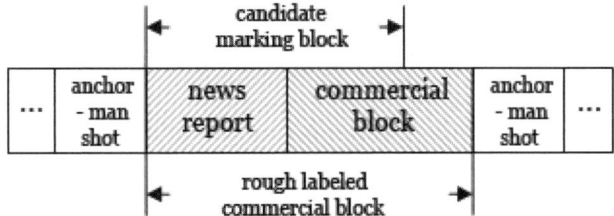

Fig. 3. The roughly labeled commercial block and the candidate marking block. The roughly labeled commercial block contains a news report and a real commercial block.

Either news report or commercial spot has its own visual styles that are represented as repeated emergence of the similar background or the same person. While a commercial block starts, the previous visual style terminates and the new style begins with high cut-frequency. That is to say it's a cut-off point of two different visual styles.

In this paper, five features are put forward to recognize this cut-off point, which include two novel features that are utilized to detect commercials in videos, for the first time.

The first two features are shot duration related. We compute two measures: duration ratio between current shot and preceding shots, duration ratio between current shot and subsequent shots. The rational lies in that before the commercial shot, it is news report and the succeeding is commercial. As we know the duration of news shot is long, while the duration of commercial shot is short. We take k=10 shots before and after the current shot to calculate the ratios: F1 and F2.

$$F1 = \frac{\text{duration of current shot}}{\text{average of the duration of preceding k shots}} \tag{2}$$

$$F2 = \frac{\text{duration of current shot}}{\text{average of the duration of subsequent k shots}} \quad (3)$$

The accumulative HSV histogram of a shot is computed by accumulating the HSV histogram over each frame of the shot and then making an average. Compared with the key-frame HSV histogram, it puts more emphasis on the overall color characteristics of the shot.

So, in order to evaluate the overall similarity between two shots well, the accumulative HSV histogram is used as the third feature in the paper to represent the color information of a shot instead of the key-frame. The corresponding histogram similarity is denoted as Fh.

Besides the shot duration ratios and histogram similarity, it is not suffice to discriminate between commercial shot and news report shot. It is of high probability that the color of a commercial shot is similar with that of a news shot. So we propose two additional features to enhance the discrminativeness.

In many commercials, the color is vivid and exaggerated to promote the products. We compute average color saturation (the S channel of the image in the HSV color space) and the ratio of high-saturated pixels with brightness greater than 0.4, which is denoted as Fs. Such features is used in [10] to detect cartoons in videos and discriminate paintings from photographs [11]. Fs is high for key frames from commercial shots.

In Ref. [11], intensity edges/color edges ratio is used to distinguish paintings from photographs. The removal of color eliminates more visual information from a commercial than from a news picture of a real scene. Canny edge detector is applied to intensity component image and the normalized RGB channels respectively. A quantitative measure is computed as follows:

$$Fe = \frac{\text{number of pixels : intensity, not color edge}}{\text{total number of edge pixels}} \quad (4)$$

We observe that Fe is smaller for the key frame of commercial shot than that of a news shot.

Fig. 4 and Fig. 5 give two pairs of frames grabbed from our test videos. It can be observed that Fs is higher and Fe is smaller for pictures from commercial than those from news report.

Fig. 4. Two images of cars. The left is from a commercial shot, while the right is from a news shot. Their (Fs,Fe)s are (0.582 0.605) and (0.017 0.979) respectively.

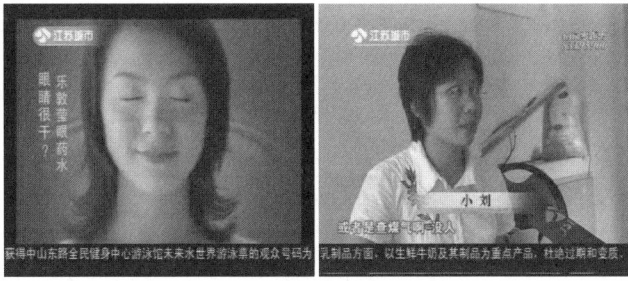

Fig. 5. Two images of ladies. The left is from a commercial shot, while the right is from a news shot. Their (Fs,Fe)s are (0.607 0.503) and (0.079 0.927) respectively.

Up to now, we get a feature vector for a candidate commercial shot: (F1, F2, Fh, Fs, Fe).

Next, we propose to use SVM[12] to classify these features, as it is one of the most competitive classifiers for small samples problem.

In the commercial /news shot classification problem, there are few commercials in a full-length news program, at most 3-4 segments in one hour. It is impractical to collect hundreds of videos to train the classifier. So we have synthesized many "news+commercial" segments from several real video programs. Every clip of news report is concatenated with a clip of commercial, thus obtaining one sample of "news+commercial" segment.

4 Experiment Results

Our experiments are carried out on the data of 4 real news video programs from our local TV station. Each news video program lasts about 50 minutes. The training data is synthesized from another 2 news video programs.

The detection results are given in Table 1.

Table 1. The detection results on four real videos

Total Frames	Commercial Blocks	Missed blocks	Wrongly marked blocks
79815	4	0	0
82499	4	0	1
80902	4	0	0
79735	4	1	0

From Table 1, it can be observed that our algorithm can locate precisely most of the commercials in the news video. The wrongly marked block in Table 1 is a short shot with a close-up of a paper contract, which differs from the rest of the news report. The missed block happens when the beginning shot of the commercial is about some mobile phone and the picture in the clip is with gray background. Fig. 6 gives the key frames of those two error cases. We believe this is rare and special which could be tackled by adding other features in our future work.

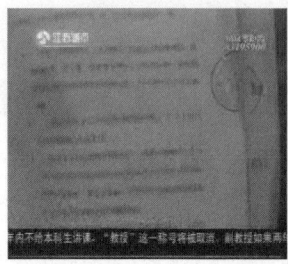 

Fig. 6. Illustration of two error cases in our experiments. The first one is a close-up of a paper contract, and the second one is a mobile phone promotion.

5 Conclusions

This paper proposes a novel method for detecting commercials in news video, which is based on temporal and visual features, such as shot duration, color, saturation and edge statistics. These features cannot be easily changed by the advertising industry. Thus, the method we proposed could be adjusted to most kinds of news program videos. The main contribution lies in that through clustering of shot duration, the statistics of news shot and commercial shot can be estimated and thus manual tuning of some thresholds can be avoided. Furthermore, we have proposed two novel visual features that are utilized to detect commercials in news videos for the first time. The performance of our algorithm is promising.

In the future, we will continue our research with the addition of some other visual features such as text and motion to improve the overall performance.

Acknowledgments. The authors would like to thank Mr. Zhang Haiyong from Jiangsu Broadcast Corporation, China, for his help providing us with the videos used in this paper.

References

1. Lienhart, R., Kuhmunch, C., Effelsberg, W.: On the detection and Recognition of Television Commercials. In: IEEE International Conference on Multimedia Computing and Systems, pp. 509–516. IEEE Computer Society, Washington, D.C. (1997)
2. Albiol, A., Fulla, M.J., Albiol, A., Torres, L.: Detection of TV commercials. In: IEEE International Conference on Acoustics, Speech and Signal Processing, vol. 3, pp. 541–544. IEEE Press, Montreal, Canada (2004)
3. Yeh, J., Chen, J., Kuo, J., et al.: TV commercial detection in news program videos. In: IEEE International Symposium on Circuits and Systems, pp. 4594–4597. IEEE Press, Japan (2005)

4. Sánchez, J.M., Binefa, X.: Local color analysis for scene break detection applied to TV commercials recognition. In: Huijsmans, D.P., Smeulders, A.W.M. (eds.) VISUAL 1999. LNCS, vol. 1614, pp. 237–244. Springer, Heidelberg (1999)
5. Duygulu, P., Chen, M.Y., Hauptmann, A.: Comparison and combination of two novel commercial detection methods. In: IEEE Int. Conf. on Multimedia and Expo, pp. 1267–1270. IEEE Press, Taipei (2004)
6. Hauptmann, A., Withrock, M.: Story segmentation and detection of commercials in broadcast news video. In: Proceedings of the Advances in Digital Libraries Conference, pp. 168–179. IEEE Computer Society, Washington, D.C. (1998)
7. Hua, X.-S., Lu, L., Zhang, H.-J.: Robust learning-based TV commercial detection. In: IEEE Int. Conf. on Multimedia and Expo, pp. 149–152. IEEE Press, Amsterdam, Netherlands (2005)
8. Duan, L., Wang, J., Zheng, Y., et al.: Segmentation, categorization, and identification of commercial clips from TV streams using multimodal analysis. In: Proceedings of the 14th Annual ACM international Conference on Multimedia, pp. 201–210. ACM Press, New York (2006)
9. Zhang, H., Kankanhalli, A., Smoliar, S.W.: Automatic partitioning of full-motion video. Multimedia System 1, 10–28 (1993)
10. Ianeva, T.I., Vries, A.P., Rohrig, H.: Detecting cartoons: a case study in automatic video-genre classification. In: IEEE International Conference on Multimedia and Expo, vol. 2, pp. 449–452. IEEE Press, Baltimore (2003)
11. Cutzu, F., Riad, H., Leykin, A.: Estimating the photorealism of images: Distinguishing paintings from photographs. In: IEEE International Conference on Computer Vision and Pattern Recognition, pp. 305–312. IEEE Press, Wisconsin (2003)
12. Burges, C.J.C.: A Tutorial on Support Vector Machines for Pattern Recognition. Data Mining and Knowledge Discovery 2, 121–167 (1998)

Salient Region Filtering for Background Subtraction

Wasara Rodhetbhai and Paul H. Lewis

Intelligence, Agents, Multimedia Group,
School of Electronics and Computer Science,
University of Southampton,
Southampton, SO17 1BJ,
United Kingdom
{wr03r,phl}@ecs.soton.ac.uk

Abstract. The use of salient regions is an increasingly popular approach to image retrieval. For situations where object retrieval is required and where the foreground and background can be assumed to have different characteristics, it becomes useful to exclude salient regions which are characteristic of the background if they can be identified before matching is undertaken. This paper proposes a technique to enhance the performance of object retrieval by filtering out salient regions believed to be associated with the background area of the images. Salient regions from background only images are extracted and clustered using descriptors representing the salient regions. The clusters are then used in the retrieval process to identify salient regions likely to be part of the background in images containing object and background. Salient regions close to background clusters are pruned before matching and only the remaining salient regions are used in the retrieval. Experiments on object retrieval show that the use of salient region background filtering gives an improvement in performance when compared with the unfiltered method.

Keywords: Background Clustering, Salient Regions, Object Retrieval.

1 Introduction

Salient regions are regions in an image where there is a significant variation with respect to one or several image features. In content-based image retrieval (CBIR), salient points and regions are used to represent images or parts of images using local feature descriptions. In [1, 2] the salient approach has been shown to outperform the global approach. Many researchers have proposed different techniques based on salient points and regions. For example, Schmid and Mohr [3] proposed using salient points derived from corner information as salient regions for image retrieval, whilst Q. Tian [4] *et al* used a salient point detector based on the wavelet transform.

Salient regions are also applied to the problem of object retrieval, for example, in the case where a specific object in a query image is required to be retrieved from the image database. Traditional CBIR based on salient regions begins with salient region detection. Each salient region is then typically represented by a feature vector extracted from the region. In the query step, there is matching between salient regions from the query image and those from images in the collection and similar images are ranked according to the quality of match.

However, one of the reasons that the accuracy of object retrieval may be less than optimal is the presence of salient regions in the retrieval process which are not located on the object of interest.

Attempts to reduce the influence of irrelevant regions have appeared in some research projects. Ling Shao and Michael Brady [5] classify the selected regions into four types before the use of correlations with the neighbouring region to retrieve specific objects. Hui Zhang [6] *et al* pruned salient points using segmentation as a filter.

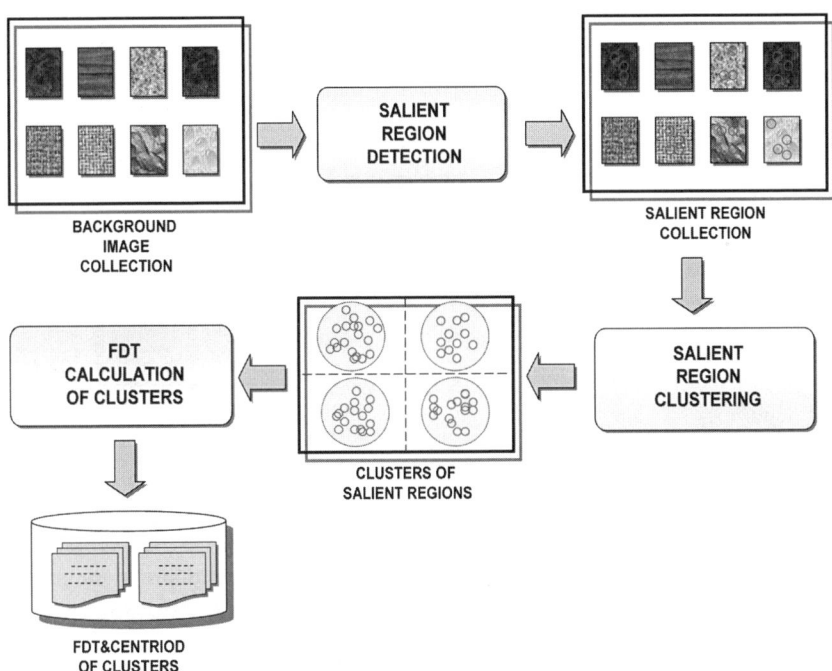

Fig. 1. Background clusters for filtering salient regions

In this paper we propose a method to filter salient regions using background information. Situations where the technique may be particularly appropriate are those where the image backgrounds are not completely arbitrary but can be characterized by a limited number of prototypes. Identifying particular objects in indoor scenes is an example.

In our approach, the system begins by creating clusters of salient regions from a collection of background only images. Thereafter, when processing images containing objects, salient regions with a high probability of belonging to a background cluster are removed before further processing. The process will be described in Section 2. It is illustrated schematically in Figure 1 and uses a distance threshold from the centre of each cluster called the fractional distance threshold (FDT).

For image retrieval, the background filtering step is applied after salient regions have been extracted. The process is illustrated schematically in Figure 2.

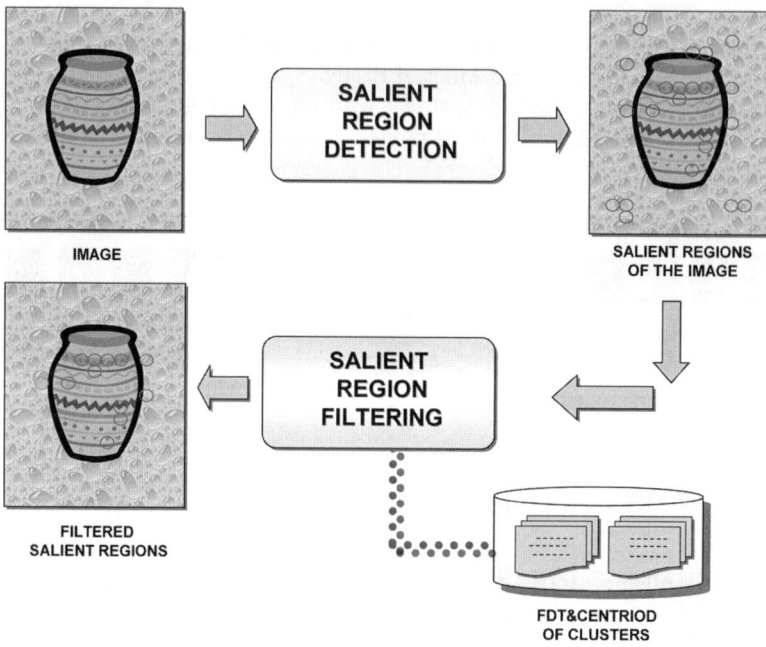

Fig. 2. The salient region filtering process

In the following sections we show that by using salient region filtering it is possible to reduce the number of unwanted salient regions and improve the precision of the retrieval process.

The paper is organised as follows. In Section 2, the methods of background clustering and calculation of the fractional distance threshold are introduced. The experimental procedure is described in Section 3. Results and discussion are presented in Section 4 and finally, Section 5 presents the conclusions and a brief discussion of future work.

2 Background Clustering

2.1 Salient Region Detection and Feature Extraction

Recently, many local detectors which can identify salient regions in an image have been described and evaluated [7, 8]. One of the popular approaches to salient region detection and representation is to use the multi-scale difference-of-Gaussian (DoG) pyramid for region location and scale estimation and the SIFT (Scale Invariant Feature Transform) from Lowe [9] to represent the detected salient regions. For each salient region, a 3D histogram of gradient locations and orientations is calculated. The SIFT descriptor has been evaluated by Mikolajczyk and Schmid in [10] to be one of the best performing local descriptors. The DoG and the SIFT approaches to salient region detection and representation are those adopted in our work.

2.2 Background Cluster Construction

One assumption of the method presented here is that salient regions from the foreground objects are reasonably distinct from background salient regions, or that any similarities involve a sufficiently small proportion of the total object salient regions to make their removal negligible.

The method begins with the detection of salient regions in a collection of background images. Since large numbers of salient regions may typically be detected in a single image, a random sample of salient regions are selected from all the background images and feature descriptors are extracted and used to cluster the salient regions into k clusters using the k-means clustering algorithm.

Since the clusters are derived from salient regions on background only images, these clusters are identified as the *background clusters*. The centroid of each cluster is calculated, essentially as a 128 element SIFT descriptor. Members in the same cluster are background salient regions that are similar to each other and dissimilar to the salient regions of other groups.

Many of the clusters are quite small in number so deriving a valid statistical model of the background clusters was not possible but to discriminate between salient regions on foreground and background, we determine an appropriate percentile distance from each cluster centroid, which we call the Fractional Distance Threshold (FDT) for each of the background clusters. The FDT of a cluster is the distance between a cluster member at a particular percentile and the centroid of that cluster. Thus FDT (90) is the distance from the centroid to a cluster member for which 90% of cluster members are nearer the centroid. The same percentile is used for all clusters and the appropriate percentile value found by experiment (see Section 3). The actual FDT and centroid for each cluster is retained for use in the retrieval process.

In the salient region filtering step, salient regions are detected and the features extracted. Any salient region (S) which has a feature distance (D) to the centroid (C) greater than the FDT (i) value for all (n) clusters, is assumed to be a salient region on the foreground (S_F). Otherwise, it is assumed to belong to a background region (S_B) as is represented by the following formula.

$$S = \begin{cases} S_F & \text{, if } \forall k \left[D(S, C_k) > FDT(i)_k \right] \\ S_B & \text{, otherwise} \end{cases} \quad (1)$$

where $k \in \{1, 2, ..., n\}$ and $i \in \{1, 2, ..., 100\}$.

3 Experiment

We separate the experimentation into 2 parts. The first part is to establish appropriate parameters for the clustering and Fractional Distance Threshold estimation and the second is to evaluate the retrieval performance using background salient region filtering.

3.1 FDT Percentile Estimation

A background only image collection, composed of 120 background only images (400 x 300 pixels) was created for 12 different backgrounds (10 images per background). Salient regions were extracted from each of the images and the number of salient regions found in each image varied between 8 and 3,503 depending on image content. Figure 3 shows some example background images from the dataset.

Fig. 3. Sample background images

In order to find appropriate values for the number of clusters in the k-means clustering, k, the number of randomly selected salient regions to use, S, and the percentile setting for the FDT calculation, a range of k and S combinations was used for clustering and the FDT estimated at each percentile from 50 to 100 in steps of 5. Each of eleven different k and S combinations were used. The resulting FDTs were used to check the percentage of correctly assigned foreground and background salient regions on a collection of object and background images. For these images, the ground truth was established by manually delineating the area covered by the object and if the centre of a salient region (SR) fell in the object area it was taken as an object SR. Otherwise, it was taken as a background SR.

Figure 4 shows examples of the decisions made by the system using some of the different FDT values. Salient regions in white circles represent foreground SRs and those with black circles represent background SRs.

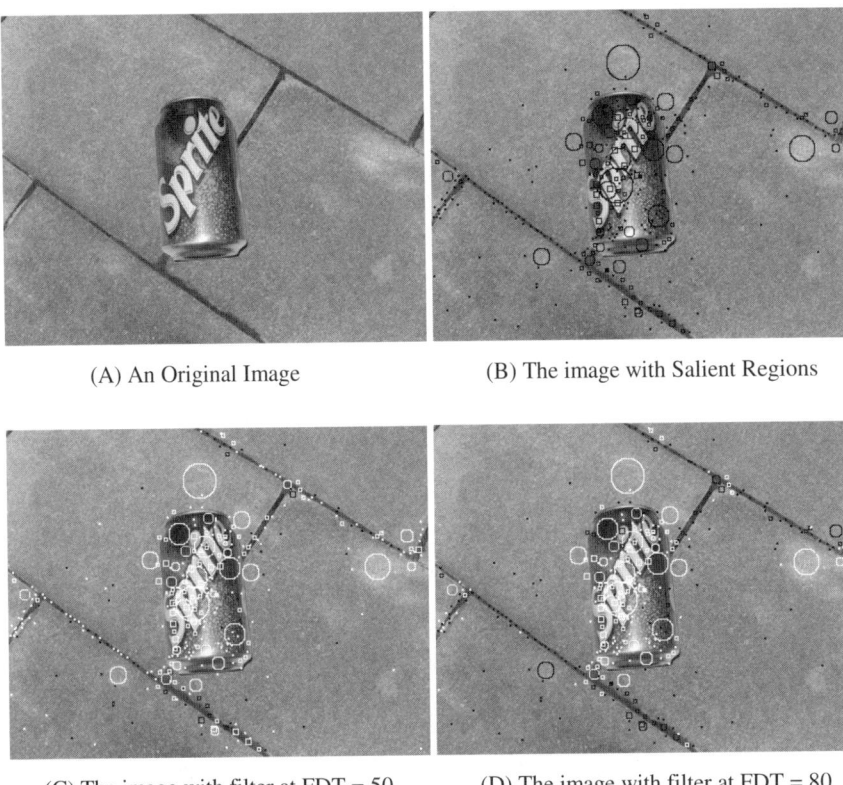

Fig. 4. Foreground (white circle) and Background (black circle) salient regions at FDT = 50 and FDT = 80

The performance of the decisions for a range of k, S and FDT values is measured via the receiver operating characteristic (ROC) space [11]. A ROC space represents the relationship of true positive rates (TP) and false positive rates (FP). Each classification produces a (TP and FP) pair corresponding to a single point in ROC space. We define

> w as the number of **correct** predictions that an instance is **Foreground SR**
> x as the number of **incorrect** predictions that an instance is **Background SR**
> y as the number of **incorrect** predictions that an instance is **Foreground SR**
> z as the number of **correct** predictions that an instance is **Background SR**

The recall or true positive rate (TP) determines the proportion of background SRs that were correctly identified, as calculated using the equation:

$$TP = \frac{z}{y+z} \qquad (2)$$

The false positive rate (FP) defines the proportion of foreground SRs that were incorrectly classified as background SRs, as calculated using the equation:

$$FP = \frac{x}{w+x} \qquad (3)$$

Figure 5 shows the ROC curve (TP against FP) as the FDT percentile is varied from 50 to 100.

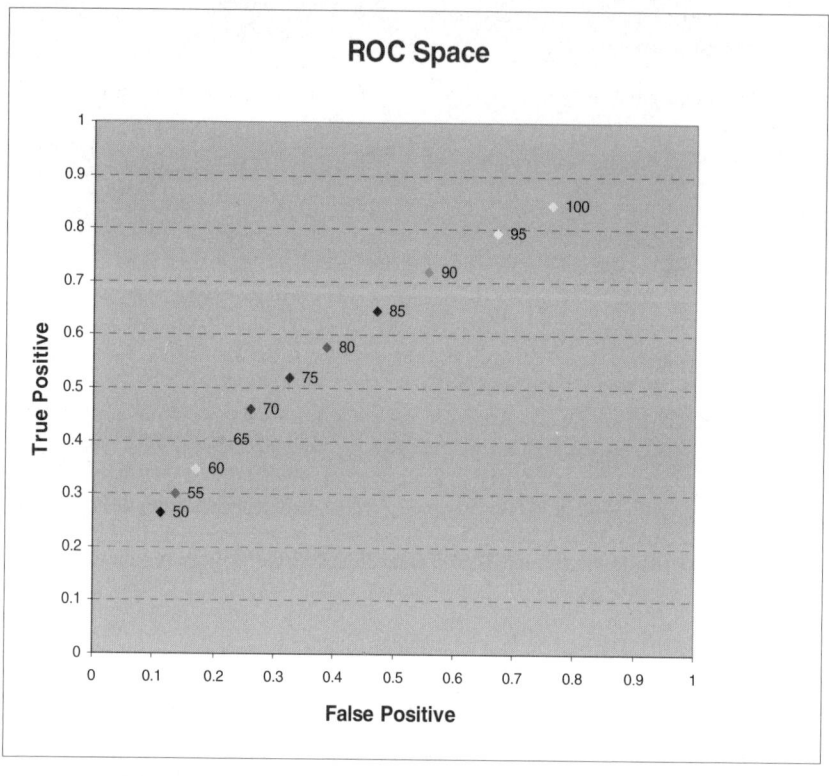

Fig. 5. The TP and FP coordinates of DFT at 50 to 100 on the ROC space

To comparing the prediction performance, distances are calculated from all points to the perfect classifier point in ROC space which is the point (0, 1). The point (0, 1) means all regions are classified correctly.

The overall results are presented in Table 1 where, for each of the k and S combinations, the table shows the distance from all of the TP and FP pairs to the point (0, 1). It illustrates how the percentage correct varies with the percentile for the FDT. It can be seen that in general a percentile of 85 gives the best results and that this is achieved with a k value of 5,000 and an S value of 50,000. These values were used in the retrieval experiments in the following section.

Table 1. The distance to (0,1) of 11 background cluster types (A-K) at the different FDT value (50 – 100). The lower the distance, the better the classifier.

Background Cluster (k - cluster, S - sample)	FDT										
	50	55	60	65	70	75	80	85	90	95	100
A (k500,S5000)	0.7528	0.7332	0.6997	0.6661	0.6336	0.6134	0.6038	0.6121	0.6408	0.7325	0.7893
B (k500,S10000)	0.6747	0.6460	0.6104	0.5874	0.5647	0.5675	0.5987	0.6381	0.6976	0.7547	0.8831
C (k500,S50000)	0.6048	0.5725	0.5474	0.5469	0.5600	0.5880	0.6332	0.7012	0.7755	0.8759	0.9778
D (k500,S100000)	0.5925	0.5614	0.5388	0.5335	0.5488	0.5809	0.6403	0.7145	0.7951	0.8967	0.9859
E (k1000,S5000)	0.8536	0.8431	0.8155	0.7786	0.7204	0.6799	0.6574	0.6285	0.5791	0.6279	0.6334
F (k1000,S10000)	0.7871	0.7635	0.7200	0.6646	0.6253	0.5931	0.5669	0.5840	0.5968	0.6954	0.7600
G (k1000,S50000)	0.6637	0.6198	0.5797	0.5466	0.5305	0.5456	0.5691	0.6295	0.7074	0.8094	0.9363
H (k1000,S100000)	0.6494	0.6040	0.5645	0.5315	0.5200	0.5374	0.5799	0.6621	0.7448	0.8612	0.9658
I (k5000,S10000)	0.9731	0.9730	0.9687	0.9556	0.9473	0.9437	0.9372	0.8842	0.7693	0.7304	0.7291
J (k5000,S50000)	0.8637	0.8439	0.7962	0.7437	0.6926	0.6214	0.5667	0.5149	0.5304	0.6075	0.6864
K (k5000,S100000)	0.8206	0.7886	0.7470	0.6902	0.6254	0.5733	0.5179	0.5194	0.5610	0.6440	0.7763

3.2 Object Retrieval

In order to test the effectiveness of background filtering, two datasets, each of 120 individual object images, were created from 10 objects on 12 different backgrounds which are not duplicated from the background training dataset. In dataset 1, the number of salient regions in these images varied between 98 and 1,728 regions. There are no scale and orientation change in each object. In dataset 2, the number of salient regions per image is between 174 and 2,349 regions. The scale and orientation is varied. From the results of the clustering experiments described earlier, the 5,000 background clusters from 50,000 salient points were used as the background clusters in the retrieval experiment and the chosen FDT value for all clusters was set to 85. Each object image was used in turn as the query image. The salient regions were extracted and background salient regions were filtered out from both the query image and the remaining object dataset images. After pruning, the strongest 50 salient regions from the remaining SRs were used to calculate the similarity between the query and dataset images and precision and recall results were obtained. The experiment was repeated without background filtering.

4 Results and Discussion

The precision and recall graphs with and without background filtering are shown in Figure 6 for dataset 1. From the graph it can be seen that the object retrieval system with background filtering outperforms the system without background filtering with an improvement in precision. The average precision with background filtering is 0.2483 and without background filtering average precision is 0.1810.

For the more challenging dataset 2, the precision and recall graph is shown in Figure 7. Again the performance is improved by using background filtering. The average precision is 0.2393 without background subtraction and is 0.2740 with background subtraction.

Looking back to Figure 5 it can be seen that best salient region classification performance was still far from the perfect classifier. The salient region filtering uses the particular differences between object areas and background areas to discriminate these regions and this is clearly not very robust.

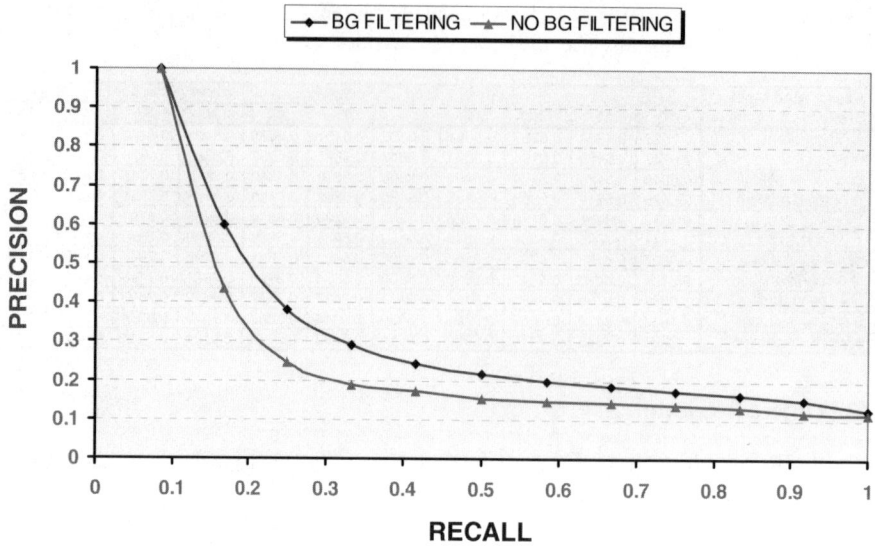

Fig. 6. Dataset 1. Precision and recall with and without background filtering.

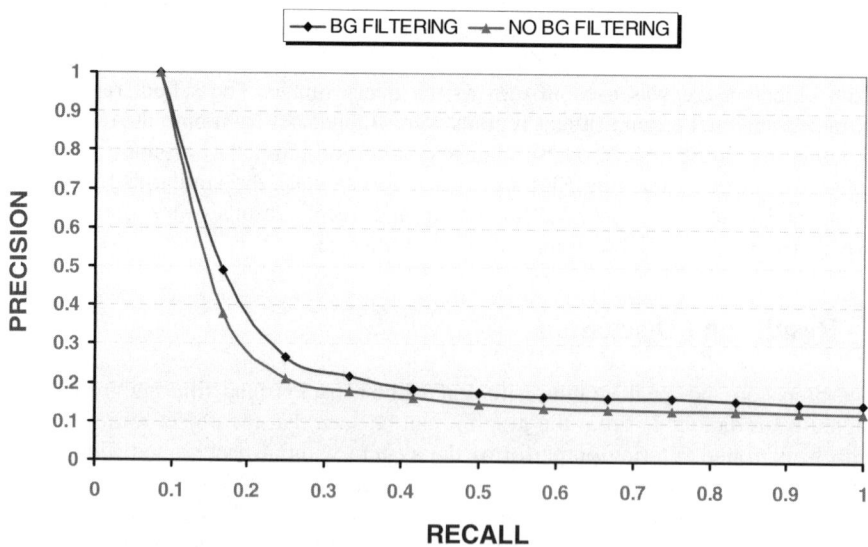

Fig. 7. Dataset 2. Precision and recall with and without background filtering.

5 Conclusion and Future Works

A novel method of filtering background salient regions for object retrieval is developed and implemented. A comparison has been made between retrieval with and without background salient region filtering and the filtering process is found to give improvements in precision. This was a rather preliminary evaluation of the technique and a more substantial evaluation is planned together with a search for a more robust way of modeling the backgrounds in terms of salient regions.

The main future work is developing a method for discriminating effectively between the object and background salient regions. More powerful feature descriptors will be incorporated to represent salient regions once identified. For the current method, more evaluation on the scale and rotation image dataset is required.

Another way to improve the performance of salient region filtering is to introduce techniques for modeling the object classes, in cases where these are known, rather than or in addition to the modeling of the background.

In summary, the background filtering method is an attempt to distinguish between the objects and the surrounding areas. Since certain types of query can benefit from using background information to filter irrelevant regions further attempts are being made to improve performance of this technique.

Acknowledgments. Our thanks to Jonathon S. Hare for assistance with the salient region extraction process system.

References

1. Sebe, N., Tian, Q., Loupias, E., Lew, M.S., Huang, T.S.: Evaluation of Salient point Techniques. In: Lew, M.S., Sebe, N., Eakins, J.P. (eds.) CIVR 2002. LNCS, vol. 2383, pp. 367–377. Springer, Heidelberg (2002)
2. Hare, J.S., Lewis, P.H.: Salient Regions for Query by Image Content. In: Enser, P.G.B., Kompatsiaris, Y., O'Connor, N.E., Smeaton, A.F., Smeulders, A.W.M. (eds.) CIVR 2004. LNCS, vol. 3115, pp. 317–325. Springer, Heidelberg (2004)
3. Schmid, C., Mohr, R.: Local Grayvalue Invariants for Image Retrieval. IEEE Transactions on Pattern Analysis & Machine Intelligence 19, 530–535 (1997)
4. Tian, Q., Sebe, N., Lew, M.S., Loupias, E., Huang, T.S.: Image Retrieval using Wavelet-based Salient Points. Journal of Electronic Imaging, Special Issue on Storage and Retrieval of Digital Media 10, 835–849 (2001)
5. Ling Shao, M.B.: Specific Object Retrieval Based on Salient Regions. Pattern Recognition 39, 1932–1948 (2006)
6. Zhang, H., Rahmani, R., Cholleti, S.R., Goldman, S.A.: Local Image Representations Using Pruned Salient Points with Applications to CBIR. In: Proc. of the 14th Annual ACM International Conference on Multimedia (ACM Multimedia), ACM Press, New York (2006)
7. Fraundorfer, F., Bischof, H.: A Novel Performance Evaluation Method of Local Detectors on Non-planar Scenes. In: Proc. CVPR. of Computer Vision and Pattern Recognition, pp. 33–33 (2005)
8. Moreels, P., Perona, P.: Evaluation of Features Detectors and Descriptors Based on 3D Objects. In: ICCV 2005. Proc. of 10th IEEE International Conference on Computer Vision, pp. 800–807. IEEE Computer Society Press, Los Alamitos (2005)
9. Lowe, D.G.: Distinctive Image Features from Scale-Invariant Keypoints. International Journal of Computer Vision 60, 91–110 (2004)
10. Mikolajczyk, K., Schmid, C.: A Performance Evaluation of Local Descriptors. IEEE Transactions on Pattern Analysis & Machine Intelligence 27, 1615–1630 (2005)
11. Fawcett, T.: ROC Graphs: Notes and Practical Considerations for Researchers. Machine Learning (2004)

A Novel SVM-Based Method for Moving Video Objects Recognition

Xiaodong Kong, Qingshan Luo, and Guihua Zeng

Laboratory of Coding and Communication Security, Shanghai Jiaotong
University, Shanghai 200240, PR China
kongxd@hotmail.com

Abstract. A novel method for moving video objects recognition is presented in this paper. In our method, support vector machine (SVM) is adopted to train the recognition model. With the trained model, the moving video objects can be recognized based on the shape features extraction. Comparing with the traditional methods, our method is faster, more accurate and more reliable. The experimental results show the competitiveness of our method.

Keywords: Video monitoring, Model recognition, Machine learning, SVM.

1 Introduction

As is known, a reliable recognition method is the key for the smart video monitoring technology [1], [2], [3], [4], [5]. In recent years, two methods are prevalent. One is the template matching method. Another is the Fisher discriminant function method. In the template matching method, every video object is described by a set of templates. When certain unknown object appears, all templates are compared with it. Finally, the most similar template is found and the unknown object will be recognized as its corresponding video object. In the Fisher discriminant function method, some parameters are worked out through the optimal formula. These parameters partition the projection space into several sections and each section corresponds to one video object. After that, project the unknown object into a point in certain section. Thus, the unknown object can be recognized as the corresponding video object of that section.

Above two methods have been applied in many fields, but they are far from a success. The primary problem is that once the number of the video objects increases, the performances of these methods will descend sharply.

To solve above problem, we presented a SVM-based recognition method. This method can maintain the recognition stabilization when the number of the video objects is increasing. At the same time, its computation speed is fast enough to satisfy the requirements of the real-time recognition. This method is introduced at length in Section 2. In Section 3, experiments are performed to demonstrate the effectiveness of the method. The conclusion is given in section 4.

2 Moving Video Object Recognition Based on SVM

At first, the basic theory for SVM is introduced briefly. And then, the specific method for moving video object recognition is shown in detail.

2.1 The Basic Theory for SVM

SVM is developed by Boser, Guyon and Vapnik in 1990s [6], [7]. After that, SVM exhibits its large advantages in the model recognition of finite samples, nonlinear and high-dimension. In addition, SVM can be extended to many other fields of machine learning.

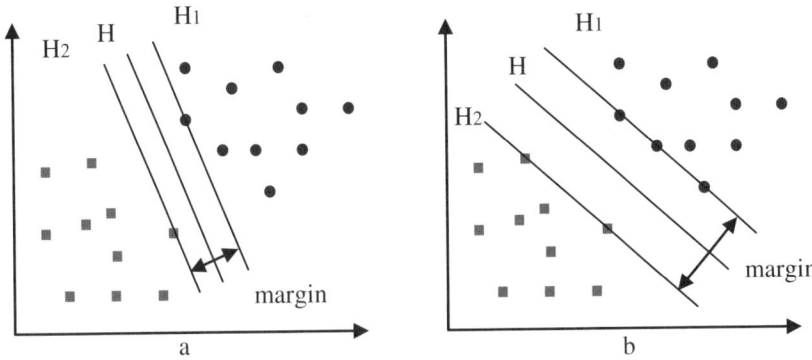

Fig. 1. Optimal Hyperplane

SVM is firstly presented for deal with the question of the optimal hyperplane. In Fig.1, the square points and the circle points respectively represent positive and negative samples. H represents the line that can classify the samples into two classes without errors. H1 (H2) represents the line, which is parallel to H and passes by the nearest points to H in negative (positive) samples set. The distance between H1 and H2 is called as margin. The optimal line is the line that can not only classify the samples into two classes without errors but also make the margin largest. The former means the experimental risk is smallest. The latter means the generalization ability is best. Extending the question to higher dimension space, the optimal line will be the optimal hyperplane.

Given the linear samples set $(\mathbf{x}_i, y_i)$, $i = 1, 2, \cdots, n$, $\mathbf{x}_i \in \mathbf{R}^d$ and $y_i \in \{+1, -1\}$. In d-dimension space, the general equation of linear decision function is shown below.

$$g(\mathbf{x}) = \mathbf{w} \cdot \mathbf{x} + b. \tag{1}$$

When $g(\mathbf{x}) = 0$, this equation will represent the optimal hyperplane. Normalize the decision function and make all samples satisfy $|g(\mathbf{x})| \geq 1$. Therefore, the nearest points to optimal hyperplane will satisfy $|g(\mathbf{x})|=1$ and the margin between H1 and H2 will be $2/\|\mathbf{w}\|$. After that, the question to find the optimal hyperplane will be transformed to make $\|\mathbf{w}\|$ smallest. Thus, all samples will satisfy the following equation:

$$y_i(\mathbf{w}\cdot\mathbf{x}_i+b)-1\geq 0, \ i=1,2,\cdots,n. \tag{2}$$

The hyperplane that can satisfy equation (2) and make $\|\mathbf{w}\|$ smallest does be the optimal hyperplane. Those points that locate in the H1 and H2 are called as support vectors.

Therefore, the question to construct the SVM model will be transformed to solve the following equations:

$$\begin{cases} \min \dfrac{1}{2}\|\mathbf{w}\|^2 \\ y_i(\mathbf{w}\cdot\mathbf{x}_i+b)-1\geq 0 \end{cases} \tag{3}$$

Because above equations both are protruding, according to the optimal theory, there is a unique global optimal solution existing for this question. So optimal hyperplane and the support vectors can be worked out and saved as the SVM model.

With the Lagrange factor method, the decision function with the optimal hyperplane can be rewritten as below:

$$f(\mathbf{x})=\mathrm{sgn}((\mathbf{w}^*\cdot\mathbf{x}+b^*))=\mathrm{sgn}(\sum_{i=1}^{n}\alpha_i^* y_i(\mathbf{x}_i\cdot\mathbf{x})+b^*). \tag{4}$$

Where α_i^* and b^* are the parameters of the optimal hyperplane. Because the value of any α_i^* corresponding to non support vector is zero, above decision function only deals with the support vectors. Since the support vectors only take a little part of all samples, the recognition speed is very fast [8].

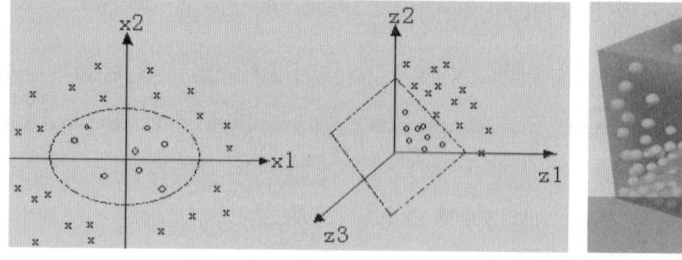
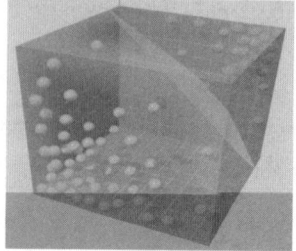

Fig. 2. Nonlinear classification [9]

For the nonlinear question, shown in Fig.2, SVM adopts the method of increasing dimensions to deal with it. Because the nonlinear samples may be transformed to linear samples in the higher dimension space, SVM firstly map the samples from the low dimension space to high space. After that, some linear methods will be taken to deal with them. Since the map is nonlinear, the nonlinear question can be solved in the higher dimension space [10], [11].

For the multi-class question, we can set $y \in \{1, 2, \cdots, k\}$ and construct the computation equation over again. Thus the SVM can deal with the multi-class question [12], [13], [14].

SVM is a very excellent method. At presents, this method has been applied in many fields, such as: manuscript recognition, face recognition, image indexing, weather forecast and so on. Although there are still many questions left to research, this method is being the most important method for model recognition [15].

2.2 The Recognition Method for Video Objects

The SVM-based recognition method for moving video objects can be divided into two parts: 1. training the SVM model based on shape features extraction; 2. recognizing the unknown moving video objects with the trained SVM model. That is to say, before video monitoring, an excellent model was achieved by training. In monitoring, the unknown objects will be recognized with this model.

2.2.1 Training the Model of SVM Based on Shape Features Extraction

Before video monitoring, the SVM model needs to be trained. That is shown in Fig.3. All images in the training sets are segmented, labeled and extracted to obtain their shape features. These features are combined with their video object label to compose the samples set $(\mathbf{x}_i, y_i)$. All these samples are taken into the equation (3) to work out the parameters for equation (4). Thus the SVM model is achieved.

There are several places left for attentions.

a. Because this question is a multi-class question, the equation (3) needs to be reconstruction according to multi-class SVM theory [16].
b. JSEG method is adopted in image segmentation [17]. Compare with other methods, the segmentation result of JSEG is more reliable. Although the speed of this method is something slow, it does not affect the performance of recognition since the segmentation is implemented before recognition.
c. The shape features employed here include wavelet descriptor, invariant moments, shape factor [18], [19], [20], [21], [22], [23], [24]. Among these features, the wavelet descriptor is a novel shape feature, which is modified based on the distance directional histogram [25]. Wavelet descriptor can describe the shape accurately. In addition, it is invariant to image transfer, rotation and scaling. Therefore, it is very important to effective recognition.

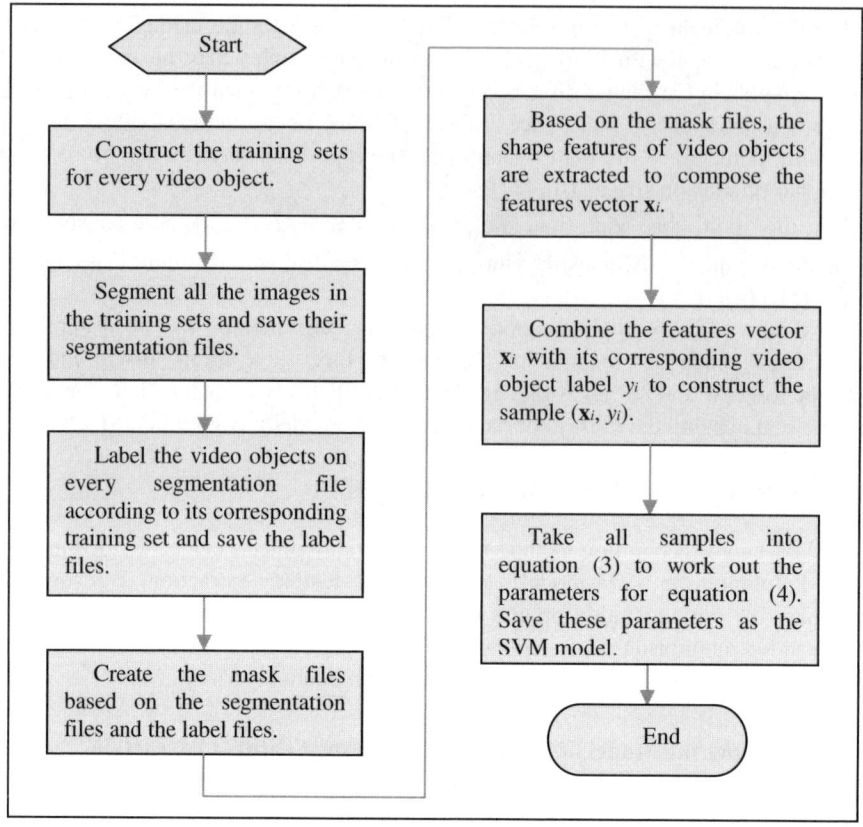

Fig. 3. The flow chart of training SVM model

2.2.2 Recognizing Moving Video Objects with Trained SVM Model

In video monitoring, the trained SVM model needs to be imported to recognize the unknown objects. At first, the unknown objects are obtained through moving video object extraction and tracking. And then, its shape features are extracted to compose the features vector x. In the final, the features vector x is taken into equation (4) to work out the recognition result.

There are several places left for attentions.

a. The particle filter method is adopted to implement the tracking of moving video object.
b. The mask file is created directly from the result of the video object extraction and tracking without segmentation or labeling.
c. The extraction speeds of the shape features are fast enough to satisfy the requirements of real-time monitoring.

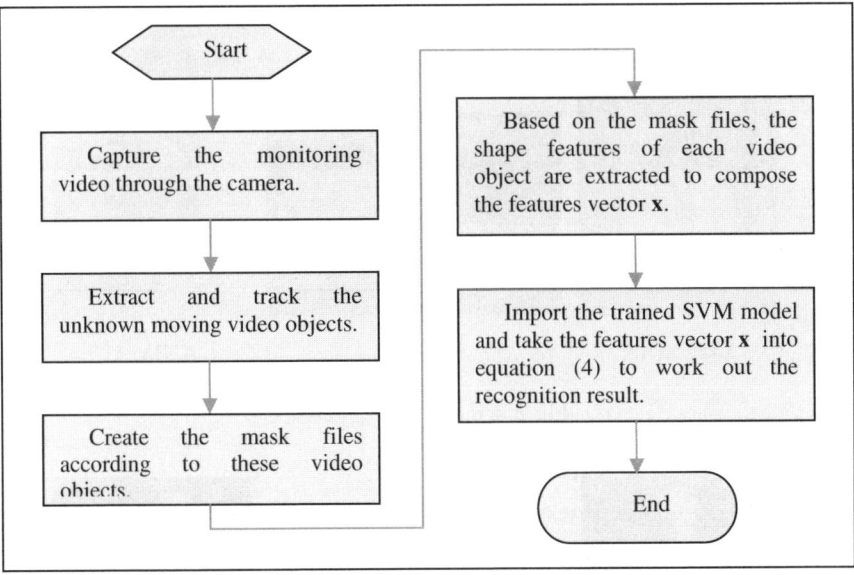

Fig. 4. The flow chart of recognition with SVM model

3 Experiments and Result Analysis

In the first experiments, three kinds of moving video objects (HUMAN, CAR and ANIMAL) are used to test our recognition method. We construct the training set for every video object. There are about 100 images in each training set. All images are segmented, labeled, extracted to train the SVM model.

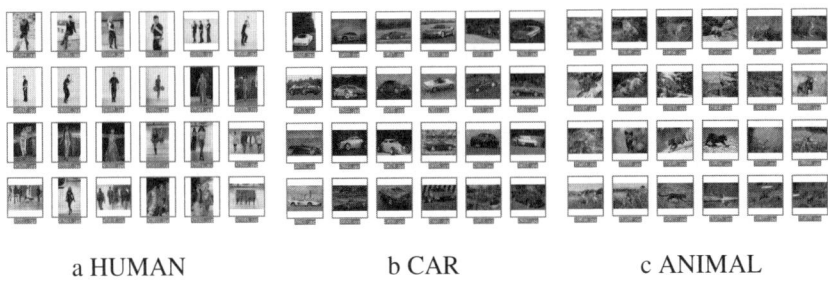

    a HUMAN   b CAR   c ANIMAL

Fig. 5. Training sets

Our experiment results show that all the recognition rates are above 97%. In addition, the mean recognition time is 94ms and can satisfy the requirements of real-time monitoring.

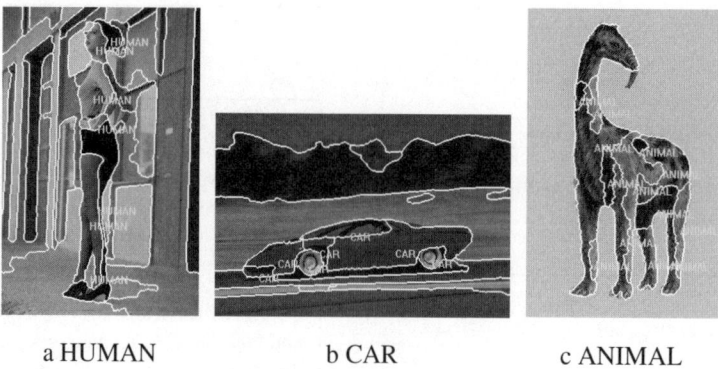

a HUMAN　　　　　　b CAR　　　　　　c ANIMAL

Fig. 6. Examples for object label

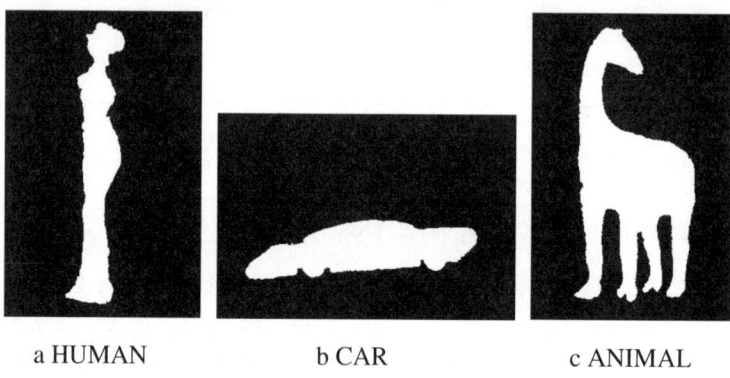

a HUMAN　　　　　　b CAR　　　　　　c ANIMAL

Fig. 7. Mask files for video object

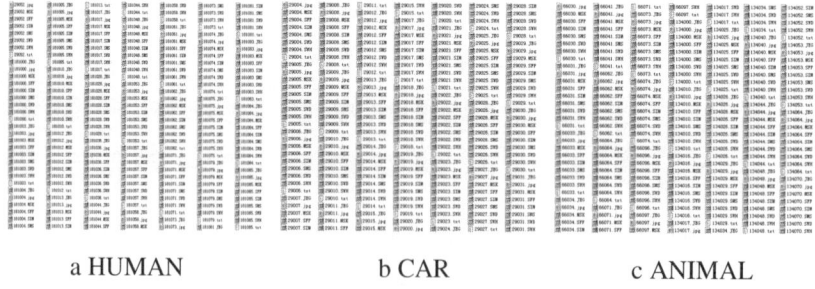

a HUMAN　　　　　　b CAR　　　　　　c ANIMAL

Fig. 8. Shape feature files

In the second experiment, we test the recognitions under different video objects number. Nine different video objects are employed to test, including animal, car, human with legs side-by-side, human with legs fork, human sitting, human lying, human bending and human riding.

Objects Number:	2	3	4	5	6	7	8	9
Recognition Rate(%):	100.00	98.65	97.97	97.96	97.96	98.02	97.56	97.59

Fig. 9. Recognition results

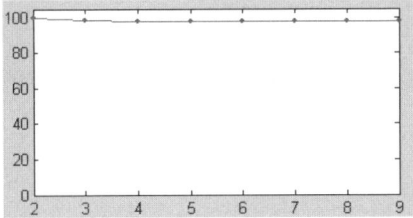

Fig. 10. Recognition rate with increasing of objects number

Just as shown in Fig.10, as the increasing of video objects number, the recognition rate descends very slowly. So, this method can overcome the disadvantages of the traditional methods in certain sense.

According to above experiments, the SVM-based method shows large advantages in computation speed, recognition rate and reliability. In addition, with the increasing of the number of the video objects, the recognition of this method maintains stabilization.

4 Conclusion

A novel SVM-based moving video objects recognition method is presented in this paper. The experimental results show that this method has large advantages in recognition rate, computation complexity and recognition reliability. Compare with the traditional methods, it is more competitive. In the future, we will continue increasing the number of recognizable video objects and extending the application fields of this method.

Acknowledgements. This work was supported in part by the National Natural Science Foundation of China (Grant No. 60472018) and supported in part by Ministry of Information and Communication (MIC) under the IT Foreign Specialist Inviting Program (ITFSIP) supervised by IIFA and ITRC supervised by IITA and International Cooperative Research Program of the Ministry of Science and Technology and KOTEF, 2nd stage BK21, Korea.

References

1. Amoid, M.W., Simone, S., et al.: Content-based image retrieval at the end of the early years [J]. IEEE Trans on Pattern Analysis and Machine Intelligence 22(12), 1349–1379 (2000)
2. Gupta, A., Jain, R.: Visual information retrieval. Communications of the ACM 40(5), 70–79 (1997)
3. Lee, J.C.M., Jain, A.K.: Special Issue on Image Database. Pattern Recognition 20(4), 537–658 (1997)
4. Marr, D.: Representing Visual Information. AI Memo 415, Mass. Inst. Technol. AI Lab., Cambridge (1977)
5. Duda, R.O., Hart, P.E.: Pattern Classification and Scene Analysis. John Wiley & Sons, Chichester (1973)
6. Vapnik, V.: Statistical Learning theory. Wiley, New York (1998)
7. Bian, Z.Q., Zhang, X.G.: Model Recognition. Tsinghua University Press, Beijing (2000)
8. Platt, J.C.: Fast Training Support Vector Machines using Sequential Minimal Optimization. In: Advances in Kernel Methods-Support Vector Learning, MIT Press, Cambridge (1998)
9. Feng, H.Z.: (2006), http://www.whihr.com.cn/htdocs/downfile/%E5%86%AF%E6%B1%89%E4%B8%AD.ppt
10. Cristianini, N., Taylor, J.S.: An Introduction to Support Vector Machines and Other Kernel-based Learning Methods. Cambridge University Press, Cambridge (2000)
11. Mangasarian, O.L., Musicant, D.R.: Successive Overrelaxation for Support Vector Machines. IEEE Trans. Neural Network 10(5), 1032–1037 (1999)
12. Bredensteiner, E.J., Bennett, K.P.: Multicategory Classification by Support Vector Machines. Computational Optimizations and Applications, pp. 53–79 (1999)
13. Westen, J., Watkins, C.: Multi-class Support Vector Machines. In: Proceedings of ESANN 1999 (1999)
14. Grammer, K., Singer, Y.: On the learnability and design of output codes for multiclass problems. In: Computational Learning Theory, pp. 35–46 (2000)
15. Osuna, E., Freund, R., Girosi, F.: Training Support Vector Machine: An Application to Face Detection. In: Proceedings of CVPR 1997 (1997)
16. Chang, C.C., Lin, C.J.: LIBSVM: A Library for Support Vector Machines (2001)
17. Deng, Y.N., Manjunath, B.S.: Unsupervised segmentation of color-texture regions in images and video. IEEE Trans. on Pattern Analysis and Machine Intelligence 23(8), 800–810 (2001)

18. Cheng, J.K., Huang, T.S.: Image registration by matching relational structures. Pattern Recognition 17(1), 149–159 (1984)
19. Mokhtarian, F., Mackworth, K.: A Theory of Multiscale, Curvature-Based Shape Representation for Planar Curves. IEEE Transaction on Pami 14(8) (1992)
20. Niblack, W., Barber, R., Equitz, W., Flicker, M., Glasman, E., Petkovic, D., Yanker, P., Faloutsos, C.: The QBIC Project: Query images by content using color texture and shape. In: SPIE Proceedings of Storage and Retrieval for Image and Video Databases, pp. 173–187 (1993)
21. Scassellati, B., Alexopoulos, S., Flickner, M.D.: Retrieving images by-2D shape: a comparison of computation methods with human perceptual judgments. In: Proceeding of SPIE Storage and Retrieval for Image and Video databases, pp. 2–9 (1994)
22. Mehtre, B.M., Kankanhalli, M.S., Lee, W.F.: Shape measures for content based image retrieval: a comparison. Information Processing & Management 33(3), 319–337 (1997)
23. Jacobs, C.E., Finkelstein, A., Salesin, D.H.: Fast multiresolution image querying. In: Computer Graphics Proceeding SIGGRAPH 1995, pp. 277–286 (1995)
24. Jain, A.K., Vilaya, A.: Image retrieval using color and shape. Pattern Recognition 29(8), 1233–1244 (1996)
25. Kong, X.D., Luo, Q.S., Guo, Y., Zeng, G.H.: A New Boundary Descriptor Based on the Directional Distance Histogram. In: Proceedings of APCC 2006 (2006)

Image Classification and Indexing by EM Based Multiple-Instance Learning*

H.T. Pao[1], Y.Y. Xu[2], S.C. Chuang[2], and H.C. Fu[2]

[1] Department of Management Science
[2] Department of Computer Science
National Chiao Tung University,
Hsin Chu, Taiwan, ROC
htpao@cc.nctu.edu.tw,{yyxu,scchung,hcfu}@csie.nctu.edu.tw

Abstract. In this paper, we propose an EM based Multiple-Instance learning algorithm for the image classification and indexing. To learn a desired image class, a set of exemplar images are selected by a user. Each example is labeled as conceptual related (positive) or conceptual unrelated (negative) image. A positive image consists of at least one user interested object, and a negative example should not contain any user interested object. By using the proposed learning algorithm, an image classification system can learn the user's preferred image class from the positive and negative examples. We have built a prototype system to retrieve user desired images. The experimental results show that for only a few times of relearning, a user can use the prototype system to retrieve favor images from the WWW over Internet.

Keywords: Multiple-Instance learning, Image retrieve, WWW.

1 Introduction

In the domain of video/image classification and indexing, color histogram has been widely used [1]. However, the position and orientation of an interested object in an image are not considered in the image indexing or classification from the color distribution method. This lack of spatial relationship makes the classification undesirable. Several Internet image searching engines also include texture of a whole image or those of a user-specified region in addition to the color distribution, such as QBIC [2] and NETRA [3]. Recently, some of the systems adapted spatial relationship as a new kind of features. For example, SaFe [4] provides a framework to search for and to compare images by the spatial arrangement of regions or objects. In the query of the SaFe system, interested objects or regions are first assigned by a user, and thus properties such as spatial

---

* This research was supported in part by the National Science Council under Grant NSC 94-2213-E009-139.

location, size and visual features are all considered. The SaFe system could find the images that best match the query.

Usually, the contents of an image are very complicated, so an image can be seen as the combination of the small subimages, in which each subimage has its own content. For example, if an image contains an interested subimage such as *"Waterfall"* and some other uninterested subimages. One would like to identify this image as *"The Image Containing a Waterfall"*. In traditional methods, the feature vector is extracted from the whole image. It is very hard to extract a suitable feature vector from the whole image just to represent *"The Image Containing a Waterfall"*. Thus, some methods first segment interested subimages from an image, and then extract feature vectors from the interested subimages. In fact, it is very difficult to segment the interested subimages precisely. Beside, the interested subimages in an image may be different for different users, as a consequence, different feature vectors may be extracted from different interested subimages when the same image are queried by different users. Thus, this approach complicates the system design and confuses users in querying the system.

In order to represent an image correctly, multiple instances (subimages) are used. The problem that using a set of instances to represent a concept is call the *Multiple-Instance learning problem* [5]. In the image indexing and classification domain, a feature vector is extracted from an instance and a concept is composed of a set of instances that is associated to the user's interested images. Since an image is represented by a set of instances, the learning method of the Multiple-Instance problem is different from traditional methods.

In [6], O. Maron and A. Lakshmi Ratan proposed the Multiple-Instance learning method to learn several instances with various Diverse Densities, and to maximize Diverse Density by Quasi-Newton method. In this paper, we propose an *EM based Multiple-Instance learning algorithm*. We intend to provide a more comprehensive treatment in deriving the maximization procedure of Diverse Density.

This paper is organized as follows. Section 2 presents the proposed EM based Multiple-Instance learning algorithm. Then, the image classification and indexing using proposed method is described in Section 3. Some experimental results are shown in Section 4. The concluding remark is presented in Section 5.

2 EM Based Multiple-Instance Learning Algorithm

In the Multiple-Instance learning, conceptual related (positive) and conceptual unrelated (negative) images are used for reinforced and antireinforced learning of a user's desired image class. Each positive training image contains at least one interested subimage related to the desired image class, and each negative training image should not contain any subimage related to the desired image class. The target of the Multiple-Instance learning is to search the optimal point of the image class in the feature space, where the optimal point is close to the intersection of the feature vectors extracted from the subimages of the positive training images and is far from the union of the feature vectors extracted from the subimages of the negative training images.

For example, if one wants to train an image class t with P_t positive images and N_t negative images. Each positive image has V_t^+ subimages, and each negative image has V_t^- subimages. We denote the kth feature vector extracted from the kth subimage of the ith positive image as $\mathbf{X}_{ik}^+$, and the kth feature vector extracted from the kth subimage of the ith negative example as $\mathbf{X}_{ik}^-$. The probability that $\mathbf{X}_{ik}^+$ belongs to class t is $P(t \mid \mathbf{X}_{ik}^+)$, and the probability that $\mathbf{X}_{ik}^-$ belongs to class t is $P(t \mid \mathbf{X}_{ik}^-)$. A measurement called Diverse Density is used to evaluate that how many different positive images have feature vectors near a point t, and how far the negative feature vectors are from a point t. The Diverse Density for a class t is defined as

$$DD_t = \prod_{i=1}^{P_t}(1 - \prod_{k=1}^{V_t^+}(1 - P(t \mid \mathbf{X}_{ik}^+))) \prod_{i=1}^{N_t}(\prod_{k=1}^{V_t^-}(1 - P(t \mid \mathbf{X}_{ik}^-))). \quad (1)$$

The optimal point of the class t is appeared where the Diverse Density is maximized. By taking the first partial derivatives of Eq.(1) with respect to parameters of the class t and setting the partial derivatives to zero, the optimal point of the class t can be obtained. Suppose the density function of the class t is a D-dimensional Gaussian mixture with uncorrelated features. The parameters are the mean μ_{tcd}, the variance σ_{tcd}^2, and the cluster prior probability p_{tc} of each cluster in the class t. The estimating parameters of the class t can be derived by $\frac{\partial}{\partial \mu_{tcd}}DD_t = 0$, $\frac{\partial}{\partial \sigma_{tcd}}DD_t = 0$, and $\frac{\partial}{\partial p_{tc}}DD_t = 0$. Thus

$$\mu_{tcd} = \left[\sum_{i=1}^{P_t} \left(\frac{\mathbb{P}_{ti}}{1-\mathbb{P}_{ti}} \right) \sum_{k=1}^{V_t^+} Q_{tc}(\mathbf{X}_{ik}^+) x_{ikd}^+ \right.$$
$$\left. - \sum_{i=1}^{N_t} \sum_{k=1}^{V_t^-} Q_{tc}(\mathbf{X}_{ik}^-) x_{ikd}^- \right]$$
$$\left/ \left[\sum_{i=1}^{P_t} \left(\frac{\mathbb{P}_{ti}}{1-\mathbb{P}_{ti}} \right) \sum_{k=1}^{V_t^+} Q_{tc}(\mathbf{X}_{ik}^+) \right. \right.$$
$$\left. \left. - \sum_{i=1}^{N_t} \sum_{k=1}^{V_t^-} Q_{tc}(\mathbf{X}_{ik}^-) \right], \quad (2)\right.$$

$$\sigma_{tcd}^2 = \left[\sum_{i=1}^{P_t}\left((\frac{\mathbb{P}_{ti}}{1-\mathbb{P}_{ti}})\sum_{k=1}^{V_t^+}Q_{tc}(\mathbf{X_{ik}^+})\|x_{ikd}^+ - \mu_{tcd}\|^2\right)\right.$$
$$\left. - \sum_{i=1}^{N_t}\sum_{k=1}^{V_t^-}Q_{tc}(\mathbf{X_{ik}^-})\|x_{ikd}^- - \mu_{tcd}\|^2\right]$$
$$\Bigg/ \left[\sum_{i=1}^{P_t}\left((\frac{\mathbb{P}_{ti}}{1-\mathbb{P}_{ti}})\sum_{k=1}^{V_t^+}Q_{tc}(\mathbf{X_{ik}^+})\right)\right.$$
$$\left. - \sum_{i=1}^{N_t}\sum_{k=1}^{V_t^-}Q_{tc}(\mathbf{X_{ik}^-})\right], \tag{3}$$

$$p_{tc} = \left[\sum_{i=1}^{P_t}\left((\frac{\mathbb{P}_{ti}}{1-\mathbb{P}_{ti}})\sum_{k=1}^{V_t^+}Q_{tc}(\mathbf{X_{ik}^+})\right)\right.$$
$$\left. - \sum_{i=1}^{N_t}\sum_{k=1}^{V_t^-}Q_{tc}(\mathbf{X_{ik}^-})\right]$$
$$\Bigg/ \left[\sum_{i=1}^{P_t}\left((\frac{\mathbb{P}_{ti}}{1-\mathbb{P}_{ti}})\sum_{k=1}^{V_t^+}(\frac{P(t|\mathbf{X_{ik}^+})}{1-P(t|\mathbf{X_{ik}^+})})\right)\right.$$
$$\left. - \sum_{i=1}^{N_t}\sum_{k=1}^{V_t^-}(\frac{P(t|\mathbf{X_{ik}^-})}{1-P(t|\mathbf{X_{ik}^-})})\right], \tag{4}$$

where

$$P(c|\mathbf{X_{ik}^\star}, t) = \frac{p_{tc}\cdot P(\mathbf{X_{ik}^\star}|c,t)}{P(t|\mathbf{X_{ik}^\star})}, \tag{5}$$

$$P(t|\mathbf{X_{ik}^\star})^{(l)} = \frac{P(\mathbf{X_{ik}^\star}^{(l)}|t)P_t}{P(\mathbf{X_{ik}^\star})}, \tag{6}$$

$$\mathbb{P}_{ti} = \prod_{k=1}^{P_i}\left(1 - P(t|\mathbf{X_{ik}^\star})\right),$$

$$Q_{tc}(\mathbf{X_{ik}^\star}) = \frac{P(t|\mathbf{X_{ik}^\star})P(c|\mathbf{X_{ik}^\star}, t)}{1 - P(t|\mathbf{X_{ik}^\star})},$$

$$P(\mathbf{X_{ik}^\star}|c,t) = \frac{1}{\prod_{d=1}^{D}(2\pi\sigma_{tcd}^2)^{\frac{1}{2}}}$$
$$\cdot \exp(-\frac{1}{2}\sum_{d=1}^{D}(\frac{x_{ikd}^\star - \mu_{tcd}}{\sigma_{tcd}})^2),$$

and the notation $\star$ represents $+$ or $-$.

According to Eqs.(2), (3), and (4), we proposed an EM based Multiple-Instance learning algorithm to learn these parameters. The EM based Multiple-Instance learning algorithm contains two steps: the expectation step (E-step) and the maximization step (M-step). The algorithm is described as follows.

1. Choose an initial point in the feature space, and let its parameters are $\mu_{tcd}^{(0)}$, $\sigma_{tcd}^{2(0)}$, and $p_{tc}^{(0)}$.
2. **E-Step:** Using the calculated model parameters $\mu_{tcd}^{(l)}$, $\sigma_{tcd}^{2(l)}$, $p_{tc}^{(l)}$, Eqs.(5) and (6), estimate $P^{(l)}(c|\mathbf{X}_{ik}^\star,t)$ and $P^{(l)}(t|\mathbf{X}_{ik}^\star)$.
   M-Step: Using the estimated $P^{(l)}(c|\mathbf{X}_{ik}^\star,t)$ and $P^{(l)}(t|\mathbf{X}_{ik}^\star)$, compute the *new* model parameters $\mu_{tcd}^{(l+1)}$, $\sigma_{tcd}^{2(l+1)}$, and $p_{tc}^{(l+1)}$ according to Eqs.(2), (3), and (4).
3. Calculate the diverse density $DD^{(l+1)}$. If $(DD^{(l+1)} - DD^l)$ is smaller than a predefined threshold ϵ, then stop the process. Otherwise, loop step 2.

3 Image Feature Extraction and Indexing

Before training the system for indexing images, multiple feature vectors are extracted from the multiple instances of several exemplar training images. Then, the system are trained according to the proposed EM based Multiple-Instance learning algorithm. Finally, the testing images are evaluated using Bayesian decision rule for indexing and classification.

3.1 Image Feature Extraction

The image features extraction we used are similar to the method proposed in [6]. First, a number of instances are randomly selected from an image. Then, the feature vectors are extracted from the instances as shown in Figure 1. The feature vector in the position (i,j) is defined as $\mathbf{X} = \{x_1, \cdots, x_{15}\}$, where

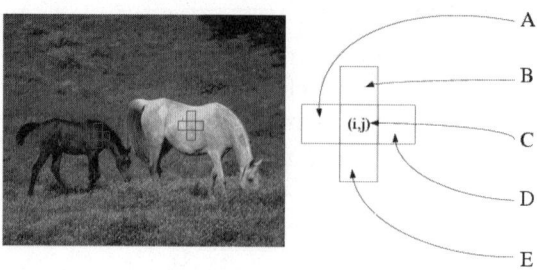

Fig. 1. Feature vectors are extracted from "+" shaped subimage (instance). The instance consists of 5 subregions: A,B,C,D and E. Each subregion is composed of 2×2 pixels. The feature vector, $\mathbf{X} = \{x_1, \cdots, x_{15}\}$, is the YUV value of C, and the difference values of C and its 4 neighbors.

- $\{x_1, x_2, x_3\}$ is the average YUV values of C.
- $\{x_4, x_5, x_6\}$ is the average YUV values of A minus average YUV values of C.
- $\{x_7, x_8, x_9\}$ is the average YUV values of B minus average YUV values of C.
- $\{x_{10}, x_{11}, x_{12}\}$ is the average YUV values of D minus average YUV values of C.
- $\{x_{13}, x_{14}, x_{15}\}$ is the average YUV values of E minus average YUV values of C.

It is clear to see that the proposed feature extraction provides not only the color information but also some of the spatial information.

3.2 Image Indexing

After the feature vectors are extracted from the subimages of the training images, the system can be trained to perform indexing and classification for the images with respect to the class t. First, a user needs to select some related and unrelated images for a class as training images, then feature vectors of these images are extracted. Once the training feature vectors are ready, the EM based Multiple-Instance learning algorithm is used to compute the mean μ_{tcd}, the variance σ^2_{tcd}, and the cluster prior probability p_{tc}. By using these parameters, the posterior probabilities $P(t|\mathbf{X_i})$ of an unindex image can be computed for each class. The unindex image is indexed to the class f if the $P(f|\mathbf{X_i})$ is the highest among all the $P(t|\mathbf{X_i})$.

4 Experimental Results

We have built a prototype system to evaluate the proposed image classification and indexing method. This system is called the *"Intelligent Multimedia Information Processing Systems"* (IMIPS)[7]. This system has been used as a video search engine over the WWW. Once a new video file is found by a video spider in IMIPS, the system will download the file and save several key frames in a database.

When a desired image class is to be trained, the positive and the negative exemplar images are selected from the stored key frames. Then, the system learns the desired image class using the proposed EM based Multiple-Instance learning algorithm. When the optimal model for desired class is trained, each key frame in the database is indexed by its posterior probability associated with the desired class.

A web-based user interface is depicted in Figure 2. Each of the title of the trained classes are display in a pulldown menu: "Select Query Type". When a user is trying to search an image of a certain class, one can use "Select Query Type" to select a class. Then, the associated images will be shown. Suppose, the class "Human" is selected, the system responds with all the images belonging to "Human" from the database in a descending order according to their computed value of posterior probability. As we can see, most of the shown images are

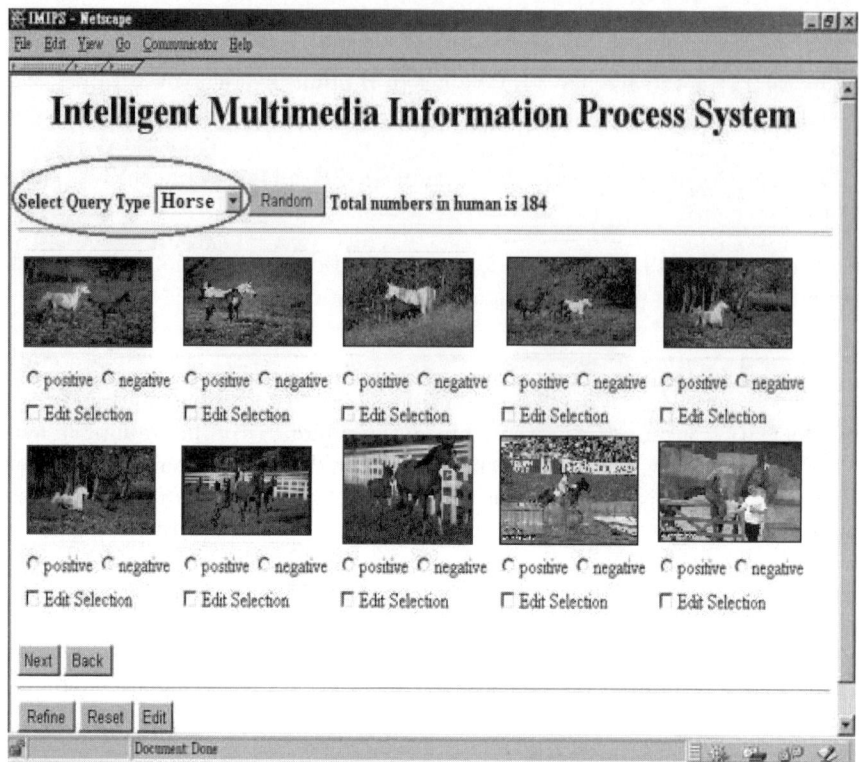

Fig. 2. When a user enters the IMIPS, one can select a class on interested of images, which are key frames of a video over the WWW. Then, the user can click on an image to view its associated video program.

human-related. If the user wants to create a new class, one can click the "Random" button, then a set of randomly selected images will be shown. One can select the "Positive" button to include conceptual related images, and select the "Negative" button to exclude conceptual unrelated images. When the "Refine" button is pressed, the system will train a new class according to the selections.

Table 1. Results of natural scenes classification. The number in each row indicates the correctly retrieved images of each classes with respect to total retrieved images.

templates	Total of Retrieve				
	10	20	30	40	50
Horse	10	20	30	39	48
Human	10	15	18	22	23
Sky	7	13	19	26	29
Star	10	20	30	35	40
Fire	9	13	N/A	N/A	N/A

In this prototype system, we have trained five classes of nature scenes: "Human", "Star", "Sky", "Horse", and "Fire". The experimental results are shown in Table 1. The number in the intersection of row "Horse" and column "10" indicates that the correctly retrieved images is 10 out of the total 10 retrieved images. The correctness of retrieve is judged according to human perception.

5 Conclusion

In this paper, we propose an EM based Multiple-Instance learning algorithm and implement a user friendly video search engine (IMIPS) over the WWW. The experimental results show that the retrieved images is quite match with human perception. How to properly determine the correct number of the clusters in the mixture Gaussian model of each class is a problem we want to solve in the future. In order to build a more powerful model, some features, such as shapes, textures, domain knowledge, etc., will also be included in the future systems.

References

1. Kasturi, R., Strayer, S.H.: An evaluation of color histogram based methods in video indexing. In: Research Progress Report CSE-96-053. Hangzhou, China, vol. 3 (1995)
2. Flickner, M., Sawhney, H., Niblack, W., Ashley, J., Huang, B.D.Q., Gorkani, M., Hafner, J., Lee, D., Petkovic, D., Steele, D., Yanker, P.: Query by image and video content: The qbic system. IEEE Computer, special issue on content based picture retrieval system 28, 23–32 (1995)
3. Ma, W.: NETRA: A Toolbox for Navigating Large Image Databases. PhD thesis, Dept. of Electrical and Computer Engineering, University of California at Santa Barbara (1997)
4. Spatial and feature query system, avaliable
   http://disney.ctr.columbia.edu/safe/
5. Dietterich, T.G., Lathrop, R.H., Lozano-Pérez, T.: Solving the multiple-instance problem with axis-parallel rectangles. Artifical Intelligence Journal 89 (1997)
6. Maron, O., Ratan, A.L.: Multiple-instance learning for natural scene classification. In: Machine Learning: Proceedings of the 15th international Conference, pp. 23–32 (1998)
7. The intelligent multimedia information processing system, avaliable
   http://imips.csie.nctu.edu.tw/imips/imips.html

Palm Vein Extraction and Matching for Personal Authentication

Yi-Bo Zhang[1], Qin Li[2], Jane You[2], and Prabir Bhattacharya[1]

[1] Institute for Information Systems Engineering, Concordia University, Quebec, Canada
prabir@ciise.concordia.ca
[2] Biometrics Research Centre, Department of Computing, The Hong Kong Polytechnic University, KLN, Hong Kong
{csqinli,csyjia}@comp.polyu.edu.hk

Abstract. In this paper, we propose a scheme of personal authentication using palm vein. The infrared palm images which contain the palm vein information are used for our system. Because the vein information represents the liveness of a human, this system can provide personal authentication and liveness detection concurrently. The proposed system include: 1) Infrared palm images capture; 2) Detection of Region of Interest; 3) Palm vein extraction by multiscale filtering; 4) Matching. The experimental results demonstrate that the recognition rate using palm vein is good.

Keywords: Palm vein, Personal identification, Liveness detection, Infrared palm images, Multiscale filtering.

1 Introduction

The personal identification using hand and palm vein has gained more and more research attentions these years [1] [2] [3] [4] [5]. There are many good properties of this kind of biometric feature: 1) the vein information can represent the liveness of an object; 2) it is difficult to be damaged and modified as an internal feature; 3) it is difficult to simulated using a fake palm. Because of these, hand and palm vein seems a better biometric feature that finger print and face.

In [1] [2], the thermal images are used to extract palm vein and obtained good results. But the infrared thermal camera is very expensive. In [5], a low cost CCD camera is used to capture near infrared image of palm. That system's recognition rate is good. But because both vein and texture information are used in [5] for recognition, that method can not ensure the liveness of a person.

In this work, we use a low cost CCD camera to capture the infrared palm images. The palm vein, rather than hand vein (back of the hand), is used by our system because it is easier to design a platform to help a user fixing his/her hand on the image capture device. And, we only use the palm vein as biometric feature without any other features such as palm texture and palm line so that this system can ensure the liveness of an object.

2 Image Capture and ROI Locating

The capture device is modified from our previous work on palmprint [6]. Fig. 1 shows part of our device. There are three poles to help a user to fix his/her hand (Fig. 1(b)). A low cost CCD camera is used in this system. In order to obtain infrared images, a set of infrared light source is installed around the camera.

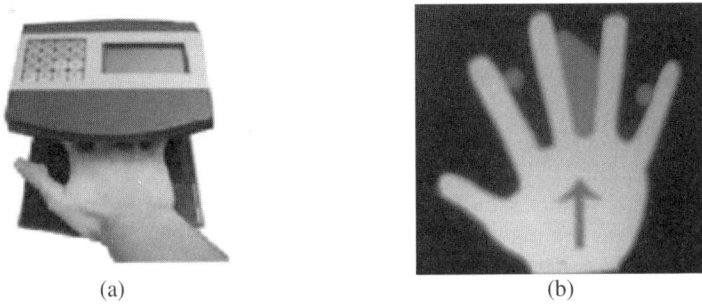

Fig. 1. Capture device. (a) outside of the device; (b) poles to fix a palm.

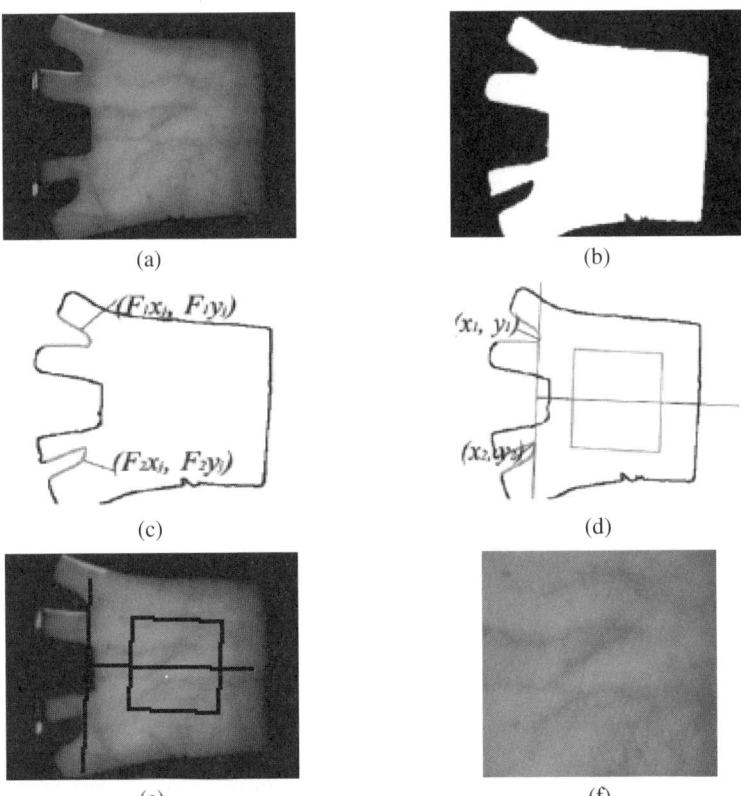

Fig. 2. Locate ROI. (a) a infrared palm image captured by our device; (b) binarized image; (c) boundaries; (d) and (e) ROI locating; (f) the subimage in ROI.

After image capture, the region of interest (ROI) is located by the same method of our previous work [6]. A small area (128*128 pixels) of a palm image is located as ROI to extract the features and to compare different palms. Using the features within ROI for recognition can improve the computation efficiency significantly. Further, because this ROI is located by a normalized coordinate based on the palm boundaries, the recognition error caused by a user who slightly rotate or shift his/her hand is minimized. Figure 2 illustrates the procedure of ROI locating:

1). Binarize the input image (Fig. 2(a) and (b));
2). Obtain the boundaries of the gaps, (Fixj;Fiyj) (Fig. 2(c));
3). Compute the tangent of the two gaps (Fig. 2(d)), use this tangent (the line connect (x1,y1) and (x2,y2)) as the Y-axis of the palm coordinate;
4). Use a line passing through the midpoint of the two points (x1,y1) and (x2,y2), which is also perpendicular to the Y-axis, as the X-axis (the line perpendicular to the tangent in Fig. 2(d));
5). The ROI is located as a square of fixed size whose center has a fixed distance to the palm coordinate origin (Fig. 2(d) and (e));
6). Extract the subimage within the ROI (Fig. 2(e) and (f)).

3 Palm Vein Extraction

By observing the cross-sections of palm veins, we found that they are Gaussian-shaped lines. Fig. 3 shows some cross-sections of the palm veins. Based on this observation, the matched filter proposed in [7] [8] can be used to detect palm veins. And, we propose a multiscale scheme to improve the performance of vein detection. This scheme includes multiscale matched filters and scale production [9] [10] [11].

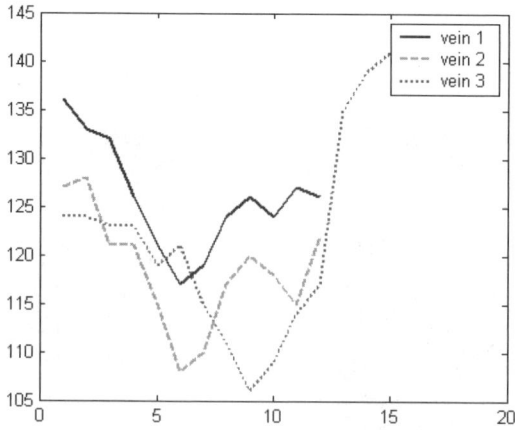

Fig. 3. Cross section evaluation

The matched filter was defined as

$$\begin{cases} g_\phi(x,y) = -\exp(-x'^2/\sigma_x^2) - m, & \text{for } |x'| \le 3\sigma_x, \; |y'| \le L/2 \\ x' = x\cos\phi + y\sin\phi \\ y' = y\cos\phi - x\sin\phi \end{cases} \quad (1)$$

where ϕ is the filter direction, σ is standard deviation of Gaussian, m is the mean value of the filter, L is the length of the filter in y direction which is set according to experience. This filter can be regarded as Gaussian filter in x direction. A Gaussian-shaped filter can help to denoise and the zero-sum can help to suppress the background pixels. Fig. 4 shows the matched filters in 1-D (a) and 2-D (b) view. Fig. 5 gives the filter response in a single scale.

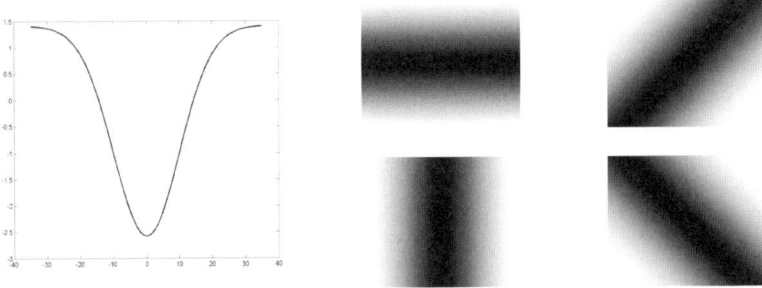

Fig. 4. Matched filters (a) 1-D and (b) 2-D

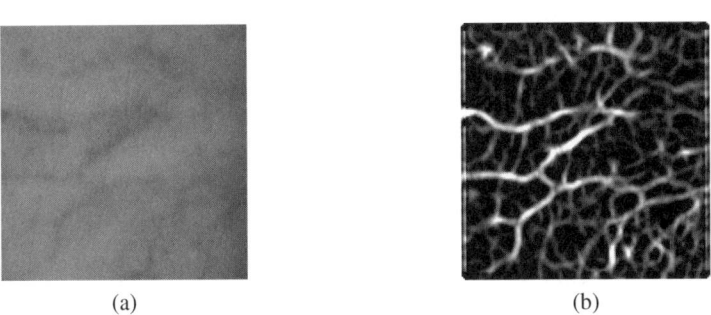

(a)   (b)

Fig. 5. (a) The subimage within ROI and (b) its single scale Matched filter response

From Fig. 5, we find that there is too much noise in the matched filter responses of infrared palm images. To gain a proper signal-noise ratio, we propose a multiscale scheme to detect the palm vein. In [12], Mallat illustrated mathematically that signals and noise have different singularities and that edge structures present observable magnitudes along the scales, while noise decreases rapidly. With this observation, we responded to those problems of edge and line detection and noise reduction by thresholding the multiscale products [9] [10] [11]. For Multiscale analysis, a scale parameter is added to equation (1) to control the filter size:

$$g_{\phi,s}(x,y) = -\exp(-x'^2/s\sigma_x^2) - m, \quad \text{for } |x'| \leq 3s\sigma_x, \ |y'| \leq sL/2 \quad (2)$$

The response of multiscale matched filter can be expressed by

$$R_g(x,y) = g_{\phi,s}(x,y) * f(x,y) \quad (3)$$

where $f(x,y)$ is the original image and $*$ denotes convolution.

The scale production is defined as the product of filter responses at two adjacent scales

$$P^{s_j}(x,y) = R_g^{s_j}(x,y) \cdot R_g^{s_{j+1}}(x,y) \quad (4)$$

Fig. 6 illustrates the multiscale line detection and scale production, where Mf1 and Mf2 are the responses at two different scales, P1,2 is their production. The noise in Mf1 and Mf2 nearly reaches the half peak of the signal response. After production, the noise becomes much smaller.

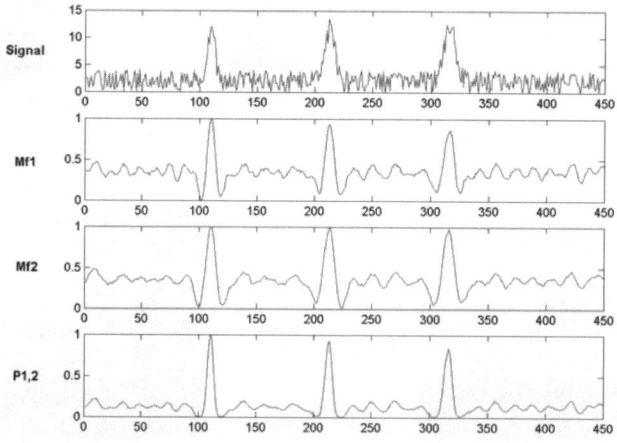

Fig. 6. Multiscale filtering and scale production: Mf1 and Mf2 are the responses at two different scales, P1,2 is their production

According to our experience, the palm vein widths in our infrared images vary from 10 pixels to 20 pixels that corresponding to Gaussian with standard derivation from 1.2 to 2.4. In order to produce strong filter responses, the standard derivation of the matched filters must be similar to the standard derivation of veins. We apply matched filters at two different scales having standard derivation 1.8 and 2.6.

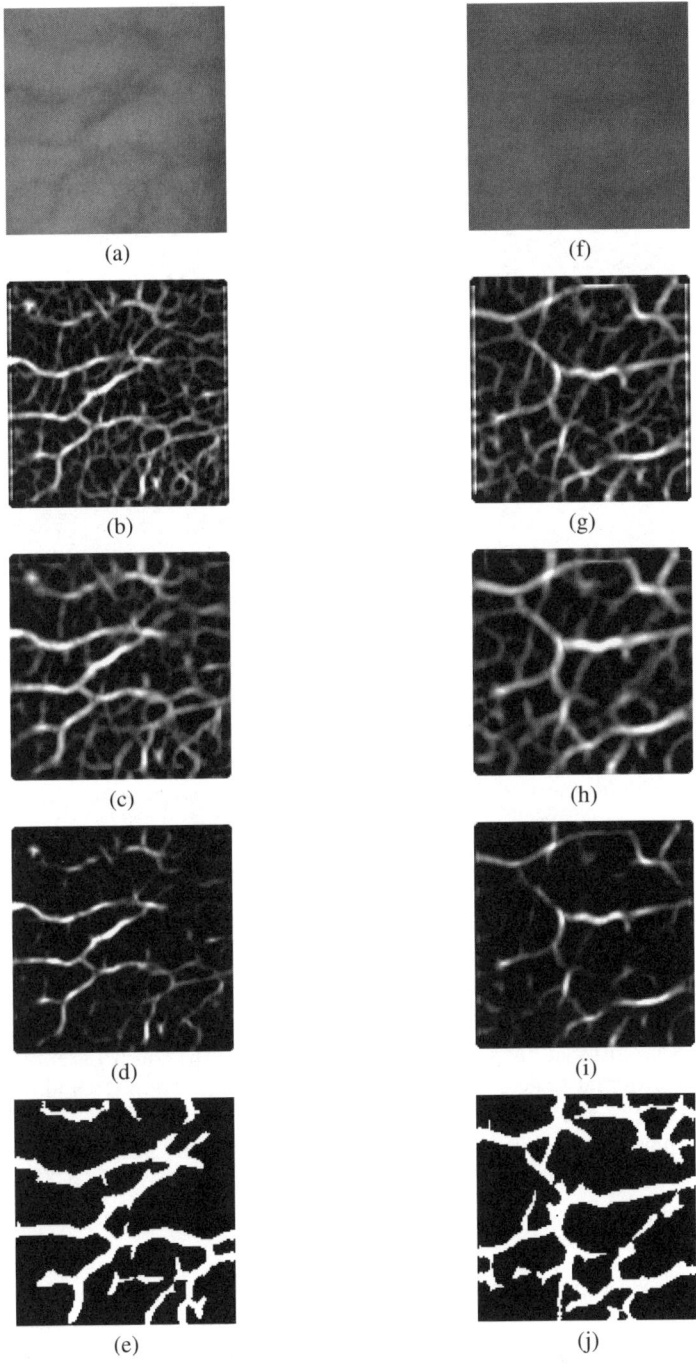

Fig. 7. Thresholding multiscale products: (a) a subimage, (b) & (c) its matched filter responses at two different scales, (d) scale production of (b) & (c), (e) binarized image; (f)~(j) corresponding images of another palm

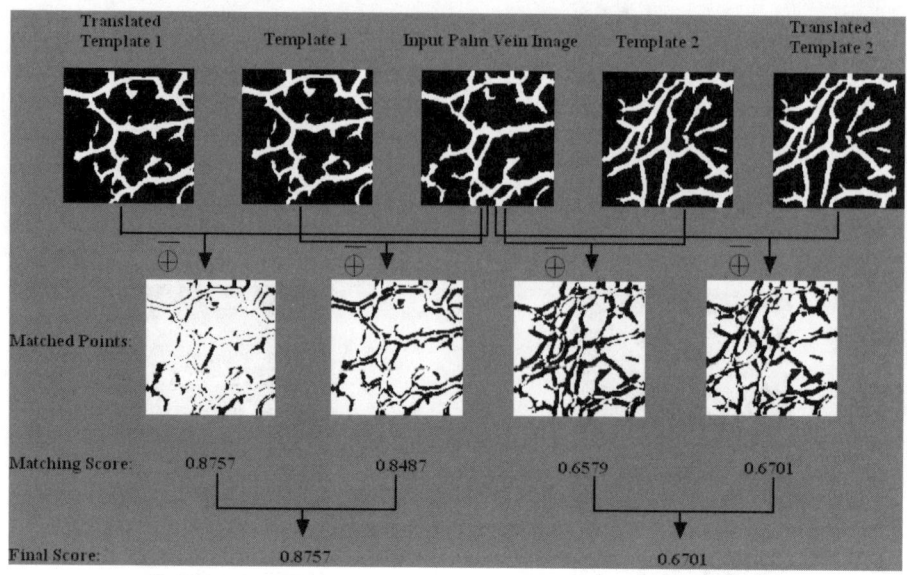

Fig. 8. Matching procedure

There are two reasons for using larger filter standard derivation: (1) Filters with larger size are better at denoising; (2) Reorganization will be more tolerance to rotation and shift. Even the ROI is located by a normalized coordinate, there is still little rotation and shift causing recognized error. In [13], Wu proposed this kind of error can be avoided by dilating the line. The convolution of a Gaussian signal whose standard derivation is σ_1 with a Gaussian filter whose standard derivation is σ_2 will produce a filter response that is a Gaussian whose standard derivation is $\sigma_3 = \sqrt{\sigma_1^2 + \sigma_2^2}$. With a larger σ_2, the vein will be "dilated".

After multiscale filtering and scale production, the filter responses will be binarized by thresholding the multiscale products [9] [10]. Fig. 7 gives some examples of thresholding multiscale products of infrared palm images. Fig. 7(a) is a subimage of an infrared palm image within ROI; Fig. 7(b) and (c) are matched filter responses at different scales; Fig. 7(d) is the scale production of (b) and (c); Fig. 7(e) is the binarized image of (d). Fig. 7(f)~(i) are the corresponding images of another palm. The lines in the binarized images, which represent palm vein, are used as biometric features in our system. This binarized image is named as palm vein image in the following paper.

4 Palm Vein Matching

Using the palm vein image as palm vein templates (only 0 and 1 in the templates), the similarity of two palm images can be calculated by template matching. Let T denote a prepared template in the database and I denote the palm vein image of a new input palm, we match T and I through logical "exclusive or" operation. The matching score of T and I is calculated as

$$S(T,I) = \frac{1}{M \times N} \sum_{i=1}^{M} \sum_{j=1}^{N} \left[\overline{T(i,j) \oplus I(i,j)} \right] \quad (5)$$

where $M \times N$ is the size of T or I (A and B must be the same size), $\oplus$ is the logical "exclusive or" operation, and $-$ is the logical "not" operation. Even we already registered palm images captured at different times at the step of ROI locating, there may still be little rotation and translation between them. To overcome this problem, we vertically and horizontally translate the template T a few points. The final matching score is taken to be the maximum matching score of all the translated positions. This matching procedure is illustrated in Fig. 8.

5 Experimental Results

There are totally 144 infrared palm images in our database. Each of 24 individuals has 6 images. Figure 9 shows images of a palm captured at different times, where the image quality is good. Figure 10 shows images of a palm captured at different times, where the image quality is bad. The probability distribution of genuine and imposter

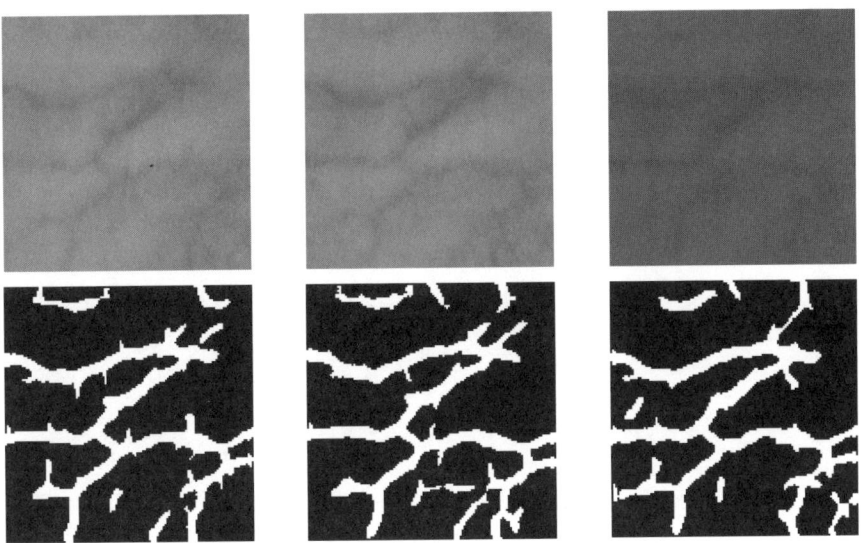

Fig. 9. Images of a palm captured at different times, where the image quality is good. The first row shows the images captured by our device. The second row shows the corresponding vessel extraction results.

is shown in Figure 11. The recognition performance is shown in Figure 12 using ROC curve. We achieved 98.8% recognition rate where the false acceptance rate is 5.5%. Most of the false recognitions were caused by the images of poor quality.

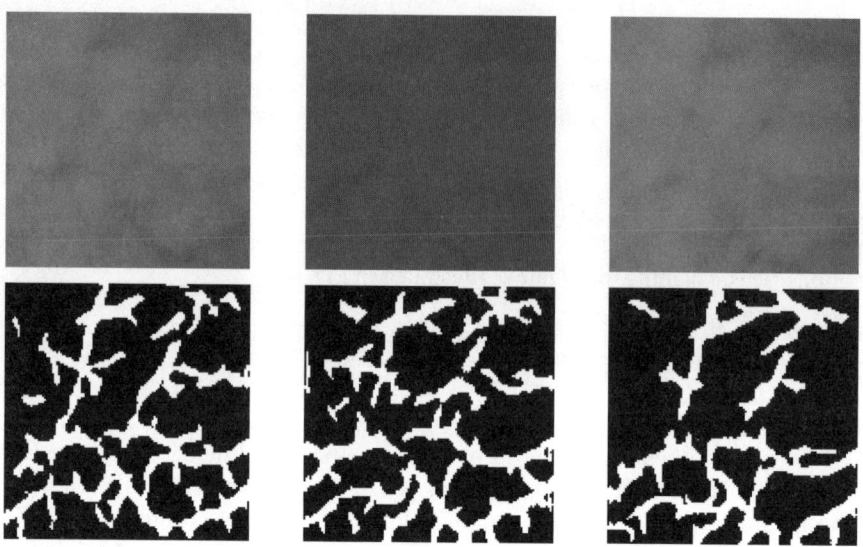

Fig. 10. Images of a palm captured at different times, where the image quality is bad. The first row shows the images captured by our device. The second row shows the corresponding vessel extraction results.

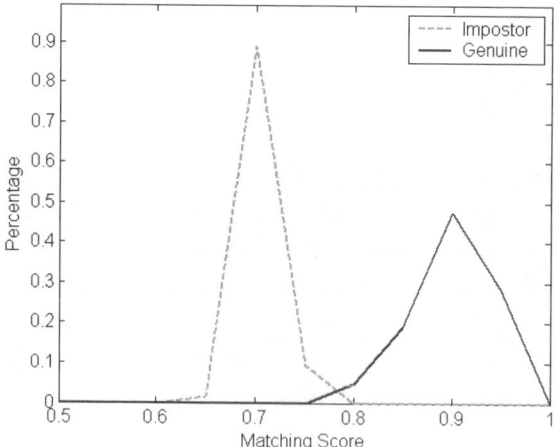

Fig. 11. The probability distribution of genuine and imposter

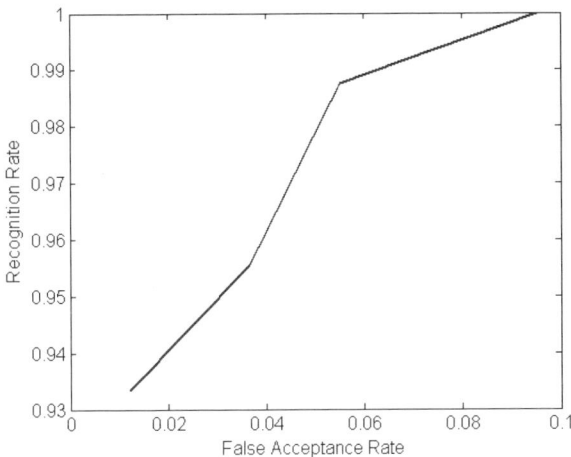

Fig. 12. System performance evaluation using ROC curve

6 Conclusion and Future Work

We proposed a personal identification system using palm vein biometrics to detect the liveness of a person. A low cost CCD camera and a set of infrared light source are used to capture the infrared palm images. A subimage is extracted by locating ROI in terms of image registration. The vein within ROI is used as biometric features to do recognition. The experimental results demonstrate that the recognition rate of our system is fine but not good enough to be a real system.

At present, our capture device is very sensitive to the outside lights. The outside lights can affect the inside infrared light source so that some images have very poor quality. If the capture device can be improved, the system performance should be better. Further, our database is too small to be convincible. More data are required to be collected for the evaluation of our system.

References

1. Lin, C.L., Fan, K.C.: Biometric Verification Using Thermal Images of Palm-Dorsa Vein Patterns. IEEE Transactions on Circuits and Systems for Video Technology 14(2), 199–213 (2004)
2. Fan, K.C., Lin, C.L: The Using of Thermal Images of Palm-dorsa Vein-patterns for Biometric Verification. In: IEEE ICPR, IEEE Computer Society Press, Los Alamitos (2004)
3. Yan, K.W., Zhang, Z.Y., Zhuang, D.: Hand Vein Recognition Based on Multi Supplemental Features of Multi-Classifier Fusion Decision. In: IEEE ICMA, IEEE Computer Society Press, Los Alamitos (2006)
4. Wang, L., Graham, L.: Near- and Far- Infrared Imaging for Vein Pattern Biometrics. In: IEEE ICAVSS, IEEE Computer Society Press, Los Alamitos (2006)

5. Toh, K., Eng, A.H.L., Choo, Y.S., Cha, Y.L., Yau, W.Y., Low, K.S.: Identity Verification Through Palm Vein and Crease Texture. In: IEEE ICB, pp. 546–553. IEEE Computer Society Press, Los Alamitos (2006)
6. Zhang, D., Kong, W.K., You, J., Wong, M.: Online Palmprint Identification. IEEE Trans. on Pattern Analysis and Machine Intelligence 25(9), 1041–1050 (2003)
7. Chaudhuri, S., Chatterjee, S., Katz, N., Nelson, M., Goldbaum, M.: Detection of blood vessels in retinal images using two-dimensional matched filters. IEEE Trans. on Medical Imaging 8, 263–269 (1989)
8. Hoover, A., Kouznetsova, V., Goldbaum, M.: Locating blood vessels in retinal images by piecewise threshold probing of a matched filter response. IEEE Trans. on Medical Imaging 19(3), 203–210 (2000)
9. Bao, P., Zhang, L.: Noise Reduction for Magnetic Resonance Image via Adaptive Multiscale Products Thresholding. IEEE Trans. on Medical Imaging 22, 1089–1099 (2003)
10. Bao, P., Zhang, L., Wu, X.L.: Canny Edge Detection Enhancement by Scale Multiplication. IEEE Trans. Pattern Analysis and Machine Intelligence 27(9) (2005)
11. Li, Q., You, J., Zhang, L., Zhang, D., Bhattacharya, P.: A New Approach to Automated Retinal Vessel Segmentation Using Multiscale Analysis. In: IEEE ICPR, IEEE Computer Society Press, Los Alamitos (2006)
12. Mallat, S., Zhong, S.: Characterization of signals from multiscale edges. IEEE Trans. Pattern Analysis and Machine Intelligence 14, 710–732 (1992)
13. Wu, X., Zhang, D., Wang, K.: Palm Line Extraction and Matching for Personal Authentication. IEEE Transactions On Systems, Man, and Cybernetics, Part A 36(5), 978–987 (2006)

A SVM Face Recognition Method Based on Optimized Gabor Features

Linlin Shen[1], Li Bai[2], and Zhen Ji[1]

[1] Faculty of Information & Engineering, Shenzhen University, China, 518060
{llshen, jizhen}@szu.edu.cn
[2] School of Computer Science & Information Technology, University of Nottingham,
UK, NG8 1BB,
bai@cs.nott.ac.uk

Abstract. A novel Support Vector Machine (SVM) face recognition method using optimized Gabor features is presented in this paper. 200 Gabor features are first selected by a boosting algorithm, which are then combined with SVM to build a two-class based face recognition system. While computation and memory cost of the Gabor feature extraction process has been significantly reduced, our method has achieved the same accuracy as a Gabor feature and Linear Discriminant Analysis (LDA) based multi-class system.

Keywords: Gabor features, Support Vector Machine, Linear Discriminant Analysis.

1 Introduction

Automatic recognition of human faces has been an active research area in recent years because it is user-friendly and unintrusive and it does not require elabrated collaboration of the users, unlike fingerprint or iris recognition. In addition to the importance of advancing research, it has a number of commercial and law-enforcement applications such as surveillance, security, telecommunications and human-computer intelligent interaction.

Gabor wavelet based recognition algorithms have been shown to be advantageous over many other methods in the literature. For example, the Elastic Bunch Graph Matching (EBGM) algorithm has shown very competitive performance and was ranked the top performer in the FERET evaluation [1]. In a recent face verification competition (FVC2004), both of the top two methods used Gabor wavelets for feature extraction. The application of Gabor wavelets for face recognition was pioneered by Lades et al.'s work since Dynamic Link Architecture (DLA) was proposed in 1993 [2]. In this system, faces are represented by a rectangular graph with local features extracted at the nodes using Gabor wavelets, referred to as Gabor jets. Wiskott et al [3] extended DLA to EBGM, where graph nodes are located at a number of facial landmarks. Since then, a large number of elastic graph based methods have been proposed [4-7]. Chung et al. [8] use the Gabor wavelet responses over a set of 12 fiducial points as input to a Principal Componet Analysis (PCA) algorithm, yielding a feature vector of 480 components. They claim to have improved the recognition rate up to 19% with this method compared to that by a raw PCA. All of these methods can be classified as analytic approaches since the local features extracted from selected points in faces are used for recognition. Recently, Gabor wavelets have also been applied in global form for face recognition [9, 10]. Liu et al. [9] vectorize the Gabor responses and then apply a downsampling by a factor of 64 to reduce the computation cost of the following subspace training. Their Gabor-based enhanced Fisher linear discriminant model outperforms Gabor PCA and Gabor fisherfaces. These holistic methods normally use the whole image after Gabor wavelets processing for feature representation. A more detailed survey on Gabor wavelet based face recognition methods can be found in [11].

Despite the success of Gabor wavelets based face recognition systems, the huge dimension of Gabor features extracted using a set of Gabor wavelets demands large computation and memory costs, which makes them impractical for real applications [11]. For the same reason, Support Vector Machine (SVM) has also seldom been applied to face recognition using Gabor features. Some works in the literature have tried to tackle this problem by (1) downsampling the images [12], (2) considering the Gabor responses over a reduced number of points [8], or (3) downsampling the convolution results [9, 10]. Strategies (2) and (3) have also been applied together [13]. However, these methods suffer from a loss of information because of the downsampling, or dimension reduction. Furthermore, the feature dimension after downsampling might still be too large for the fast training of SVM. Our works [14] have also shown that facial landmarks like eyes, nose and mouth might not be the optimal locations to extract Gabor features for face recognition.

In this paper, we propose a general SVM face recognition framework using optimized Gabor features. The most signifiant positions for extracting features for face recognition are first learned using a boosting algorithm, where the optimized Gabor responses are computed and used to train a two-class based SVM for identification. Since only the most important features are used, the two-class SVM based identification algorithm is both efficient and robust.

2 Gabor Wavelets and Feature Extraction

In the spatial domain, the 2D Gabor wavelet is a Gaussian kernel modulated by a sinusoidal plane wave [11]:

$$\begin{aligned} g(x, y) &= w(x, y)s(x, y) = e^{-(\alpha^2 x'^2 + \beta^2 y'^2)} e^{j2\pi f x'} \\ x' &= x\cos\theta + y\sin\theta \\ y' &= -x\sin\theta + y\cos\theta \end{aligned} \quad (1)$$

where f is the central frequency of the sinusoidal plane wave, θ is the anti-clockwise rotation of the Gaussian and the plane wave, α is the sharpness of the Gaussian along the major axis parallel to the wave, and β is the sharpness of the Gaussian minor axis perpendicular to the wave. To keep the ratio between frequency and sharpness constant, $\gamma = \frac{f}{\alpha}$ and $\eta = \frac{f}{\beta}$ are defined and the Gabor wavelets can now be rewritten as:

$$\varphi(x, y) = \frac{f^2}{\pi\gamma\eta} g(x, y) = \frac{f^2}{\pi\gamma\eta} e^{-(\alpha^2 x'^2 + \beta^2 y'^2)} e^{j2\pi f x'} \quad (2)$$

Fig.1 shows four Gabor wavelets with different parameters in both spatial and frequency domain.

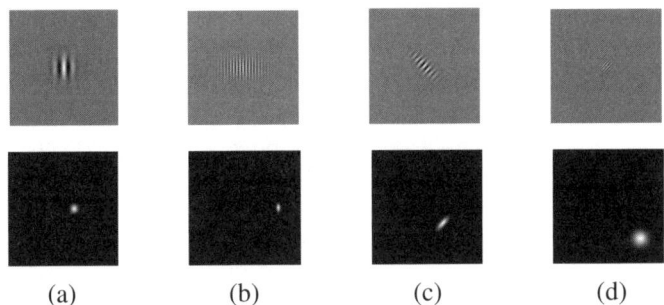

(a)   (b)   (c)   (d)

Fig. 1. Gabor filters with different parameters $\Pi(f, \theta, \gamma, \eta)$ in spatial domain (the 1st row) and frequency domain (the 2nd row), (a) $\Pi_a(0.1, 0, 1, 1)$; (b) $\Pi_b(0.3, 0, 6, 3)$; (c) $\Pi_c(0.2, 3\pi/4, 3, 1)$; (d) $\Pi_d(0.4, 3\pi/4, 2, 2)$

Once a set of Gabor wavelets have been designed, image features at different locations, frequencies and orientations can be extracted by convolving the image $I(x, y)$ with the filters:

$$O_{\Pi(f, \theta, \gamma, \eta)}(x, y) = I * \varphi_{\Pi(f, \theta, \gamma, \eta)}(x, y) \quad (3)$$

The number of scales and orientations may vary in different systems. We use in this paper a wavelet bank with 5 scales and 8 orientations to extract image features:

$$f_u = \frac{f_{max}}{\sqrt{2}^u}, u = 0, \ldots, 4 \qquad \theta_v = \frac{v}{8}\pi, v = 0, \ldots, 7 \quad (4)$$

The results S are thus the convolutions of an input image $I(x, y)$ with all of the 40 wavelets:

$$S = \{O_{u,v}(x, y) | u \in \{0, \ldots, 4\}, v \in \{0, \ldots, 7\}\} \quad (5)$$

where $O_{u,v}(x, y) = \|I * \varphi_{\Pi(f_u, \theta_v)}(x, y)\|$.

When the convolution results $O_{u,v}(x, y)$ over each pixel of the image are concatenated to form an augmented feature vector, the size of the vector could be very large. Take an image of 24×24 for example, the convolution result will give

24×24×5×8=23,040 features. To make SVM applicable to such a large feature vector, Qin and He [13] reduced the size of feature vector by including only the convolution reults over 87 manually marked landmarks. However, locating the 87 landmarks itself is a difficult problem, and the manually selected positions might not be the optimal ones for face recognition. Furthermore, wavelets with the same parameters are used at different landmarks, which is not the optimal way to feature extraction. In this paper, a boosting based feature selection process is used to choose the most useful features, which are then given as input to SVM to learn an efficient and robust face identification system.

3 The OG-SVM Classifier

Ever since its invention, SVM has been widely applied in classification and pattern recognition. One of the main reasons for the widespread applications of SVM is that its decision function is only based on the dot product of the input feature vector with the Support Vectors (SVs) [15], i.e. it has no requirements on the dimension of the feature vector. Theoretically features with any dimension can be fed into SVM for training. However in practical implementation, features with large dimension, e.g. Gabor features, could bring substantial computation and memory cost to the SVM training and classification process. In our experiments, the SVM training process did not even complete after 74 hours when a set of Gabor features of dimension 23,040 was used, due to the large computation and memory costs.

To make the SVM classifier both efficient and accurate, we propose to use optimized Gabor features for classification. As shown in Fig.2, the system starts with the Gabor feature extraction, as described in section 2. The extracted Gabor features and associated class labels for all of the training samples are then fed into the boosting algorithm to eliminate those non-discriminative features, which are not significant for classification. Once the most important positions with tuned Gabor wavelets are identified, the optimized Gabor features can be extracted and used to train the classifier, namely, the OG-SVM classifier. Using the optimized features, the boosting algorithm also learned a reasonably good classifier - Boosted Classifier (BC). However, the nonlinear OG-SVM classifier achieved further improvement on classification accuracy, with similar efficiency.

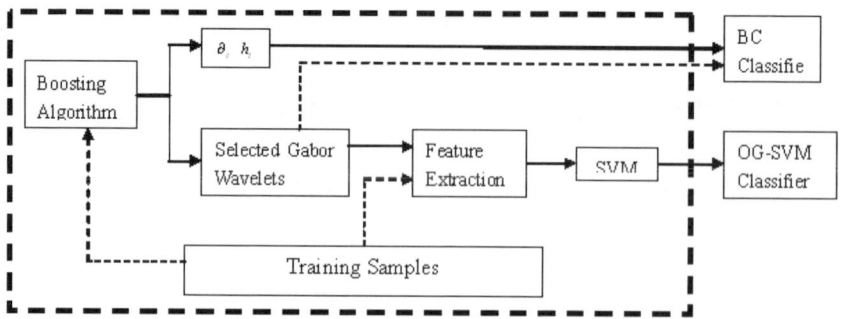

Fig. 2. Learning process of the proposed OG-SVM classifier

3.1 Boosting Based Gabor Feature Selection

Introduced by Freud and Schapire [16], boosting algorithms have been used successfully for selecting Haar-like features for general object detection [17, 18]. The essence of boosting algorithms is to select a number of 'weak' classifiers, which are then linearly combined into a single strong classifier. The algorithm operates as follows: for a two-class problem, m labelled training samples are given as $(x_i, y_i), i = 1,2,...,m$, where $y_i \in \{-1,1\}$ is the class label associated with sample $x_i \in R^N$. A large number of weak classifiers $h: R^N \to \{-1,1\}$ can be generated to form a weak classifier pool for training. In each of the iterations, the space of all possible weak classifiers is searched exhaustively to find the one that contributed the least to the overall classification error. The error is then used to update the weights associated with each sample such that the wrongly classified samples have their weights increased. The algorithm thus focuses on difficult training samples, increasing their representation in successive training sets. When a weak classifier is designed to use only a single feature to make decisions, boosting is equivalent to feature selection.

To apply the boosting algorithm to Gabor feature selection, we simplify the task of a multi-class face recognition problem to a two-class problem: selecting Gabor features that are effective for intra- and extra-person space discrimination. Such selected Gabor features should be robust for face recognition, as intra- and extra-person space discrimination is one of the major difficulties in face recognition. Two spaces, intra- and extra-person spaces are defined, with intra-person space measuring respectively dissimilarities between faces of the same person and extra-person space dissimilarities between different people. For a training set with L facial images captured for each of the K persons to be identified, $K\binom{L}{2}$ samples could be generated for class *Intra* while $\binom{KL}{2} - K\binom{L}{2}$ samples are available for class *Extra*. More details about the Gabor feature selection process can be found at [14].

Upon completion of T boosting iterations, T weak classifiers are selected to form the final strong classifier. The resulting strong classifier $H(x) = sign\left(\sum_{t=1}^{T} \alpha_t h_t(x)\right)$, called BC in this paper, is a weighted linear combination of all the selected weak classifiers, with each weak classifier using certain Gabor feature for decision. At the same time, T most significant Gabor features for face recognition have also been identified.

3.2 Support Vector Machine

Once the optimized features are selected, they can be given to SVM for classifier training. Based on an observed feature $x \in R^N$, SVM is basically a linear hyperplane classifier $f(x) = \langle w, x \rangle + b$ aimed at solving the two class problem [19]. As shown in

Fig.3a, the classifier can separate the data from two classes very well when the data is linearly separable. Since there might be a number of such linear classifiers available, SVM chooses the one with the maximal margin, which is defined as the width that the boundary could be increased by before hitting a data point. The distance between the two thin lines (boundary) in the figure thus defines the margin of the linear SVM with data points on the boundary known as Support Vectors (SV). The linear classifier $f(x)$ with maximized margin can be found using quadratic problem (QP) optimization techniques as below:

$$f(x) = sign\left(\sum \alpha_k y_k \langle x_k, x \rangle + b\right) \tag{6}$$

where $x_k \in R^N$ are the support vectors learned by SVM.

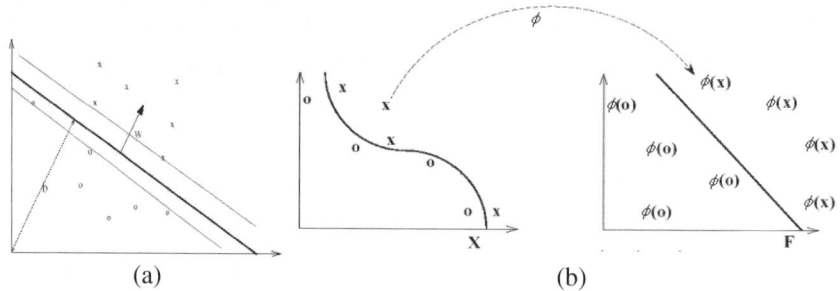

Fig. 3. A hyperplane classifier in 2-dimension feature space (a and mapping of the data (b

For non-linearly separable data, a nonlinear mapping function $\phi : R^N \to F, x \to \phi(x)$ is used to map it into a higher dimension feature space where a linear classifier can be applied. Fig.3b shows an example using the kernel method to train a non-linear SVM. Using the kernel trick [15], the non-linear SVM is now:

$$f(x) = sign\left(\sum \alpha_k y_k k(x_k, x) + b\right) \tag{7}$$

where $k(\mathbf{x}_k, \mathbf{x})$ is a kernel function, e.g., a polynomial kernel and a RBF kernel etc.

3.3 Identification

As shown in Fig.2., once the boosting iterations and the SVM learning process are completed, two classifiers, i.e. BC and OG-SVM, are created using the T selected Gabor features. Though trained to discriminate intra-person and extra-person spaces, they could also be used for recognition (identification) as follows: given a gallery $\{q_j\}$ of m known individuals and a probe p to be identified, both classifiers will first compute the Gabor feature differences $\{x_j = [d_1 \cdots d_t \cdots d_T]\}$ between the probe and each of the gallery images, and then calculate an intra-person confidence score using respective decision functions:

$$\delta_j = \begin{cases} \sum_{t=1}^{T} \alpha_t h_t(x_j), & BC \\ \sum_k a_k y_k k(x_k, x_j) + b, & SVM \end{cases} \qquad (8)$$

the probe is then identified as person j that gives the maximum confidence score δ_j.

4 Experimental Results

4.1 The Database

The FERET database is used to evaluate the performance of the proposed method for face recognition. The database consists of 14051 eight-bit grayscale images of human heads with views ranging from frontal to left and right profiles. 600 frontal face images corresponding to 200 subjects are extracted from the database for the experiments - each subject has three images of 256×384 with 256 gray levels. The images were captured at different times under different illumination conditions and contain various facial expressions. Two images of each subject are randomly chosen for training, and the remaining one is used for testing. The following procedures were applied to normalize the face images prior to the experiments:

- each image is rotated and scaled to align the centers of the eyes,
- each face image is cropped to the size of 64×64 to extract facial region,
- each cropped face image is normalized to zero mean and unit variance.

4.2 The Results

In this experiment, classification and recognition performance of the proposed two-class classifier, OG-SVM, will be tested and evaluated against that of BC and other methods, e.g. Principal Component Analysis (PCA) and Linear Discriminant Analysis (LDA). Gabor features are first selected by boosting algorithm using the training set, and then used to train BC and OG-SVM (see Fig.2. for the process). The training set thus consists of 200 intra-person difference samples and 1,600 extra-person difference samples.

Since both BC and OG-SVM are trained to discriminate intra-person and extra-person differences, we first evaluate their classification performances on the training set. Fig.4 shows the classification error of BC and OG-SVM with different kernel functions, which are computed as the ratio between the number of wrongly classified difference samples and the number of training samples. One can observe from the figure that the performances of both classifiers improve when the number of features increases. However, the performance of OG-SVM is much more stable than BC. While OG-SVM with RBF kernel achieves the lowest classification error rate (0.44%) when 140 features are used, OG-SVM with linear kernel shows similar performance.

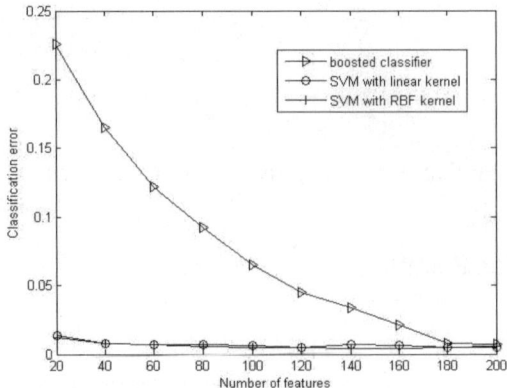

Fig. 4. Classification performances of OG-SVM and BC

The classifiers are then applied to the test set (200 images, 1 image per person) for face identification and their performances are shown in Fig.5. Similarly, OG-SVM achieves higher recognition rate than BC when different number of features are used. The highest recognition accuracy of 92% is achieved by OG-SVM with linear kernel when 120 Gabor features are used. The results also suggest that the difference of OG-SVM using RBF kernel and linear kernel is quite small, when the features selected by boosting algorithm are considered.

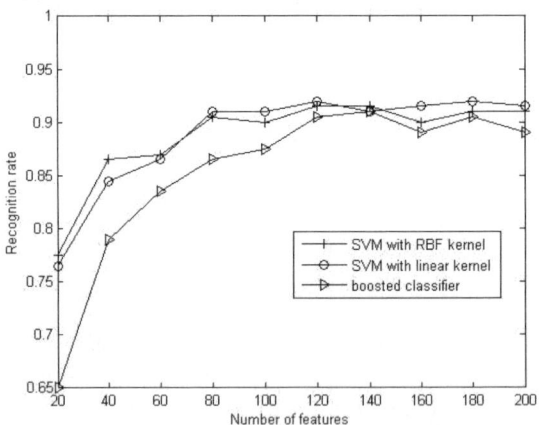

Fig. 5. Recognition performances of OG-SVM and BC

To show the efficiency and accuracy of the proposed method, we also compare its performance with other Gabor feature based approaches in Table 1. While PCA and LDA are also well known as Eigenface and Fisherface methods, details of Downsmaple Gabor + PCA and Downsample Gabor + LDA can be found in [10]. In the implementation, downsampling with rate 16 was used to reduce the dimension of extracted Gabor features before they are input to PCA, or LDA for further processing.

The table shows that the proposed OG-SVM achieved similar accuracy with Downsmaple Gabor + LDA, but with much fewer feature dimension and much less feature extraction costs. In our experiments (a normal PC with P4 3.0 GHz CPU), while it takes 100ms to train the OG-SVM classifier, the system can averagely identify 50 faces per second.

Table 1. Accuracy and efficiency of OG-SVM

Methods	Recognition Rate	No. of Convolutions for Gabor Feature Extraction	Dimension of Features
PCA	60%	N/A	64×64=4096
LDA	76%	N/A	64×64=4096
Downsample Gabor + PCA	80%	64×64×40=163,840	10,240
Downsample Gabor + LDA	**92%**	64×64×40=163,840	10,240
BC	90%	120	120
OG-SVM	**92%**	120	120

5 Conclusions

We have proposed in this paper a novel SVM face recognition method based on optimized Gabor features. While some methods in the literature consider the responses at landmark points only, our method uses a boosting algorithm to find the most significant positions and wavelet to extract features for face recognition. The features thus extracted are efficient. While downsampling could be used to reduce the dimension of features before they are fed into PCA, or LDA for further processing, it could introduce loss of important information. Furthermore, complex feature extraction process has to be used to extract high dimensional features before downsampling. By combining boosting selected Gabor features with SVM, our method not only substantially reduces computation and memory cost of the feature extraction process, but also achieves the same performance as that of Downsample Gabor + LDA, when FERET database is used for testing.

Acknowledgement. Research funded by SZU R/D Fund 200746.

References

1. Phillips, P.J., Moon, H., Rizvi, S.A., Rauss, P.J.: The FERET evaluation methodology for face-recognition algorithms. IEEE Transactions on Pattern Analysis and Machine Intelligence 22, 1090–1104 (2000)
2. Lades, M., Vorbruggen, J.C., Buhmann, J., Lange, J., Von der Malsburg, C., Wurtz, R.P., Konen, W.: Distortion invariant object recognition in the Dynamic Link Architecture. IEEE Transactions on Computers 42, 300–311 (1993)
3. Wiskott, L., Fellous, J.M., Kruger, N., von der Malsburg, C.: Face recognition by elastic bunch graph matching. IEEE Transactions on Pattern Analysis and Machine Intelligence 19, 775–779 (1997)

4. Mu, X.Y., Hassoun, M.H.: Combining Gabor features: summing vs.voting in human face recognition. In: 2003 IEEE International Conference on Systems, Man and Cybernetics, p. 737. IEEE Computer Society Press, Los Alamitos (2003)
5. Duc, B., Fischer, S., Bigun, J.: Face authentication with Gabor information on deformable graphs. IEEE Transactions on Image Processing 8, 504–516 (1999)
6. Jiao, F., Gao, W., Chen, X., Cui, G., Shan, S.: A face recognition method based on local feature analysis. In: Proc. of the 5th Asian Conference on Computer Vision, pp. 188–192 (2002)
7. Liao, R., Li, S.: Face recognition based on multiple facial features. In: Proc. of the 4th IEEE Int. Conf. on Automatic Face and Gesture Recognition, pp. 239–244. IEEE Computer Society Press, Los Alamitos (2000)
8. Chung, K.C., Kee, S.C., Kim, S.R.: Face recognition using principal component analysis of Gabor filter responses. In: International Workshop on Recognition, Analysis, and Tracking of Faces and Gestures in Real-Time Systems, Corfu, Greece, pp. 53–57 (1999)
9. Liu, C.J., Wechsler, H.: Gabor feature based classification using the enhanced Fisher linear discriminant model for face recognition. IEEE Transactions on Image Processing 11, 467–476 (2002)
10. Shen, L., Bai, L., Fairhurst, M.: Gabor wavelets and General Discriminant Analysis for face identification and verification. Image and Vision Computing 25, 553–563 (2007)
11. Shen, L., Bai, L.: A review on Gabor wavelets for face recognition. Pattern Analysis and Applications 9, 273–292 (2006)
12. Zhang, W., Shan, S., Gao, W., Chang, Y.Z., Cao, B., Yang, P.: Information fusion in face identification. In: Proceedings of the 17th International Conference on Pattern Recognition, vol. 3, pp. 950–953 (2004)
13. Qin, J., He, Z.-S.: A SVM face recognition method based on Gabor-featured key points. In: Proceedings of the 4th International Conference on Machine Learning and Cybernetics, vol. 8, pp. 5144–5149 (2005)
14. Shen, L., Bai, L.: MutualBoost learning for selecting Gabor features for face recognition. Pattern Recognition Letters 27, 1758–1767 (2006)
15. Cristianini, N., Shawe-Taylor, J.: An introduction to Support Vector Machines and Other Kernel-based Learning Methods. Cambridge University Press, Cambridge (2000)
16. Freund, Y., Schapire, R.: A short introduction to boosting. Journal of Japanese Society for Artifical Intelligence 14, 771–780 (1999)
17. Lienhart, R., Maydt, J.: An extended set of Haar-like features for rapid object detection. In: Proc. IEEE Conference on Image Processing, pp. 900–903. IEEE Computer Society Press, Los Alamitos (2002)
18. Viola, P., Jones, M.: Rapid object detection using a boosted cascade of simple features. In: Proc. of IEEE Conf. on Computer Vision and Pattern Recognition, Kauai, Hawaii, pp. 511–518. IEEE Computer Society Press, Los Alamitos (2001)
19. Burges, C.J.C.: A tutorial on support vector machines for pattern recognition. Data Mining and Knowledge Discovery 2, 121–167 (1998)

Palmprint Identification Using Pairwise Relative Angle and EMD

Fang Li, Maylor K.H. Leung, and Shirley Z.W. Tan

School of Computer Engineering, Nanyang Technological University, Singapore 639798
asfli@ntu.edu.sg

Abstract. This paper presents an efficient matching algorithm for the palmprint identification system. Line segments are extracted from an image as primitives. Each local structure is represented by a set of pair-wise angle relationships, which are simple, invariant to translation and rotation, robust to end-point erosion, segment error, and sufficient for discrimination. The Earth Mover's Distance (EMD) was proposed to match the pairwise relative angle histograms. EMD not only supports partial matching but it also establishes a neighbouring relationship in the lines information during the matching process. The system employs low-resolution palmprint images captured by normal digital camera and achieves higher identification accuracy with lower time complexity.

Keywords: line, palmprint, Earth Mover's Distance, angle, pairwise.

1 Introduction

It is commonly known that thumbprints can be used for identification purposes, due to its important property of uniqueness. In fact, bare footprints and palmprints are unique to the individual in just the same way as fingerprints. Palmprint, a new biometric, is regarded as one of the most unique, reliable and stable personal characteristics. Therefore, according to resource [1], even the UK Police Information Technology Organisation (PITO) had completed the roll-out of the palm searching capability to all police forces in England and Wales. Besides implementing palmprint matching as a forensic or security authentication tool, palmprint matching also has other areas of application, such as an Electronic Palmistry System.

A palmprint mainly composed of lines and ridges. Researchers are working on palmprint matching systems based on texture features and line features [2-4]. There are two well known techniques for extracting palmprint features. The first approach is based on the palmprint structural features while the other is on statistical features [5].

To extract structural information, like principal lines and creases, from the palm for recognition, it is a natural way to represent the feature of palmprint using line segments. Psychological studies indicate that human recognizes line drawings as quickly and almost as accurately as images [6]. These results implicate that line edge maps extracted from images can be used for object recognition or matching. In [7], the Euclidean distance between the endpoints of two line segments, the difference of angle of inclination and intercepts are used to determine the similarity between these two line segments according to three thresholds. The method is simple in concept. However, besides the thresholds determination issue, perfect alignment is required. [8] extracted a set of feature points along the prominent palm lines together with the associated line orientations. The feature points/orientation from model image and input image are matched using point matching technique. In the study, noise can affect the discriminant power because the feature point connectivity is not extracted, i.e., points on one line may match to points on multiple lines instead of one.

The statistics-based approach transforms palmprint features into other transformation domains for representation. The works that appear in the literature include eigenpalm [2][5][9], fisherpalms [5][10][11], Gabor filters [12], Fourier Transforms [13], and local texture energy [4][14]. Among statistics-based approaches, the features are extracted, represented, and compared based on the whole palmprint. The main limitation of these approaches is that the employed signatures are global measurements, and the signatures of some palmprints are very similar though they look different from human viewpoint. On the other hand, structural approaches can represent the palmprint structural features in details. However, during the extraction process of palmprint, there will always be a problem of missing or broken lines which causes difficulty in the matching process.

In this paper, we propose an algorithm capable of solving the broken line problem by introducing pairwise relative angle difference with subset division. Line is adopted as primitives to construct local structures. The inherent structural relationship between a pair of primitives will then be extracted for representation. The proposed algorithm generates histograms based on subset. The similarity of histograms is calculated using the Earth Mover's Distance (EMD). EMD not only supports partial matching but it also establishes a neighboring relationship in the lines information during the matching process. The final similarity between two images is achieved using Hausdorff distance [15]. The idea to use pairwise relative angle difference is a new research topic on palmprint matching and subset division is introduced to reinforce the algorithm. Combination of these two algorithms highly improves the matching accuracy and enables this palmprint identification system to be rotational and translational invariant. The system achieves the accuracy of 100% while using the palmprint database collected by the Hong Kong Polytechnic University [16].

The rest of this paper is organized as follows: Section 2 introduces the structure of the proposed matching system. Proposed pairwise relative angle histogram based on subset, EMD, and Hausdorff distance are discussed in details in Section 3. Section 4 presents the results. Finally, the conclusion and future work are highlighted in Section 5.

2 System Overview

Fig. 1 shows the processes of a standard palmprint matching system. Fig. 2 shows the flow chart of our proposed system. In the preprocessing module, processes such as line detection, image thresholding, thinning, contour extraction, and polygonal approximation are applied to extract Line Edge Map (LEM) [17] from a palm. Feature extraction and transformation process is described in details in [17]. The matching method is the proposed relative angle difference histogram based Hausdorff distance.

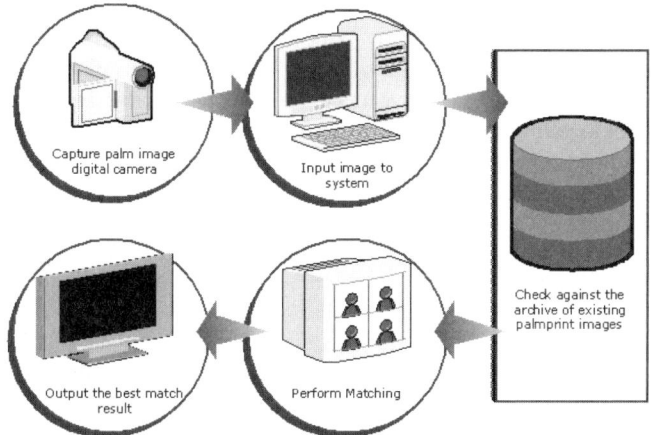

Fig. 1. Processes of standard palmprint matching system

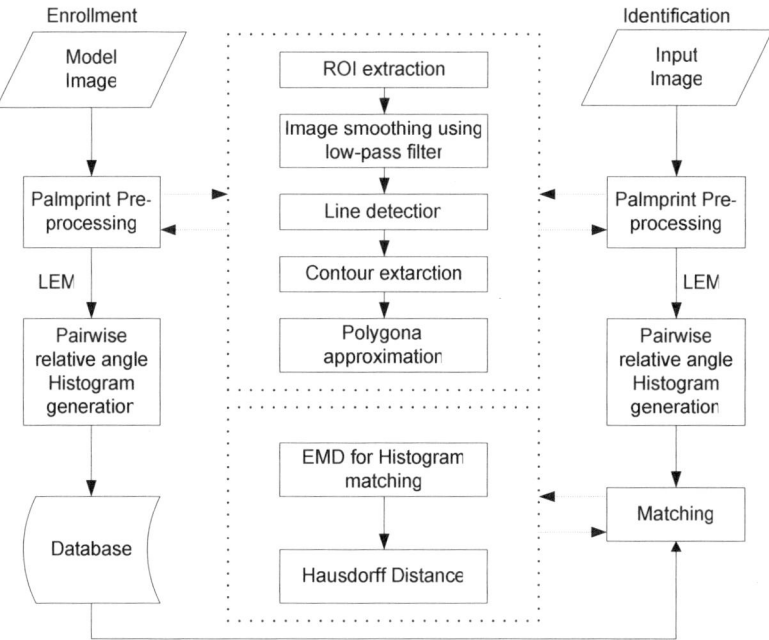

Fig. 2. Block diagram of the proposed palmprint identification system

3 Proposed Palmprint Matching Scheme

3.1 Feature Selection

Palmprint identification usually has difficulty in achieving 100% matching using line feature due to some factors such as poor quality of palmprints captured due to environments and background lighting, and alignment of input palmprints. Poor quality issue will increase the occurrence of broken and missing lines which is a known problem to palmprint matching system.

To tackle broken line problem, a few pairwise geometric attributes are investigated, such as the relative orientation between the lines, the ratio of line-segment length, the ratio of segment end-point distances, and the line-segment projection cross-ratio [18]. All the four methods are equally efficient in retrieving the pairwise geometric attributes for good quality images. However, cases of broken line are very common during feature extraction process and will pose a major problem to other methods except relative orientation.

Fig. 3 demonstrates the calculation of pairwise geometric attribute, at the same time, shows the example of broken line problem. Line *ab* and *cd* belong to same subject. In good quality image shown in Figure 3a, *cd* is a single straight line segment. However, in poor quality image captured in another time shown in Figure 3b, line *cd* is broken into 4 line segments.

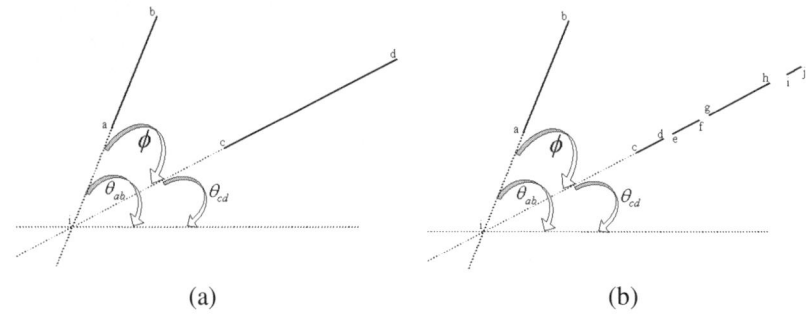

Fig. 3. The pairwise geometric attributes between line segments ab and cd

Fig. 4 is a good illustration on the handling of broken line problem. A palmprint is captured and processed twice with slight difference on the one figure with broken line display. With the use of relative angle difference, the algorithm handles the broken line problem well and from the histogram display in Fig. 4, the difference between them is very narrow and the algorithm is robust enough to produce accurate matching between the two highlighted lines. The use of Pairwise Relative Angle difference in palmprint identification is also a new proposed algorithm that has never been practice by any research group.

Another advantage for using pairwise relative angle difference is the flexibility of the input palmprints alignment due to the capability of rotational and translational invariant. The problem of perfect alignment or position of input palmprints will no longer exists.

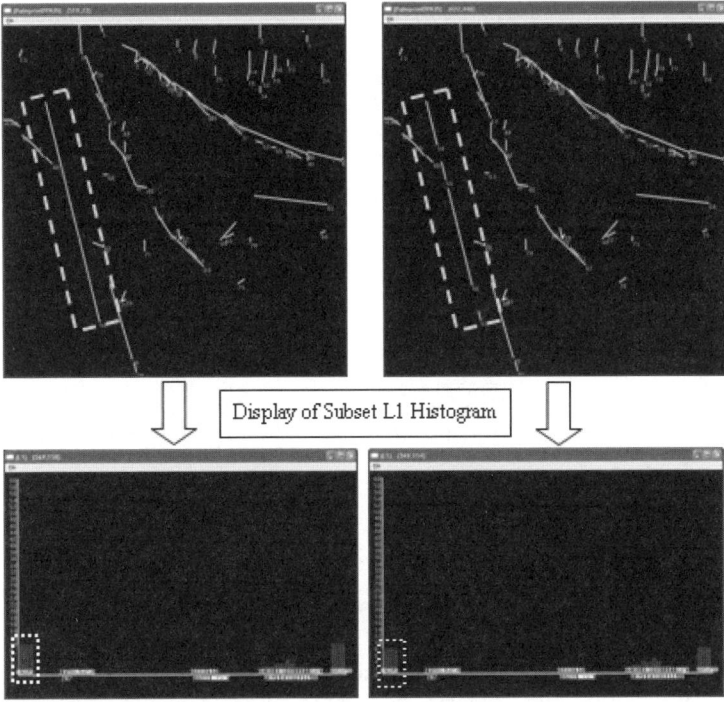

Fig. 4. Sample of solution to broken line problem

3.2 Histogram Matching

Angle offset is required due to the positioning of palmprints during the capturing process. During the stage of palmprint detection and extraction as discussed earlier, the positioning of a user palm will slightly affect the output results. If no distortion tolerance was introduced, the system will be very rigid and is only capable of matching palmprints with perfect alignment. Histogram matching methods are investigated to find a distortion tolerated one.

Simplest histogram matching is using the Euclidean Distance (ED). That is to compare the root-mean-square error between two distributions. The smaller the computed error, the closer the match is. However, this methodology is believed to be rigid because it is unable to utilize the information of relationships between the neighboring bins. Keeping these facts in mind, a flexible methodology – The Earth Mover's Distance (EMD) [19] was, hence, applied to this system to solve the bin leakage problem in pairwise relative angle histograms.

3.2.1 Histograms Versus Signatures

A histogram is used to graphically summarize and display the distribution of a process data set. It is constructed by segmenting the range of the data into equal sized bins. Data of a category allocates to the same bin. Categories without relevant data will remain as empty bins. Regardless of its bin allocation, a histogram presents its full scale binning. Thus, in cases when data set is sparsely distributed, the empty bins presented become redundant.

On the other hand, Signature is a lossless representation of its histogram in which the bins of the histogram that have value zero are not expressed implicitly [20]. Therefore, regardless of a sparse or dense distribution, a Signature presents its histogram in a more meaningful way, while retaining its initial distance relationship. Fig. 5 depicts the relationship between a histogram and a signature.

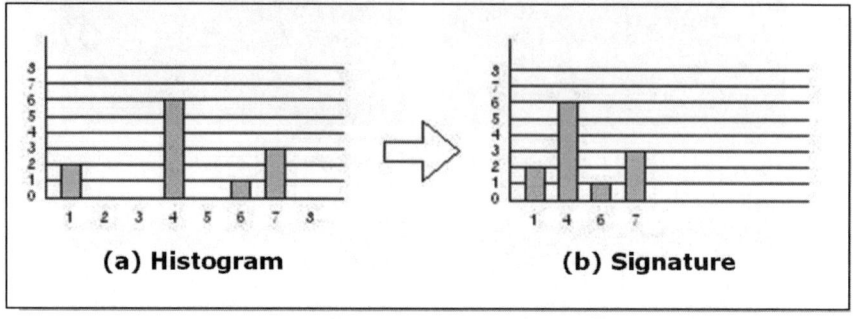

Fig. 5. Relationship between (a) Histogram and (b) Signature

3.2.2 Bin-by-Bin Versus Cross-Bin Dissimilarity Measures

Comparison between bins can be performed either by comparing exactly the corresponding bins or considering, as well, the non-corresponding bins. As the names depict, bin-by-bin comparison considers only the corresponding bin of the same index (category). This method is simple and straightforward but rigid in some sense as it takes no account of nearby bins' matching possibilities. Examples of bin-by-bin measure include Minkowski-form Distance, X2 Statistics, Kullback-Leibler Divergence, Jeffrey Divergence and Histogram Intersection (which is able to handle partial matches when areas of histograms are different) [21].

In converse, cross-bin comparison not only looks at the corresponding bins, it also contain terms that compare non-corresponding bins, compromised with a defined distance cost between the compared bins. Thus, cross-bin dissimilarity measure is more flexible but complex computationally than the former as the number of bins increases. Examples of cross-bin measure include Quadratic-form Distance, Match Distance and Parameter-based Distances.

Earth Mover's Distance (EMD) is the minimum cost incurred in transforming one signature into the other. Intuitively, given two distributions, the one with an equivalent or more amount of total weight can be deemed as piles of earth, while the other with an equivalent or less amount of total weight as a collection of holes. Therefore, there will always be sufficient amount of earth to fill all the holes to brim. Hence, EMD measures the least amount of work required in filling up all the holes with the available earth within the same space, having a unit of work corresponding to the transportation of a unit of earth by a unit of ground distance.

As shown in Fig. 6, let two discrete and finite distributions be denoted as $P = \{(p_1, w_{p1}), ..., (p_m, w_{pm})\}$, where P is the first signature with m bins indexed by p_i, and

$Q = \{(y_1, u_{q1}), ..., (y_n, u_{qn})\}$, having Q as the second signature with n bins indexed by q_j. Therefore, the ground distance between bins pi and q_j be connoted by $d_{ij} = d(p_i, q_j)$. The objective is to find a flow $F = [f_{ij}]$, with f_{ij} the flow between p_i and q_j, that minimizes the overall cost incurred in transforming P into Q [19].

$$\text{WORK}(P, Q, F) = \sum_{i=1}^{m} \sum_{j=1}^{n} d_{ij} f_{ij}$$

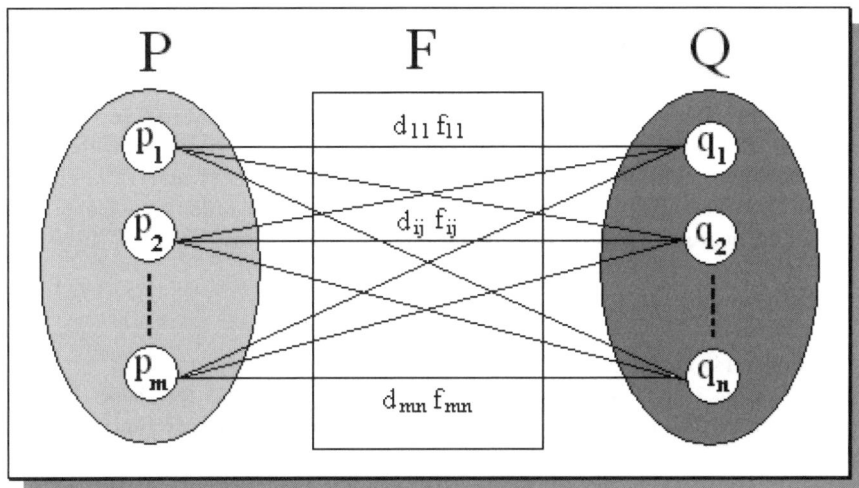

Fig. 6. EMD Definition Example

Partial EMD (EMD^γ) & Restricted EMD (τ-EMD)
A few extensions [22] were developed for the EMD and, two of which, will be briefed in this section. These include the partial EMD (EMD^γ) and the restricted EMD (τ-EMD).

To begin with, the EMD^γ also further categorizes into two types of partial matches, namely relative partial EMD and absolute partial EMD. The former takes a given fraction of the weight of the lighter distribution, while the latter uses an absolute amount of weight of the lighter distribution. Both make use of a partial (γ) of the weight of the lighter distribution, thus termed as the Partial EMD.

In contrast, the τ-EMD is a measure of the amount of feasible matching weight when ground distances for transportation are limited to a threshold τ. In other words, τ-EMD measures the maximum amount of weight that can be matched at transportation distances within τ when comparing two distributions.

Since the current Palmprint Processing System extracts line information from the entire palm image and compares the unbalanced weight distributions, the implementation of EMD to the project would be more appropriate as EMD^γ, restricting the maximum amount of matching weight to that of the lighter distribution.

Modulo-distance cost

The focus of this project was to perform matching between two distributions. In performing EMD matching, the corresponding cost between each bin had to be known. In this system, the existing histograms were sampled at a minimum sample of one degree, therefore ranging from 0° (180°) to 179°. Thus, there would be a total of 180 bins.

Sampled segments of principle lines would be allocated to the respective bins of the same turning degree. Hence, degrees that did not exist on the principle lines would remain as empty bins. With such data allocation, regardless of the number of lines extracted, the distribution would be limited to 180 bins, with a fixed sample of one degree. Therefore, the sampling interval would be taken as the unit cost of shifting earth to hole, one degree of bin difference constitutes to one unit cost of distance.

Nevertheless, it could be observed that the first (bin 0) and last bin (bin 179) had a distance of 179 − 0 = 179 units. However, from an angle's perception, the actual difference between 0° and 179° would only be 180° - 179° = 1° (1 unit). Therefore, in this project, modulo-distance cost measure was proposed in order to achieve better exploit of the nearness property of neighboring bins in EMD matching.

3.3 Image Matching

3.3.1 Hausdorff Distance

Hausdorff distance (HD) is one of the commonly used measures for shape matching and has been widely investigated [15]. It is a distance defined between two sets of points. Unlike most shape comparison methods that build a one-to-one correspondence between a model and a test image, HD can be calculated without explicit point correspondence. Hausdorff Distance (HD) is the "maximum distance of a set to the nearest point in the other set" as

$$h(A, B) = \max_{a \in A} \min_{b \in B} \| a - b \| \qquad (2)$$

where a and b are points of sets A and B respectively, where is any norm, e.g. Euclidean distance. It should be noted that Hausdorff Distance is oriented (we could say asymmetric as well). A more general definition of Hausdorff Distance would be:

$$H(A, B) = \max(h(A, B), h(B, A)) \qquad (3)$$

which defines the Hausdorff Distance between A and B, while eq.2 computes Hausdorff Distance from A to B (also called directed Hausdorff Distance). HD is one of the commonly used methods in image processing and matching applications. It measures two images' dissimilarity without explicit point correspondence.

3.3.2 Line Segment Hausdorff Distance Using Pairwise Relative Angle Histgram (LHD-h)

The original HD is a distance defined between two sets of points [15]. Proposed LHD-h extends the concept to two sets of line segments. Applying the concept on the line edge map (LEM) of an image, each line set captures the complete feature vector set of one image. Hence, we can measure two images' dissimilarity based on pairwise relative angle relationship among the reference line and neighbours using LHD-h.

4 Experiment Results

Experiments were carried out using a palmprint database from the Biometric Research Centre, The Hong Kong Polytechnic University. The palmprint images were collected from 100 individuals (6 images for each person) [16] using a special palmprint capture device. The subjects mainly consisted of volunteers from students and staffs at the Hong Kong Polytechnic University with a wide range of age distribution and different genders. 100% identification accuracy is achieved.

5 Conclusion and Future Works

The palmprint matching system has been successfully developed using the proposed algorithms of pairwise relative angle difference and subset division. In contrast to the traditional method based on local line and point feature extraction that is dependent on perfect alignment of palmprints, the system can now handle different alignment of input palmprints and is still capable to accurately identify matching palmprints due to the rotational invariant feature of the relative angle difference algorithm.

Through experiments, palmprint patterns can be very well extracted using their pairwise relative angle difference property and represented in form of histograms. Since pairwise relative angle difference algorithm is a new idea in the area of palmprint classification research and is capable of producing reliable results, future research group could build upon it further. Effort in reducing the computational complexity of this system is expected. Introducing indexing for this palmprints matching system is a promising solution. Indexing will enable the system to be scale invariant as well.

References

1. UK: Police Acquire National Palm-Print Database and Forensic Search Tool, http://ec.europa.eu/idabc/en/document/5489/5900
2. Ribaric, S., Fratric, I.: A biometric identification system based on eigenpalm and eigenfinger features. IEEE Transactions on Pattern Analysis and Machine Intelligence 27(11), 1698–1709 (2005)
3. You, J., Kong, W.K., Zhang, D., Cheung, K.H.: On hierarchical palmprint coding with multiple features for personal identification in large databases. IEEE Transactions on Circuits and Systems for Video Technology 14(2), 234–243 (2004)
4. Zhang, L., Zhang, D.: Characterization of palmprints by wavelet signatures via directional context modeling. IEEE Transactions on Systems, Man. And Cybernetics-Part B: Cybernetics 34(3), 1335–1347 (2004)
5. Connie, T., Jin, A.T.B., Ong, M.G.K., Ling, D.N.C.: An automated palmprint recognition system. Image and Vision Computing 23, 501–515 (2005)
6. Biederman, I., Gu, J.: Surface versus edge-based determinants of visual recognition. Cognitive Psychology 20, 38–64 (1988)
7. Shu, W., Rong, G., Bian, Z.Q., Zhang, D.: Automatic palmprint verification. International Journal of Image and Graphics 1(1), 135–151 (2001)

8. Duta, N., Jain, A.K.: Matching of palmprints. Pattern Recognition Letters 23, 477–485 (2002)
9. Lu, G.M., Zhang, D., Wang, K.Q.: Palmprint recognition using eigenpalms features. Pattern Recognition Letters 24, 1463–1467 (2003)
10. Connie, T., Teoh, A., Goh, M., Ngo, D.: PalmHashing: a novel approach for cancellable biometrics. Information Processing Letters 93, 1–5 (2005)
11. Wu, X.Q., Zhang, D., Wang, K.Q.: Fisherpalms based palmprint recognition. Pattern Recognition Letters 24, 2829–2838 (2003)
12. Kong, W.K., Zhang, D., Li, W.X.: Palmprint feture extraction using 2-D Gabor filters. Pattern Recognition 36, 2339–2347 (2003)
13. Li, W.X., Zhang, D., Xu, Z.Q.: Palmprint identification by Fourier transform. International Journal of Pattern Recognition and Artificial Intelligence 16(4), 417–432 (2002)
14. Dente, E., Bharath, A.A., Ng, J., Vrij, A., Mann, S., Bull, A.: Tracking hand and finger movements for behaviour analysis. Pattern Recognition Letters 27, 1797–1808 (2006)
15. Huttenlocher, D.P., Klandeman, G.A., Rucklidge, W.J.: Comparing images using the Hausdorff distance. IEEE Transactions on Pattern Analysis and Machine Intelligence 15(9), 850–863 (1993)
16. Palmprint database, Biometric Research Center, The Hong Kong Polytechnic University. Available: http://www4.comp.polyu.edu.hk/~biometrics/
17. Li, F., Leung, M.K.H.: A Two-level Matching Scheme for Speedy and Accurate Palmprint Identification. In: Cham, T.-J., Cai, J., Dorai, C., Rajan, D., Chua, T.-S., Chia, L.-T. (eds.) MMM 2007. LNCS, vol. 4352, pp. 323–332. Springer, Heidelberg (2006)
18. Heut, B., Hancock, E.R.: Line Pattern Retrieval Using Relational Histograms. IEEE Transactions on Pattern Analysis and Machine Intelligence 21 (1999)
19. Rubner, Y., Tomasi, C., Guibas, L.J.: A Metric for Distributions with Applications to Image Databases. Pattern Recognition, 59–61 (1998)
20. Serratosa, F., Sanfeliu, A.: Signature Versus Histograms: Definitions, Distances and Algorithms. Pattern Recognition, 923–925 (2005)
21. Zhao, Q.: Distance and Matching, www.soe.ucsc.edu/~taoswap/GroupMeeting/DistanceQi.ppt
22. Rubner, Y.: Perceptual Metrics for Image Database Navigation. Pattern Recognition, 39–58 (1999)

Finding Lips in Unconstrained Imagery for Improved Automatic Speech Recognition

Xiaozheng Jane Zhang, Higinio Ariel Montoya, and Brandon Crow

Department of Electrical Engineering
California Polytechnic State University, 1. Grand Ave.
San Luis Obispo, CA 93401, USA
{jzhang, hmontoya, bcrow}@calpoly.edu

Abstract. Lip movement of a speaker conveys important visual speech information and can be exploited for Automatic Speech Recognition. While previous research demonstrated that visual modality is a viable tool for identifying speech, the visual information has yet to become utilized in mainstream ASR systems. One obstacle is the difficulty in building a robust visual front end that tracks lips accurately in a real-world condition. In this paper we present our current progress in addressing the issue. We examine the use of color information in detecting the lip region and report our results on the statistical analysis and modeling of lip hue images by examining hundreds of manually extracted lip images obtained from several databases. In addition to hue color, we also explore spatial and edge information derived from intensity and saturation images to improve the robustness of the lip detection. Successful application of this algorithm is demonstrated over imagery collected in visually challenging environments.

Keywords: automatic speech recognition, visual speech recognition, lip segmentation, statistical modeling.

1 Introduction

Speech-based user interface allows a user to communicate with computers via voice instead of a mouse and keyboard. The use of speech interface in emerging multimedia applications is growing in popularity because it is more natural, easier, and safer to use. The key technology that permits the realization of a pervasive speech interface is automatic speech recognition (ASR). While ASR has witnessed significant progress in many well-defined applications, the performance of such systems degrades considerably in acoustically hostile environments such as in an automobile with background noise, or in a typical office environment with ringing telephones and noise from fans and human conversations. One way to overcome this limitation is to supplement the acoustic speech with visual signal that remains unaffected in noisy environment.

While previous research demonstrated that visual modality is a viable tool for identifying speech [1,2], the visual information has yet to become utilized in mainstream ASR systems. One obstacle is the difficulty in building a robust visual front end that tracks lips accurately in a real-world condition. To date majority of the work in automatic speechreading has focused on databases collected in studio-like environments with uniform lighting and constant background, such as CMU database [3], XM2VTS database [4], Tulips1 [5], DAVID [6], CUAVE [7], and AVOZES [8]. Hence there is a high demand for creating a robust visual front end in realistic environments. Accurately detecting and tracking lips under varying environmental conditions and for a large group of population is a very difficult task due to large variations in illumination conditions, background, camera settings, facial structural components (beards, moustaches), and inherent differences due to age, gender, and race. In particular, when we consider a real-world environment, strong illumination, uneven light distribution and shadowing, cluttered/moving background can complicate the lip identification process considerably.

In this paper we present our current progress in obtaining a robust visual front end in real-world conditions. In Section 2, we examine the use of color information and report our results on the statistical analysis and modeling of lip pixels by examining hundreds of manually extracted lip images obtained from several databases. In Section 3 we explore the effectiveness of spatial and edge information in extracting the lips. Finally Section 4 offers our conclusions and future work.

2   Color-Based Image Segmentation

2.1   Statistical Modeling and Color Space Selection

Color is an important identifying feature of an object. Prominent colors can be used as a far more efficient search criterion for detecting and extracting the object, such as the red color for identifying the lips. While color has been studied extensively in the past decade, especially in the field of face recognition, limited work was done in finding lips reliably in real-world conditions. The first critical decision for using color to find lips is to determine the best color space. RGB is the most commonly used color space for color images. However its inability to separate the luminance and chromatic components of a color hinders the effectiveness of color in object recognition. Previous studies [9] have shown that even though different people have different skin colors, the major difference lies in the intensity rather than the color itself. To separate the chromatic and luminance components, various transformed color spaces can be employed. Two such color spaces, HSV and YCbCr, were frequently used in various previous work in face and lip identification [10,11]. To determine which color space works the best for identifying the lips in unconstrained imagery, we perform statistical comparisons between the two on 421 images of human faces collected from the following three databases: 1.) Images collected using a Logitech web cam in a car or office environment with natural lightings (these images were collected by us and no compression was used); 2.) Images collected by browsing the Internet (these images were primarily face shots of celebrities so the images were most likely enhanced); 3.) Images collected from the AVICAR [12] Database, where 100 speakers were recorded in a natural automobile environment and under various driving conditions.

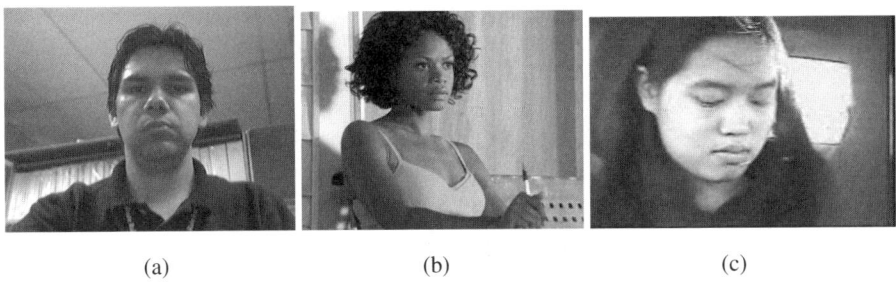

Fig. 1. Examples of Images from Three Databases

In our previous work [13], we reported the test results that were performed on all 421 images from the three databases. It was found that the Cb and Cr components offer very little distinction between the lip region pixel values and the background pixel values, while the hue component offers the best separation between the two based on the histograms shown in Figure 2. Hence the hue is better suited for distinguishing pixels between lip and non-lip regions and therefore was chosen as the primary feature for our visual front end. In Figure 2 the histogram was built by manually segmenting lip region for each of the 421 images and then normalized for equal comparison. In addition we shift the hue to the right by 0.2 because the red hue falls into two separate subsets at the low and high ends of the whole color range, as a result of the wrap-around nature of hue (hue is defined on a ring). After a right shift the red color falls in a connected region that lies at the low end close to 0. This can then be approximated by a Gaussian distribution that is shown together with the histogram in Figure 2(a). It can be shown that the hue values of the lip region are concentrated between zero and 0.6 with a mean of 0.28 and a variance of 0.0073.

Since the segmentation of the lip region will now occur in the shifted hue color space, then the histogram of the background must also be shifted to compare any overlapping sections. This shifted hue histogram together with the corresponding Gaussian model for background is shown in Figure 2(b). It is observed that the background has a much wider distribution since the background is un-controlled and varies greatly in an unconstrained environment. Note that both histograms for the lip and non-lip region are obtained by counting color pixels in manually segmented images from hundreds of images in our three databases. These histograms can well approximate the probability density functions (pdfs) for the lip and non-lips hue colors that are essential in the following lip segmentation procedure.

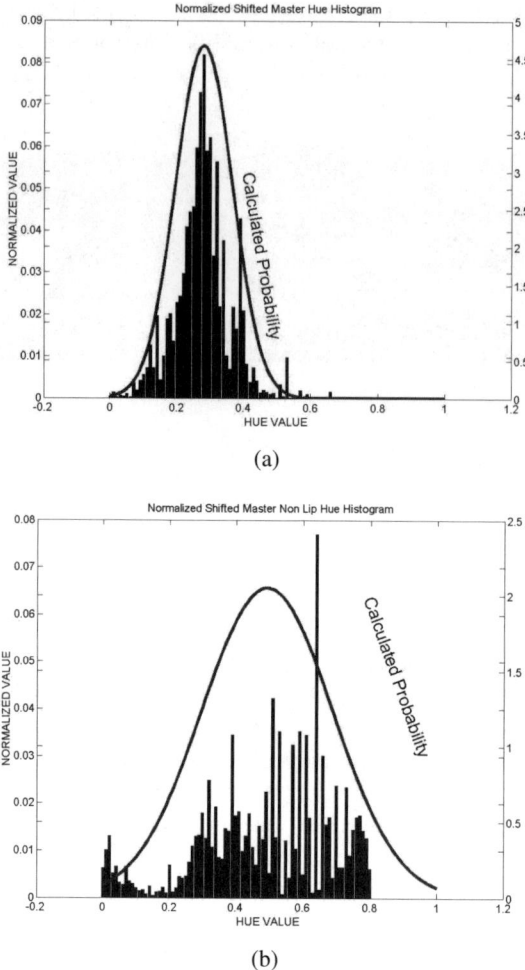

Fig. 2. Normalized Hue Histogram and Calculated PDFs for (a) Lips, (b) Background

2.2 Classification Design

To segment the lips a Bayesian classifier is employed. In Bayesian classifier a pixel x is classified as a lip pixel if the posteriori probability for the lips, $P(L \mid x)$, is larger than the posteriori probability for the background, $P(B \mid x)$, where L represents the class of lips, and B be the class for the background. By using Bayes Rule to calculate the posteriori probabilities, it can be shown that a pixel is classified as a lip pixel if

$$P[x \mid L]P[L] \geq P[x \mid B]P[B] \qquad (1)$$

In equation (1), $P[x|L]$ and $P[x|B]$ are the class conditional densities according to the Gaussian models in Figures 2(a) and (b). $P[L]$ is the a priori probability for the lip region and can be estimated by computing the percent area of the lip region averaged throughout all of our sample images. *P[B]* is the a priori probability for the background and can be estimated by computing the percent area of the background region averaged throughout all of our sample images. The implementation of the Bayesian classifier results in the following segmentation results shown in Figure 3.

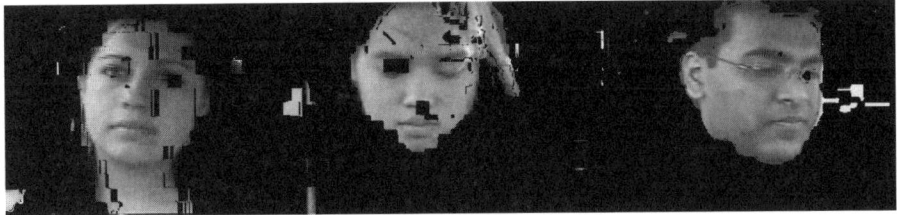

Fig. 3. Image Segmentation Results by Using Bayesian Classifier

The results of applying Bayesian classifier to our test images were that we were able to segment not only the lip region from the background, but also the entire face region. This result was neither expected nor desirable. To understand this we also calculated the hue pdf for skin around the lip region. To do so we scale the lip masks by 2 then subtract the original lip mask from the result. In Figure 4 all three hue pdfs for the lips, skin, and non-lip regions are combined in one graph for easy comparison. This figure clearly shows that there is very little hue contrast between the lips and facial skin, thus segmentation of lips from the surrounding face is difficult by using the hue color model only. However large variation of face color and the background scene allows an easy separation of both. We are able to use Bayesian classifier to successfully find faces in all images we tested. This is a significant result considering the fact that the images were collected in realistic environments that represent challenges due to large variations in illumination conditions.

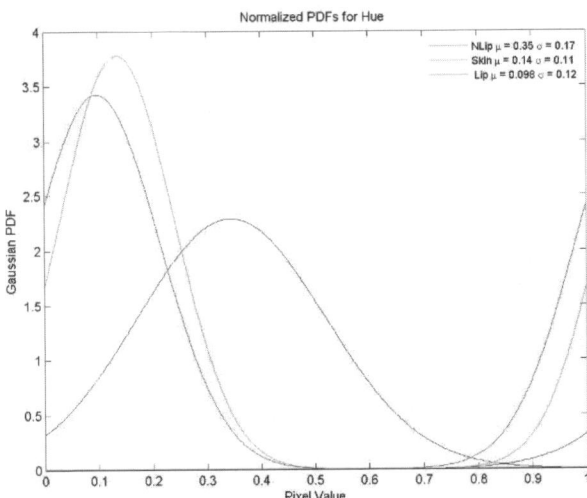

Fig. 4. Normalized Hue PDFs for Lip, skin, and non-lip regions (with no hue shift)

3 Finding Lips

To find the lip region from the face, we first reduce the search space by cropping the face image obtained from the previous module. Specifically, we remove the upper half and $1/8^{th}$ from each side. The resulting image ("half_face") is then passed to the next module.

It was observed that all inner lips are bright, near horizontal lines in the saturation color space, see Figure 5(a). At the same time the intensity values of the lips are darker than the surrounding regions. We therefore use both the saturation and value components in the HSV space to determine the lip lines. We create a mask image where a pixel is assigned to a value of 1 if its value component is less than a pre-defined threshold T_{Value} and its saturation component is greater than $T_{Saturation}$. Experimentally T_{Value} is set to be 0.3, and $T_{Saturation}$ set to be 0.45. The resulting image in Figure 5(b) is then analyzed for the component that has the longest major axis with an angle from horizontal of less than 15°, this is indicated as * in Figure 5(c). Therefore, as long as the mouth is less than 15° off axis, it should be recognized. Once this component is found, all other components are removed; the resulting image

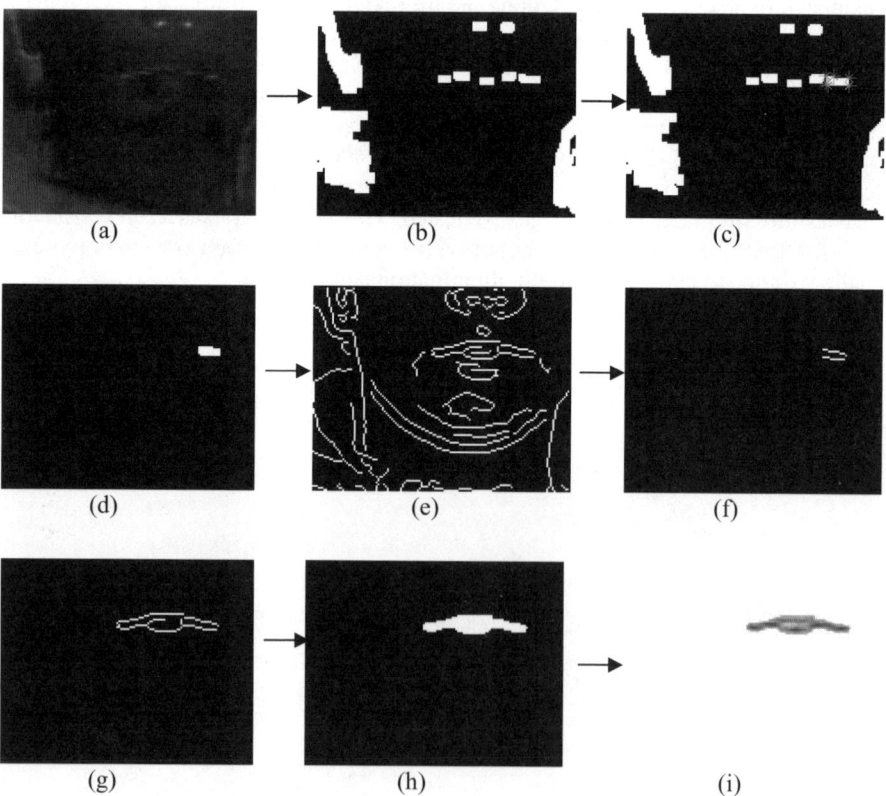

Fig. 5. Steps in Lip Finding Procedure

in Figure 5(d) is used as a mask in the subsequent step. Next a Canny edge detector is performed on the intensity values of the image and the resulting image in Figure 5(e) is multiplied by the mask image. The result is an image comprised of segments of Canny edge lines, see Figure 5(f). These segments are then analyzed for spatial location, which is then used to create a final mask image, comprised of the complete Canny components lying under the original mask. This image in Figure 5(g) is then dilated to force nearby components to connect, thus allowing filling of the region. Finally, this mask image in Figure 5(h) is inverted and then added to the original RGB image – resulting in a white background with only the lip region from the original image making it through as seen in Figure 5(i). When this procedure is applied to images in our databases, errors found are mostly due to weak edges from the lip lines extracted from Canny edge detector. Additionally, although the image was dilated prior to filling the final mask, not all regions are connected, preventing the complete mouth region from being displayed.

4 Conclusion and Future Work

Detection of lips in video sequences serves as an essential initial step towards building a robust audio-visual speech recognition system. In contrast to earlier works where the visual front end was designed based on databases collected in studio-like environment, this work seeks to find lips in imagery collected in visually challenging environment. A Bayesian classifier was developed that segments the face region in most unconstrained imagery. This was only possible after careful statistical analysis and modeling of hundreds of manually segmented lip images in various lighting and background conditions. These results demonstrate a very efficient way to narrow the search for the lip region substantially. From the detected face region, we subsequently extract lip region by incorporating spatial and edge information of the lips using both the saturation and value components in the HSV color space. From the detected lip region we can then extract physical dimensions of the lips that will be input to a recognition engine where spoken words can be classified.

In our future work we will explore additional features to further improve the robustness of our visual front end. One such feature is the motion feature. Here we notice that when a person is speaking, the motion fields around the lips are very different from those around the face. We expect this will result in improved accuracy in lip detection.

Acknowledgments. This work was sponsored by the *Department of the Navy, Office of Naval Research*, under Award # N00014-04-1-0436.

References

1. Stork, D.G., Hennecke, M.E. (eds.): Speechreading by Humans and Machines. NATO ASI Series F, vol. 150. Springer, Heidelberg (1996)
2. Potamianos, G., et al.: Audio-Visual Automatic Speech Recognition: An Overview. In: Bailly, G., Vatikiotis, E., Perrier, P. (eds.) Issues in Visual and Audio-Visual Speech Processing, MIT Press, Cambridge (2004)

3. Advanced Multimedia Processing Lab, Carnegie Mellon, Project-Audio-Visual-Processing (last accessed June 1, 2007), http://amp.ece.cmu.edu/projects/audiovisualspeechprocessing
4. The Extended M2VTS Database (last accessed June 1, 2007), http://www.ee.surrey.ac.uk/research/vssp/xm2vtsdb
5. Movellan, J.R.: Visual speech recognition with stochastic networks. In: Tesauro, G., Touretzky, D.S., Leen, T. (eds.) Advances in Neural Information Processing Systems, vol. 7, MIT Press, Cambridge, MA (1995)
6. Chibelushi, C.C., Gandon, S., Mason, J.S.D., Deravi, F., Johnston, R.D.: Design issues for a digital audio-visual integrated database. In: IEE Colloquium on Integrated Audio-Visual Processing for Recognition, Synthesis and Communication, Savoy Place, London (1996)
7. Patterson, E.K., Gurbuz, S., Tufekci, Z., Gowdy, J.N.: CUAVE: A new audio-visual database for multimodal human-computer interface research. In: Proc. ICASSP (2002)
8. Goecke, R., Millar, B.: The Audio-Video Australian English Speech Data Corpus AVOZES. In: Proc. ICSLP (2004)
9. Yang, J., Stiefelhagen, R., Meier, U., Waibel, A.: Real-time face and facial feature tracking and applications. In: Proc. of the 3rd IEEE Workshop on Applications of Computer Vision, pp. 142–147. IEEE Computer Society Press, Los Alamitos (1996)
10. Hsu, R., Abdel-Mottaleb, M., Jain, A.K.: Face detection in color images. J. IEEE Trans. Pattern Anal. Mach. Intell. 24(5), 696–706 (2002)
11. Zhang, X., Broun, C.C., Mersereau, R.M., Clements, M.A.: Automatic Speechreading with applications to human-computer interfaces. EURASIP Journal Applied Signal Processing, Special Issue on Audio-Visual Speech Processing 1, 1228–1247 (2002)
12. Lee, B., Hasegawa-Johnson, M., Goudeseune, C., Kamdar, S., Borys, S., Liu, M., Huang, T.: AVICAR: Audio-Visual Speech Corpus in a Car Environment. In: INTERSPEECH2004-ICSLP (2004)
13. Zhang, X., Montoya, H.A.: Statistical Modeling of Lip Color Features in Unconstrained Imagery. In: Proc. 11th World Multiconference on Systemics, Cybernetics and Informatic, Orlando, Florida (2007)

Feature Selection for Identifying Critical Variables of Principal Components Based on K-Nearest Neighbor Rule

Yun Li and Bao-Liang Lu

Department of Computer Science and Engineering
Shanghai Jiao Tong University
800 Dongchuan Rd, 200240 Shanghai, P.R. China
{liyun_mail,bllu}@sjtu.edu.cn

Abstract. Principal components analysis (PCA) is a popular linear feature extractor to unsupervised dimensionality reduction, and found in many branches of science including-examples in computer vision, text processing and bioinformatics, etc. However, axes of the lower-dimensional space, i.e., principal components, are a set of new variables carrying no clear physical meanings. Thus, interpretation of results obtained in the lower-dimensional PCA space and data acquisition for test samples still involve all of the original measurements. To select original features for identifying critical variables of principle components, we develop a new method with k-nearest neighbor clustering procedure and three new similarity measures to link the physically meaningless principal components back to a subset of original measurements. Experiments are conducted on benchmark data sets and face data sets with different poses, expressions, backgrounds and occlusions for gender classification to show their superiorities.

Keywords: Feature selection, Principal component analysis, K-nearest neighbor rule.

1 Introduction

In many real world problems, such as pattern recognition and data mining, reducing dimensionality is an essential step before analysis of the data [1]. Feature selection and feature extraction are two commonly adopted approaches to this issue. Feature selection refers to selecting features in the measurement space, while feature extraction technique such as Principal Component Analysis (PCA), Independent Component Analysis (ICA) and Linear Discriminate Analysis (LDA) [2], select features in a transformed space. These feature extraction methods find a mapping between the original feature space to a lower dimensional feature space. The features provided by the feature selection techniques is a subset of the original features, while the features obtained through feature extraction are a subset of new variables which have no clear practical meanings.

PCA maps data from the original measurement space to a new space spanned by a set of new variables, and principal components are linear combinations of all variables available [3]. However, these variables are not necessarily equally important to the formation of principal components. Some of the variables might be critical, but some might be redundant, irrelevant or insignificant. Motivated by this fact, various methods have been developed to attempt to link the physically meaningless principal components back to a subset of the original variables through selecting critical variables. These methods include B2 and B4 algorithms [4,5], principal variable method [6], data structure preserved (DSP) algorithm [7], Forward-least square estimation (LSE), and Backward-LSE [3]. The B2 algorithm discards variables that are highly associated with the last few principal components, while the B4 algorithm selects variables that are highly associated with the first few principal components. Since the significance of variables is evaluated individually, then redundant features might not be eliminated by the B2 and B4 algorithms. The principle variable method is to find the subset has the similar effect to the full feature set in maximizing the 'spread' of the points in the lower dimensional space achieved by PCA, retain the variation in the original space and minimize the mean square error between the predicted data to the original data. The main characteristic of the DSP algorithm is to evaluate a feature subset based on its data structural similarity to the full feature set. LSE is to evaluate a feature subset based on its capacity to reproduce sample projections on principal axes. The process of finding critical variables for LSE has the following two steps. In the first step, the principal components are derived from the full feature set using PCA, and sample projections on the principal components corresponding to the largest eigenvalues are computed. In the second step, a sequential forward-selection or backward-elimination search strategy algorithm is performed, with the objective of selecting a subset of features that has the most capacity to reproduce the data projections on the principal components. Although the LSE method simplifies the evaluation of a feature subset to an ordinary least-square estimation problem, and costs less computing time comparing with DSP. However, the complexity of LSE and DSP is still high, and they are impractical for high-dimensional data sets. In addition, Principal feature analysis (PFA) method exploits the information that can be inferred by the principal components coefficients to obtain the critical variables of principal components [8].

The rest of the paper is organized as follows. In section 2, the procedure of proposed algorithms is introduced, and three new similarity measures for row components in transformation matrix are developed. Computation complexity is also analyzed in section 2. Experiment studies are presented in Section 3, and the paper ends with concluding remarks in Section 4.

2 Algorithm Design

Assuming the training set contains N samples, $\{\mathbf{x}_l\}_{l=1}^{N}$, and each sample is represented by an n-dimensional vector $\mathbf{x}_l = [x_{l1}, x_{l2}, \cdots, x_{ln}]^T \in \mathcal{R}^n$. Feature vector $\mathbf{f}_i$ and feature matrix $\mathbf{F}$ are defined as

$$\mathbf{f}_i = [x_{1i}, x_{2i}, \cdots, x_{Ni}],$$
$$\mathbf{F} = \{\mathbf{f}_1, \mathbf{f}_2, \cdots, \mathbf{f}_n\}^T \quad (1)$$

In PCA, eigenvectors gotten from covariance matrix are sorted by eigenvalues, which represent the variance of sample distribution on eigenvectors. Normally the d largest eigenvectors $\mathbf{Q} = \{\mathbf{q}_1, \mathbf{q}_2, \cdots, \mathbf{q}_d\}$ are selected to span the PCA subspace, since they can optimally reconstruct the sample with the minimum reconstruction error. The linear transformation is denoted by transformation matrix $\mathbf{Q}$, pattern $\mathbf{x}_l$ in the new subspace is represented by

$$\mathbf{y}_l = \mathbf{Q}^T \mathbf{x}_l \quad (2)$$

where $\mathbf{y}_l = [y_{1l}, y_{2l}, \cdots, y_{dl}]^T$, and

$$\mathbf{Q} = \{\mathbf{q}_1, \mathbf{q}_2, \cdots, \mathbf{q}_d\},$$
$$\mathbf{q}_j = [q_{1j}, q_{2j}, \cdots, q_{nj}]^T, j = 1, 2, \cdots, d \quad (3)$$

where n is the number of original features, $d \leq n$ and quite often $d \ll n$. There are ten optimal properties for this choice of the linear transformation [6]. One important property is the maximization of the 'spread' of the points in the lower dimensional space which means that the points in the transformed space are kept as far apart as possible and retaining the variation in the original space. Another important property is the minimization of the mean square error between the predicted data to the original data.

2.1 Feature Selection

Consider the projection of $\mathbf{x}_l = [x_{l1}, x_{l2}, \cdots, x_{ln}]^T \in \mathcal{R}^n$ on the jth principal axis

$$y_{jl} = \mathbf{q}_j^T \mathbf{x}_l = \sum_{i=1}^{n} q_{ij} x_{li}, \quad (4)$$

As shown in (4), the projection of a sample on principal axis is a linear combination of all variables. However, some of the variables might be redundant, irrelevant or insignificant. This indicates that feature selection can be done through identifying a subset of variables whose roles are critical in determining data projections on principal axes.

The significance of variables $\mathbf{f}_i = [x_{1i}, x_{2i}, \cdots, x_{Ni}]$ can be evaluated based on the value of the corresponding parameter q_{ij}, then we can identity the critical variables of principal components through the elements q_{ij} in $\mathbf{Q}$. On the other hand, the transformation matrix $\mathbf{Q}$ also can be represented as:

$$\mathbf{Q} = \{\mathbf{v}_1, \mathbf{v}_2, \cdots, \mathbf{v}_n\}^T,$$
$$\mathbf{v}_i = [v_{i1}, v_{i2}, \cdots, v_{id}] = [q_{i1}, q_{i2}, \cdots, q_{id}],$$
$$i = 1, 2, \cdots, n \quad (5)$$

we call each vector $\mathbf{v}_i$ as row component, which represents the projection of the ith variable $\mathbf{f}_i$ to the lower dimensional PCA subspace, that is, the d elements of $\mathbf{v}_i$ correspond to the weights of $\mathbf{f}_i$ on each axis of the PCA subspace [9].

The key observation is that features that are highly correlated or have high mutual information will have similar absolute value of weight vectors (row components), and changing the sign has no statistical significance [9]. On two extreme cases, two independent features have maximally separated weight vectors; while two fully correlated features have identical weight vectors (up to a change of sign). Based on this observation, we can identify the critical variables of principal components through the choice of row component subset.

To find the best subset of features, the PFA method [8] uses the structure of the row component to find the subsets where row components are highly correlated, and then chooses one row component from each subset. The selected row components represent each group optimally in terms of high spread in the lower dimension, reconstruction and insensitivity to noise. From the selected row components, we can get corresponding original feature subset, and the feature selection is completed. The number of selected features is equal to the number of clusters. The steps of the PFA method can be briefly introduced as follows. Firstly, PCA is executed to get the transformation matrix $\mathbf{Q}$, then the row components are clustered to $p \geq d$ clusters using C-Means algorithm. For each cluster, the corresponding row component $\mathbf{v}_i$, which is closest to the mean of the cluster, is determined, and the associated original variable $\mathbf{f}_i$ is chosen as a critical variables. This step will yield the choice of p features.

In this paper, we present an approach to find the subset of row components based on k nearest neighbor (kNN) rule [10]. We first compute k nearest row components for each row component, among them the row component having the most compact subset, *i.e.* having the largest similarity to the farthest neighbor, is selected, and its k neighbors are discarded. The process is repeated for the remaining row components until all of them are considered. For determining the k nearest neighbors of row component, we assign a constant threshold τ, which is equal to the similarity of kth nearest neighbor of selected row component in the prior iteration. In the subsequent iterations, we compute the similarity between each row component and its kth nearest neighbor, and check the maximum similarity whether it is less than τ or not. If yes, then decreases the value of k. As a result, k may be changing over iterations. The proposed approach is summarized in Algorithm 1. The original features correspond to the selected row components in R are the features that have been selected and they are the critical variables of principal components.

2.2 Row Component Similarity Measures

One of key problems in our approach mentioned above is how to calculate row component similarity. Until now, various methods have been developed to compute the variable similarity such as correlation coefficient, least square regression

Algorithm 1. Row component selection using kNN rule

Step 1: Execute the PCA to get the transformation matrix $\mathbf{Q} = \{\mathbf{v}_1, \mathbf{v}_2, \cdots, \mathbf{v}_n\}^T$, where the number of row components is n, $sim(\mathbf{v}_i, \mathbf{v}_t)$ represents the similarity between row components $\mathbf{v}_i$ and $\mathbf{v}_t$, R is the reduced row component subset. Let s_i^k represent the similarity between row component $\mathbf{v}_i$ and its kth nearest-neighbor row component in R, choose an initial value of $k \leq n-1$, $R \leftarrow \mathbf{Q}$;
Step 2: Calculate a matrix $\mathbf{M}$, which is of the dimensionality n by n, and the value of element $M_{it(i \neq t)}$ is $sim(\mathbf{v}_i, \mathbf{v}_t)$ and M_{ii} ($i = 1, 2, \cdots, n$, $t = 1, 2, \cdots, n$) is zero. $\mathbf{M}$ is a symmetric matrix.
Step 3: For any row component $\mathbf{v}_i \in R$, get s_i^k according to $\mathbf{M}$;
Step 4: Retain the row component $\mathbf{v}_{i'}$, which $s_{i'}^k$ is maximum, to R and discard k nearest row components of $\mathbf{v}_{i'}$ and corresponding columns and rows in $\mathbf{M}$, then get the updated R and $\mathbf{M}$. Let $\tau = s_{i'}^k$;
Step 5: **if** $k+1$ larger than the size of R: $k = sizeof(R) - 1$.;
Step 6: **if** $k = 1$: **go to** *step 10*
Step 7: For each row component $\mathbf{v}_i \in R$, get s_i^k according to $\mathbf{M}$ and find the row component $\mathbf{v}_{i'}$, for which $s_{i'}^k$ is maximum;
Step 8: **while** $s_{i'}^k < \tau$
    $k = k - 1$
    $s_{i'}^k = \sup_{v_i \in R} s_i^k$
    if $k = 1$: **go to** *step 10*
end while
Step 9: **go to** *step 4*
Step 10: Return reduced row component subset R as the output

error, and maximal information compression index (MICI) [10]. Of course, these methods also can be used to compute the similarity of row components. Previous work has shown that MICI is superior to other methods [10].

However, ideally, the distance metric for kNN classification should be adapted to the particular problem being solved [11]. Of course, for the clustering based on kNN rule, the similarity measure also should be changed to the special issue being handled. Fortunately, , we observe that the element v_{ij} ($i = 1, 2, \cdots, n, j = 1, 2, \cdots, d$) in the transformation matrix $\mathbf{Q}$ takes value $[-1, 1]$, and changing the sign has no statistical significance [9], and the eigenvectors are independent. Therefore, we take these characteristics into account, and propose three new methods called component similarity indexes (CSI) to measure the similarity between row components. The proposed component similarity indexes are defined as follows.

Definition. For two row components, $\mathbf{v}_i = (v_{i1}, v_{i2}, \cdots, v_{id})$ and $\mathbf{v}_t = (v_{t1}, v_{t2}, \cdots, v_{td})$, v_{ij} and $v_{tj} \in [-1, 1]$, $i \neq t$, $i, t = 1, 2, \cdots, n$, and changing the sign has no statistical significance, then the similarity between $\mathbf{v}_i$ and $\mathbf{v}_t$ is calculated by:

$$CSI1 : sim(\mathbf{v}_i, \mathbf{v}_t) = 1 - \max_j(||v_{ij}| - |v_{tj}||) \qquad (6)$$

$$CSI2 : sim(\mathbf{v}_i, \mathbf{v}_t) = \frac{\sum_{j=1}^{d} \min(|v_{ij}|, |v_{tj}|)}{1/2 \sum_{j=1}^{d} (|v_{ij}| + |v_{tj}|)} \quad (7)$$

$$CSI3 : sim(\mathbf{v}_i, \mathbf{v}_t) = \frac{\sum_{j=1}^{d} \min(|v_{ij}|, |v_{tj}|)}{\sum_{j=1}^{d} \sqrt{|v_{ij}| * |v_{tj}|}} \quad (8)$$

The properties of CSI are summarized as:

- $0 \leq sim(\mathbf{v}_i, \mathbf{v}_t) \leq 1$;
- $sim(\mathbf{v}_i, \mathbf{v}_t) = 1$ if and only if $\mathbf{v}_i$ equals to $\mathbf{v}_t$;
- $sim(\mathbf{v}_i, \mathbf{v}_t) = sim(\mathbf{v}_t, \mathbf{v}_i)$.

CSI1 is based on definition of Chebychev distance, and the definition above also can be considered as an application of minimization and maximization rule, which are widely used in machine learning, such as ensemble learning and classifiers combination [12,13,14], and these methods have been successfully applied into classification of EEG [15], gender classification [16], etc. Then we use the min-max rule to get the similarity between row components. The experimental results will show its high performance. Here, we call the algorithms using the procedure described in section 2.2 and these CSIs and MICI as PCSIs ($s = 1, 2, 3$) and PMICI, respectively.

2.3 Algorithm Analysis

The characteristics of PCSIs ($s = 1, 2, 3$) can be summarized as follows:

- CSIs ($s = 1, 2, 3$) are simple and parameter free. The number of selected eigenvectors $d \leq n$, and quite often $d \ll n$, then if use the complicated model to measure the row component similarity, the estimation quality of parameter in the model is not very satisfactory.
- CSIs ($s = 1, 2, 3$) are based on the value range of element in transformation matrix $\mathbf{Q}$, and they take advantage of the independence between eigenvetors, so the proposed algorithms are professional methods for identifying critical features of principal components, and then they can achieve good performance.
- In the algorithms, k controls the size of R, since k determines the threshold τ, and then the representation of the data at different degrees of details is controlled by its choice. Then we can use this characteristic to determine the multiscale representation of critical variables of principal components.
- Clustering based on kNN rule is dynamic, partitional and nonhierarchical in nature.

Computational complexity analysis. PCSIs ($s = 1, 2, 3$), PMICI and PFA all select feature after the PCA, Therefore the difference of computation complexity among them is attributed to the difference in dealing with the transformation matrix **Q**. With respect to the number of row components (n), the proposed methods have complexity $O(n^2)$. Other search-based method for dealing with **Q** to determine the critical original feature, such as sequential forward, backward elimination, sequential floating search, orthogonal forward-selection and orthogonal backward elimination [3,17]. Among them only sequential forward and backward elimination have complexity $O(n^2)$, others have complexity higher than quadratic. On the other hand, the second factor which contributes to the speedup achieved by the PMICI and PCSIs ($s = 1, 2, 3$) is the low-computational complexity of evaluating the row component similarity. If the number of eigenvectors is d, evaluation of the similarity measure for a row component pair is of complexity $O(d)$. Thus the computation complexity of PMICI and PCSIs ($s = 1, 2, 3$) is $O(n^2 d)$.

3 Experimental Results

Without any prior knowledge, we did not know which features are more important in the data set and due to space constraints, so in order to validate the performance of the proposed methods and prove that they can select more critical variables of principal components, we conduct experiments to compare them with other algorithms on time cost and classification accuracy rate of the selected critical variables instead of listing concrete selected critical features. The whole experiment including two parts: Experiments are conducted on benchmark middle-dimensional data sets to compare PCSIs ($s = 1, 2, 3$) with PMICI, PFA and Forward-LSE. Tests are also conducted on high-dimensional data sets to test the performance of PCSIs ($s = 1, 2, 3$), PMICI and PFA in gender classification. The time cost of Forward-LSE is too high to apply in gender classification. All algorithms are implemented using Matlab compiler and run on the P4 2.8GHz PC with 2.0G-memory.

Benchmark data sets are chosen from UCI machine learning repository [18]. The details of these data sets are described in Table 1.

Table 1. Details of benchmark data sets

Data sets	Data set size	No. features	No. classes
Sonar	208	60	2
Spectf	801	44	2
Waveform	5000	21	3

For gender classification, the gallery sets used for training include 786 male samples and 1269 female samples, which have the same vector dimension of 1584 with the gabor filter [16]. The probe sets used for testing include 15 kinds of gender data with various facial views and expressions, and these probe sets are numbered as 1-15. The details of the probe sets are described in Table 2 and some examples are shown in Figure 1.

Table 2. Description of probe sets for gender classification

No. of probe set	face description	No. data
1	Front 1	1278
2	Front 2	1066
3	Down 10 degree	820
4	Down 20 degree	819
5	Down 30 degree	816
6	Smiling	805
7	Opened mouth	815
8	Closed eyes	805
9	Front with glass	813
10	Right 10 degree	814
11	Right 20 degree	815
12	Right 30 degree	805
13	Up 10 degree	819
14	Up 20 degree	816
15	Up 30 degree	816

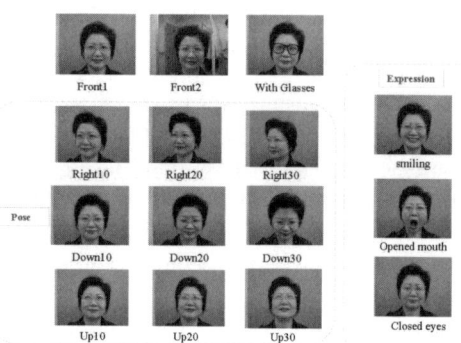

Fig. 1. Examples from face images data set

3.1 Experiments on Benchmark Data Sets

For the benchmark data sets, we use the classification accuracy of kNN classifier with the k=3 to evaluate the selected subset of features and 10-fold cross-validation is adopted. However, for Spectf data set, the original training and test data sets were used. An important issue here is the determination of a suitable value for d. This is a common problem whenever PCA is adopted. A rule of thumb is to keep eigenvalues accounting for 90% of the variance [3]. In PFA, the number of clusters equals to d, i.e., $p = d$. The experimental results are shown in Table 3.

3.2 Gender Classification

In this subsection, we present experimental results on the gender data sets for different number of selected features and various feature selection algorithms,

Table 3. Experimental results of time cost and classification accuracy for benchmark data sets

Data set	Algorithm	Accuracy	CPU time (Sec.)
Sonar	PFA	62.17	0.10
	Forward-LSE	61.99	7.80
	PMICI(k=20)	60.34	0.98
	PCSI1(k=30)	61.48	0.77
	PCSI2(k=30)	62.62	0.73
	PCSI3(k=30)	**63.18**	0.73
Spectf	PFA	79.93	0.14
	Forward-LSE	77.70	3.16
	PMICI(k=33)	74.98	0.58
	PCSI1(k=30)	79.18	0.42
	PCSI2(k=30)	79.84	0.47
	PCSI3(k=30)	**80.84**	0.41
Waveform	PFA	75.29	0.08
	Forward-LSE	77.62	28.84
	PMICI(k=6)	77.14	0.23
	PCSI1(k=8)	**80.90**	0.20
	PCSI2(k=8)	78.67	0.22
	PCSI3(k=8)	78.38	0.19

Table 4. Experimental results of maximum time cost and the highest accuracy rate for gender classification

No. selected features	Algorithm	CPU time (Sec.)	Highest accuracy
484	PFA(p=484)	136.5	82.48
	PMICI(k=1095)	509.5	85.21
	PCSIs(k=1100)	300.7	**89.05**
684	PFA(p=684)	152.9	83.29
	PMICI(k=895)	581.6	88.18
	PCSIs(k=900)	389.6	**90.14**
884	PFA(p=884)	171.6	84.60
	PMICI(k=690)	673.6	88.29
	PCSIs(k=700)	455.2	**90.61**
1084	PFA(p=1084)	190.6	83.60
	PMICI(k=491)	613.3	**91.39**
	PCSIs(k=500)	416.6	91.31

which are PFA, PMICI and PCSIs ($s = 1, 2, 3$). For space constrains, we only list the results of several numbers of selected features, which are chosen as 484, 684, 884 and 1084. Traditional Support Vector Machines (SVM) [19] is adopted, and the parameter C is set to 1. We also consider to keep eigenvalues accounting for 90% of the variance. The time cost of algorithms is shown in Table 4, and the highest accuracy rates in the fifteen probe sets for different algorithms are

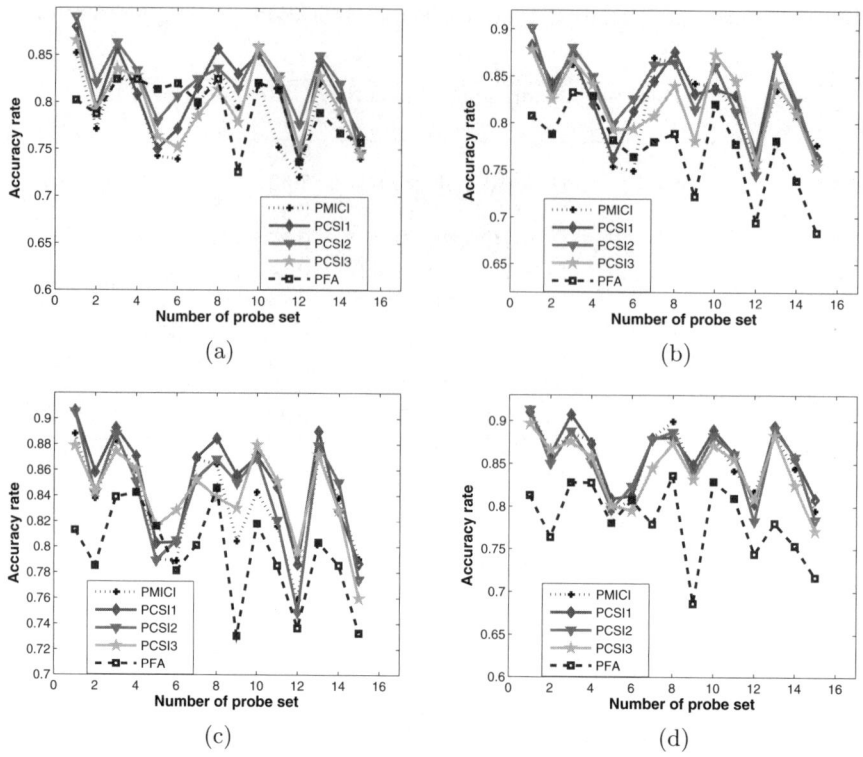

Fig. 2. The accuracy rates of fifteen probe sets for the different number of selected features (a) 484, (b) 684, (c) 884 and (d) 1084

also listed in Table 4. For PCSIs, only the maximum time cost and the highest accuracy rate among PCSI1, PCSI2 and PCSI3 are listed. The value of parameter k in PMICI and PCSIs ($s = 1, 2, 3$) is chosen to get the selected number of features. Detailed accuracy rates of different algorithms for fifteen probe sets are displayed in Figure 2. The X-axis is the No. of probe sets and Y-axis is the accuracy rate for gender classification.

3.3 Observations

From the experimental results on benchmark and gender data sets, we can obtain the following several observations.

☐ All of the component similarity indexes proposed in this paper can catch the characteristics of element in transformation matrix **Q**, and efficiently calculate the similarity between row components.

☐ PCSIs ($s = 1, 2, 3$) cost less time than PMICI and Forward-LSE. There are at least one of the algorithms PCSIs ($s = 1, 2, 3$) can get higher accuracy rate than PMICI, PFA and forward-LSE for benchmark data sets. For gender classification, there are at least one of the algorithms PCSIs ($s = 1, 2, 3$) can get higher accuracy rate than PMICI and PFA on most probe sets for the different number of selected features.

☐ In gender classification, PCSIs ($s = 1, 2, 3$) and PMICI get similar or higher accuracy rate than PFA for most probe sets. When the number of selected features is increasing, the performance of PFA is worsening or unchanging. On the contrary, PCSIs ($s = 1, 2, 3$) and PMICI are getting better. Then the proposed methods are more appropriate for feature selection from high dimensional data.

4 Conclusions and Discussions

In this paper, we present a new method for identifying critical variables of principal components. The procedure of identification is based on the k-nearest neighbors rule and an automatically changeable parameter k is used to control the size of reduced feature subset. Three efficient measures to compute the similarity of row component in transformation matrix have been proposed. These measures take advantage of the value range of element and independence of eigenvector in transformation matrix. Experiments are conducted on benchmark data sets and high-dimensional gender data set for gender classification to compare the proposed algorithms PCSIs ($s = 1, 2, 3$) with PFA, PMICI and Forward-LSE. The experimental results have shown PCSIs ($s = 1, 2, 3$) can get high recognition rate.

Acknowledgments. This research was partially supported by the National Natural Science Foundation of China via the grant NSFC 60473040, and Shanghai Jiao Tong University and Microsoft Research Asian Joint Laboratory for Intelligent Computing and Intelligent Systems. We gratefully thank OMRON Cooperation for supplying facial images.

References

1. Liu, H., Yu, L.: Toward integrating feature selection algorithms for classification and clustering. IEEE Trans. Knowledge and Data Engineering 17(4), 491–502 (2005)
2. Jain, A.K., Dulin, R.P., Mao, J.: Statistical pattern recogntion: a review. IEEE Trans. on Pattern Recognition and Machine Intelligence 22(1), 4–37 (2000)
3. Mao, K.Z.: Identifying critical variables of principal components for unsupervised feature selection. IEEE Trans. Systems, Man, and Cybernetics-part B:Cybernetics 35(2), 339–344 (2005)
4. Jolliffe, I.T.: Discarding variables in a principal component analysis I: Artificial data. Appl. Statist. 21(2), 160–173 (1972)
5. Jolliffe, I.T.: Discarding variables in a principal component analysis II: Real data. Appl. Statist. 22(1), 21–31 (1973)

6. McCabe, G.P.: Principal variables. Technometrics 26, 127–134 (1984)
7. Krzanowski, W.J.: Selection of variables to preserve multivariate data structure using principal components. Appl. Statist. 36(1), 22–33 (1987)
8. Cohen, I., Tian, Q., Zhou, X.S., Huang, T.S.: Feature selection using principal feature analysis. In: Proc. of IEEE Int'l. Conf. on Image Processing, IEEE Computer Society Press, Los Alamitos (2002)
9. Jolliffe, I.T.: Principal component analysis. Springer, New York (2002)
10. Mitra, P., Murthy, C.A., Pal, S.K.: Unsupervised feature selection using feature similarity. IEEE Trans. Pattern Recognition and Machine Intelligence 24(3), 301–312 (2002)
11. Weinberger, K.Q., Blitzer, J., Saul, L.K.: Distance metric learning for large margin nearest neighbor classification. Advances in Neural Information Processing Systems 18 (2006)
12. Kittler, J., Hatef, M., Duin, R.P.W., Matas, J.: On combing classifiers. IEEE Trans. on Pattern Recognition and Machine Intelligence 20(3), 226–239 (1998)
13. Lu, B.L., Ito, M.: Task decomposition and module combination based on class relations: a modular neural network for pattern classification. IEEE Trans. on Neural Networks 10, 1244–1256 (1999)
14. Lu, B.L., Wang, K.A., Utiyama, M., Isahara, H.: A part-versus-part method for massively parallel training of support vector machines. In: Proc. of Int'l Joint Conf. Neural Networks 2004, July 25-29, pp. 735–740 (2004)
15. Lu, B.L., Shin, J., Ichikawa, M.: Massively parallel classification of single-trial EEG signals using a min-max modular neural network. IEEE Trans. on Biomedical Engineering 51(3), 551–558 (2004)
16. Lian, H.C., Lu, B.L.: Gender recognition using a min-max modular SVM. In: Wang, L., Chen, K., Ong, Y.S. (eds.) ICNC 2005. LNCS, vol. 3611, pp. 433–436. Springer, Heidelberg (2005)
17. Mao, K.Z.: Orthogonal forward selection and backward elimination algorithms for feature subset selection. IEEE Trans. Systems, Man, and Cybernetics-part B:Cybernetics 34(1), 629–634 (2004)
18. Merz, C.J., Murphy, P.M.: UCI repository of machine learning database, http://www.ics.uci.edu/mlearn/MLRepository.html
19. Chang, C.C., Lin, C.J.: LIBSVM: a library for support vector machines, http://www.csie.ntu.edu.tw/~cjlin/papers/libsvm.ps.gz

Denoising Saliency Map for Region of Interest Extraction

Yandong Guo, Xiaodong Gu, Zhibo Chen, Quqing Chen, and Charles Wang

Thomson Corporate Research, Beijing
eemars@gmail.com, xiao-dong.gu@thomson.net

Abstract. The inherent noises can significantly degrade the accuracy of the attention area detection. In this paper, we present a novel structure of hybrid filter for suppressing noises in saliency maps which is viewed as a preliminary step towards the solution of automatic video region-of-interest determination. The filter presented in our paper makes use of the property of saliency maps and can remove almost all the Gauss noise and pepper-salt noise while preserve the details of attention area. Experimental results demonstrate the efficiency and effectiveness of our approach in extracting the region of interest.

Keywords: The saliency map, region of interest extraction, median filter, hybrid filter, suppressing noise.

1 Introduction

Visual attention analysis provides an alternative methodology to many applications such as video compression, summarization and adaptive presentation etc [1] [2] [13]. The fundamental approach for visual attention analysis is the accurate detection of attention areas. There are many methods based on the notion of saliency map proposed by Itti et al. in [3] to extract attention areas or objects automatically [4] [5].

However, all the methods mentioned above suffer from the inherent stochastic fluctuations (i.e. noise) in saliency map. Here are some examples to show the un-ideal results got from the noisy saliency maps in Fig. 4.5.6 (a) (d).

As to our knowledge, the denoising in saliency maps has not been investigated specially in the literature. We presented denoising as a promising technique to improve the reliability and correctness of the extraction of region of interest (ROI). The main difficulty is how to attenuate noises of different kinds without blurring the fine details in the saliency map.

Based on the analysis to various filters, we proposed a hybrid filter called content adaptive filter (CAF) by amelioration and combination which can get excellent denoising results. Median filter which was proposed by Turkey [6] has attracted a growing number of interests in the past few years as one of the most widely used order-statistic filters because of its intrinsic properties, edge preservation and efficient noise attenuation with robustness against impulsive-type noise. But the median filter doesn't work as well as finite-impulsive response filter (FIR) in case of Gaussian noise. FIR-median hybrid filter (FMH) introduced by Pekka [7] makes use of the desirable properties of both linear filters and median filters. In his algorithm, the input signal $x(n)$ is filtered with M linear phase FIR filters, and the output of the FMH filter is the median of the outputs of the FIR filters. The FMH does not perform well enough in the denoising of saliency sequences because of the discarding of the temporal information and the local statistic properties of saliency sequences.

We give brief discussion about the noise attenuation in regions of different statistic properties in Section 2. After that, we propose a content adaptive filter (CAF) utilizing the local statistic properties of saliency sequences to get excellent denoising results and the filter can satisfy the requirement of real-time using. The CAF is introduced in Section 3. Comparisons with other filters are presented in Section 4 and the experiment results which indicate the ability of CAF to improve the reliability and correctness of the extraction are in Section 5.

This is the first time to discuss the denoising especially for saliency maps and there are two more contributions in our approach; first, there is a novel method to form hybrid filters which makes use of the statistic properties by the different way. Second, in pervious research, analysis of saliency sequences was generally modeled for only one single frame while CAF makes use of the temporal information from the reference frames. CAF absorbs the advantages of different filters and exceeds the other filters in statistical analysis and provides denoised saliency sequences from which the region of interest can be extracted more accurately.

2 Denoising in Different Regions

We denote an image inside a video sequence as $I(x, y, t)$, where (x, y) and t indicate the spatial and temporal locations, respectively. The inputs used to estimate the value of pixel $I(x, y, t)$ are the pixels $I'(x', y', t')$ which satisfy the conditions below,

$$|x'-x|<2, |y'-y|<2, \text{ when } t'=t\pm 1,$$

$$\text{OR } |x'-x|<3, |y'-y|<3, \text{ when } t'=t; \qquad (1)$$

2.1 Denoising in EMR

Definition 1. The region belongs to the edge of objects moving drastically is defined as EMR.

When the objects in the sequences move drastically, and if the pixel being estimated belongs to the edge of objects (EMR), the temporal information cannot be considered of because the pixel at the same position in either the last frame or the next frame has little correlation with the pixel under processing. In this situation, the most authentic filter should be multistage median filter (MMF) for its predominant detail preserving ability. MMF was proposed by Ari in [8]. Take the MMF mentioned in [8] for example, the input of MMF is a set of samples inside a $(2N+1)\times(2N+1)$ window centered at the (x, y, t) location. $\{I(x+l, y+l, t): -N\leq l\leq N\}$. Define the subsets, W_1, W_2, W_3, W_4, of the square window as

$$W_1[I(x,y,t)] = \{I(x+l,y,t): -N \le l \le N\},$$
$$W_2[I(x,y,t)] = \{I(x,y+l,t): -N \le l \le N\},$$
$$W_3[I(x,y,t)] = \{I(x+l,y+l,t): -N \le l \le N\}, \qquad (2)$$
$$W_4[I(x,y,t)] = \{I(x+l,y-l,t): -N \le l \le N\},$$

Moreover, let

$$Temp_k(x,y,t) = median[I(\bullet,\bullet,\bullet) \in W_k[I(x,y,t)]] \qquad 1 \le k \le 4$$

The MMF output is described by

$$Output_{MMF}(x,y,t) = median[Output_1(x,y,t), Output_2(x,y,t), I(x,y,t)]. \qquad (3)$$

Where

$$Output_1(x,y,t) = median[Temp_1(x,y,t), Temp_2(x,y,t), I(x,y,t)]$$
$$Output_2(x,y,t) = median[Temp_3(x,y,t), Temp_4(x,y,t), I(x,y,t)]$$

Based on the analysis of root-signal it is known that the MMF preserves the details better and is computationally much more efficient than the conventional median filter, the K-nearest neighbor averaging filters and other filters mentioned in the following part. However, the MMF discards the temporal information and performs not as well as 3DPF (3D Plane Filter) and FIR (Finite Impulsive Response filter) in noise attenuation when there is additive Gaussian white noise in constant region.

2.2 Denoising in ER

Definition 2. The pixels belong to the edge of objects which do not move drastically and the pixels affected by the impulsive noise form the region called ER.

The 3D plane filter (3DPF) will output the best result when the pixel belongs to ER.

M. Bilge made observations on the root signal in binary domain based on the positive Boolean functions corresponding to the filters and derived the output distribution of 3DPF from the Boolean expressions in [9]. The results show that the 3DPF has better noise attenuation performance than MMF when the contents of the sequence do not change drastically because the temporal information which is taken into account benefits the estimation while the linear average filters have the best noise attenuation result when there are no high frequency details in the region.

The output of 3DPF is

$$Output_{3DPF}(x,y,t) = median[Output_{xy}(x,y,t), Output_{xt}(x,y,t), Output_{yt}(x,y,t)]. \quad (4)$$

Where

$$Output_{xy}(x,y,t) = median[I(x,y,t), I(x \pm 1, y, t), I(x, y \pm 1, t)].$$

$$Output_{xt}(x,y,t) = median[I(x,y,t), I(x \pm 1, y, t), I(x, y, t \pm 1)]$$

$$Output_{yt}(x,y,t) = median[I(x,y,t), I(x, y \pm 1, t), I(x, y, t \pm 1)]$$

2.3 Denoising in CR

Definition 3. When the pixel without impulsive noise influence belongs to a constant region, we say that the pixel belongs to CR.

In this situation, the FIR will take in charge because of its outstanding ability in suppressing Gaussian noise.

The average of the pixels in the (2N+1)×(2N+1)×(2N+1) window centered at the (x, y, t) location is set as the output of FIR $Output_{FIR}(x, y, t)$.

3 Content-Adaptive Filters

If there is a detector which can decide the pixel status accurately (EMR, ER, CR), we can choose the optimal filter among the three depending on the pixel status and then set the output of the optimal filter as the output of CAF. Unfortunately, it is hard to detect the pixel status accurately in a noisy image, and the inaccuracy in detection will cause faultiness (edge blurring or failure in removing impulsive noise) in the denoising process. Moreover, using outputs of different filters will cause unsmoothness on the boundary between the regions calculated by different filters.

To deal with this problem, we propose a novel method to combine the outputs of different filters to form a new hybrid filter called content-adaptive filter (CAF). The structure of CAF is shown in Fig 1.

First of all, the three filters mentioned above are used to get three outputs and each one of the three filters outperforms the others under particular condition. Second, we can get the probability of the pixels' different status by the local statistic information and then we set the expectation of the pixel value as the output of CAF by calculating weighted sum of the three filters' outputs depending on the probability of the pixel status.

$$Output_{CAF} = Output_{MMF} \times P_{EMR} + Output_{3DPF} \times P_{ER} + Output_{FIR} \times P_{CR}. \quad (5)$$

Where P_{EMR}, P_{ER}, P_{CR} denotes the probability of the pixels in EMR, ER, and CR, respectively.

We set the qualification that

$$P_{EMR} + P_{ER} + P_{CR} = 1$$

to assure the output of CAF is the unbiased estimation of the pixel value.

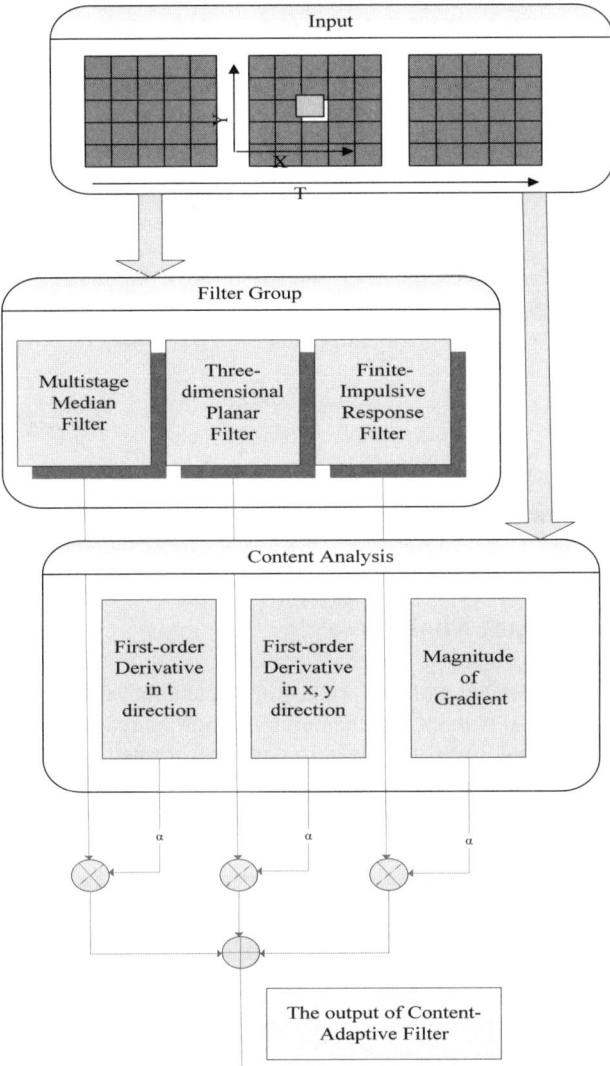

Fig. 1. The structure of content adaptive filter

There are many approaches to calculate P_{EMR}, P_{ER}, P_{CR} by statistic information. For the computational simplicity, we choose gradient magnitude to get P_{EMR}, P_{ER}, P_{CR}.

We set formulae as below because the larger magnitude of gradient is; the more possible it is that the pixel does not belong to CR. Moreover, $(|G_x|+|G_y|)$ indicates the probability which the pixel belongs to ER while the possibility that the pixel belongs to EMR is indicated by $|G_t|$. Zucker operator is adopted to calculate the magnitude of the gradient.

$$P_{EMR} = \alpha \mid G_t \mid . \tag{6}$$

$$P_{ER} = \alpha(\mid G_x \mid + \mid G_y \mid) . \tag{7}$$

$$P_{CR} = 1 - \alpha(G) . \tag{8}$$

$$G = \mid G_t \mid + \mid G_x \mid + \mid G_y \mid . \tag{9}$$

Where G_t, G_x, G_y denotes the first-order derivative in the x, y, t direction and G indicates the magnitude of gradient. α is used to normalize the gradient.

By the way, we can get the identity (11) by (7), (8) and (9)

$$0 \leq P_{EMR}, P_{ER}, P_{CR} \leq 1 . \tag{10}$$

$$\text{and,} \quad P_{EMR} + P_{ER} + P_{CR} = 1 . \tag{11}$$

which can promise that the output of CAF is the unbiased estimation of the pixel value.

4 Synthetic Data Analysis

The noises are generated with the process of calculating saliency maps, so it is hard to get the saliency map without its inherent noise. For the reason above, video sequences generated by computer were involved in our experiments to get the statistic properties of CAF. There is an example of the sequences providing a challenging environment. The top row of Fig.2 shows three consecutive original frames without noise while the

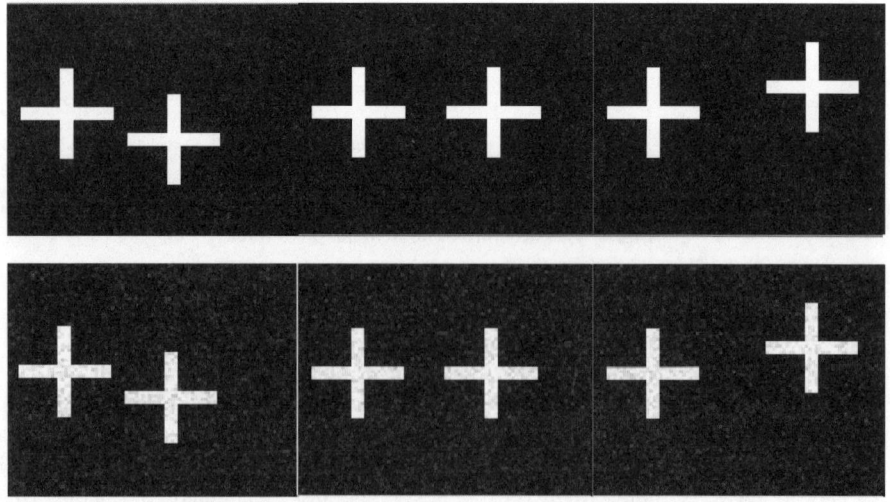

Fig. 2. Top row - original non-noisy synthetic sequence Bottom row - the sequence with additive noise

bottom row shows the frames with additive noise, respectively. We can see that the cross in left part of sequence image (LC) is stillness and the right cross (RC) moves drastically.

We use CAF to attenuate the noise in the middle frame of the noisy sequence and the first frame and the last frame of the noisy sequence are input as reference frames for their temporal information. Four other filters are involved for comparison.

Fig.3 (a) and (b) indicates that extending the mean filter (SMeanF) or median filter (SMF) directly to three-dimensional case will lead to terrible results. (c) shows that 3DPF plays well when the ROI does not move harshly (LC), but may demolish the edge when ROI moves harshly (RC). (d) shows that the frame smoothed by the MMF and the noise in background is not attenuated enough. (e) shows the frame calculated by CAF in which the edge of ROI was neither blurred nor destroyed and the noise is suppressed well.

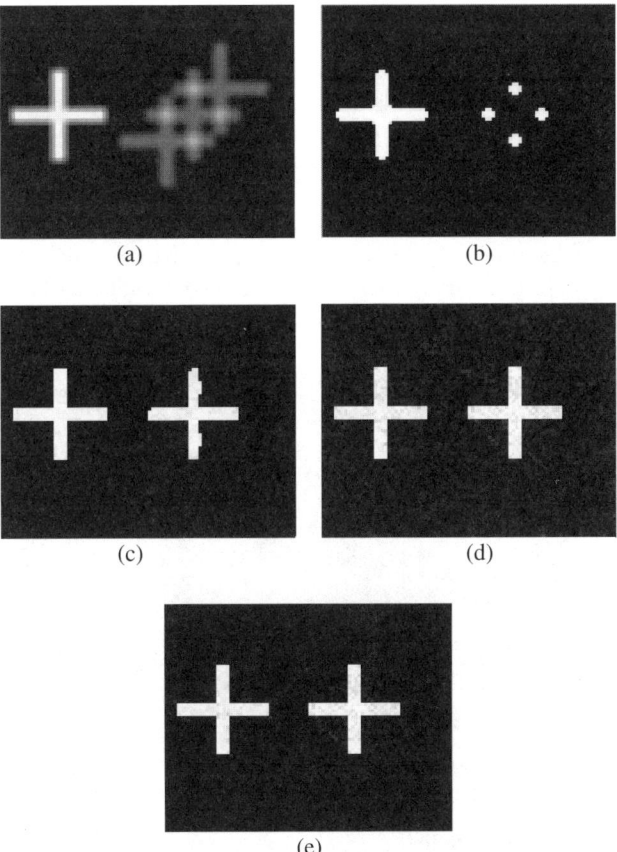

Fig. 3. (a)The saliency map got by SMeanF. (b) The saliency map got by SMF. (c) The saliency map got by 3DPF. (d)The saliency map got by MMF. (e) The saliency map got by CAF.

The PSNR and MSE of output frames of different filters are shown in Table 1.

Table 1. The statistic information of outputs using different filters

	PSNR	MSE
Noise	31.483	46.220
SMF	15.904	1669.82
SMeanF	16.397	1490.60
MMF	34.195	24.75
3DPF	24.637	223.53
CAF	36.038	16.193

5 Experiment Results

We adopt model proposed by Itti to calculate saliency maps of sequences about sports, news, and home videos. After that, we use CAF to get the new saliency sequences by attenuate the noise. At last, the approach proposed in [5] is used to draw the region of interest. The results show that the CAF benefits the extraction of attention objects, the noise on the background can be successfully removed and extracting region of interest will be more accurate.

There are some examples from a large number of our experiment results.

Fig 4, 5, 6 indicate that after the processing of CAF, the approach mentioned in [5] can get much more accurate locations of ROI.

In Fig 7 (e), the boundary of the table and the ball moving harshly are both remained after the processing of CAF. The phenomenon indicates the ability of CAF to preserve the details and objects harshly moving while suppressing noise.

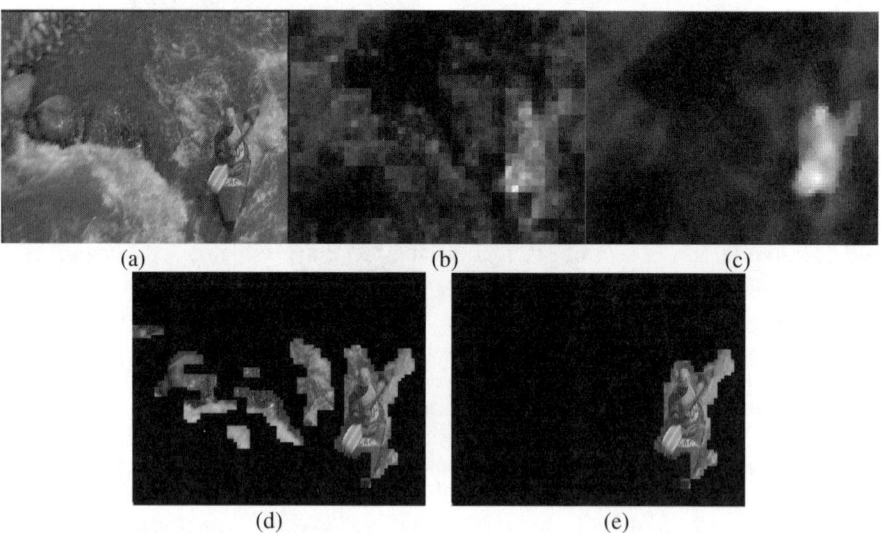

Fig. 4. (a) One frame of the real sequence. (b) The saliency map calculated by Itti's model (c) The new saliency map estimated by CAF (d) The ROI got by the original saliency map (e) The ROI got by the saliency map after filtering.

Denoising Saliency Map for Region of Interest Extraction 213

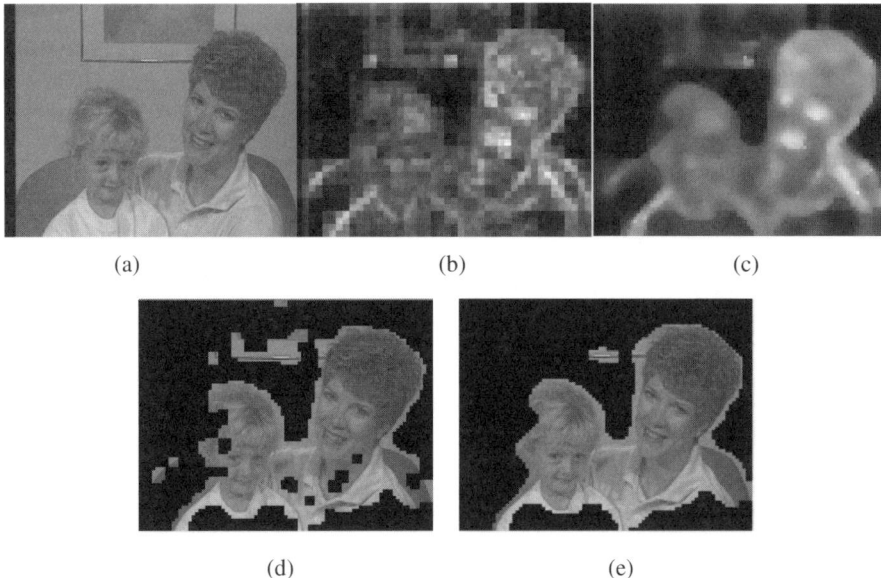

Fig. 5. (a) One frame of the real sequence. (b) The saliency map calculated by Itti's model (c) The new saliency map estimated by CAF (d) The ROI got by the original saliency map (e) The ROI got by the saliency map after filtering.

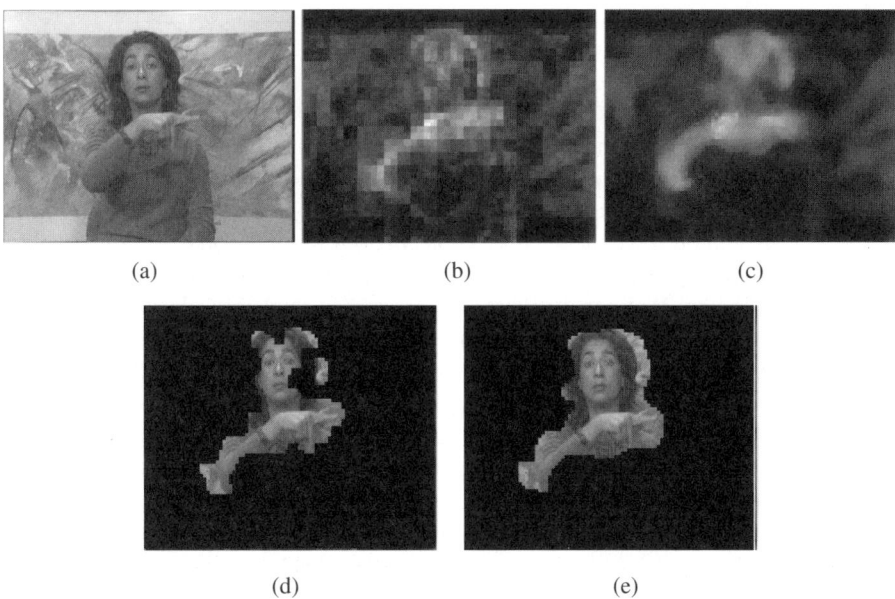

Fig. 6. (a) One frame of the real sequence. (b) The saliency map calculated by Itti's model (c) The new saliency map estimated by CAF (d) The ROI got by the original saliency map (e) The ROI got by the saliency map after filtering.

Fig. 7. (a) One frame of the real sequence. (b) The saliency map calculated by Itti's model (c) The new saliency map estimated by CAF (d) The ROI got by the original saliency map (e) The ROI got by the saliency map after filtering.

6 Conclusions

In this paper, we proposed a novel hybrid filter called CAF to denoise saliency maps which is a promising technique to improve the reliability and correctness of the extraction of ROI. The novel filter is specially designed for the saliency map and outperforms the conventional methods including SMF, SMeanF, MMF, 3DPF and FIR. The CAF can preserve the details of objects even when the objects move harshly while it also can suppress the noise well enough, which indicates that the poor performance of the traditional filters is diminished but the advantages of the traditional filters are made use of. This is attributed to the factors below; first, we define three statuses of pixels and calculate the probability of the different statuses before filtering. The output of CAF is calculated as the weighted sum of three different filters' outputs depending on the probability of the pixel status. Second, we take temporal information into account by using reference frames. At last, sufficient experimental results prove the effectiveness and necessity of our approach.

Although the paper is focused on the denoising of saliency maps, its idea and theoretical bases are not only applicable to saliency maps but also can be used for medical sequences or real sequences. In our future work, we will ameliorate the function of calculating α from the magnitude of gradient and extend the algorithm to the real sequence processing.

References

1. Ma, Y.F., Huang, X.Sh., Liu, L., Zhang, H.J.: A generic framework of user attention model and its application in video summarization. IEEE Transactions on Multimedia, 907–919 (2005)
2. Fan, X., Xie, X., Zhou, H.Q., Ma, W.Y.: Looking into video frames on small displays. In: MM 2003, Berkeley, California, USA, pp. 247–250 (2003)
3. Itti, L., Koch, C., Niebur, E.: A Model of saliency-based visual attention for rapid scene analysis. IEEE Transactions on PAMI, 1254–1259 (1998)
4. Ma, Y.F., Zhang, H.J.: Contrast-based image attention analysis by using fuzzy growing. In: Conf. ACM, pp. 374–381. ACM Press, New York (2003)
5. Li, Y., Ma, Y.F., Zhang, H.J.: Salient region detection and tracking in video. In: Proceedings International Conference ICME 2003, pp. 269–272 (July 2003)
6. Turkey, J.W.: Nonlinear methods for smoothing data. In: Conf. Rec. Eascon, p. 673 (1974)
7. Heinonen, P., Neuvo, Y.: FIR-Median hybrid filters. IEEE Transactions on Acoustic, 832–838 (1987)
8. Nieminen, A., Heinonen, P., Neuvo, Y.: A new class of detail-preserving filters for image processing. IEEE Transactions on PAMI, 74–79 (1987)
9. Bilge, M., Neuvo, Y.: 3-dimensional median filters for image sequence processing. In: Conf. ICASS, pp. 2917–2920 (1991)
10. Zucker, S.W., Hummel, R.A.: A three-dimensional edge operator. IEEE Transaction on PAMI, 324–331 (1981)
11. Canny, J.: A computational approach to edge detection. IEEE Transactions on PAMI, 679–714 (1986)
12. Marr, D.: Theory of edge detection. In: Proc. Royal Soc. London, 187–217 (1980)
13. Dhavale, N., Itti, L.: Saliency-based multi-foveated mpeg compression. In: IEEE Seventh international symposium on signal processing and its application, Paris, France, pp. 229–232. IEEE Computer Society Press, Los Alamitos (2003)

Cumulative Global Distance for Dimension Reduction in Handwritten Digits Database

Mahdi Yektaii and Prabir Bhattacharya

Concordia Institute for Information Systems Engineering,
Concordia University, Montreal, QC, H3G 1M8, Canada
{m_yektai,prabir}@encs.concordia.ca
http://www.ciise.concordia.ca

Abstract. The various techniques used to determine the reduced number of features in principal component analysis are usually ad-hoc and subjective. In this paper, we use a method of finding the number of features which is based on the saturation behavior of a graph and hence is not ad-hoc. It gives a lower bound on the number of features to be selected. We use a database of handwritten digits and reduce the dimensions of the images in this database based on the above method. A comparison with some conventional methods such as scree and cumulative percentage is also performed. These two methods are based on the values of the eigenvalues of the database covariance matrix. The Mahalanobis and Bhattacharyya distances will be shown to be of little use in determining the number of reduced dimensions.

Keywords: Principal Component Analysis, dimension reduction, separability, Mahalanobis distance, Bhattacharyya distance.

1 Introduction

An important problem occuring in the domain of pattern recognition is the *curse of dimensionality* due to the large number of features [1],[3],[8]. Some of the methods of dimension reduction have been discussed in [2],[6],[9],[11],[12]. Principal Component Analysis (PCA) (also called the *Hotelling Transform* or the *Karhunen-Loève Transform*), is one of the most popular methods in dimension reduction and data compression [1],[3]. Essentially PCA is a way of finding the basic directions along which the data are more concentrated. Ideally, the reduced number of features corresponds correctly to the *intrinsic dimensionality* [5] of the original classification problem. The extra dimensions address the representation noise or the variations of the data elements from a center point. An explanatory example could be the different views of the same face or different lighting conditions.

For many practical applications, the inherent dimension would not be very clear. However since subsequent pattern classification and object recognition techniques depend on reduced number of dimensions -designated as L in this paper, this number is of significant importance. In PCA, the dimension of the feature space after the reduction process is usually chosen in an ad-hoc manner [9],[7]. These methods are usually based on the sorted (descending order) eigenvalues of the covariance matrix. An often used method, known as *cumulative percentage*, is to obtain the reduced dimension L such that the ratio of the sum of the first L eigenvalues and the sum of all the eigenvalues is greater than a certain threshold. Another method known as *scree diagram* is to plot the eigenvalues and define L where the value of eigenvalue is *relatively* small. If the initial dimensionality of data is much higher than the number of available samples, M, then an upper bound for reduced dimensions would be same as M.

The classical statistical distance measures such as the *Mahalanobis* and *Bhattacharyya* distances as measures of separability of distributions [3],[4] could not be used to determine the number of features [13]. The simulations in this paper done on the handwritten digits database will show the deficiency of the two measures and that of scree and cumulative percentage methods for determining L. We use a new method of defining the reduced dimensions first presented in [13] and named as *cumulative global distance* (*cgd*).

The idea of [13] is to apply the PCA to the training data and compute a separability measure *cgd* among all the classes of the problem. It then needs *cgd* values to be plotted versus different values of L, the number of reduced dimensions. If this curve saturates (that is, does not change significantly after some value of L), the data classes have a separability (above that specific value of L). In this paper, this method is used for the MNIST handwritten digits database[10].

2 Background

2.1 Principal Component Analysis

Let the $N \times N$ matrix $\boldsymbol{S}$ be the covariance matrix of the training data: $\boldsymbol{S} = \frac{1}{M} \sum_{i=1}^{M} (\boldsymbol{x}_i - \boldsymbol{\mu})^T (\boldsymbol{x}_i - \boldsymbol{\mu})$ in that M is the number of data points, $\boldsymbol{x}_i \in R^N$ is the ith data point and $\boldsymbol{\mu}$ denotes the mean of those data points. In order to make the covariance matrix statistically meaningful and representative of the data to be classified, it is crucial to have enough data points in its computation. The dimension, N, of the data points is being reduced to $L(< N)$. L eigenvectors $\boldsymbol{V}_i (1 \leq i \leq L)$ corresponding to the L largest eigenvalues of $\boldsymbol{S}$ are then used to define a linear transformation given by: $\boldsymbol{\Phi} = [\boldsymbol{V}_1^T, \boldsymbol{V}_2^T, \cdots, \boldsymbol{V}_L^T]^T$ that reduces the data by transferring them to the R^L space. There are several ad-hoc methods available for selecting L [7],[9]. The eigenvalues λ_i of $\boldsymbol{S}$ are sorted in the descending order. Then a straightforward method is to plot λ_i versus i. Known as *scree*, the plot is studied in search of a saturation or an "elbow" after which the slope of the curve reduces significantly. Another approach is to consider the ratio of the sum of the first (largest) L eigenvalues and the sum of all eigenvalues - this ratio is called the *cumulative percentage* or C_p. By plotting C_p versus different values of L, we may seek a saturation effect.

2.2 Statistical Distance Measures

The Mahalanobis and the Bhattacharyya distances between two random distributions with means $\boldsymbol{\mu}_i$ and $\boldsymbol{\mu}_j$ and covariance matrices $\boldsymbol{S}_i$ and $\boldsymbol{S}_j$ are defined respectively as [3], [4]:

$$MD(i,j) = (\boldsymbol{\mu}_i - \boldsymbol{\mu}_j)^T (\frac{\boldsymbol{S}_i + \boldsymbol{S}_j}{2})^{-1} (\boldsymbol{\mu}_i - \boldsymbol{\mu}_j) \qquad (1)$$

$$BD(i,j) = \frac{1}{8} MD(i,j) + \frac{1}{2} \ln \frac{|\frac{\boldsymbol{S}_i + \boldsymbol{S}_j}{2}|}{\sqrt{|\boldsymbol{S}_i||\boldsymbol{S}_j|}} \qquad (2)$$

The *Bhattacharyya cumulative distance* (BCD) for a collection of c classes is defined as the sum of the pair-wise Bhattacharyya distance between all the classes [13]: $BCD = \sum_{i=2}^{c} \sum_{j=1}^{i-1} BD(i,j)$. [13] similarly defines *Mahalanobis cumulative distance* (MCD) as: $MCD = \sum_{i=2}^{c} \sum_{j=1}^{i-1} MD(i,j)$.

3 The Cumulative Global Distance [13]

Assume there are $c \geq 2$ distinct classes in the training data and $\boldsymbol{\mu}_i (1 \leq i \leq c)$ and $\boldsymbol{V}_k (1 \leq k \leq L)$ denote respectively the means of those classes and the eigenvectors corresponding to the L largest eigenvalues λ_k of the covariance matrix of all training data points. The cumulative global distance (*cgd*) is defined as a descriptive distance measure among several classes of the training data. Consider the following definition that would represent a possible measure of the distance between the two class means $\boldsymbol{\mu}_i$ and $\boldsymbol{\mu}_j$ along a specific eigenvector $\boldsymbol{V}_k$ (corresponding to eigenvalue λ_k):

$$d(i,j,k) = (\lambda_k (\boldsymbol{\mu}_i - \boldsymbol{\mu}_j)^T \boldsymbol{V}_k)^2 \qquad (3)$$

Equation (3) shows how "distant" the two means are in the orthogonal coordinate system created by the covariance matrix eigenvectors. In case of more than two classes of data, a *global* distance measure (*gd*) could be used:

$$gd(k) = \sum_{i=2}^{c} \sum_{j=1}^{i-1} d(i,j,k) \qquad (4)$$

When the problem is to define number of dimensions for a reduction process the *cumulative global distance* (*cgd*) is introduced:

$$cgd(L) = \sum_{k=1}^{L} gd(k) \qquad (5)$$

In that L is the dimensionality of the reduced data. The *cgd* is an increasing function of L, that is, its value will increase monotonically if more and more eigenvectors are used. If a *saturating effect* is observed in the graph of *cgd* versus L, then the corresponding L could be seen as the right dimensionality. According to [13] the Bhattacharyya and Mahalanobis distances will *not* show the saturation effect and then could not be used for dimension reduction purposes.

4 Finding Optimum Dimension of Reduced Feature Space

To obtain the optimal dimensionality using cgd, some steps of a PCA process are done [13]: The covariance matrix of the training data and its eigenvectors and eigenvalues computed. For each new dimensionality L, the cgd is computed and a graph of the cgd values is plotted. The proposed algorithm to find the number of dimensions is described below [13]:

Algorithm 1. Finding Optimum Dimension of Reduced Feature Space

---

1. Set $L \leftarrow 1$, decide a maximum dimensionality for reduction process (Max).
2. Compute $cgd(L)$ and $L \leftarrow L+1$, if $L < Max$ repeat this step, otherwise plot cgd.
3. Obtain the optimum L by fitting cgd values to the curve $y = A(1 - e^{-ax})$ where the slope starts to be less than 0.001 (this choice can be customized to take any small value).

By applying the above algorithm to the training data, the behavior of the plot in Step 2 is analyzed in order to seek the saturating behavior. Once the saturating effect is observed, the start of that saturating region will be found using the (L, cgd) pairs as mentioned in Step 3 of the above algorithm. The optimal value of L will be denoted as L^*. Using a dimension larger than L^*, would unnecessarily increase the cost of the classification process.

5 Implementation

The cgd is applied for feature reduction in handwritten digits database. The handwritten character database MNIST of the national institute of standards and technology (NIST) has been used in this work. The database contains 60000 training samples and 10000 test samples in ten classes. 24000 samples (2400 for each digit) of the whole training set were used to train the system (covariance matrix computation) and 4000 samples of the test set (400 for each digit) were used to be classified after each reduction. Each sample in this database is a 28×28 image and is already raster scanned into a vector and stored in the database. Here the vectors are normalized to one. The mean vectors of each class are also normalized after being computed. Figure 1 shows several samples of the database. The efficiency of the nearest neighbor classifier increases by increasing the number of dimensions. The graph of the cgd together with that of correctly classified samples versus number of dimensions are shown in Figure 2. As could be seen from Figure 2, there is a strong correlation between the two graphs in terms of saturation.

By applying the algorithm described in Section 4, the value of L^* (which is defined at the end of Section 4) is found to be 11 for the saturating slope of 0.001.

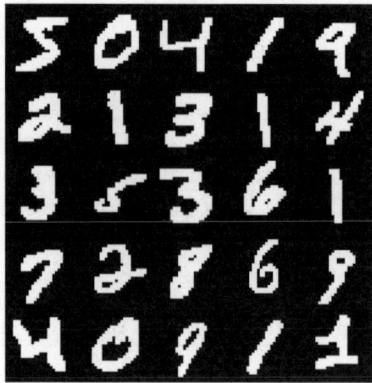

Fig. 1. Some samples of the digits in MNIST database [10]

With this number of features, the classification efficiency is around 0.94% of the case with using 50 features. The whole process requires less than 30 minutes on a Pentium IV machine with 1536 MB of RAM. In fact much of the time is used for computing the eigenvalues and eigenvectors of the covariance matrix.

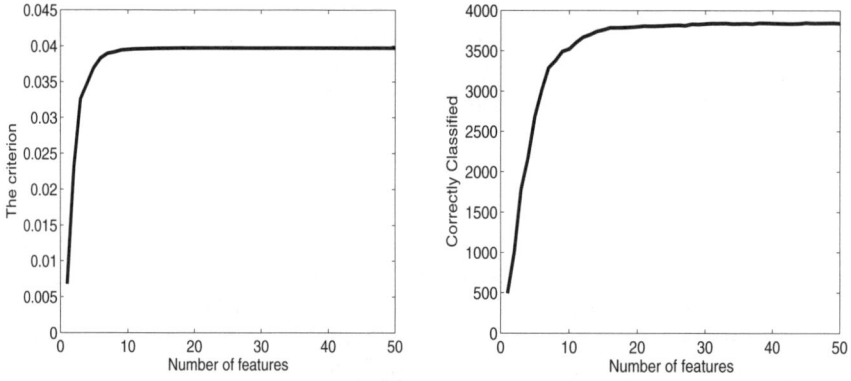

Fig. 2. Left: *cgd*, right: correctly classified samples

5.1 Comparisons

In order to show the effectiveness of the *cgd* criterion, the scree diagram and the cumulative percentage (see Section 2.1) have been tested on the database discussed in this section. The results of these two criteria for the MNIST database are shown in Figure 3. From Figure 3, the saturation effect could not be seen even for large dimensions (Compare with Figure 2).

In order to observe deficiency of BCD and MCD (see section 2) in determining the number of dimensions in PCA, the two measures have been computed and plotted in Figure 4. As could be seen from the figure, both measures increase with L but do not show a strong tendency toward saturation.

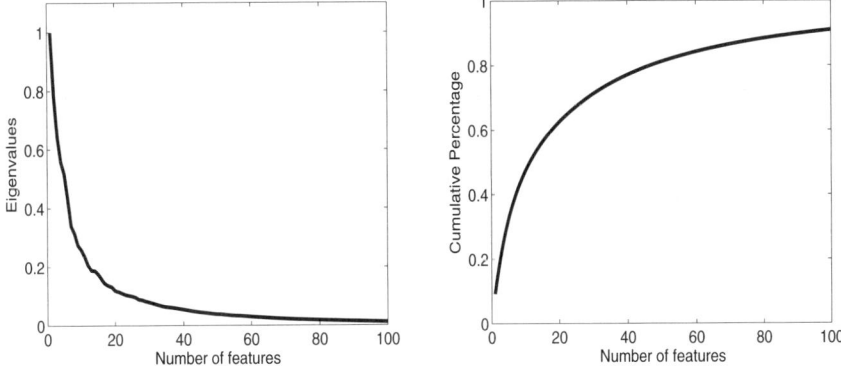

Fig. 3. Left: Scree Diagram, right: Cumulative Percentage

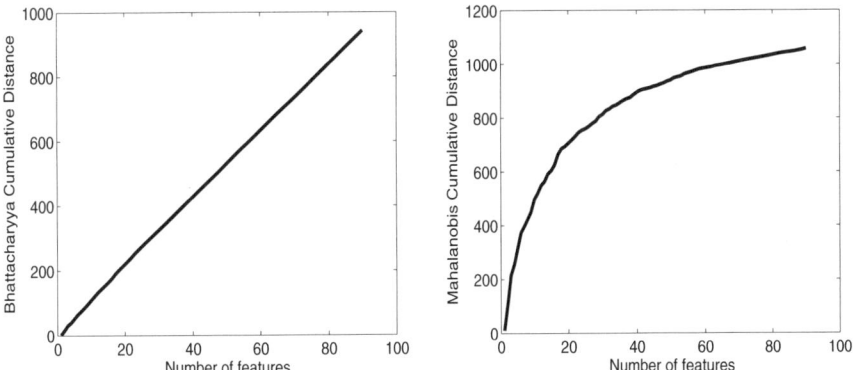

Fig. 4. Left: Bhatatcharyya Cumulative Distance, right: Mahalanobis Cumulative Distance

6 Conclusion

We used the newly introduced cumulative global distance (cgd) in reducing the dimensionality of the samples of MNIST handwritten digits database and found out that cgd is a good criterion for this purpose. It gives a lower bound on the number of features to be selected. A comparison with conventional scree diagram and the cumulative percentage was also performed. cgd shows a better performance than these conventional approaches. The modified statistical distance measures for several distributions, i.e. Mahalanobis and Bhattacharyya cumulative distances were also tested for the mentioned database. The result is that they cannot efficiently be used for this purpose.

Acknowledgments. We wish to acknowledge with thanks the use of the MNIST handwritten digits database [10].

References

1. Bishop, C.M.: Pattern Recognition and Machine Learning. Springer, New York (2006)
2. Cangelosi, R., Goriely, A.: Component retention in principal component analysis with application to cDNA microarray data. Biology Direct 2(2) (2007), available from http://www.biology-direct.com/content/2/1/2
3. Duda, R.O., Hart, P.E., Stork, D.G.: Pattern Classification, 2nd ed. John Wiley, New York (2001)
4. Fukunaga, K.: Introduction to Statistical Pattern Recognition, 2nd edn. Academic Pr., New York (1990)
5. Fukunaga, K., Olsen, D.R.: An Algorithm for Finding Intrinsic Dimensionality of Data. IEEE Trans. Comp. 20(2), 176–183 (1971)
6. Hadsell, R., Chopra, S., LeCun, Y.: Dimensionality Reduction by Learning an Invariant Mapping. In: IEEE Conf. Comp. Vision and Pattern Recog., pp. 1735–1742. IEEE Computer Society Press, Los Alamitos (2006)
7. Jackson, J.E.: A User's Guide to Principal Components. John Wiley, New York (2003)
8. Jain, A.K., Chandrasekaran, B. (eds.): Dimensionality and Sample Size Considerations in Pattern Recognition Practice, in Handbook of Statistics, pp. 835–855. North Holland, Amsterdam (1982)
9. Jolliffe, I.T.: Principal Component Analysis, 2nd edn. Springer, Berlin (2002)
10. LeCun, Y., Bottou, L., Bengio, Y., Haffner, P.: Gradient-based learning applied to document recognition. Proceedings of the IEEE 86(11), 2278–2324 (1998)
11. Rayner, M.L., Punch, W.F., Goodman, E.D., Kuhn, L.A., Jain, A.K.: Dimensionality Reduction Using Gene tic Algorithms. IEEE Trans. Evolutionary Computation 4(2), 164–171 (2000)
12. Tipping, M.E., Bishop, C.M.: Probabilistic Principal Component Analysis. J. Roy. Stat. Soc. 61(3), 611–622 (1999)
13. Yektaii, M., Bhattacharrya, P.: A Criterion for Measuring the Separability of Clusters and Its Applications to Principal Component Analysis, Concordia Institute for Information Security Institute (CIISE) internal report (March 2007)

A New Video Compression Algorithm for Very Low Bandwidth Using Curve Fitting Method

Xianping Fu[1,2], Dequn Liang[2], and Dongsheng Wang[1]

[1] Research Institute of Information Technology, Tsinghua University, Beijing China
[2] School of Computer Science and Technology, Dalian Maritime University
1 LingHai Road, Dalian, China
fxp@dl.cn

Abstract. A new video object encoding algorithm based on the curve fitting trajectory of video object moving edges pixels is proposed in this paper. This algorithm exploits the fact that, under certain circumstances where the objects in the video are not moving quickly as in video conferencing and surveillance, the only significant video object information are the edges of the moving objects. Other object information remains relatively constant over time. This algorithm is modified from the standard video compression algorithms. Simulation results show that, under standard test sequences, this encoding algorithm has a lower bit rate than the classical DCT method.

Keywords: curve fitting, video coding, low bit rate, videoconference.

1 Introduction

Although rapid progress has been made in network and digital communications, demand for data transmission bandwidth is still beyond the capabilities of available technologies [1]. Further research needs to be done to discover new ways to take advantage of the available bandwidth using low bit rate encoding. For example, low bit rate video conferencing over the PSTN and over wireless media is of particular interest [2-3].

We show that the standard video compression algorithms can be modified under certain circumstances where the objects in the video are not moving quickly as in video conferencing and surveillance[4-5]. New compression algorithms based on moving edges are therefore developed, which are based on the modification of the standard compression algorithms to achieve higher compression ratio and simultaneously to enhance visual performance[6].

Instead of using the block-based transform, the key idea of this algorithm is to use the edges of moving objects for video encoding since these edges are the most important part of each frame[7]. The other parts which do not move remain constant over time and do not need to be considered. Therefore, less information is transformed in the video encoding process. Instead of using discrete block information transform coefficients as in the DCT method, the parameters of the curve which contains the pixel values along the trajectory of the motion over consecutive frames are considered and sent through the network. In this way the system can make good use of available network resources to achieve optimal performance without losing important information.

The proposed new video coding scheme med-filter the input sequence images before pre-processing. The difference of every successive frame is accumulated to form the moving regions of the video objects. The regions contain the edges that have the same moving characteristic. Then every moving region of the moving objects is labeled. The labels of the regions guarantee these regions have the edges with the same curve fitting coefficients. The object edges are calculated using a logical operation AND between the moving region and the edge detected image. Thus a new coding method using motion detection and curve fitting based algorithm is described as follows, the proposed scheme is based on the motion of the same edges in eight previous frames. The block diagram is illustrated in Fig.1.

The algorithm in this paper includes the detection, the labeling and the curve fitting of the moving edges and the non-motion region update.

2 The Detection of Moving Edges

The changed edges are based on the moving region segmentation [8], which is obtained from the sum of frames differences processed by the morphological operations. The video object edges are calculated using a logical operation AND between the moving region and the edge image. These edges are fitted by a conic curve and the parameters of fitting are transferred, meanwhile the non-motion regions are extracted from the background image which is updated frequently by motion estimation and compensation.

Edge information that denotes with the spatial gradients of image is the basis to distinguish the different objects. The Canny method finds edges by looking for local maxima of the gradient of image. The gradient is calculated using the derivative of a Gaussian filter. This method is less likely than the others to be fooled by noise, and more likely to detect true weak edges. Therefore the Canny edge detector is used in this paper to find the object edges; the result of edge image is $I_{edge}(x, y)$.

To calculate the difference of image sequence $I_n(x, y)(1 \leq n \leq N)$ that processed by the median filtering, setting the difference of previous n frames is $D_{n-1}(x, y)$, and the difference of previous n+1 frames is $D_n(x, y)$, then

$$D_n(x, y) = D_{n-1}(x, y) + |I_{n+1}(x, y) - I_n(x, y)| \qquad (1)$$

where $2 \leq n \leq N-1$, and $D_1(x, y) = I_2(x, y) - I_1(x, y)$.

The results of edge detection and difference of frames are combined by a logical operation AND to get the final moving edges. The equation for the detection of moving edges M_{edge} is

$$M_{edge}(x, y) = I_{edge}(x, y) \& D_n(x, y) \qquad (2)$$

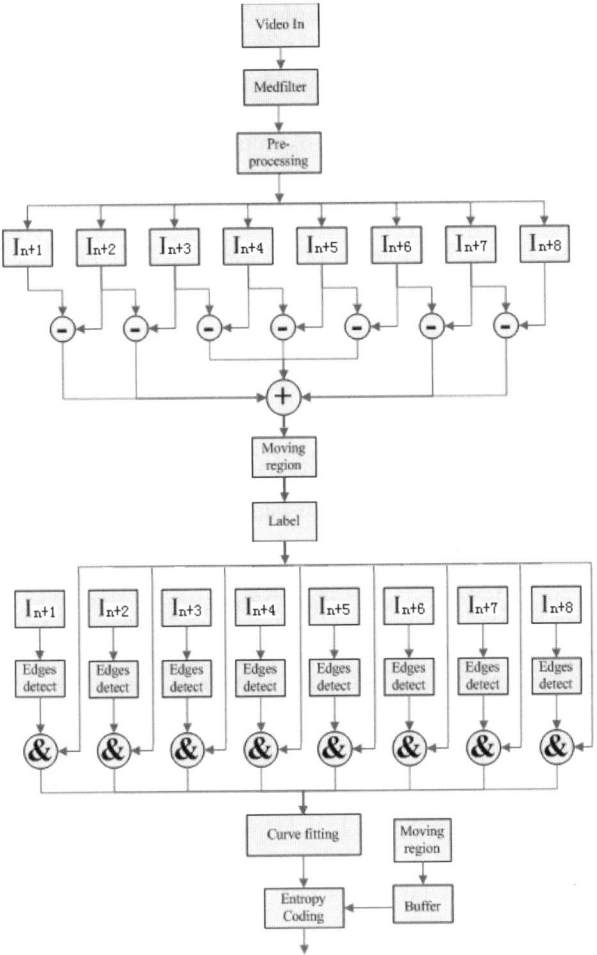

Fig. 1. Block diagram of the curve fitting-based video encoding scheme

3 Labeling Moving Edges

The detected moving edges are labeled, the purpose of label is to separate the different video object edges and count the changed edges, so the result can be used to fit each labeled edges along the temporal moving trajectory. The label is based on the moving edges that are obtained from (2), and guarantee the same edge in the successive frames to be coded using curve fitting method. The same edges in the successive frames should have the same label numbers, the curve fitting of the moving edges on the temporal direction is based on the same label number.

4 Curve Fitting Moving Edges Pixels

The key idea of curve fitting is that a moving edge consists of many pixels each of which may have a different pixel intensity in consecutive frames. The different values of an individual pixel can be curve fitted. To represent the different values of that pixel, only the fitted curve's parameters need to be coded. In the case where the objects are not moving quickly, only a small number of the pixels of a moving edge need to be coded.

Based on the moving intensity of video objects, there are two kinds of curve fitting methods. The first one is based on the background and video objects; the other is based on the video objects only. Because of the heavy calculated load and complexion of the first method, the second method is adopted in our approach.

The type of the curve that best matched the value of the moving edges is selected automatically using the minimum of SAD between the actual value and curve fitted according to the moving intensity in the video. The parameter of curve fitting should be calculated automatically. A simple conic arithmetic model is chosen to fit the moving trajectory of video object edges in this paper.

The conic equation used in this paper is

$$I(x_i) = a_2 x_i^2 + a_1 x_i + a_0 \tag{3}$$

Where x_i is the value of pixel on the same moving edges M_{edge}, i=1, 2, 3...n is the serial number of frames, the n equal 8 in our experiment, that is the curve fitting is calculated on every 8 frames.

Set I_i' is the value of pixel on the edge of the № i frame, and

$$E = \sum_{1}^{n} (I(x_i) - I_i')^2 \tag{4}$$

Equation (4) is the sum squared difference of frames, if the value is infinite, then

$$\frac{\partial E}{\partial a_k} = 0 \quad (k=0,1,2) \tag{5}$$

and

$$\sum_{1}^{n} (a_2 x_i^2 + a_1 x_i + a_0 - I_i') \bullet x_i^k = 0 \tag{6}$$

Where k =0, 1, 2, the curve fitting parameter can be got from (6).

5 The Non-motion Region Update

In the consecutive frames, some regions of each frame are static and have no changes, and other regions are changed because of motion. So the non-motion regions in frames can be considered as background images, which have basically smooth edges. The several frames that can be used to encode using curve fitting method generally have the similarity background. So the each frame can be divided into two kind regions, one is moving edges and the other is background image. The moving edges can be coded by the curve fitting method, and the other regions can be update using the background buffer. In other words, assuming that both the encoder and the decoder have the same previous eight frames, motion estimation is performed between these frames resulting in a dense motion vector field of the previous frames. Then, based on the assumption of constant motion along its trajectory, one can determine the location of the moving areas by using the previous buffer information only. The background buffer is more efficient in this case than the general images.

6 Experimental Results

The proposed scheme is implemented in the standard test color video sequences of Akiyo. During the coding, the edges of the video objects are calculated using curve fitting on every 8 frames. So there are 8 frames in buffer when coding. The compress ratio is about 8.01 times without entropy coding. The experimental results are showed in Fig.2, in which (a) is the original image of frame 1, (b) is the sum of the difference of previous 8 frames, (c) is the reconstructed image of frame 1.

This algorithm is compared with the classic DCT method. In this paper, a two-dimensional DCT of 8-by-8 blocks is calculated for an Akiyo sequence image. It discards (sets to zero) all but 10 of the 64 DCT coefficients in each block and then reconstruct the image using the two-dimensional inverse DCT of each block. Although there is some loss of quality in the reconstructed image, it is clearly recognizable. The compress ratio is about 6.4 times without entropy coding. The mask matrix is shown as below:

$$mask = \begin{bmatrix} 1 & 1 & 1 & 1 & 0 & 0 & 0 & 0 \\ 1 & 1 & 1 & 0 & 0 & 0 & 0 & 0 \\ 1 & 1 & 0 & 0 & 0 & 0 & 0 & 0 \\ 1 & 0 & 0 & 0 & 0 & 0 & 0 & 0 \\ 0 & 0 & 0 & 0 & 0 & 0 & 0 & 0 \\ 0 & 0 & 0 & 0 & 0 & 0 & 0 & 0 \\ 0 & 0 & 0 & 0 & 0 & 0 & 0 & 0 \\ 0 & 0 & 0 & 0 & 0 & 0 & 0 & 0 \end{bmatrix}$$

The most commonly used measure PSNR is given as an indication of the image quality of the reconstructed frame. It is defined as PSNR=10lg (255^2/MSE), where the MSE is the mean squared error between the original and decoded frame.

The proposed video coding has obtained the averaged PSNR 41.68dB and the DCT algorithm averaged PSNR 30.41dB for the test sequence Akiyo. The PSNR of some frames are shown in Table 1. We can see that the PSNR of the reconstructed frame is relatively constant. The original and reconstructed frame № 1 are shown in Fig.2. As it can be seen, the proposed video coding method in this paper has higher PSNR than DCT and has no blocking effects. However, we can detect some artifacts and smearing effects due to the curve fitting error.

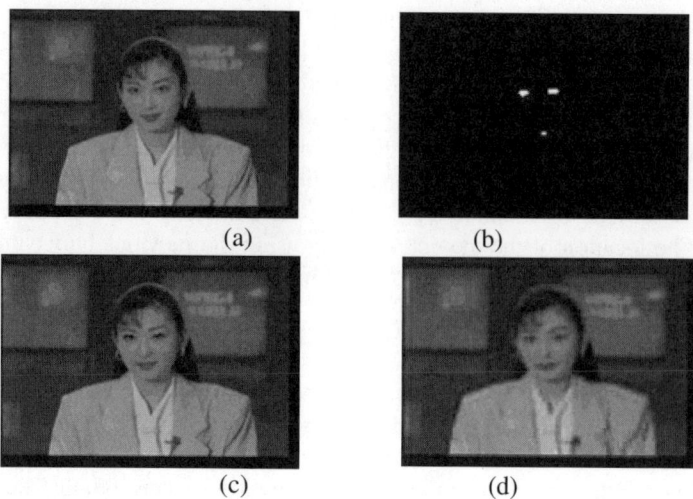

Fig. 2. (a) the original frame № 1. (b) The sum of the difference of previous 8 frames, (c) The reconstructed image of frame 1 using proposed method. (d) The reconstructed DCT frame № 1.

Table 1. PSNR of some frames of Akiyo

Frame№	Proposed method	DCT
1	44.07	30.46
2	43.63	30.42
3	43.12	30.43
4	40.64	30.43
5	40.18	30.40
6	40.94	30.40
7	40.43	30.39
8	40.41	30.37

7 Conclusions and Future Work

In this paper we presented a new approach to video encoding using the correlation of adjacent frames. The trajectory of the edge pixels in the sequence is coded using prediction of curve fitting method. With curve characteristic on temporal, the edge detection and region segmentation are used to get better coding efficiency. The experiment result on the image that has sample background and not active motion have a good subjective image quality and very low bit rate.

Future research efforts will concentrate on three items. Firstly we aim to increase the compression ratio itself by improving the encoding performance. Secondly, we will adopt a new model-based method to bring down the bit rate. Finally, the work on a good segmentation algorithm is in progress for locating objects within a head-shoulder scene.

References

1. Chen, M., Fowler, M.L.: The Importance of Data Compression for Energy Efficiency in Sensor Networks. In: Proceedings of Conference on Information Sciences and Systems, pp. 12–14. Johns Hopkins University (March 2003)
2. Fu, X., Wang, Z.: Video coding of model-based at very low bit rates. In: Proceedings of SPIE - The International Society for Optical Engineering, Lugano, Switzerland, vol. 2, pp. 1224–1231 (2003)
3. Richardson, I.E.G.: H.264 and MPEG-4 Video Compression: Video Coding for Next-generation Multimedia, pp. 159–170. John Wiley & Sons Ltd, England (2003)
4. Alsaqre, F.E., Baozong, Y.: Moving object segmentation for video surveillance and conferencing applications. In: ICCT 2003, 2003th edn. Proceedings International Conference on Communication Technology, pp. 1856–1859 (1859)
5. Mann, S.: Headmounted wireless video: computer-supported collaboration for photojournalism and everyday use Communications Magazine. IEEE 36(6), 144–151 (1998)
6. Chang, N.Y.-C., Chang, T.-S.: Combined Frame Memory Motion Compensation for Video Coding? Circuits and Systems for Video Technology. IEEE Transactions 16(10), 1280–1285 (2006)
7. Wang, C., Brandstein, M.: Robust head pose estimation by machine learning. In: Image Processing, 2000, Septmber 10–13, vol. 3, pp. 210–213 (2000)
8. Fu, X., Wang, Z., Liang, D.: Segmentation of moving objects in image sequence based on orientation information measure, Chongqing, China, pp. 969–975. World Scientific Publishing Co. Pte Ltd, Singapore (2004)

The Influence of Perceived Quality by Adjusting Frames Per Second and Bits Per Frame Under the Limited Bandwidth

Huey-Min Sun[1], Yung-Chuan Lin[1], and LihChyun Shu[2,*]

[1] Dept. of Information Management, Chang Jung Christian University, Taiwan, ROC
prince@mail.cjcu.edu.tw
[2] Dept. of Accounting, National Cheng Kung University, Taiwan, ROC
shulc@mail.ncku.edu.tw

Abstract. Under the limited bandwidth, MPEG-4 video coding stream with Fine Granularity Scalability can be flexibly dropped by very fine granularity to adapt to the available network bandwidth. Therefore, we can either reduce the frame rate, i.e., reduce the frames per second(FPS), by dropping partial frames to keep the spatial sharpness of an image or reduce the bits per frame(BPF) to keep the temporal continuity of a video. We attempt to understand that different content characteristics for the above two schemes affect the visual perceived quality when the bandwidth is limited. In this paper, the double stimulus continuous quality evaluation(DSCQE) is used as our subjective measurement. In our experiment, the subjects assess the scores of perceived quality by comparing the reference sequences with the test sequences for different content characteristics. We find that video contents with low motion characteristic suit to low frame rate and video contents with high motion characteristic suit to high frame rate under the limited bandwidth. The perceived quality of the spatial sharpness for the detailed texture sequences is influenced more than the easy texture sequences when the bit rate is increased.

Keywords: content characteristic, frame rate, bit rate, subjective measurement.

1 Introduction

The problem of maximizing the visually perceived quality under the limited network resource is emerging from the service of video on demand for streaming. Different video content characteristics, such as motion and texture, may prefer different frame rates and bit rates. Two methods can be employed to observe the change of perceived quality. One is an objective quality assessment based on the mathematical formula to quantify the visual perception. The other is a subjective

---

* Corresponding author. LihChyun Shu was supported in part by NSC grant NSC-2221-E-006-277.

quality assessment based on the rating-scale of subjects to measure the quality of perception. The former usually uses the peak-signal-noise ratio(PSNR) to measure an image quality. The latter adopts the mean opinion score(MOS) to assess an perceived quality. In the paper, we will use both the PSNR and the MOS to analyze the differences of different video contents by varying the frame rates and the bit rates.

Yadavalli et al.[2] consider only a motion characteristic to classify the video contents into low, medium, and high motions. Schaar and Radha[3] consider both a motion and a texture for the video sequences to analyze the temporal and the signal-to-noise scalabilities in MPEG-4 FGS. Gulliver and Ghinea[4] show that the higher frame rates, although resulting in a better perceived level of quality and enjoyment, across different video contents, do not significantly increase the level of a user information assimilation. Cuetos et al.[5] apply the evaluation framework based on MPEG-4 FGS to investigate the rate-distortion optimized streaming at different video frame aggregation levels. The video quality is related to the frames per second(FPS) and the bits per frame(BPF), especially on the constraint bandwidth. Therefore, the factors of influence on video quality are the frame rate, the bit rate, and the content.

For a constrained bandwidth network, MPEG-4 video coding stream with Fine Granularity Scalability (FGS) can be flexibly dropped by very fine granularity to adapt to the available network bandwidth. Under the limited bandwidth, we can either reduce the frames per second(FPS), by dropping partial frames to keep the spatial sharpness of an image or reduce the bits per frame(BPF) to keep the temporal continuity of a video. However, we don't know whether different content characteristics for the above two schemes affect the visual perceived quality when the bandwidth is limited. We want to discover the trade-off between the FPS and the BPF by varying the bit rate for the different content characteristics.

In our experiment, we take 15 video DVDs to edit 233 clips as our test databases. The clips are classified into four types for the motion and the texture characteristics by the cluster analysis method. Our experimental process refers to [14]. At the same time, we implement the assessment interface for the DSCQE to quickly collect the responses of the questionnaires. We find that the video contents with low motion characteristic prefer low frame rate and the sequences with high motion characteristic prefer high frame rate. In addition, the perceived quality for different content characteristic has significant and partial significant differences under the different bit rates and the different frame rates, respectively. We also analyze the mean opinion scores of 212 subjects by some statistic tools to understand the influence levels of different content characteristics such as the low motion and the easy texture, the high motion and the easy texture, the low motion and the detailed texture, and the high motion and the detailed texture. The results and findings indicate that the frame rate and the bit rate variables influence the different levels on the spatial sharpness, the temporal continuity, and the integral satisfaction.

2 Related Works

Multimedia contents are very diversifying, for example, sports footage, talk show, distance learning, and news. Apteker et al. [1] explore the relationship between video acceptability and frame rate by different content. They show that users perceive a reduced frame rate for a continuous-media stream differently, depending on the content. The multimedia contents are categorized by three dimensions: (1) the temporal nature of the data, (2) the importance of the auditory, and (3) visual components to understanding the message. Yadavalli et al. [2] consider only the motion characteristic to classify the video contents into low, medium, and high motions. Schaar and Radha[3] consider both the motion and the texture of the video sequences to analyze the temporal and the signal-to-noise scalabilities in MPEG-4 FGS. Gulliver and Ghinea[4] show that higher frame rates, although resulting in a better perceived level of quality and enjoyment, across different video contents, do not significantly increase the level of user information assimilation. Cuetos et al. [5] apply the evaluation framework based on MPEG-4 FGS to investigate the rate-distortion optimized streaming at different video frame aggregation levels. The video quality is related to the frames per second(FPS) and the bits per frame(BPF), especially on the constraint bandwidth. Therefore, the factors of influence on video quality are frame rate, bit rate, and content.

Most image processing researches adopt the objective quality measurement such as the peak signal to noise ratio(PSNR) and the mean squared error(MSE). Aeluri et al.[15] combine the four parameters: motion characteristic, encoder, frame rate, and bit rate based on the MSE to assess the video quality. Although some researchers try to propose new objective quality metrics, the acquired process of related information in coding is too complicated. The ANSI National Telecommunications and Information Administration (NTIA) General Model for measuring video quality adopts seven parameters including the loss of spatial information, the shift of edges from horizontal and vertical to diagonal, the shift of edges from diagonal to horizontal and vertical, the spread of the distribution of two-dimensional color samples, the quality improvement from edge sharpening or enhancements, the interactive effects from spatial and temporal, and the impairments from the extreme chroma. The seven parameters are based on four constructs including Spatial Alignment, Processed Valid Region, Gain and Level Offset, and Temporal Alignment [9,10]. Yao et al.[16] employ the visual quality scores based on a combination of three objective factors: visually masked error, blurring distortion and structural distortion.

The choice of the PSNR is motivated by Video Quality Expert Group (VQEG)[17], which states that none of the objective measures performs better than the computationally very simple PSNR in predicting the scores assigned by humans. However, the method is hard to understand the perceived quality of service for the consecutive video frames. Zink et al. [11] show that the PSNR is not an appropriate metric for variations in layer-encoded video. They conduct a subjective assessment on variations in layer-encoded video with the goal to assess the appropriateness of existing quality metrics. The quality of perception (QoP) and the user-level QoS are presented in [6] and [7]. QoP involves not only users' satisfaction but also their ability to perceive, synthesize and analyze multimedia information. The authors examine the relationship between application-level QoS, users' understanding and perception on multimedia clips by empirical experiment. Ghinea and Thomas[6] find that significant reductions in frame rate and color depth do not result in a significant QoP degradation. Nam et al.[8] present visual content adaptation techniques considering the users visual perception characteristics. They address how the visual properties of image and video content are adapted according to two types of visual accessibility characteristics: color vision deficiency and low-vision impairment. Therefore, the subjective quality measurement complements the objective quality measurement. The subjective quality evaluation has been defined the double-stimulus impairment scale(DSIS), the double-stimulus continuous quality evaluation(DSCQE), the single-stimulus(SS), stimulus-comparison(SC), the single stimulus continuous quality evaluation (SSCQE), and the simultaneous double stimulus for continuous evaluation (SDSCE) methodologies by ITU-R Recommendation BT.500 [12,13].

The DSIS method is cyclic in that the assessor is first presented with an unimpaired reference, then with the same picture impaired. Following this, the assessor is asked to vote on the second, keeping in mind the first. In sessions, which last up to half an hour, the assessor is presented with a series of pictures or sequences in random order and with random impairments covering all required combinations. The unimpaired picture is included in the pictures or sequences to be assessed. At the end of the series of sessions, the mean score for each test condition and test picture is calculated. The DSCQE method is also cyclic in that the assessor is asked to view a pair of pictures, each from the same source, but one via the process under examination, and the other one directly from the source. He is asked to assess the quality of both. The assessor is presented with a series of picture pairs (internally random) in random order, and with random impairments covering all required combinations. At the end of the sessions, the mean scores for each test condition and test picture are calculated. In the SS method, a single image or sequence of images is presented and the assessor provides an index of the entire presentation. In the SC method, two images or sequences of images are displayed and the viewer provides an index of the relation between the two presentations. In the SSCQE method, it is useful for the subjective quality of digitally coded video to be measured continuously, with subjects viewing the material once, without a source reference. The SDSCE has been developed starting from the SSCQE, by making slight deviations concerning the way of presenting the images to the subjects and concerning the rating scale. The SS and the SC methods don't be considered in the video test but to assess image quality.

In our experiment, the adopted DSCQE is suited to compare with the objective assessment for the PSNR. This is because our material must be with both test and reference sequences. In these subjective assessment methods, the mean opinion score(MOS) is used as the user-level QoS parameter or quality of perception. In order to measure MOS values, the rating-scale method[14], where an experimental subject classifies objects into some categories, is used. Each category is assigned a score to represent from bad to excellent. The rating scale is presented electronically. The range of the electronic rating scale may be between 0 and 100, between -3 and +3, or between 1 and 7.

3 Research Architecture

Our research architecture shown in Fig. 1 illustrates the relationship between three independent variables and the dependent variable, the quality of perception. We use the motion characteristic and the texture characteristic to classify the video contents into four types. We denote that Type 1 represents the low motion and the easy texture characteristics, Type 2 the low motion and the detailed texture characteristics, Type 3 the high motion and the easy texture characteristics, and Type 4 the high motion and the detailed texture characteristics. The second independent variable is a frame rate which is varied by 10, 15, 20, and 25 fps. The third independent variable is a bit rate which is varied by 200, 400, 600, 800, and 1000 kbps.

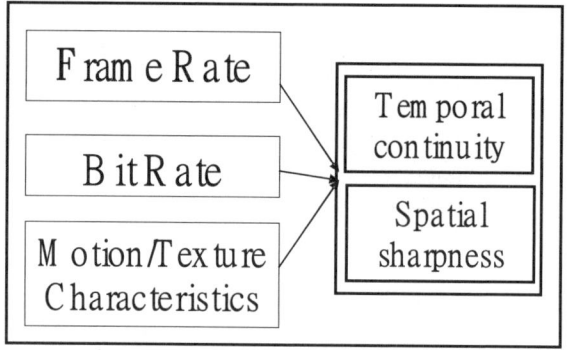

Fig. 1. Research architecture

In the dependent variable, we use two indicators to evaluate the performance. One is the temporal continuity of a sequence, which is the quality of consecutive frames. The other is the spatial sharpness of an image, which is the average quality of all the individual frames.

We employ two video quality assessment methods to measure the effect of a video quality. One is an objective measurement, which computes the average PSNR of luminance component for the spatial sharpness of an image. We make use of the average PSNR difference between a previous frame and a successive frame, denoted as $\triangle$PSNR, to evaluate the temporal continuity of consecutive frames. The choice of PSNR is motivated by Video Quality Expert Group(VQEG)[17], which states that none of the objective measures performs better than the computationally very simple PSNR in predicting the scores assigned by humans.

The other is a subjective measurement which performs the DSCQE test. Each trial consists of two presentations, one termed the "reference" (typically original source material) and one termed the "test" (typically processed material). The source material for both presentations is identical in content. The processed material shows the material with the same bit rate after alteration by varying the frames per second and the bits per frame. Subjects provide quality ratings for both the reference and test presentations. Subjects are not informed which of the two presentations is the reference and which is the test. The order of the sequences in each trial presentation is randomized. That is, the reference presentation is presented first, but in other trials the test presentation appears first. Quality ratings are made using the rating scale shown in Fig. 2. The observers assess the quality twice for the reference and the test presentations, evaluating the fidelity of the video information by moving the slider of a voting function. The assessment scale is composed of 5 levels, which include bad, poor, fair, good and excellent. The left end of the slider is 0 point, whereas the right end of the slider is 100 points. Each level steps 20 points. We will analyze the mean opinion scores (MOSs) for the result of all the subjective assessments.

Fig. 2. The rating scale for the both presentations

4 Experimental Design

In our experiments, we edit 233 sequences from 15 videos as test databases including (1) "Australia's Great Barrier Reef", (2) "Raging Planet – Volcano", (3) "Raging Planet – Tidal Wave", (4) "Crossing the Alps on a Hot-Air Balloon", (5) "The Invisible World", (6) "Koyaanisqatsi – Life Out of Balance", (7) "Maintaining a Megalopolis", (8) "Fearless", (9) "Jurassic Park", (10) "The Passion of the Christ", (11) "Bulletproof Monk", (12) "Paycheck", (13) "Collateral", (14) "The Transporter", and (15) "Tim Burton's Corpse Bride".

We use Microsoft MPEG-4 software encoder/decoder with FGS functionality [18] to encode the sequences using the frame rates and the bit rates with the CIF (352 × 288 pixels) format. The Group of Pictures (GoP) structure is set to $IBBPBB$ having 12 pictures. Every sequence has 300 frames, i.e. 10 seconds and is processed in the YUV format(Y is the luminance component; U and V are color components of a frame). The sampling rate of the base layer is 3, i.e., one frame per three frames is encoded for the base layer. The sampling rate of the enhancement layer is 1.

We adopt the DSCQE to evaluate the perceived quality. The DSCQE method is cyclic in that the assessor is asked to view a pair of pictures, each from the same source, but one via the process under examination, and the other one directly from the source. He is asked to assess the quality of both. The assessor is presented with a series of picture pairs (internally random) in random order, and with random impairments covering all required combinations. At the end of the sessions, the mean scores for each test condition and test picture are calculated.

In the subjective assessment method, the mean opinion score(MOS) is used as the user-level QoS parameter or quality of perception. In order to measure MOS values, the rating-scale method[14], where an experimental subject classifies objects into some categories, is used. Each category is assigned a score to represent from bad to excellent. The rating scale is presented electronically. The range of the electronic rating scale may be between 0 and 100.

5 Experimental Results

We computed the average bits of all the P and B frames for the motion characteristic and the average bits of all the I frames for the texture characteristic. We find out the centers of the motion and texture characteristics by the K-means of a cluster analysis method. Then, we adopt the following classification function shown in the equation (1) based on the discriminant analysis to classify all the clips into four types for the motion and texture characteristics.

$$c_i = \mu_i' \Sigma^{-1} x - \frac{1}{2}\mu_i' \Sigma^{-1} \mu_i + \ln p_i \qquad (1)$$

where μ_i is the mean vector of the ith group, Σ is the variance-covariance matrix, and p_i is the ith group prior probability. According to the classification function for the motion and texture characteristics, we make use of a statistical tool, called Statistica 6.0, to get two discriminant values 3400 and 31000. The clips are classified into the high motion and detailed texture type if the average bits of the frames are larger than 3400 and 31000, and vice versa. Table 1 illustrates the centers of four types.

We first use the objective quality assessment to measure the video quality by varying the frame rate and the bit rate. Fig. 3 illustrates the sequences with

Table 1. The cluster analysis

Cluster	low/easy	high/detailed
The average bits for the motion characteristic	2102.46	4819.67
The average bits for the texture characteristic	19926.56	47486.11

the detailed texture charactcristic having the low PSNR and the high standard deviation(stddev). Fig.4 illustrates the sequences with the high motion or the detailed texture characteristics having the low $\triangle PSNR$ and the high standard deviation.

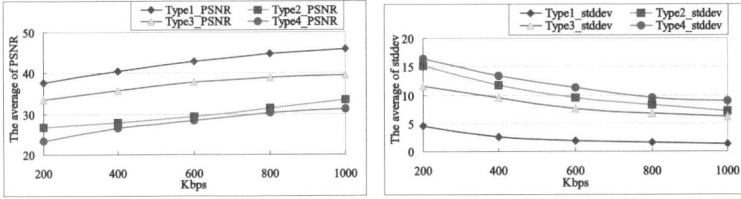

Fig. 3. The effect of varying the bit rate

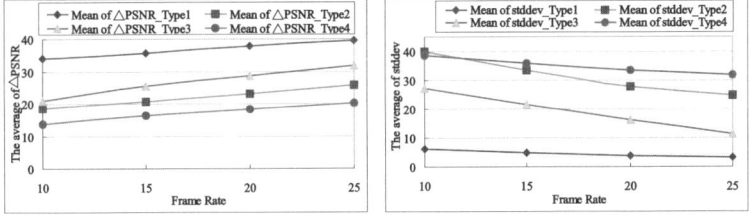

Fig. 4. The effect of varying the frame rate

5.1 Statistic Analysis

In the subjective assessment, 212 viewers took part in the test. The ages of the subjects range between 19 and 26. The subjects view the reference sequence and the test sequence by random sampling from our test databases. The mean opinion score(MOS) is used as the quality ratings.

To observe the effect levels of perceived quality on the spatial sharpness and the temporal continuity, we add an indicator, integral satisfaction, to the dependent variable. We compute the correlation coefficients, $R(-1 \leq R \leq 1)$, denoted as the equation (2) to understand the influences of the spatial sharpness and the temporal continuity to the integral satisfaction.

$$R = \frac{Cov(V_1, V_2)}{\sigma_1 \sigma_2} \quad (2)$$

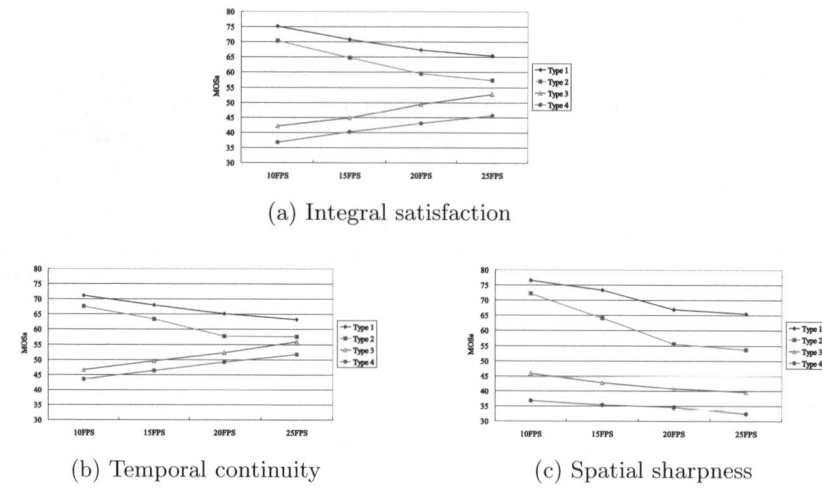

Fig. 5. The MOSs analysis for 4 Types

Table 2. The correlation analysis

Correlation	Type 1	Type 2	Type 3	Type 4
Temporal Continuity	0.733	0.811	0.725	0.720
Spatial Sharpness	0.771	0.862	-0.657	-0.175

where $Cov(V_1, V_2)$ is the covariance of V_1 and V_2, and σ_1 and σ_2 are the standard deviations. Table 2 for Type 1(low motion and easy texture) shows that high correlation($R = 0.733$ and 0.771) between the integral satisfaction, the temporal continuity, and the spatial sharpness. Fig. 5 shows that the subjects of Type 1 sequences prefer low frame rate in the average of all the bit rates. For the individual bit rate, the correlation coefficient between the integral satisfaction and the spatial sharpness is very high($R = 0.933$) at 200 kbps. This means that the subjects of Type 1 sequences prefer low frame rate when the bandwidth is not sufficient.

The subjects of Type 2(the low motion and the detailed texture) sequences also prefer a low frame rate when the bandwidth is not sufficient. The result is the same as Type 1.

High correlation between the integral satisfaction and the temporal continuity for Type 3(the high motion and the easy texture) sequences across all the bit rates means that the subjects prefer a high frame rate. The correlation coefficient of spatial sharpness is negative. It means that the bandwidth resource should be poured into the temporal continuity to increase the perceived quality.

High correlation between integral satisfaction and temporal continuity for Type 4 (the high motion and the detailed texture) sequences is the same as Type 3. However, the correlation coefficient between the integral satisfaction and the spatial sharpness is close to 0. It means that the bandwidth resource should be poured into the temporal continuity to increase the perceived quality.

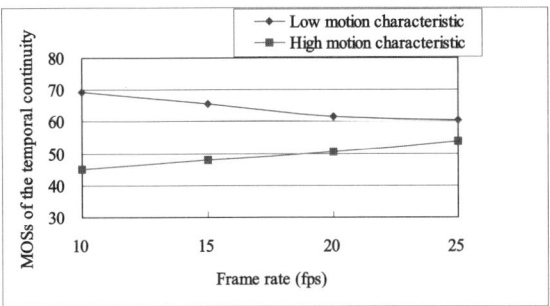

Fig. 6. The MOSs analysis of temporal continuity for frame rate

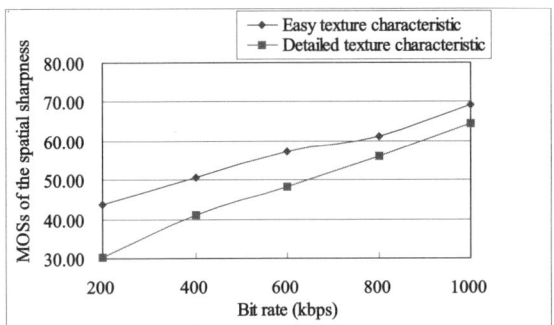

Fig. 7. The MOSs analysis of spatial sharpness for bit rate

Fig. 6 shows that the sequences of the high motion characteristic for the quality of the temporal continuity prefer a high frame rate. Whereas, the sequences of the low motion characteristic for the quality of the temporal continuity prefer a low frame rate. This is because the bits per frame are increased when the frame rate is reduced. Fig. 7 shows that the perceived quality of the spatial sharpness for the detailed texture sequences is influenced more than the easy texture sequences on varying the bit rate.

6 Conclusion

We have shown the results by the subjective assessment methodology. In this experiment, the test databases are composed of 233 sequences edited from 15 DVDs including different content characteristics such as the low motion and the easy texture, the low motion and the detailed texture, the high motion and the easy texture, and the high motion and the detailed texture. To clearly know the factors of the effect for the perceived quality of viewers, we divided the perceived quality into the spatial sharpness, the temporal continuity, and the integral satisfaction.

In the statistic analysis, we described the correlation of the integral satisfaction for the spatial sharpness and the temporal continuity. In summary, the quality of the spatial sharpness for Type 1 sequences is more important than the temporal continuity. When the bandwidth is not sufficient, we should reduce the frame rate to increase the bits of frame. Most of the results for Type 2 are the same as Type 1 except that the effect level of the bit rate variable for Type 1 is stronger than Type 2. Due to that, the perceived quality for the easy texture sequences is significantly improved by increasing the bit rate. The quality of the temporal continuity for Type 3 and Type 4 sequences is more important than the quality of the spatial sharpness.

Acknowledgement

This work was supported by the National Science Council of the R.O.C under Contract NSC 95-2221-E-309-014.

References

1. Apteker, R.T., Fisher, J.A., Kisimov, V.S., Neishlos, H.: Video acceptability and frame rate. IEEE Multimedia 2(3), 32–40 (1995)
2. Yadavalli, G., Masry, M., Hemami, S.S.: Frame rate preferences in low bit rate video. International Conference on Image Processing, Barcelona, vol. 1, pp. 441–444 (September 2003)
3. van der Schaar, M., Radha, H.: Temporal-SNR rate-control for Fine-Granular Scalability. In: International Conference on Image Processing, vol. 2, pp. 1037–1040 (2001)
4. Gulliver, S.R., Ghinea, G.: Changing Frame Rate, Changing Satisfaction? In: IEEE International Conference on Multimedia and Expo, vol. 1, pp. 177–180. IEEE Computer Society Press, Los Alamitos (2004)
5. de Cuetos, P., Reisslein, M., Ross, K.W.: Evaluating the Streaming of FGS-Encoded Video with Rate-Distortion Traces, Institut Eurecom Technical Report RR-03-078 (June 2003)
6. Ghinea, G., Thomas, J.P.: Quality of perception: user quality of service in multimedia presentations. IEEE Trans. on Multimedia 7(4), 786–789 (2005)
7. Ito, Y., Tasaka, S.: Quantitative assessment of user-level QoS and its mapping. IEEE Trans. on Multimedia 7(3), 572–584 (2005)
8. Nam, J., Ro, Y.M., Huh, Y., Kim, M.: Visual content adaptation according to user perception characteristics. IEEE Trans. on Multimedia 7(3), 435–445 (2005)
9. T1.801.03- 2003, American National Standard for Telecommunications – Digital transport of one-way video signals - Parameters for objective performance assessment, American National Standards Institute
10. Pinson, M.H., Wolf, S.: A New Standardized Method for Objectively Measuring Video Quality. IEEE Transactions on Broadcasting (3) (September 2004)

11. Zink, M., Schmitt, J., Steinmetz, R.: Layer-encoded video in scalable adaptive streaming. IEEE Trans. on Multimedia 7(1), 75–84 (2005)
12. Alpert, T., Evain, J.-P.: Subjective quality evaluation-The SSCQE and DSCQE methodologies. ITU-R Recommendation BT.500-7 (271), 12–20 (1997)
13. ITU-R Recommendation BT.500-11: Methodology for the subjective assessment of the quality of television pictures. ITU, Geneva, Switzerland (2002)
14. Hands, D.S.: A Basic Multimedia Quality Model. IEEE Transaction on Multimedia 6(6), 806–816 (2004)
15. Aeluri, P.K., Bojan, V., Richie, S., Weeks, A.: Objective quality analysis of MPEG-1, MPEG-2, and Windows Media video. In: The 6th IEEE Southwest Symposium on Image Analysis and Interpretation, March 28-30, pp, pp. 221–225. IEEE Computer Society Press, Los Alamitos (2004)
16. Yao, S., Lin, W., Lu, Z., Ong, E., Etoh, M.: Objective quality assessment for compressed video. In: ISCAS 2003. Proceedings of the 2003 International Symposium on Circuits and Systems, May 25-28, vol. 2, pp. 688–691 (May 2003)
17. Rohaly, A.M., et al.: Video Quality Experts Group: Current Results and Future Directions. In: Proc. of SPIE Visual Communications and Image Processing, Perth, Australia, vol. 4, pp. 742–753 (June 2000)
18. Microsoft, ISO/IEC 14496 Video Reference Software, Microsoft - FDAM1 - 2.3 - 001213

An Evolutionary Approach to Inverse Gray Level Quantization

Ivan Gerace[1], Marcello Mastroleo[1], Alfredo Milani[1], and Simona Moraglia[2]

[1] Dipartimento di Matematica e Informatica, Università degli Studi di Perugia, via Vanvitelli 1, I-06123 PG, Italy
[2] Dipartimento di Ingegneria Industriale, Università degli Studi di Perugia, via Duranti 67, I-06125 Perugia, Italy

Abstract. The gray levels quantization technique is used to generate images which limit the number of color levels resulting in a reduction of the image size, while it preserves the quality perceived by human observers. The problem is very relevant for image storage and web distribution, as well as in the case of devices with limited bandwidth, storage and/or computational capabilities. An efficient evolutionary algorithm for the inverse gray level quantization problem, based on a technique of dynamical local fitness evaluation, is presented. A population of blur operators is evolved with a fitness given by the energy function to be minimized. In order to avoid the unfeasible computational overhead due to the fitness evaluation calculated on the entire image, an innovative technique of dynamical local fitness evaluation has been designed and integrated in the evolutionary scheme. The sub–image evaluation area is dynamically changed during evolution of the population, and the evolutionary scheme operates a form of machine learning while exploring subarea which are significatively representative of the global image. The experimental results confirm the adequacy of such a method.

Keywords: evolutionary algorithms, image compression, machine learning.

1 Introduction

Gray levels quantization is a technique that allows to produce an image with a finite number of gray-scale levels that gives a human observer the illusion of continuous-tone of gray-scale levels. The main advantage of this technique is the reduction of the size of data to transmit or to store an image. It is thus a very relevant issue for images which are distributed over the web, stored in large image database or in all those cases in which memory, computational power and/or bandwidth are restricted such as with PDAs, webcam and mobile terminals.

When a human being obseves a quantized image he/she has the illusion of continuous-tones image, due to the fact that human eyes average the gray levels in a neighborhood of the observed points, this process is mathematically formalized by a blur operator. The problem of inverse gray levels quantization consists in estimating both the original continuous-tones image and the blur mask that best fits the human eyes illusion, given the quantized image [16].

This problem is ill-posed in the sense of Hadamard. Using both deterministic and probabilistic alternative approaches it is possible to obtain the solution of this problem as the minimum of an energy function [1,8]. Many authors, for different ill-posed visual problem, proposed the use of binary elements, called *line variables*, to recover image-discontinuities [1,2,8]. However, the characteristics of the scene are better recovered imposing smoothness constrains of order two or three, in order to obtain images locally planar or quadric [8,12]. Thus the solution, knowing the blur operator, is given by the argument of the minimum of an energy function, which incorporated the piecewise smoothness constrains. In such a way, the gray levels quantization problem is reduced to find the performing blur operator.

In this paper a method based on the genetic evolutionary scheme is proposed, the basic idea is that the fitness function is taken to be given from the energy function and the population is composed by blur masks. Gray levels quantization problem can then be solved by GA techniques evolving the blur masks population in order to minimize the fitness. The selection operator is realized by tournament method proportionally to fitness. Crossover is implemented by a random exchange the entries of the blur masks. In order to improve the computational performance of the algorithm we propose to evaluate the fitness function locally, on a sub-image, instead of evaluating it on the whole one. Anyway a wrong choice of sub-image can lead to a local minimum plateau, to overcome this problem we have introduced the dynamically change of the fitness function by iterating the sub-image selection phase after some generations. The experimental results confirm the adequacy of such a technique, as it is possible to check by a comparison between the images reconstructed by the genetic technique and the ones obtained using the same blur mask involved in the quantization process [6].

2 The Inverse Gray Levels Quantization Problem

Let y be a $n \times n$ gray levels quantized image. The inverse gray levels quantization problem consists in estimate a $n \times n$ continuous-tone image x from y. To this aim is also necessary to estimate a blur operator A such that applied to y returns an image close to x. The blur operator A has to simulate the continuous-tone illusion that human eye has believing to see a unique gray level in a neighborhood of a point.

It is possible to formally define the blur as the substitution of the intensity value of a pixel with the value of the weighted average of the intensity values of its adjacent pixels. These weights can be represented as the entries of a matrix $M \in \mathbb{R}^{(2z+1) \times (2z+1)}$, $z \in \mathbb{N}^+$, called *blur mask*. In this paper the range of the entries of the blur mask is restricted to a finite set $\chi_\mu = \{0, \ldots, \mu\}$, where μ is a positive fixed integer.

The inverse dithering problem is ill-posed in the sense of Hadamard. Thus, *regularization techniques* [1] are necessary to obtain a stable solution.

To improve the quality of the results, we introduce some binary elements, called *line variables* [1]. These elements have to take in account of the discontinuities present in the gray levels image x, indeed, the ideal images present discontinuities in correspondence with edges of different objects.

We define a *clique c* as the set of points of a square grid on which the second order finite difference is defined. We indicated with b_c the boolean line variable associated to clique c; in particular, the one value corresponds to a discontinuity of the involved image in c. The vector b is the set of all line variables b_c.

We denote the 2-th order finite difference operator of the vector x associated with the clique c by $D_c^2 x$. Namely, if $c = \{(i,j), (h,l), (r,q)\} \in C_2$, then

$$D_c^2 x = x_{i,j} - 2x_{h,l} + x_{r,q}. \tag{1}$$

The solution of inverse gray levels quantization problem can be defined as the minimum of the following *primal energy function*

$$E(x, b, A) = \|x - Ay\|^2 + \sum_{c \in C} \left[\lambda^2 (D_c^2 x)^2 (1 - b_c) + \alpha b_c\right] \tag{2}$$

where λ^2 is a *regularization parameter*, that allows us to regulate an appropriate degree of smoothing in the solution. The positive parameter α is used in order to avoid to have too many discontinuities in the restored image.

To reduce the computational cost of the minimization of the primal energy function we define the *dual energy function* through the minimization of the primal energy function respect to the line variables b_c. Namely, (cf. [1,2,8,7]):

$$E_d(x, A) = \inf_b E(x, b, A). \tag{3}$$

2.1 GNC Deterministic Algorithms

A classical deterministic technique for minimizing the dual energy function is the GNC (*Graduated Non-Convexity*) algorithm [2,3,13]. In general, E_d is not convex. The results of the algorithms for minimizing a non-convex function depend on the choice of the starting point. In order to have a suitable choice of this point, the GNC technique requires to find a finite family of approximating functions $\{E_d^{(p_i)}\}_{i \in \{1,\ldots,\bar{i}\}}$, such that the first $E_d^{(p_1)}$ is convex and the last $E_d^{(p_{\bar{i}})}$ is the original dual energy function, and then to apply the following algorithm:

```
Algorithm 1
Instance: (A, x^(0))
Output: x^(i-1)

    i = 1;
    while i ≠ ī do
```

$x^{(i+1)}$ is equal to the stationary point among the steepest descent direction of $E_d^{(p_{i+1})}$, starting from $x^{(i)}$;
$i = i + 1;$

In particular we use the family of approximations $\{E_d^{(p_i)}\}_{i \in \{1,\ldots,\bar{\imath}\}}$ proposed in [4] for convex data term and we will call the result of the algorithm as $\tilde{x}(A, x^{(0)})$. To reach the stationary point involved in the GNC algorithm the NLSOR (*Non-Linear Successive Over-Relaxation*) is used.

3 Blur Mask Estimation

In the gray levels reconstruction problem the blur operator, representing the human eyes filtering, is not known and as to be estimated in order to improve the quality of the results.

Formally we have to find

$$(A^*, x^*) = \arg\min_{(A,x)} E_d(A, x) \tag{4}$$

The conjunct minimization of the energy function is not an easy task because it is strongly non linear. To overly this problem in literature some methods have been proposed [17].

A classical technique to solve the problem is to minimize alternatively the dual energy function with respect to x and A.

Note that for small values of λ the algorithm has a trivial fixed point close to (I, y), where I is the identity matrix belonging to $\mathbb{R}^{(2z+1) \times (2z+1)}$. It is possible to overcome this problem starting from an initial guess close to the solution. Thus it is necessary to have a good estimation of the ideal blur operator.

For this reason we propose a different approach to the problem. It is possible to see the solution (4) as the solution of the following nested minimization problem

$$A^* = \arg\min_A E_d(A, x(A)) \tag{5}$$

where

$$x(A) = \arg\min_x E_d(A, x) \tag{6}$$

and

$$x^* = x(A^*). \tag{7}$$

Computing (6) means solving the problem assuming that A as blur operator, so we estimate the solution with the technique proposed in §2. The main problem remains to compute the solution of (5) for which we propose to use a GA (*Genetic Algorithm*).

3.1 Genetic Gray Levels Reconstruction

The use of GAs in order to minimize non convex function is well established in the literature and classical results [11] have proved that they converge in probability to the absolute minimum.

As we have seen in the previous paragraphs, the gray levels reconstruction problem consists in finding a blur mask which minimizes the energy function E_d in (5). In this section we formalize a GA scheme to this aim. Our basic idea lies in evolving a blur operators population through the application of classical genetics operators.

Chromosomes and genes representation. Chromosomes are here blur operators, represented by the associated blur matrix $M \in \chi_\mu^{(2z+1)\times(2z+1)}$. We have firstly considered the possibility of a binary representation, but finally we adopt an high level codification, in which blur mask entries $M_{i,j}$ are the genes. In fact, each entries $M_{i,j}$ conveys a kind of spatial information which is independent one each other.

Fitness. Initially we assume the fitness to be provided by the function

$$F(A) = Max_E - E_d(\boldsymbol{x}(A), A), \qquad (8)$$

where Max_E is an estimation of the maximum value of the dual energy. $\boldsymbol{x}(A)$ is determined by the minimization of the energy function E_d, as required in (3), and it is here approximated by the result $\tilde{\boldsymbol{x}}(A, \boldsymbol{x}^{(0)} = \boldsymbol{y})$ of Algorithm 1. Note that the computational cost of Algorithm 1 for large images is very expansive, being proportional to image size $N \times N$. In the following, to overcome this drawback, we propose an alternative fitness.

Assumed that appropriate definitions for crossover and mutation have been established, then a *straightforward GA* scheme for the blind restoration problem is given by

```
Algorithm 4
Instance: the fitness function F
Output: the last population P^(gen-1)

    inizialize randomly the population P^(0)
    while gen ≤ MG or the stop condition is false do
        evaluate the fitness F(P_id^(gen)) of each individual id;
        select the individuals with respect to their fitness to
        populate P^(gen+1);
        compute the crossover of some randomly selected
        individuals in P^(gen+1);
        mutate some randomly selected individuals in P^(gen+1);
        gen = gen + 1;
```

Let us examine in more details the features and drawbacks of the *straightforward GA* algorithm.

One of the basic element of any GA is the *selection* operator which characterizes the phase in which genes from individuals are passed into the next generation. Different techniques have been proposed [15,9] to pass on information into subsequent generations, and this topic is still an open problem. In this work the operator *selection* is used to find the individuals which constitute the *intermediate generation*, before crossover and mutation phases. Then this operator must be conceived in such a way that the survival probability of individual in a population depends on its blur mask fitness. The proposed selection operator is realized by a *tournament selection* method. In details, a tournament runs among two individuals randomly chosen from the population, higher is the fitness and higher is the probability that an individual is selected as tournament winner.

Crossover. Crossover is a process in which new individuals, called *offspring* or *children*, are generated from the intermediate population. By this technique each member of the new population would inherit the most useful traits from its parents. The here proposed GA crossover operator starts from two children, which initially assume the values of their two parents, then for each couple of indices (i,j) a random number $r \in [0,1]$ is generated, if r is less than a threshold, the entries (i,j) of the two sons are switched.

Mutation. Mutation introduces genetic diversity into the population, ensuring the exploration of a large part of solution space. Without mutation the algorithm is incomplete, since the part of the solution space, containing the optimal solution, could possibly be unexplored. A typical mutation operator guarantees that whole solution space is examined with some probability. The here proposed mutation operator exploits the features of the matrix representation of blur operators. In short each matrix entry can be mutated by randomly choosing a value in the admissible interval. First, a blur bask is randomly chosen, then for each pair of indices (i,j) of the blur mask, a random number is chosen and if its value is less than a given threshold, then the (i,j) entry is substituted with a random number $x \in \chi_\mu$. This procedure is repeated for a fixed number of times.

Termination Condition. The ideal case is that GAs should typically run until they reach a solution which is close to the global optimum. In this paper the stop condition is assumed to be an homogeneity measure of the population. Namely, we established that if the standard deviation of the population is less than a threshold then the algorithm is stopped. To avoid too long computations the process is usually terminated by the former technique or after a fixed number of generation MG. Moreover, in order to improve the convergence speed of the proposed GA we make use of the *elitist individual rule*, such that the best individual of each generation is directly transferred to the next one without modifications.

As we have already said the fitness cost term makes the algorithm computationally unfeasible for large images. The fact that the algorithm does not scale well for large images has lead us to design an original strategy for reducing the fitness evaluation.

3.2 Genetic Inverse Gray Levels Quantization with Dynamical Local Evaluation

Our goal is to find a method which is able to guarantee at the same time the accuracy of fitness calculation and a considerably low computational cost. The basic idea of the method, we have developed, consists in calculating an estimate of the fitness E_d of the image which can be computed very fast. The key point is that the estimate is not obtained by an approximation of E_d on the entire image, but it is obtained by the exact value of E_d *locally computed* in a subarea of the input image. The sampled subarea represents a *local* image to which all individuals in the current population refer for the evaluation of their current blur operators. Thus, A $\tilde{N} \times \tilde{N}$ sub–image $\tilde{y}$ is selected from the data y.

It is clear that by the minimization of the energy on a subarea of the image, the GA can produce a very good blur estimation just for the local sampled subarea, which is not necessary a good one for the whole image. In order to overcome this problem we have introduced in the local evaluation scheme the dynamical change of the sampled subarea, which is periodically activated after a certain number *par* of generations. The purpose of dynamical change is to realize a complete coverage of the whole image which is distributed over the time of GA generations, instead of realizing a complete coverage of the image at each generation, by evaluating E_d on the whole image as in the *straightforward* GA scheme.

The idea is that the blur operators, i.e. individuals, in the population would maintain a memory of their past optimization activity on other samples of the same input image, and mutate, merge and pass this information, i.e. the blur operators, to the next generations. In the end, a sequence of subarea sampling, which covers enough significantly different parts of the image, would produce the same results of a sequence of entire image evaluations, but at a considerable reduced cost.

What it is really needed in the problem, it is not a complete image coverage but a coverage of a sufficient number of significant numbers of subarea of the input image. Since real images are largely characterized by this homogeneity property, then a complete coverage of the whole image features requires a number of subarea samples which do not sum up to the entire image. For this reason the sampling strategy we adopted for dynamical change is not deterministic but it is randomized.

The main advantage of the dynamical local evaluation technique is the dramatic cost reduction. Moreover the randomized sampling strategy provides a further reduction in term of the total area which is necessary to examine to guarantee an effective coverage of blur features.

Independently on the sub–image choice problem, local minimum can not be reached due to a wrong GNC starting point. To overcome this drawback it is possible to pick a new random starting point at each generation, so the fitness function changes dynamically during the algorithm process.

Summarizing, the proposed algorithm for Genetic Inverse Gray Levels Quantization with Dynamical Local Evaluation can be described by the following scheme:

```
Genetic Inverse Gray Levels Quantization
Instance: the data image y
Output: the reconstructed image x
```

$\quad$ initialize randomly the population $P^{(0)}$
$\quad$ while $gen \leq MG$ or the population is not homogeneous do
$\quad\quad$ if $gen \equiv 0 \mod par$ then pick up a sub--image $\tilde{y}$;
$\quad\quad$ generate a random starting point $\tilde{x}^{(0)}$;
$\quad\quad$ evaluate the *fitness* $F(P_{id}^{(gen)})$ of each individual id with respect to $\tilde{y}$ and $\tilde{x}^{(0)}$;
$\quad\quad$ Put the elitest of $P^{(gen)}$ in $P_0^{(gen+1)}$;
$\quad\quad$ *select* the individuals by a tournament to populate the remain $P^{(gen+1)}$;
$\quad\quad$ compute the *crossover* of some randomly selected individuals in $P^{(gen+1)}$;
$\quad\quad$ *mutate* some randomly selected individuals in $P^{(gen+1)}$;
$\quad\quad$ $gen = gen + 1$;
$A = P_0^{(gen+1)}$;
$x = \tilde{x}(A, y)$;

4 Experimental Results

In this section we present the results obtained by experimenting the proposed algorithm. We have quantized three different images using the GNC algorithm proposed in [6]. In this algorithm the result image is obtained by an a priori blur mask M. We consider

$$M = \begin{pmatrix} 0 & 1 & 0 \\ 1 & 4 & 1 \\ 0 & 1 & 0 \end{pmatrix}. \quad (9)$$

In order to have a comparison we have reconstructed the quantized images both by the blur mask obtained by the genetic algorithm and by M.

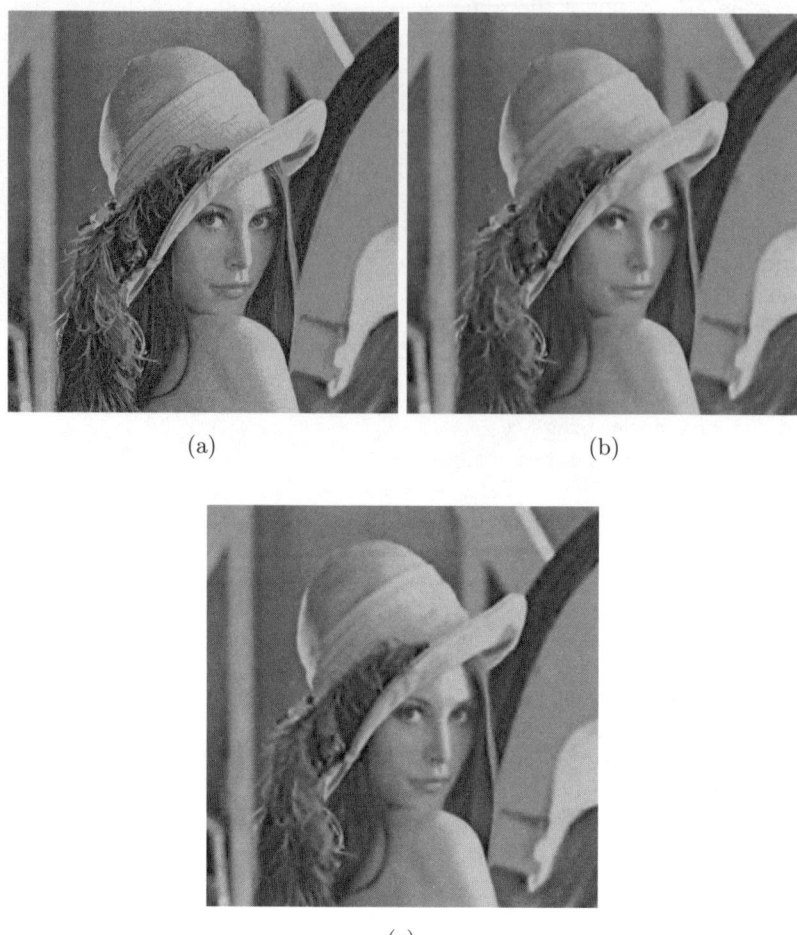

Fig. 1. (a) quantized image with 5 levels of gray for each RGB component; (b) gray level reconstruction with the a priori mask M; (c) genetic gray level reconstruction

We have empirically chosen the free parameters involved in the reconstruction, α and λ. In any case, in the literature there exist different methods to determine a good estimation of these parameters [5,10,14].

First, we have tested our algorithm on the 512×512 image presented in Figure 1 (a) quantized in 5 levels of gray for each RGB components, then on the 512×512 one presented in Figure 2 (a) quantized in only 8 colors i.e. black and white RGB components. For the first test the free parameters of GNC algorithm are $\lambda = 3$ $\alpha = 1000$, while for the second one they are $\lambda = 0.5$ $\alpha = 100$. In our last experimental result we consider the gray levels image presented in Figure 3 (a). This image is quantized in three gray levels. The reconstruction by M is shown

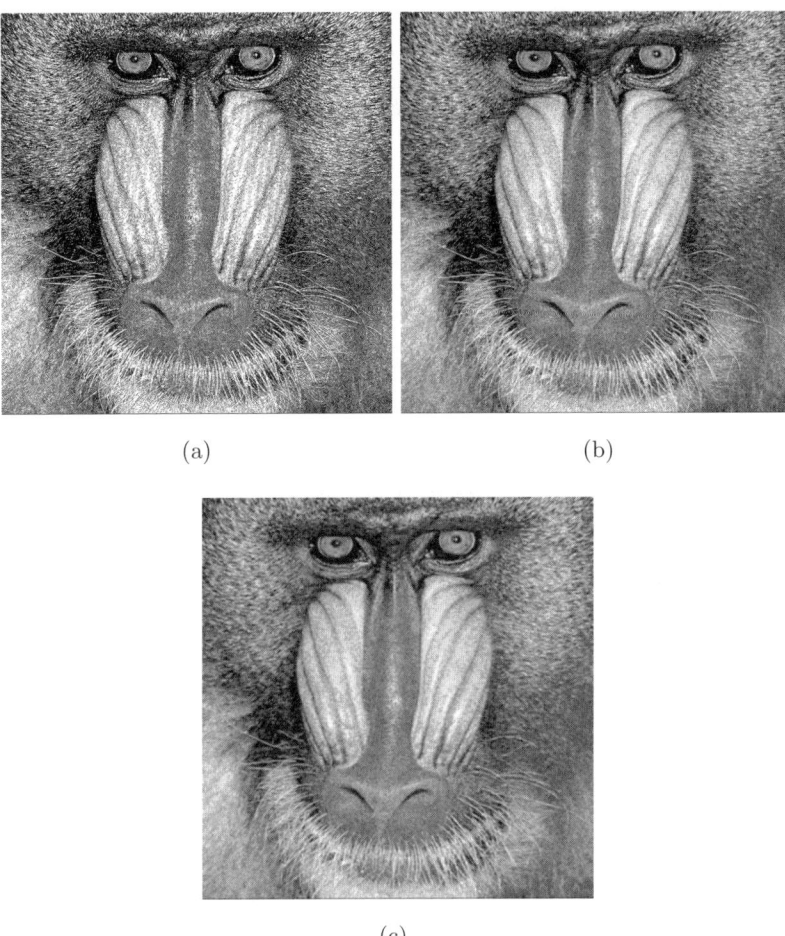

Fig. 2. (a) quantized image with black an white RGB components; (b) gray level reconstruction with the a priori mask M; (c) genetic gray level reconstruction

in Figure 3 (b), while the genetic result is given in Figure 3 (c). In this case we use $\lambda = 0.5$ $\alpha = 100$. In each test we have fixed the population size $MP = 100$, the maximum generation $MG = 2000$, the maximum blur mask entry $\mu = 15$ and the sub–image extraction period $par = 10$. In all presented experimental the better quality of the genetic reconstruction, which is evidently perceivable, has been confirmed by a formal computation of the Mean Square Errors (MSE) between the original image with no quantization and the images obtained by using M (M) and evolutionary approach (E) respectively ($MSE_{M1b} = 7.4$ vs $MSE_{E1c} = 7.3$, $MSE_{M2b} = 7.1$ vs $MSE_{E2c} = 7.0$ and $MSE_{M3b} = 8.2$ vs $MSE_{E3c} = 6.8$, where indices 1, 2 and 3 are referring to respective figures).

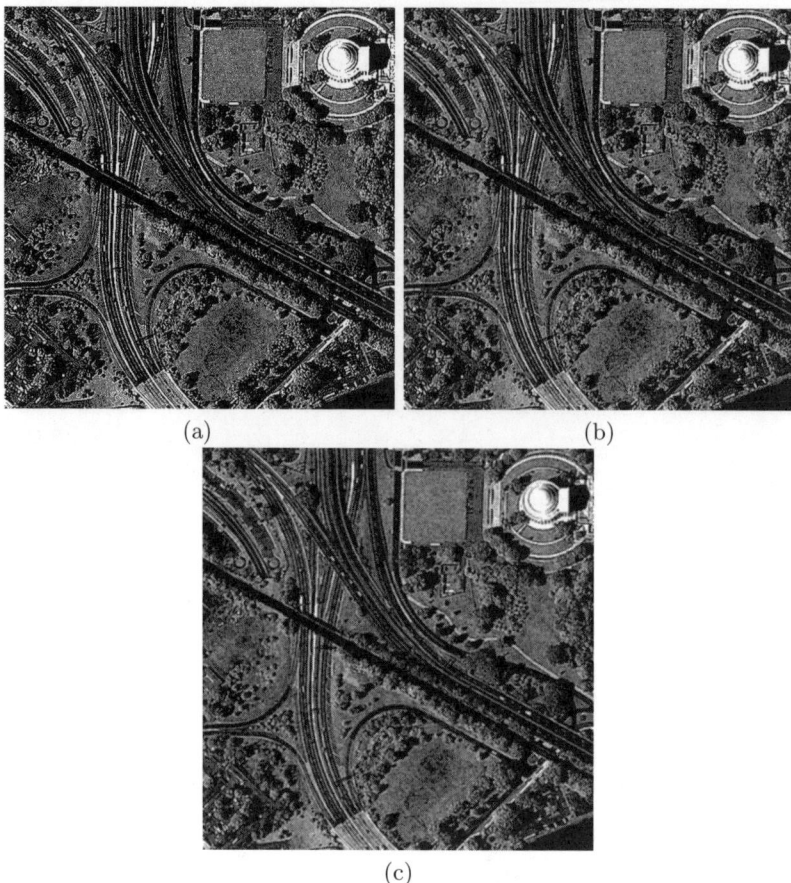

Fig. 3. (a) quantized image with black an white RGB components; (b) gray level reconstruction with the a priori mask M; (c) genetic gray level reconstruction

5 Conclusion

We have presented an evolutionary algorithm for inverse gray level quantization problem which uses an innovative technique of dynamical local fitness evaluation. Positive results in terms of computational time reduction and sensible image quality improvement have been obtained by experiments and allow the applicability of the technique in contexts, such as web large image database, PDA and mobile devices where bandwidth, computational and storage capabilities are limited. The main advantage of the dynamical local evaluation technique is the dramatic cost reduction. It is easy to see that the dynamical local evaluation technique will offer a quadratic computational cost reduction, because it depends on the area of the sample. The randomized sampling strategy also provides a further reduction in term of the total area which is necessary to examine to guarantee an effective coverage of blur features. The underlying hypothesis is that the set of explored subarea are significatively representative of the global image. The evolutionary approach is realizing a form of machine learning by which the relevant features discovered in some sub–images are maintained through the generations by the population of blur matrices.

References

1. Bedini, L., Gerace, I., Salerno, E., Tonazzini, A.: Models and Algorithms for Edge-Preserving Image Reconstruction. Advances in Imaging and Electron Physics 97, 86–189 (1996)
2. Bedini, L., Gerace, I., Tonazzini, A.: A Deterministic Algorithm for Reconstruction Images with Interacting Discontinuities. CVGIP: Graphical Models Image Process 56, 109–123 (1994)
3. Blake, A.: Comparison of the Efficiency of Deterministic and Stochastic Algorithms for Visual Reconstruction. IEEE Trans. Pattern Anal. Machine Intell. 11, 2–12 (1989)
4. Blake, A., Zisserman, A.: Visual Reconstruction. MIT Press, Cambridge, MA (1987)
5. Gerace, I., Martinelli, F., Sanchini, G.: Estimation of the free parameters in th problem of edge-preserving image reconstruction by a shooting method. In: SMMSP 2006. The 2006 International TICSP Workshopo on Spectral Methods and Multirate Signal Processing, Florence, Italy, pp. 205–212 (2006)
6. Gerace, I., Pandolfi, R., Pucci, P.: A new GNC Algorithm for Spatial Dithering. In: SMMSP2003. proceedings of the 2003 International Workshop on Spectral Methods and Multirate Signal Processing, pp. 109–114 (2003)
7. Gerace, I., Pandolfi, R., Pucci, P.: A new estimation of blur in the blind restoration problem. In: ICIP 2003. proceeding of IEEE International Conference on Image Processing, p. 4. IEEE Computer Society Press, Los Alamitos (2003)
8. Geman, D., Reynolds, G.: Constrained Restoration and the Recovery of Discontinuities. IEEE Trans. Pattern Anal. Machine Intell. 14, 367–383 (1992)
9. Goldberg, D., Debb, K.: A comparative analysis of selection schemes used in Genetic Algorithms. In: Rawlins, G.J.E. (ed.) Foundation of genetic Algorithms, Morgan Kaufman, San Francisco (1991)
10. Hansen, C.: Analysis of Discrete Ill-Posed Problems By Means of the L-Curve. SIAM Review 34, 561–580 (1992)
11. Holland, S.H.: Adaptation in natural and artificial systems. The University of Michigan press, Ann Arbor, MI (1975)
12. Li, S.Z.: Roof-Edge Preserving Image Smoothing Based on MRFs. IEEE Trans. Image Process 9, 1134–1138 (2000)
13. Nikolova, M.: Markovian Reconstruction Using a GNC Approach. IEEE Trans. Image Process 8, 1204–1220 (1999)
14. Reginska, T.: A Regularization Parameter in Discrete Ill-Posed Problems. SIAM J. Sci. Comput. 17, 740–749 (1996)
15. Sokolov, A., Whitley, D.: Unbiased tournament selection. In: Proceeding of GECCO 2005, Washington, DC, USA, pp. 1131–1138 (2005)
16. Stevenson, R.L.: Inverse Halftoning via MAP Estimation. IEEE Trans. Image Process 6, 574–583 (1997)
17. Tonazzini, A.: Blur Identification Analysis in Blind Image Deconvolution Using Markov Random Fields. Pattern Recogn. and Image Analysis 11, 669–710 (2001)

Mining Large-Scale News Video Database Via Knowledge Visualization

Hangzai Luo[1], Jianping Fan[2], Shin'ichi Satoh[3], and Xiangyang Xue[4]

[1] Software Engineering Institute, East China Normal University, Shanghai, China
memcache@gmail.com
[2] Department of Computer Science, UNC-Charlotte, Charlotte, USA
jfan@uncc.edu
[3] National Institute of Informatics, Tokyo, Japan
satoh@nii.ac.jp
[4] Department of Computer Science, Fudan University, Shanghai, China
xyxue@fudan.edu.cn

Abstract. In this paper, a novel framework is proposed to enable intuitive mining and exploration of large-scale video news databases via knowledge visualization. Our framework focuses on two difficult problems: (1) how to extract the most useful knowledge from the large amount of common, uninteresting knowledge of large-scale video news databases, and (2) how to present the knowledge to the users intuitively. To resolve the two problems, the interactive database exploration procedure is modeled at first. Then, optimal visualization scheme and knowledge extraction algorithm are derived from the model. To support the knowledge extraction and visualization, a statistical video analysis algorithm is proposed to extract the semantics from the video reports.

Keywords: Multimedia Mining, Knowledge Visualization.

1 Introduction

The broadcast video news has extensive influence. Different organizations and individuals use it for different purposes. The government can boost up the morale by publishing positive reports. However, the terrorists can also use it to display their announcement (e.g., Al Jazeera), formulate their plans, raise funds, and spread propaganda. The intelligence agents can collect secrets by watching these news reports. Therefore, analyzing and mining these international television news videos acts as an important role in enhancing the Nation's counter terrorism capabilities and translating raw video data into useful information. The national security community has a growing interest and compelling need of more effective techniques for automatic video analysis and knowledge discovery from large sets of international news video programs to allow the government to learn of terrorist's plans, disrupt their plots, and gain early warning about impending attacks. In addition, strategic investment decision-makers can evaluate the political, economic, and financial status with the information in video news reports. The general audiences can also build a thorough understanding for interesting events by watching news reports from different sources.

Due to the large amount of video news reports generated every day, discovering and analyzing news stories of interest is becoming an increasing problem. In news analysis, for example, it is becoming untenable to hire people to manually process all available news videos and produce summarization for them. Manual analysis of large-scale news video reports is too expensive, and it may take long time for response. Therefore, there is an urgent demand for achieving intuitive exploration of large-scale video news databases. However, it still suffers from the following challenging problems.

The first problem is how to *extract the most useful knowledge* from the large-scale video news database. Because the total amount of information (knowledge) for a large-scale video news database is very large (e.g., many thousands of hours of video), most of the information is irrelevant to the point of interest. If all information is delivered to the analysts or audiences, they may easily get lost and miss the important information. For example, "Bush is the president of the USA" is a piece of well-known information. Disclosing this information to an analyst does not make sense. Abnormal information is more useful and interesting for the users. The association rules are widely used in knowledge discovery and data mining applications to extract knowledge from databases. A-priori algorithm [1] and its variations are used to extract association rules based on fixed or adaptive support and confidence value. However, neither the support nor confidence is directly relevant to the interest of the users. Thus, there is an **interest gap** between the underlying information collection and the user's interest. How to extract interesting knowledge is still a problem.

The second problem is how to *present the knowledge to the users intuitively*. The amount of knowledge of large-scale video news databases is very large. It is untenable to ask the users read it one by one. Knowledge visualization is a potential solution. Several algorithms have been proposed to visualize text documents, such as InSpire [2], TimeMine [3], and ThemeRiver [4]. However, all of these visualization systems cannot directly provide the knowledge to the users. They disclose all information to the users. The users must "mine" the information of interest with the provided tools. Although these tools disclose different distribution structures of the database, most of the distribution structures are common and uninteresting. How to couple visualization techniques with the large-scale video news database to achieve intuitive exploration is still an open problem.

The third problem is how to *extract the underlying semantics* to support the knowledge extraction and visualization. Before the system can provide automatic video news exploration services, it must understand the underlying semantics of the input video clips. However, there is a big **semantic gap** [5,6] between the low-level visual features and the high-level semantic video concepts. Existing video exploration systems can only support services based on low-level visual features [7]. The users, nevertheless, can only express their information needs via high-level semantics and concepts [6,8]. Semantic video classification approaches can extract limited video semantics from video clips, but they can hardly satisfy the requirements of semantic video news database exploration. With a limited video semantics, how to provide intuitive applications for video news database mining and exploration is still an open problem.

In addition, all existing algorithms address the above problems separately. Therefore, different components of the system may optimize the solution for different target. On the one hand, the semantic video classification algorithms are generally optimized for keyword-based video retrieval applications [9]. As a result, the semantic video concepts implemented may be suitable for search but not suitable for visualization. On the other hand, the visualization approaches focus on providing new techniques for information representation [10] and assume the information is available all the time. However, the most useful information can only be extracted by using state-of-the-art semantic analysis algorithms. By addressing these problems together, these mismatches can be avoided and the overall performance can be improved significantly.

The Informedia Digital Video Library project is the only existing system that tries to addresses these problems in the same system. It has achieved significant progresses on several areas. A detailed survey is in [11]. Several applications have been reported, such as keyword-based video retrieval, and query results visualization. Unfortunately, the above problems are treated separately, and they did not report their solution for large-scale video database mining and exploration. To help analysts and general audiences find the news stories of interest efficiently, these problems must be addressed jointly.

To resolve the above problems, we offer here a knowledge visualization framework that can integrate achievements on semantic video analysis, information retrieval, knowledge discovery, and knowledge visualization. The framework is introduced in Section 2. Sections 3 and 4 introduce algorithms to implement different components of the framework. Finally we conclude in Section 5.

2 Knowledge Visualization Framework

Based on the above observations, a solution can only be achieved by integrating achievements on *semantic video analysis, information retrieval, knowledge discovery,* and *knowledge visualization*. In addition, all components must be optimized toward a single aim to achieve intuitive and intelligent exploration of large-scale video databases. Based on this understanding, the workflow of our framework is shown in Fig. 1. First, the semantic interpretation is extracted from raw video clips via semantic video analysis techniques. Second, the knowledge interpretation is extracted by weighting the semantic interpretation according to an interestingness measurement. Third, visualization techniques are adopted to represent and interpret the knowledge intuitively. In addition, the semantic interpretation and the knowledge interpretation can be improved through the user input received via the visualization interface.

To establish the best visualization design, we first model the database exploration process mathematically and resolve it to obtain the optimal design.

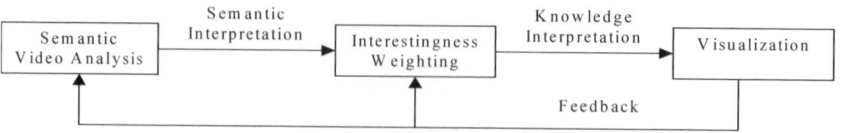

Fig. 1. The workflow of the framework

2.1 Modeling the Database Exploration Process

For a given video database D, we represent its all knowledge as K_D. The utmost goal of our system is to deliver as much knowledge in K_D to the users as possible. It is important to note that the amount of knowledge delivered by the system is not equal to the amount of knowledge received by the users. Only the **interesting knowledge** can be effectively delivered to the users, uninteresting knowledge may be simply ignored. As a result, we need to model the amount of **received knowledge** by the users.

In this paper we use U to represent the user who is using the system. The exploration procedure is an interactive iteration. Then, the amount of knowledge that U receives from the system can be modeled as $K_U = \sum_{t=1}^{T_U} K(r_t)$. Where t is the iteration counter, r_t is the t-th query submitted by U, $K(r_t)$ is the knowledge delivered to U from the system in response to r_t, T_U is the time that U is willing to use the system continuously, and $K_U \subseteq K_D$ is the total amount of knowledge U received from the system. The purpose of the knowledge visualization framework is to maximize K_U. It can be achieved by maximizing $K(r_t)$, T_U or both. Because U wants to know as much information as possible, the probability that U is willing to submit more inputs can be maximized by maximizing $K(r_t)$ with respect to the query r_t, i.e., T_U can be maximized by maximizing $K(r_t)$. As a result, K_U can be maximized by resolving:

$$\widehat{K}(t) = \max_{D,U} \{K(r_t)\} \tag{1}$$

To resolve Eq. (1), we model the knowledge as a set of items: $K_D = \{k_1, k_2, ..., k_N\}$. Consequently, we need to find a subset $K_D(U, r_t) \subseteq K_D$ and visualize it properly to enable U to receive as much knowledge of $K_D(U, r_t)$ as possible. Based on this model, $K(r_t)$ can be represented as $K(r_t) = Q(K_D(U, r_t)) = \sum_{i=1}^{N_t} q_U(k_i, r_t)$. Where $Q(*)$ is the visualization algorithm, k_i is a knowledge item, $q_U(k_i, r_t)$ is the amount of knowledge from k_i that is received by U under Q, and N_t is the number of knowledge items displayed. The most important two factors affecting $q_U(k_i, r_t)$ are: (1) the efficiency $\alpha(k_i)$ of on-screen presentation, and (2) the user's "interestingness" toward k_i. The interestingness is related to two factors. First, it is proportional to the correlation, $\psi(k_i, r_t)$, between k_i and r_t. Second, the users may have different inherent preferences $w_U(k_i)$ for different k_i. For large-scale databases, a large number of knowledge items may have the same relevance with a query r_t. As a result, the user preference $w_U(k_i)$ is also very important. By integrating these factors, $K(r_t)$ can be represented as:

$$K(r_t) = \sum_{i=1}^{N_t} q_U(k_i) = \sum_{i=1}^{N_t} \alpha(k_i) \times \psi(k_i, r_t) \times w_U(k_i) \tag{2}$$

Eq. (2) breaks the knowledge received by U into three independent parts. The representation efficiency $\alpha(k_i)$ is related to the visualization, $\psi(k_i, r_t)$ is related to the query, and the interestingness $w_U(k_i)$ is related to the content. This model breaks the large-scale video database exploration and mining problem into three smaller ones: (1) how to extract the knowledge interpretation, i.e. extracting k_i and $w_U(k_i)$, (2) how to measure the relevance between the query and the content, i.e. computing $\psi(k_i, r_t)$, and (3) how to optimally visualize the extracted knowledge, i.e. computing $\alpha(k_i)$.

Computing $\psi(k_i, r_t)$ is analog to an information retrieval problem. Sophisticated models can be built by integrating natural language processing, ontology, and other techniques. To focus on the large-scale database exploration problem, we do not cover this problem in depth. We set $\psi(k_i, r_t) = 1$ when $r_t \subseteq k_i$ and $\psi(k_i, r_t) = 0$ otherwise. It means we select only knowledge items that are explicitly related to the query for visualization. If the users do not have specific interest, $r_0 = \emptyset$ is used to inquire an initial visualization. As a result, our system can provide a reasonable initial visualization before the user input the first query. Because $\psi(k_i, r_t)$ can only be 0 or 1 in our system, it can be eliminated from Eq. (2) to simplify the visualization optimization:

$$K(r_t) = \sum_{i=1}^{N'_t} \alpha(k_i) \times w_U(k_i) \quad (3)$$

Next we need to extract the knowledge representation and compute the optimal visualization. As the knowledge representation must be optimized toward the requirements of visualization, we discuss the visualization problem in this section. The next section will introduce how to extract the knowledge interpretation.

2.2 Optimizing the Knowledge Visualization

To optimize the visualization, we assume all interestingness weights $\{w_U(k_i)\}$ have already been computed. As a result, knowledge items must be presented on the screen to enable efficient and intuitive exploration. All knowledge items in the database will compete for the limited display place. Therefore, the optimal visualization algorithm should allocate appropriate display area to knowledge items, so that U can learn as much knowledge as possible.

Several properties of presentation techniques must be considered. First, one unit of display area can only disclose a certain amount of information. If the underlying content carries more information than this limit, information loss may happen. Second, k_i may carry only limited information. If it is presented in too large an area, it may not efficiently utilize the display area. Therefore, if we use $A(k_i)$ to represent the display area allocated to k_i, and $B(k_i)$ to represent the maximal display area that k_i can fully utilize, we have the following relations:

$$\alpha(k_i) \propto A(k_i) \leq B(k_i) \propto w_U(k_i), \quad A(k_i) \geq 0, \quad \sum_{i=1}^{N_t} A(k_i) = c \quad (4)$$

where c is the total display area for visualization. Furthermore, because we concern only the ratios among $\alpha(k_i)$, we can replace $\alpha(k_i)$ with $A(k_i)$ in Eq. (3). As a result, the optimal visualization is to resolve $\widehat{A}(k_i)$ by maximizing $K(r_t)$. By using mathematical induction, it can be resolved as in Eq. (5).

$$\widehat{A}(k_i) = \underset{A(k_i)}{\arg\max} \left\{ \sum_{i=1}^{N_t} A(k_i) \times w_U(k_i) \right\} \Rightarrow \begin{cases} \widehat{A}(k_i) = \begin{cases} B(k_i) & 1 \leq i \leq \widehat{N_t} \\ 0 & \widehat{N_t} < i \leq N_t \end{cases} \\ \forall (1 \leq i < j \leq N_t): w_U(k_i) \geq w_U(k_j) \\ \sum_{i=1}^{\widehat{N_t}} B(k_i) \leq c, \quad \sum_{i=1}^{\widehat{N_t}+1} B(k_i) > c \end{cases} \quad (5)$$

It suggests a "winners take" strategy: the first winner takes as much as it can, then the second winner. The procedure stops until all resource is taken. With this mathematical foundation, we can implement the optimal visualization.

2.3 Implementing the Knowledge Visualization

Although there is only one mathematical solution, we may have different implementations of the visualization for different types of knowledge items. In our system, the first type of units for knowledge interpretation is keyframes of video shots and keywords of closed caption, videotext and ASR script. Keyframes and keywords are good representation for the underlying semantics. The users are able to catch rough idea of news reports by watching a sequence of keyframes and keywords. By reorganizing them with proper visualization, the users are able to capture the underlying semantics and hidden knowledge intuitively. However, the reorganization of keyframes and keywords may lose their sequence information. Because each keyframe or keyword can carry very limited information, the visualization may not disclose enough information to users. Some kind of correlations among keyframes and keywords must be preserved and visualized to deliver high-level semantics to the users. To resolve this problem, we propose to use binary relations of keyframes and keywords to represent the correlation information. A pair of keyframes and/or keywords is used as used as a relation if they occur in a closed caption or ASR sentence simultaneously.

As we assign interestingness to all knowledge items of K_D, each keyframe, keyword, and binary relation will also be weighted by the interestingness. The interestingness of the knowledge items not only filters out weak items but also suppress uninteresting and useless items.

For different types of knowledge items, different visualization may be adopted to enable the users receive as much information as possible. The visualization organization algorithm relies on the properties of $B(k_i)$. First, some knowledge items may not be suitable for variable display area, or their maximal display area may not be computed accurately. Relations are this type of knowledge items. In this situation, we assign a single constant value to all $B(k_i)$. It means we just pick up a preset number of most interesting items and represent them equally on the screen. One example of relation visualization is given in Fig. 2 and 3. To validate our proposed visualization interface,

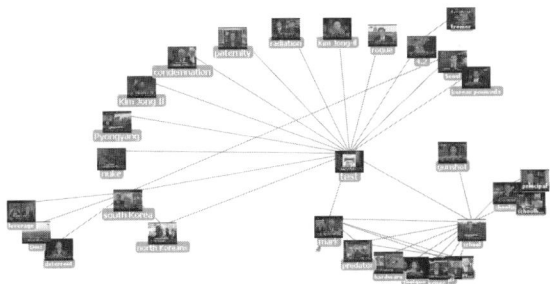

Fig. 2. Links of the semantic item "test" disclose details of the event and response of the international community during the North Korean nuclear weapon test

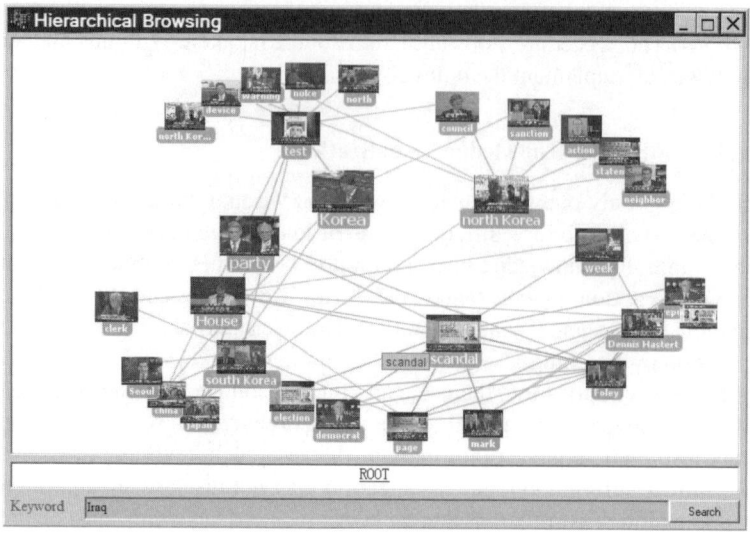

Fig. 3. North Korean nuclear weapon test event in the global context of all US news reports

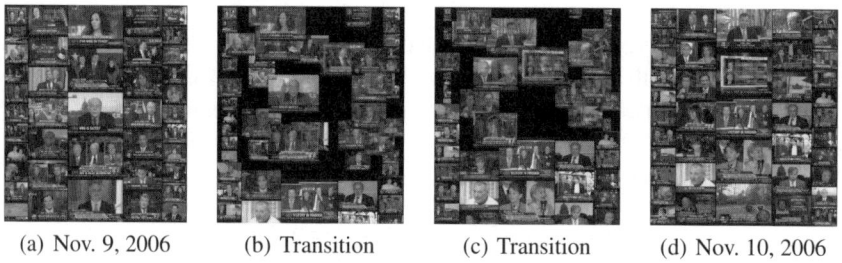

(a) Nov. 9, 2006   (b) Transition   (c) Transition   (d) Nov. 10, 2006

Fig. 4. An example of video news visualization

we have evaluated it on a TB scale video database. The online demo can be found at http://webpages.uncc.edu/~hluo/relation/Relation.html.

If the knowledge items can be easily changed in display size, variable sized visualization should used. According to Eq. (5), the actual display size of k_i equals to $B(k_i)$. Therefore, we need to resolve $B(k_i)$. The solution of $B(k_i)$ must satisfy two conditions: (1) $B(k_i) \propto w_U(k_i)$, and (2) any item should not be displayed in a size larger than its original size. Condition (2) is easy to understand. Condition (1) is tenable because $w_U(k_i)$ is directly related to U's interest of k_i. As a result, the expected amount of information of k_i is proportional to $w_U(k_i)$ for user U. Therefore, k_i must be visualized in a size proportional to $w_U(k_i)$ to satisfy the user's expectation.

We take this strategy for global overview visualization because the global overview is organized by keyframes that can be easily scaled. An example is given in Fig. 4. Because news topics change quickly over time, we also use an animation to present the dynamic trend over time of the global overview. An example of animation can be downloaded at http://webpages.uncc.edu/~hluo/NewsDemo.avi.

3 Knowledge Interpretation Extraction

To implement the above visualization, the knowledge interpretation $\acute{K}_D$ must be extracted. As discussed in previous section, an appropriate knowledge interpretation for the proposed visualization is composed of a set of knowledge items and their interestingness, i.e., $\acute{K}_D = \{(k_i, w_U(k_i)) \mid 1 \leq i \leq N)\}$. An important factor of our knowledge interpretation is the interestingness of knowledge items: $w_U(k_i)$. The interestingness enables our system to emphasize interesting knowledge and suppress uninteresting knowledge. It relieves the users from the burdensome exploration of uninteresting or useless knowledge and enables the users to focus on interesting knowledge that needs immediate process. The interestingness is subjective thus very difficult to compute. The best solution is to ask a group of experts to annotate all video reports and the system computes the average. However it's impractical as discussed before. Furthermore, it's unnecessary because the "expert annotations" are already there. The news producers have a weight for each news report. This weight is used to guide the producing of news reports, such as story selection, sequence, length of each story, etc. Therefore, we can use this weight as the expert annotation for the news reports. The remaining problem is how to quantify this weight.

There are two factors can be quantified. The first one is the frequency of knowledge items, $\delta(k_i)$, because more important news stories may be repeated time by time and channel by channel. In addition, the users may not be interested in knowledge that is already known. It implies that the interestingness $w_U(k_i)$ is inverse proportional to the user's prior knowledge, $\mu_U(k_i)$. Based on above observations, $w_U(k_i)$ can be modeled by integrating the two quantities:

$$\begin{aligned} w_U(k_i) &\propto \delta(k_i) \\ w_U(k_i) &\propto \frac{1}{\mu_U(k_i)} \end{aligned} \Rightarrow w_U(k_i) = \gamma \frac{\delta(k_i)}{\mu_U(k_i)} \qquad (6)$$

where γ is for normalization. Because the visualization algorithm uses only the relative ratios among $w_U(k_i)$, γ can be to optimize other purpose.

In Eq. (6), $\delta(k_i)$ can be computed by statistical analysis on K_D: $\delta(k_i) = P_{K_D}(k_i)$. Where $P_{K_D}(k_i)$ is the probability of k_i in K_D. To compute $\mu_U(k_i)$, the approaches that U builds her priori knowledge model must be discussed. Although U may learn k_i in many different ways, the dominant case is that k_i is rooted from a broadcast report. Therefore, the priori knowledge of U is built from broadcast news reports directly or indirectly. The effects of broadcast reports to a given user may have large difference due to many reasons. But their effects to the public may be smoothed from the individual biases due to statistic average on the large number of the audiences. For large-scale video database exploration and investigation applications, general knowledge that reflects the thinking of a general user is more interesting than the personalized knowledge extracted for a particular investigator. As a result, a broadcast news report database, $\acute{D}$, can be used to characterize the user's priori knowledge $\mu_U(k_i)$. $\acute{D}$ must cover all reports that may be learned by a general user directly or indirectly.

A knowledge database $K_{\acute{D}}$ can be extracted from $\acute{D}$ as before. As a result, $\mu_U(k_i)$ is computed by using $K_{\acute{D}}$ based on the above observation: $\mu_U(k_i, t) = P_{K_{\acute{D}}}(k_i)$. Then, the interestingness can be computed as:

$$w_U(k_i) = \gamma \frac{P_{K_D}(k_i)}{P_{K_{\hat{D}}}(k_i)} \tag{7}$$

To simplify the post process and fusing with other factors γ is selected to normalize $w_U(k_i)$ to the range of $[0, 1]$.

Except the frequency information, video production rules can also imply the importance assigned by news producers. To enable more efficient visualization of large-scale news video collections, these special visual features should be considered. Unlike frequency information, video production rules generally do not have the "prior knowledge" problem. However these video production rules are difficult to extract because they are at high-level semantics. In addition, the keyframes, keywords and their relations are all semantic information. To have a complete large-scale news video database exploration system we need to extract semantic interpretation of the database. Next section introduces the algorithm for semantic interpretation extraction.

4 Semantic Interpretation Extraction

Semantic video analysis and understanding are still very challenging for current computer vision technologies. The problem is caused by the semantic gap between the semantics of video clips and the low-level features [5]. Nevertheless, supporting semantic video analysis plays an important role in enabling more efficient exploration of large-scale news videos. Without extracting the semantics from large-scale news video collections, it is very difficult to visualize them effectively. Based on this observation, we have developed novel algorithms to extract the multi-modal knowledge items (i.e., video, audio, text) and video production rules automatically. Weights are assigned automatically with a statistical video analysis algorithm.

4.1 Semantic Video Analysis

The basic unit for news video interpretation is the video shot. Unlike the keywords of text documents, a video shot may contain abundant information (i.e., an image is more than a thousand words). This specific property of the video shot makes it difficult to effectively achieve statistical analysis of its visual properties and assign importance weights for news video visualization. To overcome this, we have developed a novel framework for statistical video analysis.

There are three types of semantic units that are critical to determine the importance weights for the corresponding video shots. The first one is the statistical properties of the shots. The second one is the special video objects that appear in the shots. The last one is the semantic concepts that are associated with the shots. Because these three types of semantic units have different properties, different algorithms are needed to extract such multi-modal semantic items.

Statistical Property Analysis of Physical Video Shots. The physical video shot is the basic unit for news video interpretation. Therefore, it can be used as a semantic item. However, unlike the keywords in text documents, the repetition of physical video shots cannot be detected automatically by using simple comparison between the shots. New techniques are needed for detecting the repeat of video shots in news videos [12].

News producers may repeat a certain shot in several ways. By detecting the repeat pattern of shots, we can infer the interestingness weights assigned by news producers. Consequently, we need to discriminate these patterns and assign appropriate weights to them. Through experiments, we found that most repeated shots can be weighted by an intra-program repetition weights and an inter-program repetition weights:

$$w_{intra}(i) = e^{-\left(\frac{r_{intra}(i)-2}{2}\right)^2}, \quad w_{inter}(i) = e^{-\left(\frac{r_{inter}(i)-5}{8}\right)^2} \quad (8)$$

where $r_{intra}(i)$ is the intra-program repeating number of shot i, and $r_{inter}(i)$ is the inter-program repeating number. More details can be found in [12].

Video Objects Detection. For news videos, text areas and human faces may provide important clues about news stories of interest. Text lines and human faces in news videos can be detected automatically by computer vision techniques [12]. Then the objects can be used to quantify the weight:

$$w_{textArea}(i) = \frac{1}{1+e^{-\frac{\max\{\alpha_t(i)-\nu_t, 0\}}{\lambda_t}}}, \quad w_{faceArea}(i) = \frac{1}{1+e^{-\frac{\max\{\alpha_f(i)-\nu_f, 0\}}{\lambda_f}}} \quad (9)$$

where α is the ratio that the object is in the frame, ν and λ are parameters determined by experiments. More details can be found in [12].

By performing face clustering, face objects can be clustered to groups and the human objects can be identified. The human object is similar to the knowledge items: too frequent items may not be interesting, such as the anchor person. Consequently, the same weighting algorithm introduced in Section 3 is adopted to compute the weight:

$$w_{face}(i) = \begin{cases} \max_{x \in FACE(i)} \{w_U(x)\} & FACE(i) \neq \emptyset \\ 0.5 & FACE(i) = \emptyset \end{cases} \quad (10)$$

where $FACE(i)$ is the set of face objects of shot i, and $w_U(x)$ is the weight of the face object x computed by using Eq. (7).

Semantic Video Classification. The semantic concepts of video shots can provide valuable information to enable more efficient and effective visualization and retrieval of large-scale news video collections. Semantic video classification is one method that helps detect the semantic concepts for the video shots. To incorporate this capability, we adopt a principal video shot-based semantic video classification algorithm [6] in our system.

Two types of information about semantic concepts can be used for weight assignment. First, the users may have different preferences for different semantic concepts. Therefore, a priori weight can be assigned to each semantic concept according to the user preference. We adopt a scheme that approximates the preference of the public, as assigned in Table 1. Where $C(i)$ is the semantic concept of i, and $w_c(C(i))$ is

Table 1. Semantic Concept Importance

Concept ($C(i)$)	$w_c(C(i))$	Concept ($C(i)$)	$w_c(C(i))$	Concept ($C(i)$)	$w_c(C(i))$
Announcement	0.9	Report	0.3	Gathered People	1
Sports	0.5	Weather	0.5	Unknown	0.8

the weight. Second, semantic concepts are similar to the knowledge items thus can be weighted by the algorithm of Section 3. Finally, the weight of semantic concept is determined by:

$$w_{concept}(i) = w_c(C(i)) \times w_U(C(i)) \quad (11)$$

where $w_c(C(i))$ is looked up from Table 1, and $w_U(C(i))$ is the weight of $C(i)$ computed by using Eq. (7).

Multi-modal Data Fusion. To enable more efficient visualization of large-scale news video collections, an overall weight is assigned with each video shot based on the weights described above. Our purpose of weighting is to detect the existence of some visual properties and emphasize those shots with interesting visual properties. The existence of one visual property may be indicated by different visual patterns. For example, the repeat property may be represented by w_{intra} or w_{inter}. To ensure we detect the existence of interesting visual properties and capture the patterns we are looking for, we first use max operation to fuse weights for the same visual property:

$$\begin{aligned} w_{repeat}(i) &= \max\{w_{intra}(i), w_{inter}(i)\} \\ w_{object}(i) &= \max\{w_{faceArea}(i), w_{textArea}(i)\} \\ w_{semantics}(i) &= \max\{w_{face}(i), w_{concept}(i)\} \end{aligned} \quad (12)$$

Where w_{repeat} measures the visual property of physical video shot repetition, w_{object} measures the visual property of salient objects, and $w_{semantics}$ measures the visual property of visual concepts.

Then the overall visual importance weight for a given video shot is determined by the geometric average of the three weights:

$$w_{video}(i) = \sqrt[3]{w_{repeat}(i) \times w_{object}(i) \times w_{semantics}(i)} \quad (13)$$

4.2 Audio and Text Keywords Extraction

The keywords can be extracted from closed caption and ASR scripts. Advanced natural language processing techniques, such as named entity detection, coreference resolving, and part-of-speech (POS) parsing are used in our system to extract appropriate keywords for shots. More details can be found in [12]. Finally each shot is associated a set of keywords. The keyword weight of a shot is computed by:

$$w_{keyword}(i) = \max_x \{w_U(x) | x \text{ is a keyword of } i\} \quad (14)$$

In Eq. (14) we use the proposed provider behavior model to weight each keyword.

With the keyword weight and the visual weight computed above, the overall weight for a given video shot is determined by averaging w_{video} and $w_{keyword}$:

$$w(i) = \gamma \times w_{video}(i) + (1-\gamma) \times w_{keyword}(i) \qquad (15)$$

In our current experiments, we set $\gamma = 0.6$.

5 Conclusions

In this paper, a large-scale video database mining and exploration system is proposed by integrating novel algorithms of visualization, knowledge extraction and statistical video analysis. By optimizing all components toward a single target, the proposed system achieves more effective and intuitive video database mining and exploration.

To implement the proposed system, a knowledge visualization model is developed to optimally deliver knowledge to the users. A knowledge interpretation is extracted to implement the proposed visualization. The proposed knowledge interpretation is able to suppress uninteresting knowledge. As a result, the users may find news reports of interest without the burdensome of mining large volume of uninteresting and useless reports. The proposed knowledge interpretation is able to bridge the **interest gap**. The proposed knowledge interpretation is extracted by using the proposed statistical video analysis algorithm, which integrates our previous achievements on semantic video analysis. Therefore, our system is also able to bridge the **semantic gap**.

Future work involves implementing personalized knowledge extraction under current framework, and exploration over long period and hundreds of geographically dispersed broadcast channels. This work will also be applied to other types of media.

References

1. Agrawal, R., Imielinski, T., Swami, A.: Mining association rules between sets of items in large databases. In: SIGMOD. vol. 22, pp. 207–216 (1993)
2. Wise, J.A., Thomas, J.J., Pennock, K., Lantrip, D., Pottier, M., Schur, A., Crow, V.: Visualizing the non-visual: Spatial analysis and interaction with information from text documents. In: IEEE InfoVis, pp. 51–58. IEEE Computer Society Press, Los Alamitos (1995)
3. Swan, R., Jensen, D.: Timemines: Constructing timelines with statistical models of word. In: ACM SIGKDD, pp. 73–80. ACM Press, New York (2000)
4. Havre, S., Hetzler, B., Nowell, L.: Themeriver: Visualizing theme changes over time. In: IEEE InfoVis, pp. 115–123. IEEE Computer Society Press, Los Alamitos (2000)
5. Smeulders, A.W., Worring, M., Santini, S., Gupta, A., Jain, R.: Content-base image retrieval at the end of the early years. IEEE Trans. on Pattern Analysis and Machine Intelligence 22(12), 1349–1380 (2000)
6. Fan, J., Luo, H., Elmagarmid, A.K.: Concept-oriented indexing of video database toward more effective retrieval and browsing. IEEE Trans. on Image Processing 13(7), 974–992 (2004) (IF: 2.715. Google Cite: 12. SCI Cite: 6)
7. Flickner, M., Sawhney, H., Niblack, W., Huang, J.A.Q., Dom, B., Gorkani, M., Hafner, J., Lee, D., Petkovic, D., Steele, D., Yanker, P.: Query by image and video content: The qbic system. Computer 28(9), 23–32 (1995)

8. Fan, J., Gao, Y., Luo, H.: Multi-level annotation of natural scenes using dominant image components and semantic image concepts. In: ACM Multimedia, pp. 540–547. ACM Press, New York (2004) (Best paper runner-up. Accept rate: 17)
9. Dimitrova, N., Zhang, H., Shahraray, B., Sezan, L., Huang, T., Zakhor, A.: Applications of video-content analysis and retrieval. IEEE Trans. on Multimedia 9(3), 42–55 (2002)
10. van Wijk, J.J.: Bridging the gaps. Computer Graphics and Applications 26(6), 6–9 (2006)
11. Hauptmann, A.G.: Lessons for the future from a decade of informedia video analysis research. In: Leow, W.-K., Lew, M.S., Chua, T.-S., Ma, W.-Y., Chaisorn, L., Bakker, E.M. (eds.) CIVR 2005. LNCS, vol. 3568, Springer, Heidelberg (2005)
12. Luo, H., Fan, J., Yang, J., Ribarsky, W., Satoh, S.: Exploring large-scale video news via interactive visualization. In: IEEE Symposium on Visual Analytics Science and Technology, pp. 75–82. IEEE Computer Society Press, Los Alamitos (2006)

Visualization of the Critical Patterns of Missing Values in Classification Data

Hai Wang[1] and Shouhong Wang[2]

[1] Sobey School of Business, Saint Mary's University, Canada
hwang@smu.ca
[2] Charlton College of Business, University of Massachusetts Dartmouth, USA
swang@umassd.edu

Abstract. The patterns of missing values are important for assessing the quality of a classification data set and the validation of classification results. The paper discusses the critical patterns of missing values in a classification data set: missing at random, uneven symmetric missing, and uneven asymmetric missing. It proposes a self-organizing maps (SOM) based cluster analysis method to visualize the patterns of missing values in classification data.

Keywords: Data visualization, missing values, classification, cluster analysis, self-organizing maps, data mining.

1 Introduction

Classification is one of the important techniques of data engineering [6,7]. Commonly, one can rarely find a sample data set for classification that contains complete entries of each observation for all of the variables. The possible reasons of missing values could be numerous, including negligence, deliberate avoidance for privacy, damage, and aversion. The extent of damage of missing data is unknown when it is virtually impossible to return the data source for completion. The incompleteness of data is vital to data quality for classification [9,12]. Indeed, missing data has been an important debatable issue in the classification field [1,2,16,18,19].

There have been several traditional approaches to handling missing values in statistical analysis, including eliminating from the data set those records that have missing values [11] and imputations [4]. However, these traditional approaches can lead to biased classification results and invalid conclusions. Research [17] has pointed out that the patterns of missing values must be taken into account in analyzing incomplete data.

This paper is to examine the patterns of missing values in classification data that are critical for assessing the data quality and validating classification results, and proposes a model to detect and visualize critical patterns of missing values for classification data based on self-organizing maps (SOM) [10]. An experiment on real-world data is employed to demonstrate the usefulness of the proposed model.

2 Critical Patterns of Missing Values in Classification Data

There are three critical patterns of missing values in classification data. These patterns verify the data quality and influence the validation strategy on classification results.

2.1 Missing at Random (MAR)

Missing values often randomly distributed throughout the sample space. There is no particular assumption on the reason of value missing. Few correlations among the missing values can be observed if the missing values have the MAR pattern. Since values are missing at random, the missing values are not distributed unequally towards a particular class in the classification context. Any imputation analysis would be invalid in this case. The quality of the entire classification data set is determined by the degree of data incompleteness.

2.2 Uneven Symmetric Missing (USM)

Data are often missing in some variables more than in others. Also, missing values in those variables can be correlated. It is difficult to use statistical techniques to detect multi-variable correlations of missing values. An implication of uneven missing for classification is that the data in the subpopulation with those missing data is under representative. In the two-class classification context, uneven missing values may not be biased towards a particular class. In other words, uneven missing values might distribute symmetrically in both classes. The quality of classification data with this pattern of missing values is less homogeneous than that with MAR. Applications of the general classification results based on the complete data set should be cautious for the subpopulation with this pattern.

2.3 Uneven Asymmetric Missing (UAM)

Uneven missing values can be biased towards a particular class. Similar to USM, UAM makes the quality of classification data less homogeneous than that with MAR. More importantly, the general classification results based on the complete data set are highly questionable for the subpopulation with this pattern.

When the data set is high-dimensional, statistical tests for the critical missing patterns might become powerless due to the curse of dimensionality. On the other hand, the interest of a data pattern depends on the purpose of the data analysis and does not solely depend on the estimated statistical strength of the pattern [15].

3 Self-Organizing Maps for Visualizing and Detecting Patterns of Missing Values

3.1 Self-Organizing Maps (SOM)

Cluster analysis, which reveals abnormal attributes of the measurements in the multivariate space, is often used to discover patterns of the data [7]. A definition of clusters for our purposes is: a cluster consists of observations that are close together in terms of Euclidian distance and the clusters themselves are clearly separated [5]. Although there are three general types of statistical clustering methods, namely, hierarchical [8,13], partitioning [8] and overlapping [14], only partitioning is strictly compatible with the definition of clustering in this study. In a typical partitioning clustering method such as the k-means algorithm, an initial specification of cluster "centers" is required. Then observations are assigned to the clusters according to their nearest cluster centers. Cluster "centers" are refined and observations are reallocated. The procedure continues until some type of steady-state is reached. The best known partitioning procedure is the k-means algorithm [8]. A major difficulty in using the k-means algorithm is the selection of parameters for the algorithm a priori. Also, the k-means algorithm acts as a black-box and provides little comparison of the clusters.

The self-organizing maps (SOM) method based on Kohonen neural network [10] has become one of the promising techniques in cluster analysis. SOM-based cluster techniques have advantages over statistical methods in cluster analysis to discover patterns of missing values in classification data. First, classification often deals with very high-dimensional data. The SOM method does not rely on any assumptions of statistical tests, and is considered as an effective method in dealing with high-dimensional data. Secondly, data for classification often do not have regular multivariate distributions, and then the traditional statistical cluster analysis methods have their limitations in these cases. On the other hand, the SOM method has demonstrated its flexibility and usefulness in cluster analysis because of the relaxation of statistical assumptions [10,3]. Thirdly, the SOM method provides a base for the visibility of clusters of high-dimensional data. This feature is not available in any other cluster analysis methods. It is extremely important for our study since we are interested in not only individual clusters, but also the global pattern of clusters such as no-cluster pattern.

SOMs are used to reveal certain useful patterns found in their input data through the unsupervised (competitive) learning process. The SOM method is a typical artificial intelligence technique of cluster analysis. It maps the high-dimensional data onto low-dimensional pictures, and allows human to view the clusters. The neural network depicted in Figure 1 is the two-layer SOM used in this research. The nodes at the lower layer (input nodes) receive inputs presented by the sample data points. The nodes at the upper layer (output nodes) will represent the one-dimensional organization map of the input patterns after the unsupervised learning process. Every low layer node is connected to every upper layer node via a variable connection weight. The lateral interconnections are not included in this study.

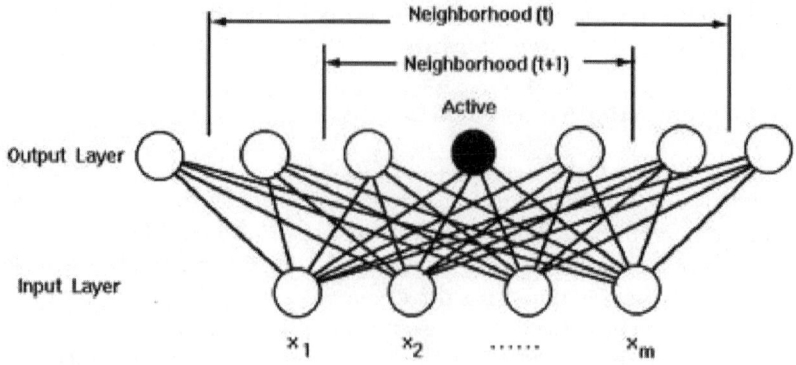

Fig. 1. Two Layer Self-Organizing Maps

The unsupervised learning process in SOM can be briefly described as follows. The connection weights are assigned with small random numbers at the beginning. The incoming input vector presented by a sample data point is received by the input nodes. The input vector is transmitted to the output nodes via the connections. The activation of the output nodes depends upon the input. In a "winner-take-all" competition, the output node with the weights most similar to the input vector becomes active. In the learning stage, the weights are updated following Kohonen learning rule [10]. The weight update only occurs for the active output node and its topological neighbors (Figure 1). In this one-dimensional output case, we assume a linear neighborhood. The neighborhood starts large and slowly decreases in size over time. Because the learning rate is reduced to zero, the learning process will eventually converge. The weights will be organized such that nodes that share a topological resemblance are sensitive to inputs that are similar. The output nodes in SOM will thus be organized and represent the real clusters in the self-organizing map. The reader is referred to [10] for a more detailed discussion.

3.2 Visualizing and Detecting Patterns of Missing Values

The method of visualizing and detecting patterns of missing values is described as follows. The first step of this method is to convert observations with missing values into shadow observations by assigning a value 1 for the data item with missing value and 0 for the data item without missing values. For example, suppose we have an observation with five data items [2, 4, M, 1, 4] where M indicates a missing value. Its shadow observation is [0, 0, 1, 0, 0].

Shadow observations are then presented to the SOM and they are self-organized on the map represented by the output nodes of the SOM. The map is then depicted graphically. The one-dimensional map is almost the same as histogram in statistics. The horizontal axis represents locations of output nodes of the SOM. The height of a bar indicates the number of data points which activate the output node of the SOM at the corresponding location. To determine whether

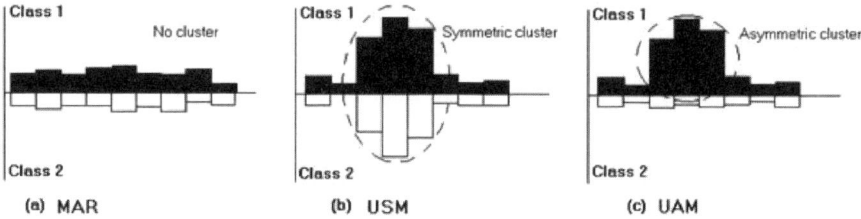

Fig. 2. The Critical Patterns of Missing Values in Classification Data

missing values are biased towards a particular class, we depict the map in two parts, assuming without loss of generality that two classes are involved in the classification of the data set. The upper part represents patterns for one class, and the lower part represents those for the other class.

The three types of critical patterns of missing values in classification data can be illustrated in Figure 2. Figure 2(a) shows a case where no significant cluster is generated by the SOM. Apparently, this pattern shows MAR. The cluster circled in Figure 2(b) represents the pattern USM. It indicates that a substantial number of observations have missing values in certain variables concurrently. The cluster distributes almost evenly in the two spaces corresponding to the two classes, indicating it does not have a significant bias. If one traces these shadow observations back to their original multivariate data, specific variables that are correlated in missing values can be found. The cluster circled in Figure 2(c) represents the pattern UAM since the cluster appears in only one class.

To validate the patterns of missing values visualized from SOM, one might make changes to the parameters of the SOM network, including the neighborhood function and its change rate, and the learning rate and its change rate, to regenerate several self-organizing maps. The conventional split-half validation method can also be used to validate the patterns detected. Using this pattern discovery method, the experiment process and validation are conducted simultaneously, and an adequate result can always be obtained.

4 An Experiment with Real-World Data

Student evaluation of instruction survey methods are widely used at universities to evaluate the teaching performance of instructors. The data used in this experiment came from a student evaluation of instruction survey at a Canadian university. In this case, twenty questions describe the characteristics of an instructor's performance. Each question is rated on a five-point scale for students to answer. A high mark for a question indicates a positive answer to the question. One question of the survey is related to the classification of the instructors into two groups (i.e., effective and ineffective) based on the median value of the entire survey data on this question.

Sample data of 1893 incomplete observations were collected for this experiment. These observations with missing values were converted into shadow

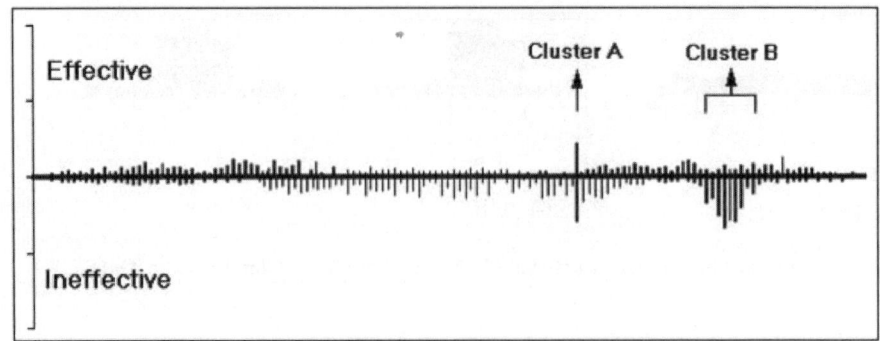

Fig. 3. The Self-Organizing Map for the Missing Values

Table 1. Summary of the Experimental Results

Patterns of Missing Values	Variables Involved	Questions	Interpretations
UAM (Cluster B)	x_{14}	Tests and assignments are reasonable measures of student learning	The four variables were correlated in missing values. The observations with concurrent missing values in these variables were biased towards the ineffective class.
	x_{15}	Where appropriate, student work is graded promptly	
	x_{16}	Where appropriate, helpful comments are provided when student work is graded	
	x_{18}	Tests and assignments provide adequate feedback on student progress	
USM (Cluster A)	x_{20}	The textbook(s) and course material are useful	Missing values in this variable were unbiased. They has little correlation with missing values in other variables

observations for SOM. In making trials of SOM, we started with the extremely large amount of output nodes in accordance with the number of the training samples (i.e., 1893 in this case), and then gradually reduced the number of output nodes to find a clear feature map. Our SOM program was implemented in C++, and had a data conversion interface with Microsoft Excel for graphics visualization. Figure 3 shows the result using the SOM with 200 output nodes, 200 nodes for the initial neighborhood, the initial learning rate of 0.01, and 2000 learning iterations. Since the map presented clear clusters, it was employed for further analysis.

Using the validated cluster analysis result on the SOM, two large clusters were identified, as shown in Figure 3. The original observations corresponding to the shadow observations of the clusters were extracted, and the nature of missing values in the survey and the hidden correlations of missing values were then identified, as summarized in Table 1. As shown in Table 1, the critical patterns of missing values revealed by the two clusters were UAM and USM. Notably, four variables of the twenty had a correlation of missing values which were biased towards the ineffective class.

5 Conclusions and Discussion

In this paper, we have discussed the critical patterns of missing values in the context of classification. SOM has been considered a useful technique in visualized clustering analysis when statistical clustering methods are difficult to apply. This paper proposes a SOM-based visualized pattern discovery method for missing values in classification data. Through the real-world case, it has been shown the usefulness of this method.

Compared with other traditional methods of treating missing values in classification data, the proposed method has certain advantages. This method provides visual presentations of patterns of missing values so that the user of the method is allowed to detect the quality of data and determine the strategy for validating classification results based on the particular problem domain. It requires no statistical assumptions in comparison with other statistical clustering analysis methods, but can effectively support statistical classification analysis.

Acknowledgements

The authors would like to thank an anonymous reviewer for helpful comments and suggestions. The first author is supported in part by Grant 312423 from the Natural Sciences and Engineering Research Council of Canada (NSERC).

References

1. Bello, A.L.: Imputation techniques in regression analysis: Looking closely at their implementation. Computational Statistics and Data Analysis 20, 45–57 (1995)
2. Chan, P., Dunn, O.J.: The treatment of missing values in discriminant analysis. Journal of the American Statistical Association 6, 473–477 (1972)
3. Deboeck, G., Kohonen, T.: Visual Explorations in Finance with Self-Organizing Maps. Springer, London, UK (1998)
4. Dempster, A.P., Laird, N.M., Rubin, D.B.: Maximum likelihood from incomplete data via the EM algorithm. Journal of the Royal Statistical Society B39(1), 1–38 (1997)
5. Gnanadesikan, R., Kettenring, J.R.: Discriminant analysis and clustering. Statistical Science 14(1), 34–69 (1989)
6. Hand, D.J.: Discrimination and Classification. Wiley, New York (1981)
7. Hand, D.J.: Data mining: Statistics and more? The American Statistician 52(2), 112–118 (1998)

8. Hartigan, J.A.: Clustering Algorithms. Wiley, New York, NY (1995)
9. Kalton, G., Kasprzyk, D.: The treatment of missing survey data. Survey Methodology 12, 1–16 (1986)
10. Kohonen, T.: Self-Organization and Associative Memory, 3rd edn. Springer, Heidelberg (1989)
11. Little, R.J.A., Rubin, D.B. (eds.): Statistical Analysis with Missing Data, 2nd edn. John Wiley and Sons, New York (2002)
12. Mundfrom, D.J., Whitcomb, A.: Imputing missing values: The effect on the accuracy of classification. Multiple Linear Regression Viewpoints 25(1), 13–19 (1998)
13. Romesburg, H.C.: Cluster Analysis for Researchers, Robert E. Krieger: Malabar, FL (1990)
14. Seber, G.A.F.: Multivariate Observations. Wiley, New York, NY (1984)
15. Silberschatz, A., Tuzhilin, A.: What makes patterns interesting in knowledge discovery systems. IEEE Transactions on Knowledge and Data Engineering 5(6), 970–974 (1996)
16. Troyanskaya, O., Cantor, M., Sherlock, G., Brown, P., Hastie, T., Tibshirani, R., Bostein, D., Altman, R.B.: Missing value estimation methods for DNA microarrays. Bioinformatics 17(6), 520–525 (2001)
17. Wang, H., Wang, S.: Data mining with incomplete data, in Encyclopedia of Data Warehousing and Mining. In: Wang, J. (ed.), Idea Group Inc. Hershey, PA, pp. 293–296 (2005)
18. Yang, Q., Ling, C., Chai, X., Pan, R.: Test-cost sensitive classification on data with missing values. IEEE Transactions on Knowledge and Data Engineering 18(5), 626–638 (2006)
19. Zhang, S., Qin, Z., Ling, C., Sheng, S.: "Missing is useful": Missing values in cost-sensitive decision trees. IEEE Transactions on Knowledge and Data Engineering 17(12), 1689–1693 (2005)

Visualizing Unstructured Text Sequences Using Iterative Visual Clustering

Qian You[1], Shiaofen Fang[1], and Patricia Ebright[2]

[1] Department of Computer and Information Science, Indiana University- Purdue University, 723 W. Michigan Street, SL 280, Indianapolis, 4620
[2] Department of Adult Health, School of Nursing, Indiana University, 1111 Middle Dr., NU442, Indianapolis, IN 46202
{qiyou, sfang}@cs.iupui.edu, prebrigh@iupui.edu

Abstract. This paper presents a keyword-based information visualization technique for unstructured text sequences. The text sequence data comes from nursing narratives records, which are mostly text fragments with incomplete and unreliable grammatical structures. Proper visualization of such text sequences can reveal patterns and trend information rooted in the text records, and has significant applications in many fields such as medical informatics and text mining. In this paper, an Iterative Visual Clustering (IVC) technique is developed to facilitate multi-scale visualization, and at the same time provide abstraction and knowledge discovery functionalities at the visualization level. Interactive visualization and user feedbacks are used to iteratively group keywords to form higher level concepts and keyword clusters, which are then feedback to the visualization process for evaluation and pattern discovery. Distribution curves of keywords and their clusters are visualized at various scales under Gaussian smoothing to search for meaningful patterns and concepts.

Keywords: Text and document visualization, Nursing data processing.

1 Introduction

In recent years, interactive visualization of non-visual information, combined with carefully designed user interaction tools, has started to play a major role in many data analysis and data mining applications. The integration of human intuition and knowledge with data analysis can potentially provide a faster and more reliable way for data mining and knowledge discovery. This requires a more dynamic and exploratory visualization environment that takes user input as guidance in directly the visualization structure and flow to obtain desired visual information. Text Information such as medical records, real time narratives, and daily notes, are readily available in non-structured fashion as sequences of text, with incomplete and grammatically inaccurate sentences. This type of text data is very difficult and often impossible to analyze using natural language parsing or traditional text analysis methods, as the language structures are unreliable or non-existing. It is, therefore, critical to be able to have quick and intuitive methods to navigate and explore unstructured text data to perceive and discover potential patterns and trends information that may be hidden in the vast quantities of unstructured information. The problem is particularly important for health care applications as most clinical and patient care records and narratives are unstructured text sequences. In this paper, we specifically target unstructured text data from nursing records and narratives.

In the following, we will first describe in Section 2 some background information on the datasets and related nursing applications. Some previous work will be discussed in Section 3. In Section 4, A simple keyword based visualization method will be described. This basic visualization technique will be extended to a user guided concept visualization technique using Iterative Visual Clustering in Section 5. We will conclude the paper in Section 6 with additional remarks and future work.

2 Nursing Data and Applications

Most studies about registered nurse (RN) work focused on specific care procedures and management of single patient clinical case studies; or were time studies with data collected through direct observation to identify percentage of time spent on specific activities such as direct and indirect patient care, and documentation.[5,7,9]. However, recent research suggested that RN performance is influenced not only by knowledge specific to patient care needs, but also by factors characteristic of complex environments, including time pressures, uncertain information, conflicting goals, high stakes, stress, and dynamic conditions [1,2,13]. What had not been explored until recently were the numerous activities engaged in by RNs to manage the environment and patient flow, including the invisible cognitive work necessary to create and maintain safety. The challenge for disseminating findings regarding the complex nature of RN work is in representing the work in a way that demonstrates the dynamic, multi-tasking features and patterns that characterize RN work performance and decision making. Previous to discovering the complexity of RN work, what an RN did was captured by job descriptions and lists of tasks.

A procedure for manual recording of direct observations of RN work was developed. Observation data was recorded on legal pads, line by line, using an abbreviated shorthand method. For example, each entry started with a verb representing the RN activity (e.g., adjusting the IV flow rate; assisting patient out of bed) or the preposition "to" to indicate travel (e.g., to the supply room…; to the med cart). No pre-coded categories were used so that all actual RN activities in the sequence encountered were captured continuously over three-hour periods. Although times were recorded frequently next to data to provide an estimate of elapsed time for situations, the timing of individual activities was not the focus. A segment of an sample session of nursing tasks is as shown below:

Walks to Pt #1 Room
Gives meds to Pt #1
Reviews what Pt #1 receiving and why
Assesses how Pt#1 ate breakfast
Teaches Pt#1 to pump calves while in bed
Explains to Pt#1 fdngs- resp wheezes

Reinforces use of IS Pt#1
Positions pillow for use in coughing Pt#1
Listens to tech about new pain in Pt#1
Assesses Pt#1 for new onset pain in shoulder
Explains to Pt#1 will call physician
To hallway
Returns to Pt#1 for if BP has changed
To station to call
....

Analysis of the patterns and trends in data through visualization help to convey the complexity and dynamic aspects of RN work. What to the outside casual observer appears to be single task elements becomes a much more complicated array of overlapping functions with inter-related patterns and trends. The following are but a few of the anticipated applications to RN work: (1) identification of specific non-clinical work in the midst of clinical care that could be redesigned for efficiency and safety; (2) identification of work patterns that could be supplemented and enhanced with informatics technology; (3) identification of work patterns requiring skills and knowledge needing to be addressed in basic nursing curriculum but not presently addressed; and (4) staffing and assignment implications based on patterns in work across time. We will not attempt to directly address these analysis goals in this paper, as that requires a lot more datasets than what we have now and more extensive domain studies. Instead we aim to develop a general visualization framework that allows the user to guide the visualization procedures based on a specific analysis goal to obtain visual information of a certain context.

The notion of time in this type of datasets is an issue that requires some attention. Although the start and stop times are recorded in each session with a set of varying number of activities, the time each activity takes is not quantified, and the activities across different sessions are not synchronized or aligned either. This makes comparisons between different sessions difficult. Since no information is available regarding time distributions, we will simply normalize the sessions and equally distribute the activities within the standard time interval. This inevitably creates misalignments of activities across sessions. To alleviate this problem, we will apply a Gaussian filter to the keyword distribution curves to provide a smoothing effect that can better preserve the global shape and pattern information of these distributions.

3 Related Work

Since a meaningful mapping from text information to a visual space is difficult, there have been relatively few text visualization techniques compared to other information visualization domains. The problem becomes even more difficult when the text is unstructured, which makes semantic analysis extremely difficult or impossible. There are a large number of document visualization techniques in the literatures, but these are usually concerned with the summary and categorical information of each document, and do not directly visualize or represent the information from the text. There are, however, several text visualization techniques dealing with text information directly that are related to the work presented here.

The CareView [10] visualization was developed for the purpose of visualizing integrated trends in clinical records of patients which consist of both quantitative and qualitative data and are recorded as narratives. Quantitative values are plotted along the vertical axis with respect to the timeline in a 2D space and a color-scheme is used to indicate the intensity of a particular data range. The qualitative data are decomposed into numeric values and similarly plotted on the vertical axis. CareView was found to significantly reduce the amount of time required to identify trends in patients' conditions, as opposed to a tabular interface. LifeLines [11, 12] describes a visualization environment which provides an overview of a series of events such as personal histories, medical records, and court records. The varying facets of the records are displayed as individual time lines in two-dimensional space. Thickness of the lines and different colors are used to encode various attributes of the records. In addition, discrete events within the records are visualized using icons that are indicative of the nature of events. ThemeRiver [6] uses the natural image of a flowing river as a visual metaphor to illustrate changes of themes over time. The vertical width of the river indicates the total collective strength of the selected theme. It was found that the ThemeRiver visualization was much easier and more productive to use than the other forms of visualization they had been using. 3D ThemeRiver [8] extends the work of 2D ThemeRiver by adding one more attribute in the Z axis, there by giving a three dimensional flow of the river.

Mining Text using Keyword Distributions [4] explores and addresses the problem of knowledge discovery in large collections of unstructured text documents using distributions of keywords to label the documents. The distributions are determined with respect to the occurrence frequency of keywords in the documents. The system first labels the documents with a predefined set of keywords from a restricted vocabulary. Then the keywords are reorganized into a structured hierarchy. A set of customized knowledge discovery operations on those hierarchies are created to find patterns in text documents.

Our earlier work in [3] employed a keyword-based visualization method that used icons and a technique similar to Theme-river to represent the keyword distributions in the time sequence data. In this paper we present a different visualization approach based on a user guided keyword clustering technique. The concept of integrating interactive visual feedback within the visualization structures for knowledge discovery is not present in previous text visualization techniques.

4 Keywords Distribution View

4.1 Text Data Preprocessing

The lack of semantic structures in the text information indicates that keyword is the only available information unit that can be processed and visualized. A set of keywords are first extracted from the text data. The frequencies of occurrence of each keyword form a distribution curve over the time of each session. As in all forms of communication, written, and spoken, the frequency and quantity of specific words provide two specific perceptions for the observer. First, the frequency or number of times a word occurs over the course of a dialogue often provides indication of a theme or trend. Second, the number of times a word is referenced within one event, such as each sentence within a dialogue, often indicates the importance or weight of a given topic.

To start with, each text document is broken up into a list of relevant keywords to enable the visualization of keyword occurrences in time and the relative distribution of keywords. Then each line of text sequences is denoted as an event. Each line or event is assigned a numerical number known as an event position. For each event, the various relevant keywords are broken up from that line of text and logged as keywords having the same event position. Thus, all the keywords in the same line of text would have the same event position number.

4.2 Keyword Distribution Curves

While the keyword distribution can be viewed individually within each dataset (one sequence), it is more informative to combine the distributions of each keyword in all available datasets. Patterns and trends, if present, will more likely to appear in a cumulative view. The distribution shows the aggregate pattern over the specified period for each keyword in the form of either a distribution curve, or a filled graph plot [3], as shown in Figure 1.

Each plot contains and visually represents the frequencies at which a particular set of events (keywords) occurred over a time interval. It represents what the individual data set trends look like when they are collapsed together in one cohesive display. In the filled graph plot, the thickness of each solid graph girth represents the frequency of each particular event over time. It is similar to ThemeRiver [6] except that the data plot is anchored at the bottom by a uniform flat starting point as the base, which is easier to compare total height of the stacked plot representation over the length of the display. In the visual pattern recognition process in this paper, however, the similarities of the distribution curves for different keywords need to be visually identified to generate keyword clusters. This shape information is easier to identify using simple curve representations in separate views (Figure 1(b)).

By combining the distribution curves in all datasets, we make the assumption that all events are uniformly distributed within the 3-hour interval of each session, and these sessions are then accumulated according to the events' relative positions on the timeline of each session. When the events are not evenly distributed on the timeline, there are misalignments in the cumulative distribution curve. While we do not expect that minor misalignments would alter the shape patterns of the curves, a proper smoothing filter can help recover and maintain the global shapes of the curves. In our experiment, a Gaussian filter of varying scales is applied to each curve to generate a shape pattern that is simple and easy compare.

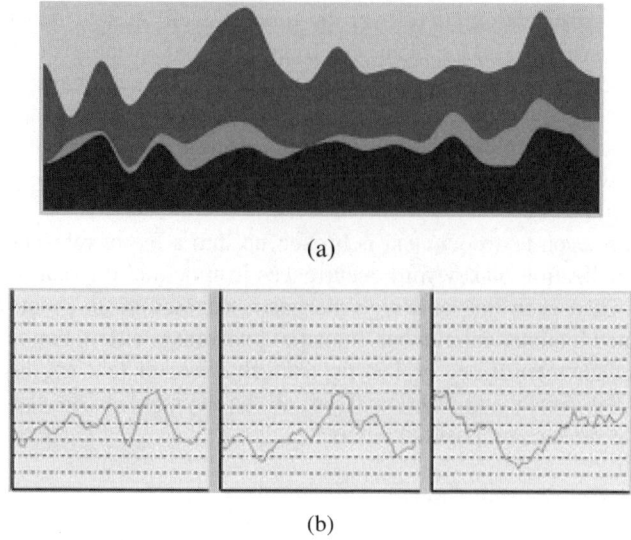

Fig. 1. (a) Filled graph plots; (b) Distribution curves

5 Iterative Visual Clustering

5.1 Design Goal of Iterative Visual Clustering

The goal of this text visualization technique is to let the user to visually identify patterns and similarities among various groups of keywords and explore the potential trends and knowledge hidden in the datasets. Generally speaking, this visual analytics method is more effective for this type of data since there is not a clear analysis goal – the goal depends on what patterns show up in the visualization process. It is also difficult to determine and extract features from the keyword distribution curves because the concepts of similarity and pattern are very vague and less deterministic in nature.

The Iterative Visual Clustering approach we developed in this paper employs a hierarchical keyword merging process that allows the user to visually determine:

Is there an interested pattern present in a certain representation?
Are two patterns similar?
Is a pattern become stronger by merging a new group of keywords?

Making these decisions by visually examining the keyword distribution curves is relatively easy for human but extremely difficult and unreliable by automatic algorithms.

In our approach, the user will first screen the keyword list and manually remove trivial keywords and merge words that represent the same or highly similar meanings. The distribution curves of the filtered keywords will be visualized in an interactive interface (as shown in Figure 2) that contains multiple pages of keyword distribution curves.

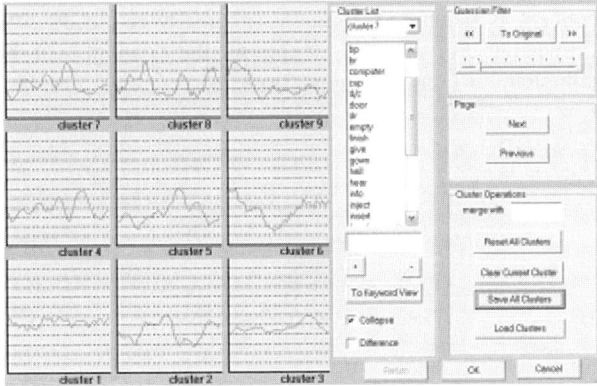

Fig. 2. The visualization interface

5.2 Iterative Visual Clustering Algorithm

In the Iterative Visual Clustering algorithm, the user first select a number of distinct patterns that are represented by the initial keyword groups, and will then attempt to merge each of the remaining keywords with these initial patterns. The merging starts with a group that is the most similar. The merge will stay if one of the following two scenarios occurs:

The original pattern is strengthened; or
A new stronger pattern appears.

If a keyword cannot be merged with any of the existing groups, it will stay by its own as a new cluster.

In the second round, the clusters generated in the first round will be merged in pairs, based on again the above two criteria. This process will continue iteratively with as many rounds as needed until no more merging can be made. In this process, if a cluster cannot be merged in two consecutive rounds, this cluster will be set aside and become a stand-alone cluster. This is to avoid the large differences in size when merging two clusters. The strength of the curve of a large cluster will overcome the influence of a small cluster, and can sometimes incorrectly hide a meaningful stand-alone small cluster through merging. Figure 3 shows some sample screens of several rounds of merging with 14 sets of nursing record datasets.

This process is based entirely on the visual appearances of the curves without references to the meanings of the words. So not all of the final clusters represent meaningful concepts. But some of these clusters do reveal patterns of certain general concepts. For example, Cluster #1 clearly represents a concept related to "interaction with patients", as it involves mostly keywords such as "ask", "call", "listen", "observe", "procedure", "request", etc. These are often involved in descriptions such as: "Listens to pt ask re: how control K+ better?", "Observes as pt approaches", and "Asks Pt#1 if understands procedure". The curve's pattern for cluster #1 indicates that there is a consistently high level of patient interactions throughout the entire sessions.

Once such a concept is generated, user can manually re-cluster several keywords based on the knowledge. For instance, "explain" and "patient" can be added to the first cluster to complete the concept "interaction".

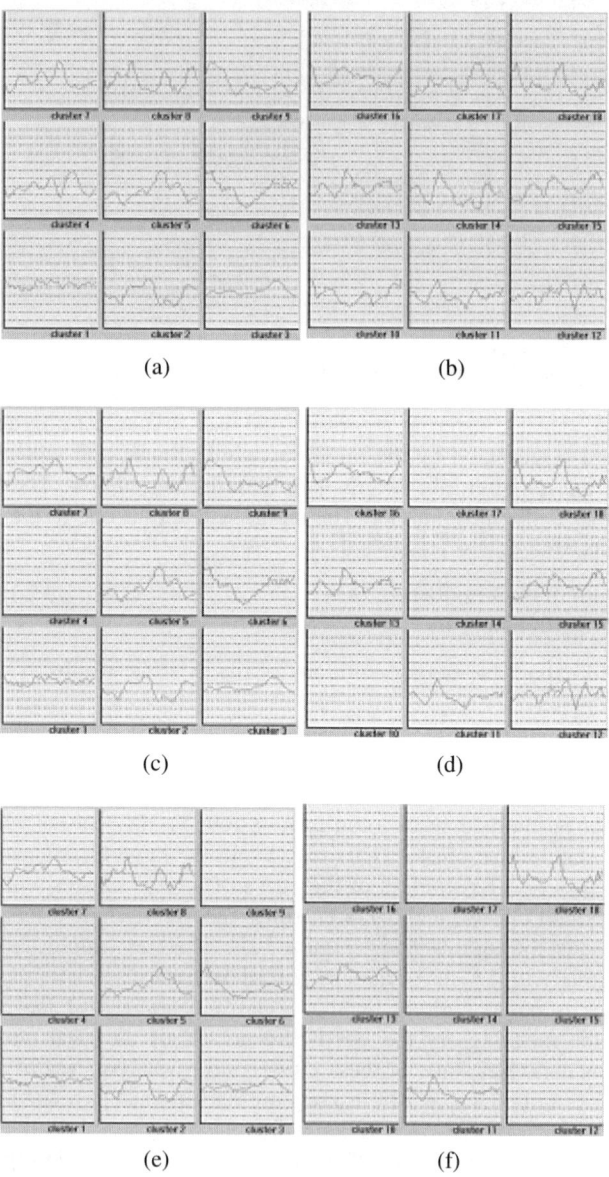

Fig. 3. (a) (b) After first round; (c): Merged 4 and 7; (d) merged 11 and 14; (e) Merged 6 and 9; (f) some of the final clusters

6 Conclusion

Unstructured text sequences contain insufficient semantic information, thus many traditional text and document analysis techniques are not effective. Interactive visualization allows the users to intuitively and visually define and generate similar and interested patterns in an iterative and interactive manner. This effectively captures the user's design intentions and analysis goals through an interactive visualization process without formal representations of objective functions or a formal knowledge representation scheme, which are often too difficult to formulate. The framework described in this paper provides such an environment for users to explore and navigate the possible feature space and to search for interested concepts and patterns that match the user's analysis goals. The iterative visual clustering process is designed to form subgroups of the keywords that exhibit good patterns in their distribution curves. While not all of these groups represent meaningful concepts, it does substantially narrow down the search space.

In the future, we would like to implement operators (e.g. intersection, subtraction, order, etc.) into the keyword groups to form a more sophisticated concept representation. Additional testing and user studies in the nursing field will be necessary to evaluate the effectiveness of this approach.

References

1. Ebright, P.R., Patterson, E.S., Chalko, B.A., Render, M.L.: Understanding the complexity of registered nurse work in acute care settings. Journal of Nursing Administration 33(12), 630–638 (2003)
2. Ebright, P.R., Urden, L., Patterson, E.S., Chalko, B.A.: Themes surrounding novice nurse near miss and adverse event situations. Journal of Nursing Administration 34(12), 531–538 (2004)
3. Fang, S., Lwin, M., Ebright, P.R.: Visualization of Unstructured Text Sequences of Nursing Narratives. In: Proc. ACM Symposium on Applied Computing, Dijon, France, pp. 239–243 (2006)
4. Feldman, R., Dagan, I., Hirsh, H.: Mining Text Using Keyword Distributions. Journal of Intelligent Information Systems: Integrating Artificial Intelligence and Database Technologies, pp. 281–300 (1998)
5. Hansten, R., Washburn, M.: Why don't nurses delegate? Journal of Nursing Administration 26(12), 24–28 (1996)
6. Havre, S., Hetzler, E., Whitney, P., Nowell, L.: ThemeRiver: Visualizing Thematic Changes in Large Document Collections. Visualization and Computer Graphics, IEEE Transactions 8, 9–20 (2002)
7. Hughes, M.: Nursing workload: an unquantifiable entity. Journal of Nursing Management 7(6), 317–322 (1999)
8. Imrich, P., Mueller, K., Imre, D., Zelenyuk, A., Zhu, W.: Interactive Poster: 3D ThemeRiver. In: IEEE Information Visualization Symposium, IEEE Computer Society Press, Los Alamitos (2003)
9. Kumarich, D., Biordi, D.L., Milazzo-Chornick, N.: The impact of the 23-hour patient on nursing workload. Journal of Nursing Administration 20(11), 47–52 (1990)

10. Mamykina, L., Goose, S., Hedqvist, D., Beard, D.: CareView: Analyzing Nursing Narratives for Temporal Trends. In: Proceedings of the SIGCHI Conference on Human Factors in Computing Systems, pp. 1147–1150 (2004)
11. Plaisant, C., Milash, B., Rose, A., Widoff, S., Shneiderman, B.: LifeLines: Visualizing Personal Histories. In: Proceedings of the SIGCHI conference on Human factors in computing systems, p. 221 (1996)
12. Plaisant, C., et al.: LifeLines: Using Visualization to Enhance Navigation and Analysis of Patient Records. In: Proceedings of American Medical Informatics Association Conference (1998)
13. Potter, P., Wolf, L., Boxerman, S., Grayson, D., Sledge, J., Dunagan, C., Evanoff, B.: Understanding the cognitive work of nursing in the acute care environment. JONA 35(7/8), 327–335 (2005)

Enhanced Visual Separation of Clusters by M-Mapping to Facilitate Cluster Analysis*

Ke-Bing Zhang[1], Mehmet A. Orgun[1], and Kang Zhang[2]

[1] Department of Computing, Macquarie University, Sydney, NSW 2109, Australia
{kebing, mehmet}@ics.mq.edu.au
[2] Department of Computer Science, University of Texas at Dallas
Richardson, TX 75083-0688, USA
kzhang@utdallas.edu

Abstract. The goal of clustering in data mining is to distinguish objects into partitions/clusters based on given criteria. Visualization methods and techniques may provide users an intuitively appealing interpretation of cluster structures. Having good visually separated groups of the studied data is beneficial for detecting cluster information as well as refining the membership formation of clusters. In this paper, we propose a novel visual approach called *M-mapping*, based on the projection technique of HOV^3 to achieve the separation of cluster structures. With M-mapping, users can explore visual cluster clues intuitively and validate clusters effectively by matching the geometrical distributions of clustered and non-clustered subsets produced in HOV^3.

Keywords: Cluster Analysis, Visual Separability, Visualization.

1 Introduction

Cluster analysis is an iterative process of clustering and cluster verification by the user facilitated with clustering algorithms, cluster validation methods, visualization and domain knowledge to databases. The applications of clustering algorithms to detect grouping information in real world applications are still a challenge, primarily due to the inefficiency of most existing clustering algorithms on coping with arbitrarily shaped distribution data of extremely large and high-dimensional databases. Moreover, the very high computational cost of statistics-based cluster validation methods is another obstacle to effective cluster analysis.

Visual presentations can be very powerful in revealing trends, highlighting outliers, showing clusters, and exposing gaps in data [18]. Nowadays, as an indispensable technique, visualization is involved in almost every step in cluster analysis. However, due to the impreciseness of visualization, it is often used as an observation and rendering tool in cluster analysis, but it has been rarely employed directly in the precise comparison of the clustering results.

HOV^3 is a visualization technique based on hypothesis testing [20]. In HOV^3, each hypothesis is quantified as a measure vector, which is used to project a data set for

---

* The datasets used in this paper are available from
http://www.ics.uci.edu/~mlearn/Machine-Learning.html

investigating cluster distribution. The projection of HOV^3 is also proposed to deal with cluster validation [21]. In this paper, in order to gain an enhanced visual separation of groups, we develop the projection of HOV^3 into a technique which we call M-mapping, i.e., projecting a data set against a series of measure vectors.

We structure the rest of this paper as follows. Section 2 briefly introduces the current issues of cluster analysis. It also briefly reviews the efforts that have been done in the visual cluster analysis and discusses the projection of HOV^3 as the background of this research. Section 3 discusses the M-mapping model and its several important features. Section 4 demonstrates the effectiveness of the enhanced separation feature of M-mapping on cluster exploration and validation. Finally, section 5 concludes the paper with a brief summary of our contributions.

2 Background

2.1 Cluster Analysis

Cluster analysis includes two processes: clustering and cluster validation. Clustering aims to distinguish objects into partitions, called clusters, by a given criteria. The objects in the same cluster have a higher similarity than those between the clusters. Many clustering algorithms have been proposed for different purposes in data mining [8, 11]. Cluster validation is regarded as the procedure of assessing the quality of clustering results and finding a fit cluster scheme for a specific application at hand. Since different cluster results may be obtained by applying different clustering algorithms to the same data set, or even, by applying a clustering algorithm with different parameters to the same data set, cluster validation plays the critical role in cluster analysis.

However, in practice, it may not always be possible to cluster huge datasets by using clustering algorithms successfully. As Abul *et al* pointed out "In high dimensional space, traditional clustering algorithms tend to break down in terms of efficiency as well as accuracy because data do not cluster well anymore"[1]. In addition, the very high computational cost of statistics-based cluster validation methods directly impacts on the efficiency of cluster validation.

2.2 Visual Cluster Analysis

The user's correct estimation of the cluster number is important for choosing the parameters of clustering algorithms in the pre-processing stage of clustering, as well as assessing the quality of clustering results in the post-processing stage of clustering. The success of these tasks heavily relies on the user's visual perception of the distribution of a given data set. It has been observed that visualization of a data set is crucial in the verification of the clustering results [6].

Visual cluster analysis enhances cluster analysis by combining it with visualization techniques. Visualization techniques are typically employed as an observational mechanism to understand the studied data. Therefore, instead of contrasting the quality of clustering results, most of the visualization techniques used in cluster analysis focus on assisting users in having an easy and intuitive understanding of the cluster structure in the data. Visualization has been shown to be an intuitive and effective method used in the exploration and verification of cluster analysis.

Several efforts have been made in the area of cluster analysis with visualization: OPTICS [2] uses a density-based technique to detect cluster structures and visualizes clusters in "Gaussian bumps", but it has non-linear time complexity making it unsuitable to deal with very large data sets or to provide the contrast between clustering results. H-BLOB [16] visualizes clusters into blob manners in a 3D hierarchical structure. It is an intuitive cluster rendering technique, but its 3D and two stages expression limits its capability in an interactive investigation of cluster structures.

Kaski et. al [15] uses Self-organizing maps (SOM) to project high-dimensional data sets to 2D space for matching visual models [14]. However, the SOM technique is based on a single projection strategy and it is not powerful enough to discover all the interesting features from the original data.

Huang et. al [7, 10] proposed several approaches based on FastMap [4] to assist users on identifying and verifying the validity of clusters in visual form. Their techniques are good on cluster identification, but not able to deal with the evaluation of cluster quality very well. Moreover, the techniques discussed here are not very well suited to the interactive investigation of data distributions of high-dimensional data sets. A recent survey of visualization techniques in cluster analysis can be found in the literature [19].

Interactive visualization is useful for the user to import his/her domain knowledge into the cluster exploration stage on the observation of data distribution changes. Star Coordinates favors to do so with its interactive adjustment features [21]. The M-mapping approach discussed in this paper has been developed based on the Star Coordinates and the projection of HOV^3 [20]. For a better understanding of the work in this paper, we briefly describe them next.

2.3 The Star Coordinates Technique

The Star Coordinates is a technique for mapping high-dimensional data to 2D dimensions. It plots a 2D plane into n equal sectors with n coordinate axes, where each axis represents a dimension and all axes share the initials at the centre of a circle on the 2D space. First, data on each dimension are normalized into [0, 1] or [-1, 1] interval. Then the values of all axes are mapped to orthogonal X-Y coordinates which share the initial point with Star Coordinates on the 2D space. Thus, an n-dimensional data item is expressed as a point in the 2D plane.

The most prominent feature of Star Coordinates and its extensions such as VISTA [4] and HOV^3 [20] is that their computational complexity is only in linear time (that is, every n-dimensional data item is processed only once). Therefore they are very suitable as visual interpretation and exploration tools in cluster analysis. However, it is inevitable to introduce overlapping and bias by mapping high-dimensional data to 2D space. For mitigating the problem, Star Coordinates based techniques provide some visual adjustment mechanisms, such as axis scaling (called α-*adjustment* in VISTA), footprints, rotating axes angles and coloring data points [12].

- **Axis Scaling**

The purpose of axis scaling in Star Coordinates is to adjust the weight value of each axis dynamically and observe the changes to the data distribution under the newly weighted axes. We use the Iris, a well-known data set in machine learning area as an example to demonstrate how axis scaling works as follows.

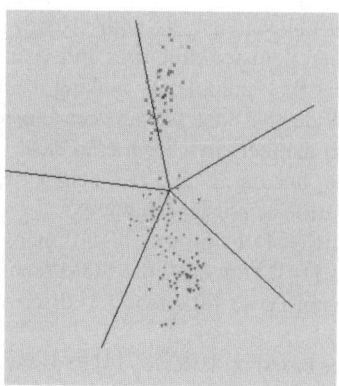 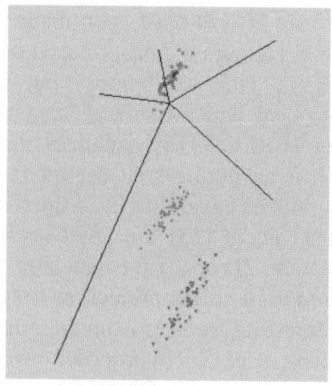

Fig. 1. The initial data distribution of clusters of Iris produced by k-means in VISTA

Fig. 2. The tuned version of the Iris data distribution in VISTA

Iris has 4 numeric attributes and 150 instances. We first applied the K-means clustering algorithm to it and obtained 3 clusters (with k=3), and then tuned the weight value of each axis of Iris in VISTA [4]. The diagram in Fig.1 shows the original data distribution of Iris, which has overlapping among the clusters. A well-separated cluster distribution of Iris is illustrated in Fig. 2 by a series of axis scaling. The clusters are much easier to be recognized than those in the original one.

- **Footprints**

To observe the effect of the changes to the data points under axis scaling, Star Coordinates provides the "footprints" function to reveal the trace of each point [12]. We use another data set *auto-mpg* to demonstrate this feature. The data set *auto-mpg* has 8 attributes and 397 items. Fig. 3 presents the footprints of axis scaling of attributes "weight" and "mpg", where we may find some points with longer traces, and some with shorter footprints.

However, the imprecise and random adjustments of Star Coordinates and VISTA limits them to be utilized as quantitative analysis tools.

Fig. 3. Footprints of axis scaling of "weight" and "mpg" attributes in Star Coordinates [12]

2.4 The HOV³ Model

The HOV3 Model improves the Star Coordibates model. Geometrically, the difference of a matrix D_j (a data set) and a vector M (a measure) can be represented by their inner product, $D_j \cdot M$. Based on this idea Zhang *et al* proposed a projection technique, called HOV³, which generalizes the weight values of axes in Star Coordinates as a hypothesis (measure vector) to reveal the differences between the hypotheses and the real performance [20].

The Star Coordinates model can be simply described by the Euler formula. According to the Euler formula: $e^{ix} = \cos x + i \sin x$, where $z = x + i.y$, and i is the imaginary unit. Let $z_0 = e^{2\pi i/n}$; we see that $z_0^1, z_0^2, z_0^3, \ldots, z_0^{n-1}, z_0^n$ (with $z_0^n = 1$) divide the unit circle on the complex plane into n equal sectors. Thus Star Coordinates mapping can be simply written as:

$$P_j(z_0) = \sum_{k=1}^{n}[(d_{jk} - \min d_k)/(\max d_k - \min d_k) \cdot z_0^k] \quad (1)$$

where $\min d_k$ and $\max d_k$ represent the minimal and maximal values of the kth coordinate respectively, and m_k is the kth attribute of measure M. In any case equation (1) can be viewed as mappings from $R^n \to C^2$.

Then given a non-zero measure vector M in R^n and a family of vectors P_j, the projection of P_j against M according to formula (1), in the HOV^3 model [18], is given as:

$$P_j(z_0) = \sum_{k=1}^{n}[(d_{jk} - \min d_k)/(\max d_k - \min d_k) \cdot z_0^k \cdot m_k] \quad (2)$$

As shown above, a hypothesis in HOV^3 is a quantified measure vector. HOV^3 not only inherits the axis scaling feature of Star Coordinates, but also generalizes the axis scaling as a quantified measurement. The processes of cluster detection and cluster validation can be tackled with HOV^3 based on its quantified measurement feature [20, 21]. To improve the efficiency and effectiveness of HOV^3, we develop the projection technique of HOV^3 further with *M-mapping*.

3 M-Mapping

3.1 The M-Mapping Model

It is not easy to synthesize hypotheses into one vector. In practice, rather than using a single measure to implement a hypothesis test, it is more feasible to investigate the synthetic response of applying several hypotheses/predictions together to a data set. For simplifying the discussion of the M-mapping model, we give a definition first.

Definition 1 (Poly-multiply vectors to a matrix). The inner product of multiplying a series of non-zero measure vectors $M_1, M_2, \ldots, M_S$ to a matrix A is denoted as

$$A \cdot * \prod_{i=1}^{S} M_i = A \cdot * M_1 \cdot * M_2 \cdot * \ldots \cdot * M_S.$$

A simple notation of HOV^3 projection as $\mathcal{D}_p = \mathcal{H}_C(\mathcal{P}, M)$ was given by Zhang et al [20], where $\mathcal{P}$ is a data set; $\mathcal{D}_p$ is the data distribution of $\mathcal{P}$ by applying a measure vector M. Then the projection of *M-mapping* is denoted as $\mathcal{D}_p = \mathcal{H}_C(\mathcal{P}, \prod_{i=1}^{S} M_i)$. Based on equation (2), *M-mapping* is formulated as follows:

$$P_j(z_0) = \sum_{k=1}^{n}[(d_{jk} - \min d_k)/(\max d_k - \min d_k) \cdot z_0^k \cdot \prod_{i=1}^{S} m_{ik}] \quad (3)$$

where m_{ik} is the kth attribute (dimension) of the ith measure vector M_i, and $s \geq 1$. When $s=1$, the equation (3) is transformed into equation (2) (the HOV³ model).

We may observe that instead of using a single multiplication of m_k in formula (2), it is replaced by a poly-multiplication of $\prod_{i=1}^{s} m_{ik}$ in equation (3). Equation (3) is more general and also closer to the real procedure of cluster detection. It introduces several aspects of domain knowledge together into the process of cluster detection by HOV³. Geometrically, the data projection by M-mapping is the synthesized effect of applying each measure vector by HOV³. In addition, the effect of applying M-mapping to datasets with the same measure vector can enhance the separation of grouped data points under certain conditions. We describe this enhanced the separation feature of M-mapping below.

3.2 The Features of M-Mapping

For the explanation of the geometrical meaning of the M-mapping projection, we use the real number system. According to the equation (2), the general form of the distance σ (i.e., weighted Minkowski distance) between two points a and b in HOV³ plane can be represented as:

$$\sigma(a,b,m) = \sqrt[q]{\sum_{k=1}^{n} |m_k(a_k - b_k)|^q} \quad (q>0) \quad (4)$$

If $q = 1$, σ is Manhattan (city block) distance; and if $q = 2$, σ is Euclidean distance. To simplify the discussion of our idea, we adopt the Manhattan metric. Note that there exists an equivalent mapping (bijection) of distance calculation between the Manhattan and Euclidean metrics [13]. For example, if the distance ab between points a and b is longer than the distance $a'b'$ between points a' and b' in the Manhattan metric, it is also true in the Euclidean metric, and vice versa.

In Fig 4, the orthogonal lines represent the Manhattan distance and the diagonal lines are the Euclidean distance (red for $a'b'$ and blue for ab) respectively. Then the Manhattan distance between points a and b is calculated as in formula (5).

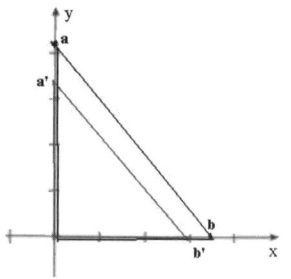

Fig. 4. The distance representation in Manhattan and Euclidean metrics

$$\sigma(a,b,m) = \sum_{k=1}^{n} |m_k(a_k - b_k)| \quad (5)$$

According to the equations (2), (3) and (5), we can present the distance of M-mapping in Manhattan distance as follows:

$$\sigma(a,b,\prod_{i=1}^{s} m_i) = \sum_{k=1}^{n} |\prod_{i=1}^{s} m_{ki}(a_k - b_k)| \quad (6)$$

Definition 2 (The distance representation of M-mapping). The distance between two data points a and b projected by *M-mapping* is denoted as $\overset{s}{\underset{i=1}{M}}\sigma ab$. If the measure vectors in an M-mapping are the same, $\overset{s}{\underset{i=1}{M}}\sigma ab$ can be simply written as $M^s\sigma ab$; if each attribute of M is 1 (no measure case), the distance between points a and b is denoted as σab.

For example, the distance between two points a and b projected by M-mapping with the same two measures can be represented as $M^2\sigma ab$. Thus the projection of HOV^3 of a and b can be written as $M\sigma ab$.

- **Contracting Feature**

From the equations (5) and (6), we may observe that the application of M-mapping to a data set is a contracting process of data distribution of the data set. This is because, when $|m_{ki}|<1$ and $\sigma ab \neq 0$, we have $\sum_{k=1}^{n}|m_k(a_k - b_k)| < \sum_{k=1}^{n}|(a_k - b_k)|$ $\Rightarrow$ $\sigma(a, b, m) < \sigma ab$. In the same way, we have $\sigma(a, b, m^2) < \sigma(a, b, m)$ and $\sigma(a, b, m^{n+1}) < \sigma(a, b, m^n)$, $n \in \mathbb{N}$.

Hinneburg at el proved that a contracting projection of a data set could strictly preserve the density of the dataset [9]. Chen and Liu also proved that in the Star Coordinates 2D space, the original closed data points are also more closed relatively in the newly produced data distribution by axis scaling [4]. Thus, the relative geometrical position of data points within a cluster in a data set would be closer by applying M-mapping to the data set.

- **Enhanced Separation Feature**

If the measure vector is changed from M to M', and $|M\sigma ab - M\sigma ac| < |M'\sigma ab - M'\sigma ac|$ then

$$\frac{|M'\sigma ab - M'\sigma ac| - |M'^2 \sigma ab - M'^2 \sigma ac|}{|M'\sigma ab - M'\sigma ac|} > \frac{|M\sigma ab - M\sigma ac| - |M'\sigma ab - M'\sigma ac|}{|M\sigma ab - M\sigma ac|}.$$

Due to the space limitation, the detailed proof of this property can be found in [22].

This inequality shows that if the difference of the distance ab and the distance ac are increased by scaling axes from M to M' (which can be observed by the footprints of points a, b and c, as shown in Fig 3), then after applying M-mapping to a, b and c, the distance variation rate of distances ab and ac would be enhanced. In other words, if it is observed that several groups of data points can be roughly separated (where ambiguous points exist between groups) by projecting a measure vector in HOV^3 to a data set, then the application of M-mapping with the measure vector to the data set would lead to the groups being more contracted, i.e., they will be a good separation of the groups.

These two features of M-mapping are significant for identifying the membership formation of clusters in the process of cluster exploration and cluster verification. This is because the contracting feature of M-mapping keeps the data points within a cluster relatively closer, i.e., grouping information is preserved. On the other hand, the enhanced separation feature of M-mapping can extend the distance of far data points relatively further.

- **Improving Accuracy of Data Point Selection with Zooming**

External cluster validation [19] refers to the comparison of previously produced cluster patterns with newly produced cluster patterns to evaluate the genuine cluster structure about a data set. However, due to the very high computational cost of statistical methods on assessing the consistency of cluster structures between the subsets of a large database, achieving this task is still a challenge.

Let us assume that, *if two sampling subsets of a dataset have similar data distributions and a measure vector is applied to them by HOV^3, the similarity of their data distributions should still be high*. Based on this assumption, Zhang et al [21] proposed a visual external cluster validation approach with HOV^3. Their approach uses a clustered subset from a database and a same-sized unclustered subset as an observation. It then applies several measure vectors that can separate clusters in the clustered subset. Thus each cluster and its geometrically covered data points (called *quasi-cluster* in their approach) based on a given threshold distance are selected. Finally, the overlapping rate of each cluster and quasi-cluster pair is calculated; and if the overlapping rate approaches 1, this means that the two subsets have a similar cluster distribution. Compared to statistics-based external validation methods, their method is not only visually intuitive, but also more effective in real applications [21].

However, separating a cluster from lots of overlapping points manually is often time consuming. We claim that the enhanced separation feature of HOV^3 can provide improvements not only in efficiency but also in accuracy in dealing with external cluster validation by the proposed approach [21].

As mentioned above, the application of M-mapping to a data set is a contracting process. In order to avoid the contracting effect causing pseudo data points being selected, we introduce a zooming feature with M-mapping. According to equation (2), zooming in HOV^3 can be understood as projecting a data set with a vector, which "has the same attribute values, i.e., each m_k in equation (2) has the same value. Then we choose $\min(m_k)^{-1}$ as the zooming vector values, where $\min(m_k)$ is the non-zero minimal value of m_k. Thus the scale of patterns in HOV^3 is amplified by applying the combination of M-mapping and zooming. This combination is formalized in equation (7).

$$P_j(z_0) = \sum_{k=1}^{n} [(d_{jk} - \min d_k)/(\max d_k - \min d_k) \cdot z_0^k \cdot (m^s{}_k \cdot \min(m_k)^{-s})] \qquad (7)$$

Because $|m_k|<1$, and $\min(m_k)$ is the non-zero minimal value of m_k in a measure vector, thus $|(m_k)^S \cdot \min(m_k)^{-S}|>1$ if there exists $|m_k|>|\min(m_k)|$. With the effect of $|(m_k)^S \cdot \min(m_k)^{-S}|$, M-mapping enlarges the scale of data distributions projected by HOV^3. With the same threshold distance of the data selection proposed by Zhang et al [21], M-mapping with zooming can improve the precision of geometrically covered data point selection.

4 Examples and Explanation

In this section we present several examples to demonstrate the efficiency and the effectiveness of M-mapping in cluster analysis.

4.1 Cluster Exploration with M-Mapping

Choosing the appropriate cluster number of an unknown data set is meaningful in the pre-clustering stage. The enhanced separation feature of M-mapping is advantageous in the identification of the cluster number in this stage. We demonstrate this advantage of M-mapping by the following examples.

- **Wine Data**

The Wine data set (*Wine* in short) has 13 attributes and 178 records. The original data distribution of *Wine data* in 2D space is shown in Fig. 5a, where no grouping information can be observed. Then we tuned axes weight values randomly and had a

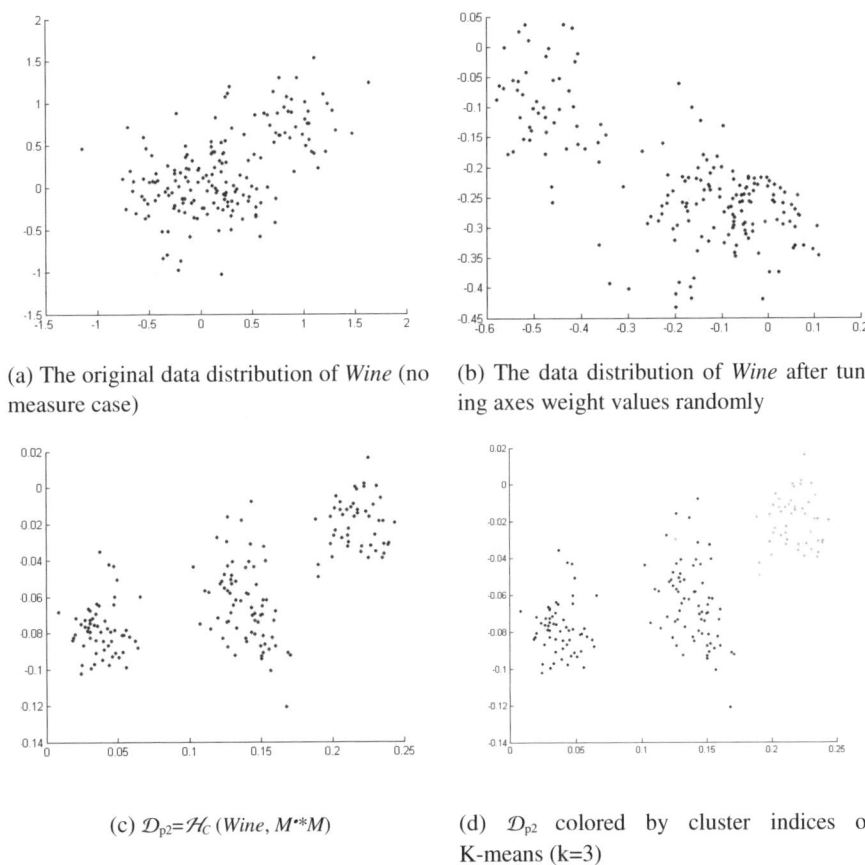

(a) The original data distribution of *Wine* (no measure case)

(b) The data distribution of *Wine* after tuning axes weight values randomly

(c) $\mathcal{D}_{p2} = \mathcal{H}_C(Wine, M^{\cdot*}M)$

(d) $\mathcal{D}_{p2}$ colored by cluster indices of K-means (k=3)

Fig. 5. Distributions of *Wine* data produced by HOV^3 in MATLAB

roughly separated data distribution (what looks like two groups) of *Wine*, as demonstrated in Fig 5b; we recorded the axes values of *Wine* as M = [-0.44458, -0.028484, -0.23029, -0.020356, -0.087636, 0.015982, 0.17392, 0.21283, -0.11461, 0.099163, -0.19181, 0.34533, 0.27328].

Then we employed M^2 (inner dot) as a measure vector and applied it to *Wine*. The newly projected distribution $\mathcal{D}_{p2}$ of *Wine* is presented in Fig 5c. It has become easier to identify 3 groups of *Wine* in Fig 5c. Thus, we colored the *Wine* data with cluster indices that were produced by the K-means clustering algorithm with k=3. The colored data distribution $\mathcal{D}_{p2}$ of *Wine* is illustrated in Fig 5d. To demonstrate the effectiveness of the enhanced separation feature of M-mapping, we contrast the statistics of $\mathcal{D}_{p2}$ of *Wine* (clustered by their distribution, as shown in Fig 5c) and the clustering result of *Wine* by K-means (k=3). The result is shown in Table 1, where left side of Table1 (C_H) is the statistics of clustering result of *Wine* by M-mapping, and the right side of Table1 (C_K) is the clustering result by K-means (k=3).

By comparing the statistics of these two clustering results, we may observe that the quality of clustering result by the distribution of $\mathcal{D}_{p2}$ of *Wine* is slightly better than that produced by K-means according to their "variance" of clustering results. Observing the colored data distribution in Fig 5d carefully, we may find that there is a green point grouped in the brown group by K-means.

Table 1. The statistics of the clusters in wine data produced by M-mapping in HOV[3] and K-means

C_H	Items	%	Radius	Variance	MaxDis	C_k	Items	%	Radius	Variance	MaxDis
1	48	26.966	102.286	0.125	102.523	1	48	27.528	102.008	0.126	102.242
2	71	39.888	97.221	0.182	97.455	2	71	39.326	97.344	0.184	97.579
3	59	33.146	108.289	0.124	108.497	3	59	33.146	108.289	0.124	108.497

By analyzing the data of these 3 groups, we have found that, group 1 contains 48 items and with "Alcohol" value 3; group 2 has 71 instances and with "Alcohol" value 2; and group 3 includes 59 records with "Alcohol" value 1.

- **Boston Housing Data**

The Boston Housing data set (simply written as *Housing*) has 14 attributes and 506 instances. The original data distribution of *Housing* is given in Fig. 6a. As in the above example, based on observation and axis scaling we had a roughly separated data distribution of *Housing*, as demonstrated in Fig 6b; we fixed the weight values of each axis as M = [0.5, 1, 0, 0.95, 0.3, 1, 0.5, 0.5, 0.8, 0.75, 0.25, 0.55, 0.45, 0.75].

By comparing diagrams of Fig. 6a and Fig. 6b, we can see that the data points in Fig. 6b are constricted to possibly 3 or 4 groups. Then M-mapping was applied to *Housing*. Fig.6c and Fig. 6d show the results of M-mapping with M.*M and M.*M.* M correspondingly. It is much easier to observe the grouping insight from Fig. 6c and Fig. 6d, where we can identify the group members easily. We believe that with the domain experts getting involved in the process, the M-mapping approach can perform better in real world applications of cluster analysis.

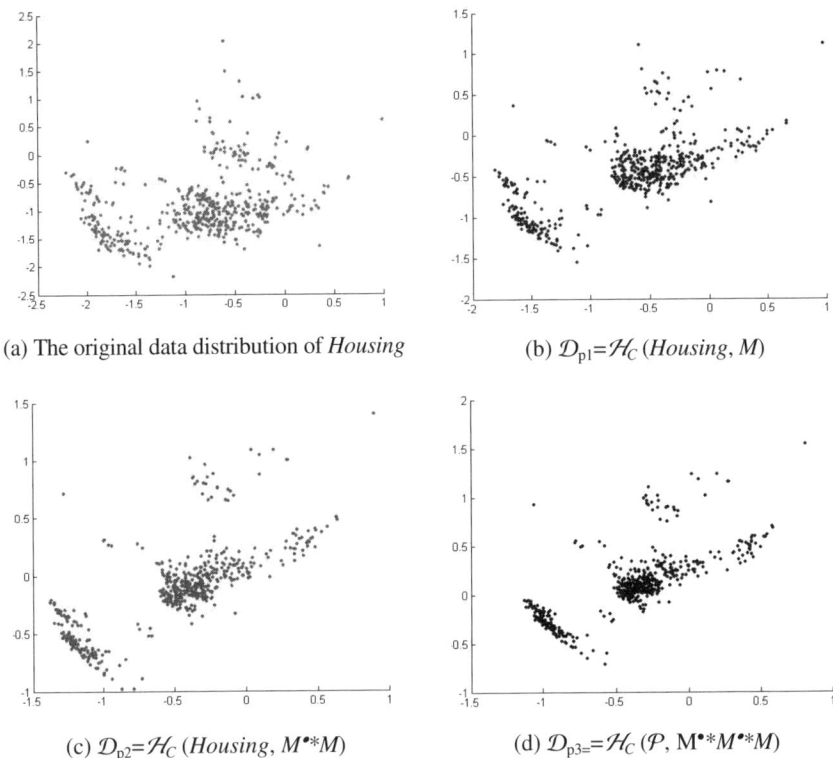

Fig. 6. The enhanced separation of the data set *Housing*

4.2 Cluster Validation with M-Mapping

We may observe that the data distributions in Fig 6c and Fig 6d are more contracted than the data distributions in Fig 6a and Fig 6b. To ensure that this contracting

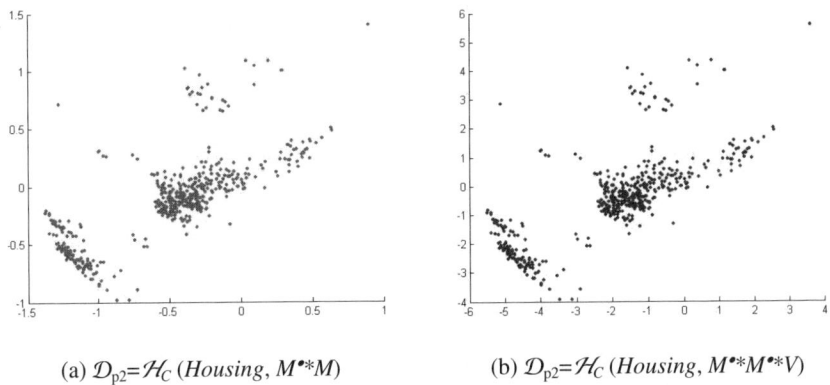

Fig. 7. The distributions produced by M-mapping and M-mapping with Zooming

process does not affect data selection, we introduce zooming in the M-mapping process. For example, in the last example, the non-zero minimal value of the measure vector M is 0.25. We then use V = [4, 4, 4, 4, 4, 4, 4, 4, 4, 4, 4, 4, 4, 4, 4] (4=1/0.25) as the zooming vector. We discuss the application of M-mapping with zooming below. It can be observed that the shape of the patterns in Fig.7a is exactly the same as that in Fig. 7b, but the scale in Fig. 7b is enlarged. Thus the effect of combining M-mapping and zooming would improve the accuracy of data selection in external cluster validation by HOV^3 [21].

5 Conclusions

In this paper we have proposed a visual approach, called M-mapping, to aid users to enhance the separation and the contraction of data groups/clusters in cluster detection and cluster validation. We have also shown that, based on the observation of data footprints, users can trace grouping clues, and then by applying the M-mapping technique to the data set, they can enhance the separation and the contraction of the potential data groups, and therefore find useful grouping information effectively.

With the advantage of the enhanced separation and contraction features of M-mapping, users can identify the cluster number in the pre-processing stage of clustering efficiently, and they can also verify the membership formation of data points among the clusters effectively in the post-processing stage of clustering by M-mapping with zooming.

References

1. Abul, A.L., Alhajj, R., Polat, F., Barker, K.: Cluster Validity Analysis Using Subsampling. In: proceedings of IEEE International Conference on Systems, Man, and Cybernetics, vol. 2, pp. 1435–1440. Washington DC (October 2003)
2. Ankerst, M., Breunig, M.M., Kriegel, S.H.P.J.: Ordering points to identify the clustering structure. In: Proc. of ACM SIGMOD Conference, pp. 49–60. ACM Press, New York (1999)
3. Baumgartner, C., Plant, C., Railing, K., Kriegel, H.-P., Kroger, P.: Subspace Selection for Clustering High-Dimensional Data. In: Perner, P. (ed.) ICDM 2004. LNCS (LNAI), vol. 3275, pp. 11–18. Springer, Heidelberg (2004)
4. Chen, K., Liu, L.: VISTA: Validating and Refining Clusters via Visualization. Journal of Information Visualization 13(4), 257–270 (2004)
5. Faloutsos, C., Lin, K.: Fastmap: a fast algorithm for indexing, data mining and visualization of traditional and multimedia data sets. In: Proc. of ACM-SIGMOD, pp. 163–174 (1995)
6. Halkidi, M., Batistakis, Y., Vazirgiannis, M.: Cluster validity methods: Part I and II, SIGMOD Record, 31 (2002)
7. Huang, Z., Cheung, D.W., Ng, M.K.: An Empirical Study on the Visual Cluster Validation Method with Fastmap. In: Proc. of DASFAA01, pp. 84–91 (2001)
8. Han, J., Kamber, M.: Data Mining: Concepts and Techniques. Morgan Kaufmann, San Francisco (2001)

9. Hinneburg, K.A., Keim, D.A., Wawryniuk, M.: Hd-eye: Visual mining of high-dimensional data. Computer Graphics & Applications Journal~19(5), 22--31 (1999)
10. Huang, Z., Lin, T.: A visual method of cluster validation with Fastmap. In: Proc. of PAKDD-2000 pp. 153–164 (2000)
11. Jain, A., Murty, M.N., Flynn, P.J.: Data Clustering: A Review. ACM Computing Surveys 31(3), 264–323 (1999)
12. Kandogan, E.: Visualizing multi-dimensional clusters, trends, and outliers using star coordinates. In: Proc. of ACM SIGKDD Conference, pp. 107–116 (2001)
13. Kominek, J., Black, A.W.: Measuring Unsupervised Acoustic Clustering through Phoneme Pair Merge-and-Split Tests. In: 9th European Conference on Speech Communication and Technology (Interspeech'2005), Lisbon, Portugal, pp. 689–692 (2005)
14. Kohonen, T.: Self-Organizing Maps, 2nd edn. Springer, Berlin (1997)
15. Kaski, S., Sinkkonen, J., Peltonen, J.: Data Visualization and Analysis with Self-Organizing Maps in Learning Metrics. In: Kambayashi, Y., Winiwarter, W., Arikawa, M. (eds.) DaWaK 2001. LNCS, vol. 2114, pp. 162–173. Springer, Heidelberg (2001)
16. Sprenger, T.C, Brunella, R., Gross, M.H.: H-BLOB: A Hierarchical Visual Clustering Method Using Implicit Surfaces. In: Proc. of the conference on Visualization 2000, pp. 61–68. IEEE Computer Society Press, Los Alamitos (2000)
17. Shneiderman, B.: Inventing Discovery Tools: Combining Information Visualization with Data Mining. In: Jantke, K.P., Shinohara, A. (eds.) DS 2001. LNCS (LNAI), vol. 2226, pp. 17–28. Springer, Heidelberg (2001)
18. Seo, J., Shneiderman, B.: A Knowledge Integration Framework for Information Visualization. In: Hemmje, M., Niederée, C., Risse, T. (eds.) From Integrated Publication and Information Systems to Information and Knowledge Environments. LNCS, vol. 3379, Springer, Heidelberg (2005)
19. Vilalta, R., Stepinski, T., Achari, M.: An Efficient Approach to External Cluster Assessment with an Application to Martian Topography, Technical Report, No. UH-CS-05-08, Department of Computer Science, University of Houston (2005)
20. Zhang, K-B., Orgun, M.A., Zhang, K.: HOV3, An Approach for Cluster Analysis. In: Li, X., Zaïane, O.R., Li, Z. (eds.) ADMA 2006. LNCS (LNAI), vol. 4093, pp. 317–328. Springer, Heidelberg (2006)
21. Zhang, K-B., Orgun, M.A., Zhang, K.: A Visual Approach for External Cluster Validation. In: CIDM2007. Proc. of the first IEEE Symposium on Computational Intelligence and Data Mining, Honolulu, Hawaii, pp. 577–582. IEEE Computer Press, Los Alamitos (2007)
22. Zhang, K-B., Orgun, M.A., Zhang, K.: A Prediction-based Visual Approach for Cluster Exploration and Cluster Validation by HOV3. In: ECML/PKDD 2007, Warsaw, Poland, September 17-21. LNCS, vol. 4702, pp. 336–349. Springer, Heidelberg (2007)

Multimedia Data Mining and Searching Through Dynamic Index Evolution

Clement Leung and Jiming Liu

Department of Computer Science,
Hong Kong Baptist University,
Kowloon Tong, Hong Kong
{clement, jiming}@comp.hkbu.edu.hk

Abstract. While the searching of text document has grown relatively mature on the Internet, the searching of images and other forms of multimedia data significantly lags behind. To search visual information on the basis of semantic concepts requires both their discovery and meaningful indexing. By analyzing the users' search, relevance feedback and selection patterns, we propose a method which allows semantic concepts to be discovered and migrated through an index hierarchy. Our method also includes a robust scoring mechanism that permits faulty indexing to be rectified over time. These include: (i) repeated and sustained corroboration of specific index terms before installation, and (ii) the ability for the index score to be both incremented and decremented. Experimental results indicate that convergence to an optimum index level may be achieved in reasonable time periods through such dynamic index evolution.

Keywords: Dynamic Indexing, Evolution Strategy, Information Recovery, Multimedia Data Mining, Relevance Feedback, Semantic Search, Web Intelligence.

1 Introduction

In Internet search activities, the PageRank algorithm and its variants focus on ranking query results based on scoring and the link structure of the Web [2, 3, 4, 5, 10, 11, 13, 14]. In this way, the relevance and relative importance of Web pages may be quantified and ranked accordingly. On the other hand, visual information search is significantly more difficult compared with the relatively well-developed algorithms of text-oriented web page searches, and has become increasingly common in many Internet search activities. The reason is that, unlike text-based documents, the features and characteristics of multimedia objects are often not automatically extractable.

Research in image retrieval has, so far, been divided between two main categories: "concept-based" image retrieval, and "content-based" image retrieval [1, 6, 7, 8, 12, 15, 16, 17, 18]. The former focuses on higher-level human perception using words to retrieve images (e.g. keywords, captions), while the latter focuses on the visual features of the image (e.g. colour, texture). Efficient indexing and retrieval of multimedia data are necessary for a successful multimedia system, and should measure object similarity in a manner consistent with human perception. However, with current technological limitations, it is not possible to have machines automatically extracting semantic content and some form of manual indexing is necessary. Compared with automatic machine processing, the discovery and inclusion of new indexing terms is always costly and time-consuming.

Our goal in this paper is to provide a methodology that supports semantic visual information search through selection scoring and incremental indexing. In addition, our methodology is able to incorporate index evolution which allows the system to respond dynamically to changing usage patterns.

2 Basic Elements and Structure

In this study, we concentrate on the indexing of *semantic contents* of multimedia objects and will exclude metadata from consideration, since indexing by metadata is relatively straightforward. For example, indexing and retrieving a photograph by date or GPS coordinates is relatively trivial as many cameras already extract such information automatically, and these can be searched to identify particular photographs before they are displayed.

Here, we consider a set of data objects $\{O_j\}$, which are typically multimedia objects where automatic machine extraction of their characteristics or properties is not possible (such as video, music, or images). Each particular object O_j has an *index set* which consists of a number of elements:

$$I_j = \{e_{j1}, e_{j2}, \ldots, e_{jM_j}\}. \tag{1}$$

Each index element is a triple

$$e_{jk} = (t_{jk}, s_{jk}, o_j), \tag{2}$$

where t_{jk} is an index term, s_{jk} is the score associated with t_{jk}, and o_j represents the ObjectID. The value of the score induces a partition of the associated ndex term into N levels $L_1, L_2, \ldots, L_N$ according to a set of parameters $P_1, P_2, \ldots, P_N$ (see Fig. 1). For a given index term with score x, the index term will be placed in level L_i if

$$P_i \leq x < P_{i+1}, \quad i = 1, \ldots, N-1, \tag{3}$$

and will be placed in level N if

$$P_N \leq x. \tag{4}$$

The index sets of all the objects in the database are collectively referred to as the *index hierarchy*. In general, terms in the higher levels are more significant than those in the lower levels, and thus the top level will be searched first; in some situations, the lower levels may not need to be probed at all.

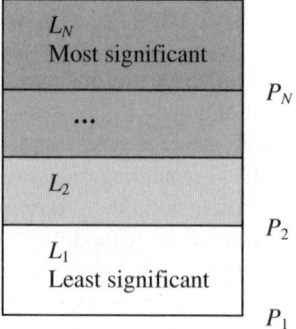

Fig. 1. Partitioning of the Index Set

3 Object Ranking

In this study, we concentrate on the indexing of *semantic contents* of multimedia objects and will exclude metadata from consideration, since indexing by metadata is relatively straightforward. For example, indexing and retrieving a photograph by date or GPS coordinates is relatively trivial as many cameras already extract such information automatically, and these can be searched to identify particular photographs before they are displayed.

Suppose a user enters a series of search terms $T_1, T_2, \ldots, T_n$ in a query $Q(T_1, T_2, \ldots, T_n)$, and furthermore suppose a series of corresponding weights $W_1, W_2, \ldots, W_n$ are also specified, which indicate for this particular query the relative importance of the different features. We calculate the *query score* $S(Q|O_j)$ for a particular object O_j as follows. We first look up the score for a particular term T_i in the index hierarchy for that object. In SQL notation, this will be

> SELECT Score
> FROM Index_Hierarchy
> WHERE Object = "ObjectID" and Index_Term = "T_i"

In cases where an index term is absent for a particular object, the score is taken to be zero. Each score s_{jk} so obtained will then be weighted by the corresponding query weight W_i and then aggregated over all index terms:

$$S(Q|O_j) = \sum_{k=1}^{n} W_k \, s_{jk} \qquad (5)$$

and the results will be ranked according to $S(Q|O_j)$.

If the query weights are not specified, we may take them to be the same and set $W_i = 1, \forall \, i$, and the above reduces to

$$S(Q|O_j) = \sum_{k=1}^{n} s_{jk} \,. \qquad (6)$$

In returning objects resulting from query processing, the returned objects will be ranked in accordance with the computed query score.

4 Score Updating Algorithms

Incrementing the Score

Suppose a user issues a query including a query term T. Then typically all objects having that term in their index hierarchy will be included in the retrieval results, ranked according to a the query score computed above, though not all of them will be considered relevant to this query by the user. In typically multimedia object retrieval, the user will go through a number of iterations using relevance feedback (Fig. 2).

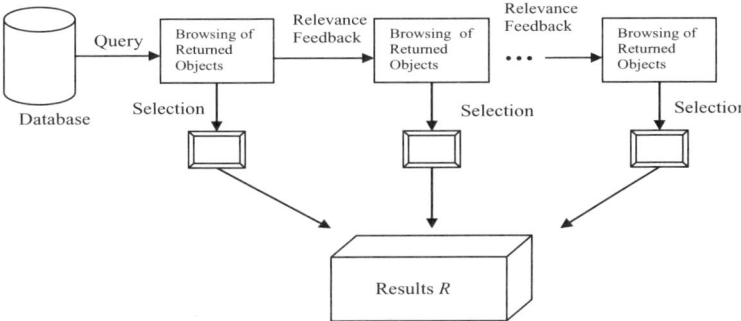

Fig. 2. Iterative Retrieval with Relevance Feedback

Thus, the final results R may include three sets:

1. S_2, a set of objects having the index term T is present in the top level L_N of its index hierarchy,
2. S_1, a set of objects having the index term T is present in one of the levels $L_1, L_2, \ldots, L_{N-1}$ of its index hierarchy,
3. S_0, a set of objects where the index term T is absent from any of the levels of the index hierarchy.

Thus we have,

$$R = S_2 \cup S_1 \cup S_0, \quad \text{with} \quad S_i \cap S_j = \emptyset, \quad \text{for } i \neq j. \tag{7}$$

In the cases of S_2 and S_1 the score of T will be incremented by a pre-determined amount Δ. This may cause the promotion of T to the next higher level. In the case of S_0, the index term T will be added to the lowest level of the index hierarchy with an initial score of P_1. (Note: Here for simplicity, we use a fix amount of increase Δ. In general cases, different amounts may be incremented depending on which stage the image is selected for inclusion in the final results. Maximum increment will occur in the first stage of selection.)

Decrementing the Score
Let U_T be the set of objects having the term T at some level in its index hierarchy. Now, some objects of U_T may not be included in the final selected results R. This suggests that term T may not be very useful as an index term for that object. Possible reasons may be that the particular feature described by term T is not sufficiently prominent or significant in the multimedia object (e.g. a particular actress, although included in the index of a movie, is given an insignificant role in that movie). Thus, denoting the score of T by $|T|$, for objects in the set $U_T - R$, the following can occur:

1. $|T| \geq P_1 + \Delta$ in which case $|T|$ is decremented by the amount Δ, and the updated $|T|$ is still greater than P_1 after the decrement, and hence remains in the index hierarchy,
2. $|T| < P_1 + \Delta$ in which case $|T|$ is decremented by the amount Δ, and the updated $|T|$ is less than P_1 after the decrement, and hence it will be dropped from the index hierarchy.

We shall refer to an increment as *positive feedback*, and a decrement as *negative feedback*. We shall illustrate the case where $N = 3$, $P_1 = 0$, $P_2 = 10$, $P_3 = 25$, $\Delta = 3$ (Fig. 3). We physically separate the three levels of the index hierarchy into three tables as follows (other physical organizations are possible such as using a single table to store all three levels). The primary key in all three tables will be the composite key (Index Term, ObjectID). All scores in the Main Index Table will have

Main Index Table

Index Term	Score	ObjectID
T1	27	#20
T2	42	#167
T1	30	#39
...	...	...

Pre-Index Table

Index Term	Score	ObjectID
T19	24	#71
T2	12	#277
T78	18	#71
...	...	...

Potential Index Table

Index Term	Score	ObjectID
T5	9	#20
T7	0	#167
T11	3	#61
...	...	...

Fig. 3. Physical Implementation of the Index Hierarchy

score ≥ 25; those in the Pre-Index Table will have scores less than 25 but ≥ 10, and those in the Potential Index Table will have scores less than 10 but ≥ 0. Here, a positive feedback to T5 will cause it to be deleted from the Potential Index Table and added to the Pre-Index Table (since 9+3 ≥ 10). Conversely, a negative feedback to T1 for object #20 will cause it to be deleted from the Main Index Table and added to the Pre-Index Table (since 27-3 < 25). Furthermore, a negative feedback to T7 will cause it to be dropped from the index hierarchy altogether.

5 Minimal Indexing and Index Evolution

Definition. An object which is *minimally indexed* is defined as one having the following properties:

(i) it has only a single index term T, and
(ii) T is *simple* (i.e. T consists of only a single word).

An index term which is not simple is referred to as a *compound* index term. An object for which there is no index term associated with it is referred to as *unindexed*. In most multimedia object collections, it is often minimally indexed (or slightly better) from the automatic keyword extraction of some kind of caption information provided alongside the object. Sometimes, however, it is completely unindexed as no caption information is available.

Augmentation of an Existing Index Term
Consider a particular object J, which is minimally indexed with term $T = w$ (i.e. T consists of the single word w, e.g. "swan"). Suppose a user enters a query using $w \oplus u$, which signifies a situation where $w \oplus u$ forms a single meaningful compound index term (e.g. "black swan"). In general, u may be added before or after w and serves to further qualify w (hence narrowing down the search possibilities). From this query, all data objects with w in its index will be returned – there may be some objects with $w \oplus u$ or $w \oplus u'$ (e.g. white swan or black swan). After some browsing through these returned objects, the user finally selects a particular object. As a result of this selection, it is inferred that the selected object will have the property represented by the more complete index term $w \oplus u$ (i.e. black swan). Thus the term $w \oplus u$ will be added to the index hierarchy of this object, and if this happens many times, the score of $w \oplus u$ will increase and this will cause $w \oplus u$ to be installed along with w in the index of the object so that the retrieval results will be much more accurate. The original index term w will remain in the index hierarchy, and may be retained or gradually dropped in accordance with the score updating algorithm. In general, an object which is minimally indexed with term $T = w$ may be successively augmented to $w \oplus u_1 \oplus ... \oplus u_k$ through this mechanism.

Addition of New Index Term
Consider a particular object J, which is indexed with term T (which may, for example, be extracted from its caption), where T may be simple or compound. When a query is entered using the term T, many objects of different types will be returned, all of which

will have the term T in their index. Among this results set is the required object J, and this is the only relevant object for the user. Other objects that have been returned are not relevant for this user query. Due to the volume of objects returned, it will take considerable time and effort for the user to navigate through them to get to the target object J, which is inefficient.

Next, suppose the same object J can have another index term T' which is not yet included in the current index but represents an additional property of J. Occasionally, users would include both terms T and T' in a query with a view of increasing the precision of the search. Since T' is not indexed, the initial search results will still be the same as before. As the user is interested in J and not other data objects, he will eventually select J from among the objects returned. This suggests that T' may be a potential index term and will enter the index hierarchy at the lowest level with a score of P_1. Once it has been installed in the index hierarchy, repeated search using it and subsequent selection will serve to raise its score and may eventually elevate it to the main index. In so doing, an object which is indexed with a single term may subsequent have a number of index terms associated with it.

In general, an object having an initial set of index terms $T_1, T_2, \ldots, T_n$, may have new index terms $T_1'\ T_2', \ldots, T_m'$ added to it through repeated usage.

Example

Consider the searching of the movie "The Prize" which starred Paul Newman. Let us assume that only the term "Prize" has been indexed for this particular multimedia data object. When a query is entered using the term "Prize", many objects of different types will be returned, all of which will have the term "Prize" in their index. Among this results set is the required movie "The Prize", and this is the only relevant object for the user. However, some queries may be more specific, with both "Prize" and "Paul Newman" specified, but the initial search results will still be the same as before as "Paul Newman" has not been indexed. The user will eventually select this movie, and this suggests that the term "Paul Newman" may also be included in the index of this movie. Thus "Paul Newman" would be included in the index hierarchy for this movie. Thus, every time the terms "Prize" and "Paul Newman" are both specified in a query, and if the user subsequently selects that movie, the index score of this new term term "Paul Newman" will be increased. When this score reaches the required threshold, then "Paul Newman" will be installed as a proper index term of this movie. Similarly, other terms may be added to the index in a dynamic way.

Attaining Intelligence through Index Evolution

Using a combination of the above two mechanisms, the number and precision of index terms of a data object will evolve and be enriched as usage progresses. This will increase search effectiveness as time goes on. The main principle is that human users – through their considerable time spent in interacting with the system and their visual judgment – has progressively transferred and instilled their intelligence into the system so that the index of data objects is gradually enriched, which cannot be achieved by purely automatic means as current technology does not allow image objects to be meaningfully recognized. If the data objects reside collectively on the web, then the intelligence of the web will be enhanced through such process.

In this way, richer data object semantics, particularly *entities* and *relationships*, may be incorporated into the query processing algorithms of multimedia data objects.

For example, an image with a boy riding on an elephant is initially minimally indexed with only the term "elephant" (an entity). As time goes on, using the mechanisms indicated above, "boy" (a further entity) will be added to the index of the image. Still later, the term "riding" (relationship) will also be added. Thus, after progressive usage, the three terms "boy", "riding", "elephant" (representing entities and relationship) will all be part of the index of this particular image, even though at the beginning, only "elephant" is in the index. Through the incorporation of all three terms in the index, and through the inclusion of all three terms in a query, a much more precise retrieval can occur, reducing time and effort in navigating a large number of irrelevant multimedia objects.

Maximal Indexing

For a given multimedia data object, which may for example be unindexed or minimally indexed, the number of its index terms will increase through usage. As time goes on, the number of index terms associated with it will converge to a given level and will remain there after which no further increase will take place. Such an index will be referred to as a *stable maximal index*. In more general situations, even the maximal index will change over time and is dictated by changing usage patterns and user preference; this is referred to as an *unstable index*. In our study, we shall be mostly concerned with stable index.

Probability of Object Recovery

In relation to information recovery, where one wishes to retrieve a specific known data object, let us consider the recovery rate ρ, which gives the probability of successful object recovery, for a situation where $z\%$ of the collection is maximally indexed. Here, we suppose that an object will be retrieved when some or all of its indexed terms match some or all of the terms specified in a query. Assuming all the object accessed activities are uniformly distributed (i.e. each query is equally likely to target any of the objects with no special preference for any particular ones), then for a given query Q, there are four cases. Denoting the set of maximally indexed terms of a given object by X, the set of indexed terms included in the query Q by X_q, where we assume $X_q \neq \emptyset$, and the set of indexed terms included in the target object by X_o, then the four cases are:

1. $X_q = X_o = X$, i.e. the target object is maximally indexed, and the query Q specifies all the maximally indexed terms
2. $X_q \subset X_o = X$, i.e. the target object is maximally indexed, but the query Q specifies only some, but not all, of the maximally indexed terms (unlike "$\subseteq$", here "$\subset$" excludes improper subset, i.e. X_q and X_o are not permitted to be equal)
3. $X_o \subset X_q = X$, i.e. the target object is not maximally indexed, but the query Q specifies all the maximally indexed terms
4. $X_o \subset X$, and $X_q \subset X$, i.e. the target object is not maximally indexed, and Q does not specify all the maximally indexed terms.

Cases 1 and 2 will result in the successful retrieval of the target object. Since $z\%$ of the objects are maximally indexed, and since all objects sustain the same amount of access activity, these two cases will account for $z\%$ of the queries. Thus, for $z\%$ of the queries, the recovery rate $\rho = 100\%$. Hence, even assuming the recall of other queries is 0%, the recovery rate averaged over all queries is $z\%$.

However, the recovery rate of Cases 3 and 4 are not necessarily 0%. In Case 3, since the query is maximally indexed, and if the target object is not unindexed, i.e. $X_o \neq \emptyset$, then matching of indexed terms will take place between the query and the target object (i.e. $X_q \cap X_o \neq \emptyset$), and so successful retrieval of the target object will result. That is, Case 3 will result in unsuccessful retrieval only if $X_o = \emptyset$. In Case 4, if there are coincidences between the terms given in the query and the indexed terms of the object (i.e. $X_q \cap X_o \neq \emptyset$), then matching will take place, and the target object will be retrieved. If $X_q \cap X_o = \emptyset$, then no matching will take place and the target object is not retrieved. Thus, Case 4 will yield 0% recovery only if there is no overlap between the query terms and the indexed terms of the target object. Thus, we conclude that the recovery rate $\rho \geq z\%$, and most likely strictly greater than $z\%$.

6 Experimental Results

Experiments have been carried out to determine the evolution time and feasibility of such indexing, and the following cases are considered.

Case A
Initial number of index terms per object = 0 (unindexed). Maximal number of index terms per object = 3. Database size = 10,000 images. Access rate follows a Poisson distribution with parameter λ= 10 accesses per hour [9]. Each access has a probability of 0.05 of incrementing the index set by one index term. Fig. 4 shows the time for the database to achieve 80% maximal indexing capability (i.e. 8,000 images in the database will have reached 3 index terms).

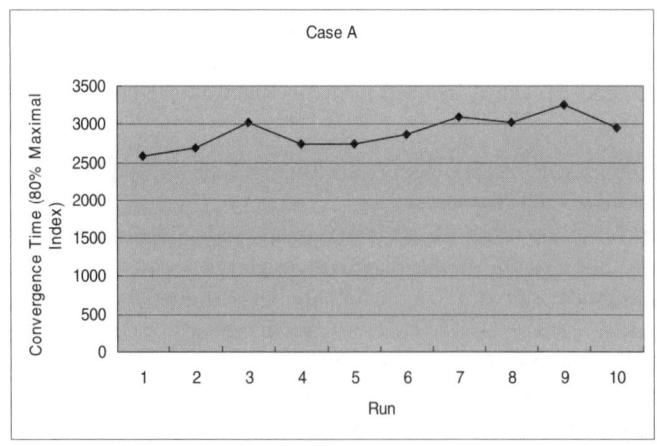

Fig. 4. Indexing Convergence Time (Hours) for Case A

Here, the mean time to achieve 80% maximal indexing capability is 2897.75 hours which is about 17 weeks. The standard deviation is 209.31 hours or about 8.7 days.

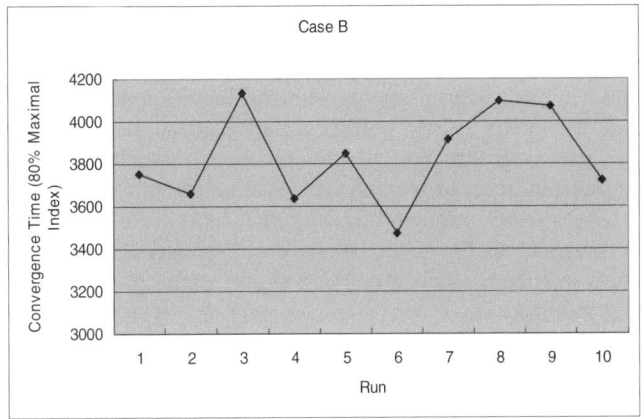

Fig. 5. Indexing Convergence Time (Hours) for Case B

Case B
Initial number of index terms per object = 3. Maximal number of index terms per object = 8. Database size = 10,000 images. Access rate follows a Poisson distribution with parameter λ = 10 accesses per hour. Each access has a probability of 0.05 incrementing the index set by one index term. Fig. 5 shows the time for the database to achieve 80% maximal indexing capability (i.e. 8,000 images in the database will have reached 8 index terms).

Here, the mean time to achieve 80% maximal indexing capability is 3830.86 hours which is about 23 weeks. The standard deviation is 221.45 hours or about 9.2 days. Thus, even allowing two standard deviations from the mean value (23 weeks + 18.4 days), there is a high probability – over 95% using the Central Limit Theorem approximation to a normal distribution [9] – that 80% maximal indexing capability can be achieved within six months.

7 Conclusions

The efficient identification of web documents in the past has made use of user retrieval patterns and access characteristics. We have proposed a method whereby the indexing of images, which can be web-based or non web-based, can be done by similarly keeping track of the users querying behaviour. By analyzing the users' search, relevance feedback and results selection patterns, our method allows semantic concepts to be gradually discovered and migrated through an index hierarchy. Our method also includes a robust scoring mechanism which enables faulty indexing to be rectified over time. This includes the following properties: (i) repeated and sustained corroboration of specific index terms is required before their installation, and (ii) the flexibility not only for the index score to be incremented but also for it to be decremented (which may lead to the eventual deletion of the index term).

Given that the automatic recognition of semantic multimedia object contents, and hence their automatic indexing, is not possible, such a semi-automatic evolutionary approach will allow human intelligence and judgment to be progressively transferred to that of the underlying system and will bring substantial benefits. In particular, this will obviate the need to perform time-consuming manual indexing which has shown to be costly and, if done by a small unrepresentative group, can also produce a biased and subjective indexing structure. Although such indexing is not one-off or immediate, a competent level of semantic retrieval performance may be achieved over a reasonable time period through updating the score. We have shown that, by judiciously tracking and analysing the behavior of user accesses over time, a database of 10,000 images may be semantically indexed and attain good object recovery rates after three to six months.

Acknowledgements

The authors are grateful to Li Xin for carrying out the experimental and programming works of this research.

References

[1] Azzam, I., Leung, C.H.C., Horwood, J.: Implicit concept-based image indexing and retrieval. In: Proceedings of the IEEE International Conference on Multi-media Modeling, Brisbane, Australia, pp. 354–359 (January 2004)
[2] Chakrabarti, S., Dom, B., Gibson, D., Kleinberg, J., Raghavan, P., Rajagopalan, S.: Automatic Resource Compilation by Analyzing Hyperlink Structure and Associated Text. In: Proc. Seventh Int'l World Wide Web Conf. (1998)
[3] Chakrabarti, S., Joshi, M.M., Punera, K., Pennock, D.M.: The Structure of Broad Topics on the Web. In: Proc. 11th Intl World Wide Web Conf. (2002)
[4] Diligenti, M., Gori, M., Maggini, M.: Web Page Scoring Systems for Horizontal and Vertical Search. In: Proc. 11th Int'l World Wide Web Conf. (May 2002)
[5] Dwork, C., Kumar, R., Naor, M., Sivakumar, D.: Rank Aggregation Methods for the Web. In: Proc. 10th Int'l World Wide Web Conf. (2001)
[6] Finkelstein, L., Gabrilovich, E., Matias, Y., Rivlin, E., Solan, Z., Wolfman, G., Ruppin, E.: Placing Search in Context: The Concept Revisited. In: Proc. 10th Int'l World Wide Web Conf. (2001)
[7] Funkhouser, T., Min, P., Kazhdan, M., Chen, J., Halderman, A., Dobkin, D., Jacobs, D.: A Search Engine for 3D Models. ACM Transactions on Graphics 22(1), 1–28 (2003)
[8] Gevers, T., Smeulders, A.V.M: Image search engines An Overview. In: Emerging Topics in Computer Vision, pp. 1–54. Prentice-Hall, Englewood Cliffs (2004)
[9] Ghahramani, S.: Fundamentals of Probability with Stochastic Processes, 3rd edn. Prentice-Hall, Englewood Cliffs (2005)
[10] Haveliwala, T.H.: Topic-Sensitive PageRank. In: Proc. 11th Int'l World Wide Web Conf. (May 2002)

[11] Haveliwala, T.H.: Topic-Sensitive PageRank: A Context-Sensitive Ranking Algorithm for Web Search. IEEE Transactions on Knowledge and Data Engineering 15(4), 784–796 (2003)
[12] Hawarth, R.J., Buxton, H.: Conceptual-Description from Monitoring and Watching Image Sequences. Image and Vision Computing 18, 105–135 (2000)
[13] Jeh, G., Widom, J.: Scaling Personalized Web Search. In: Proc. 12th Int'l World Wide Web Conf. (May 2003)
[14] Kamvar, S.D., Haveliwala, T.H., Manning, C.D., Golub, G.H.: Extrapolation Methods for Accelerating PageRank Computations. In: Proc. 12th Int'l World Wide Web Conf. (May 2003)
[15] Müller, H., et al.: Performance Evaluation in Content–Based Image Retrieval: Overview and Proposals. Pattern Recognition Letters 22(5), 593–601 (2001)
[16] Over, P., Leung, C.H.C., Ip, H., Grubinger, M.: Multimedia retrieval benchmarks. IEEE Multimedia 11(2), 80–84 (2004)
[17] Tam, A., Leung, C.H.C.: Structured natural-language descriptions for semantic content retrieval of visual materials. J. American Society for Information Science and Technology , 930–937 (2001)
[18] Venkat, N., Gudivada, Raghavan, V.V.: Modeling and Retrieving Images Content System. Information Processing and Management 33(4), 427–452 (1997)

Clustering and Visualizing Audiovisual Dataset on Mobile Devices in a Topic-Oriented Manner

Lei Wang, Dian Tjondrongoro, and Yuee Liu

Faculty of Information Technology, Queensland of University of Technology,
George Street, GPO BOX 2434, Brisbane, Queensland 4001, Australia
{wangl3, dian}@qut.edu.au,
y53.liu@student.qut.edu.au

Abstract. With the significant enhancement of telecom bandwidth and multimedia-supported mobile devices occupying the market, consuming audiovisual contents on the move is no longer a hype. A lot of telecom operators are now porting traditional TV service to PDAs, 3G cell phones. However, several surveys suggest that direct migration of service from large screen to small screen may not comply with mobile users' consuming behaviors. In this paper, we first elaborate existing surveys together with our survey result to understand consumer interests in terms of consuming audiovisual contents on mobile devices. Based on the findings, we propose a novel solution to help user locate, gather, and cross-relate topics scattered in an array of contents across various domains. A web-based demo application has been implemented for PDA. The system evaluation indicates that this paradigm of organizing and presenting multimedia archives is very welcomed.

Keywords: Mobile, Audiovisual Content, Visualization, Topic-oriented.

1 Introduction

In the last five years, mobile devices have been undergoing a booming prosperity in our every day life. More handheld devices with audio/video playback functionalities are available in the market with a reasonable price. The advent of third generation communication networks further enables telecom operators to provide better mobile multimedia service, such as smoother streaming time and higher quality of video resolution [1] [2]. Both client and server ends are significantly improved to promote the era of consuming multimedia resources on the move.

Currently, several telecom operators have already kicked off their mobile multimedia services [3] [4]. Among them, most can be categorized into two classes, namely: Mobile live TV and Push/Stream [5]. A study [6] found that while current mobile TV services offer a different experience to that of traditional TV, they still cannot deliver a true TV experience anytime, anywhere due to limitations of wireless network and mobile devices' capability. In both of these services, users are only entitled limited choices of contents.

The launch of mobile TV services with both solutions did not bring expected revenue [7] [8] due to issues such as price, reliability, and quality [9] [10]. Other survey projects [11] [12] [2] show that most users only spend fifteen (15) to twenty (20) minutes on mobile TV and mostly during the waiting time. The programs appealing to them are short, yet complete video clips such as news, sports highlights, whereas lengthy programs like full soccer match are often not welcomed given the download expenses. Therefore, it is concluded that traditional TV viewing methods which are originally designed for static viewing do not fit in a mobile multimedia context.

With realization of what current services fall short, researchers pointed out some possible solutions. Motely [13] suggests that mobile multimedia service should give customers choice and allow more interactivity. Wiggin [14] predicts that mobile TV should formulate a compelling service by delivering an array of content to customers and enable them to find favorite shows with ease. Hollink et al. [15] studies user behavior in video retrieval and found video browsing requires a visualization schema to provide a quick overview of the dataset and give users insight into the structure of the dataset, the visualization scheme allows effective layout and efficient navigation on mobile device [16].

Given the comprehensive literature review, we have drawn two hypotheses a) Users intend to view interesting segments from an array of audiovisual dataset in which intended topics may be scattered in multiple audiovisual clips b) From the perspective of human cognition, visualization of audiovisual dataset will help users quickly locate interested segments with ease.

A preliminary survey with fifty-five (55) students helped us to validate the prompted assumptions. Ninety-six percent (96%) students strongly prefer to browse between different genres for the same topics. Additionally, majority students agreed that table-based EPG (Electronic Program Guide) is more effective for locating a certain program, whereas graph-based map is desired when navigating through an array of interlinked contents.

Based on these user behavior studies, we propose a novel method for organizing and presenting audiovisual dataset. This method will apply two strategies being a) Topicalization: Given the existing audiovisual metadata, reorganize the metadata dataset into a common topics-based layer. Thus, relevant partial segments scattered in multiple clips can be clustered. By presenting topics at this fine granularity level, skimming/searching time is saved, and viewers' continuous access desire can be retained through this chain of interlinked topics, b) Visualization: Visualize the topic-based layer into graph format and illustrate relationships among topics. In this paradigm, viewers can browse audiovisual dataset across the boundary of media documents and various genres in an intuitive and productive way.

The remainder of the paper is organized as follows: some related works are introduced in section 2. In section 3, how to formulate this common topic-based layer is described. Section 4 presents how the proposed visualization scheme will help users to find favorite audiovisual contents beyond document and genre boundary. Section 5 details system implementation. System valuation result and future works are outlined in section 6.

2 Related Work

In the following sections, we will examine how state-of-art video retrieval systems help users to find favorite shows.

Currently, video browsing by key frame is the major solution for navigating through audiovisual dataset. Most video services allow users to browse the video collection for a destination clip. Upon finding the clip, users can view it with a VCR-like control (play, pause and rewind). Such services only provide limited interaction with users on a program level. YouTube[1] is the typical example. In YouTube, video clip can be associated with other relevant clips. Browsing on the program level is at a coarse granularity. YouTube treats entire video document as unit of search, thus viewing related partial segments scattered among multiple clips will incur additional searching, gathering and skimming time. From this perspective, current solution is not suitable for mobile video content consumers since it requires high transaction overhead to locate and gather intended clips.

Unlike other static medium, audio and video are simultaneously presented. VCR-like control allows user to linearly navigate, however requires high transaction overhead to find the intended segment(s). Researchers have already noticed the importance of amplifying user's cognition of the audiovisual item's structure, so that non-linear navigation through a single clip at a fine granularity can be achieved. Currently, most approaches are to break down the video document into frames, and browsing is achieved based on the structure of these frames [17]. Some researchers utilize the tree hierarchy to show the full video table of content [18]. Some researchers also list video segments in a tree view, however they are presented in a timeline style [19].

Be it navigating through audiovisual dataset or navigate through a single clip, the common limitation is that they are based on visual shot segments in a temporal sequence, rather than meaningful objects or topics. From this perspective, semantic object-based search is difficult to be achieved [20] [21]. As an alternative, hierarchical searching is not always an appropriate way for searching [22] since the relations between media items are invisible. On top that, the complexity of current video browsing mechanisms are not applicable for mobile device as users can only afford few clicks.

Having information "at your fingertips" is a crucial issue. There is an urgent need for cognitive tools aiming at presenting, structuring and retrieving the information cognitively so as to make users easily access the large data and knowledge set [23]. Jacques Bertin [24], Herman [25] and Keim [26] reveals that the graph more approaches the human perception of the real world information, because human often construct an invisible "graph" in their brain to associate what they have seen. The graphical display may make information indexing more efficient and facilitate the perceptual inference [27]. Brend [28] pointed out the design issues when applying data visualization concept to mobile devices. To our best knowledge, solution for visualizing audiovisual dataset on mobile device is yet available.

---

[1] http://www.youtube.com/

3 Topicalization: Building Topic-Based Layer for Clustering

The common topic-based layer is a universal medium on top of various formats of multimedia metadata. By constructing this layer, following aspects will benefit: a) Various description metadata can be unified into one single format, thus multimedia content described in different schemes can be integrated seamlessly; b) Temporal sequence can be topicalized into abstract object. For example, a goal event from frame 20 to frame 98 could be a topic; c) Topics can be cross-related. For example, a goal event is related to a soccer player, the soccer player is related to a news interview. In this section, we will present the solution that formulates this topic-based layer.

3.1 Topic-Based Knowledge Modeling: Topic Maps

Following the semantic web concept, an upper layer is needed to unify various description syntaxes so that knowledge can be exchanged. Knowledge model determines how the resources are organized and presented to the user. As for now, there are two candidates for knowledge modeling being Topic Maps[2] and RDF[3]. Topic maps are an ISO standard for the representation and interchange of structured knowledge models. It is able to be utilized as an algorithm to organize information in a way that is optimized for navigation. Topic maps are good at bridging topics with links such as glossaries, cross-references, thesauri and catalogs. It can also merge structured or unstructured information. The Resource Description Framework (RDF) is developed under W3C, it is an infrastructure that enables the encoding, exchange, and reuse of structured metadata.

Topic Maps and RDF share a very similar concept (if not the same) which is to model complex metadata and ontology as multi-dimensional graphs using XML as the interchange format [29]. However, there are some subtle differences between these models which lead to the choice according to the application context. Based on Garshol's findings [30], the major difference between these two models is how they represent association between two "things". RDF is a bit redundant and ambiguous while Topic Maps deliver a more complex structure than RDF statements. Furthermore, Topic Maps also wrap more information in the form of role types. As for our application context which focuses on locating, collating and cross-relating audiovisual contents, we identified that Topic Map is more appropriate for multimedia knowledge modeling. Another reason we adopt Topic Maps as the knowledge model is that it organizes information in a way that is optimized for navigation. With the choice of Topic Maps, audiovisual items can be broken down into topics and clustered together. After visualization of XML-based Topic Maps into a MetaMap, those users with vague exploration goal would have some visual clues, and are able to explore other associated topics.

3.2 Topic Maps for Audiovisual Dataset

To demonstrate how Topic Maps organize the audiovisual dataset and cluster relevant topics, we present a mockup dataset with the following diagram. The lower section

---

[2] http://www.topicmaps.org/
[3] http://www.w3.org/RDF/

indicates the structural metadata of media items from various genres. The upper section explains how related topics are gathered after conversion from various formats of structural metadata to Topic Maps.

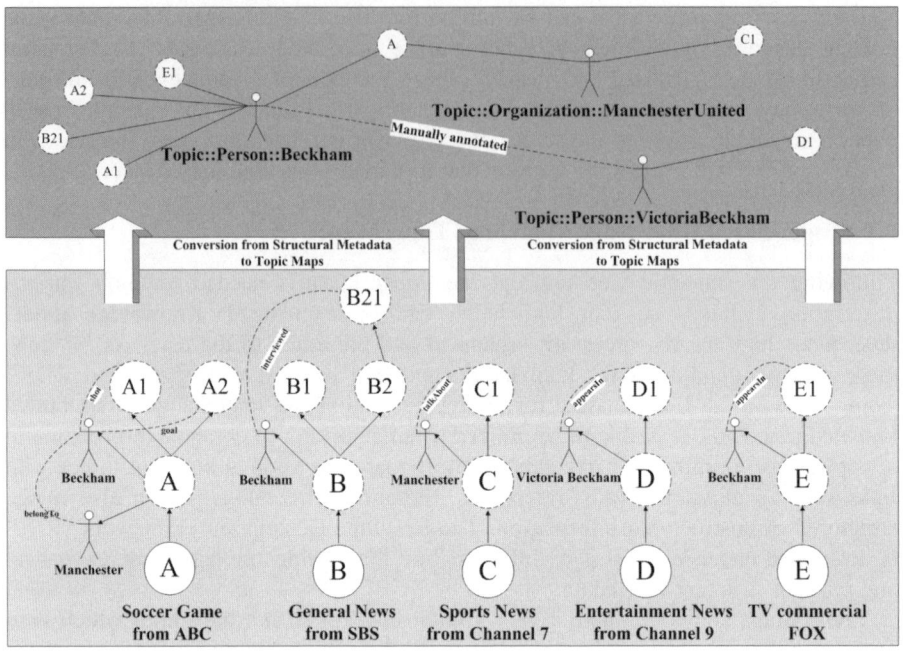

Fig. 1. Using Topic Maps to Organize Audiovisual Dataset

In Fig. 1, David Beckham is a main topic appearing in several media items across different genres. After Topic Maps which represents different sources merged, the MetaMap automatically clustered his occurrences in different occasions by the role he played (e.g. soccer player or actor). In addition, "David Beckham" can be cross-related to other external resources via intermediate topics such as "Manchester United" or "Victoria Beckham", and viewers can keep digging on the intermediate topics if they found them interesting.

3.3 Construct Topic-Based Layer Automatically

Manually constructing the topic-based layer is not feasible due to the large amount of metadata attached with media items. Thus, a proper metadata scheme and an intelligent conversion algorithm are the corner stones to success. This sub-section will present SEO (Segment-Event-Object) modeling scheme and its conversion to topic maps (i.e. SEO2TopicMaps) algorithm for mass generation of the common topic-based layer. It should be acknowledged that content description using SEO is only one possible case, the conversion algorithm should not constraint to SEO2TopicMaps but also other algorithms such as MPEG7toTopicMaps algorithm. The system should allow various conversion components to be plugged in with ease.

SEO, A domain-neutral, semantic-rich modeling scheme. To have a comprehensive common topic-based layer, the modeling scheme should provide structural and semantic information. SEO highly abstracts a common ground from all kinds of audiovisual content. It identified that all audiovisual contents are structured by segment, event and objects. Segments can contain low-level (e.g. color, texture) and mid-level features (e.g. replay, face), while event is a specialized segment which contains generic, specific and further tactical semantics of a certain domain. An object could be any entity appears in the audiovisual content which will be linked to the relevant segment or event. Using this well generalized structure, SEO can be applied to different domains as they all share the same structure. However, different domains keep their own semantic rules. For example, an object David Beckham, who is interviewed in a news program, can be annotated as an interviewee, whereas when he appears in a soccer match should be annotated as a soccerPlayer. SEO facilitates accurate description for segment, event and object depending on its domain context. The following is the SEO example considering the domain context it appears in:

```
<SEO:person type="interviewee">
<SEO:objectAlias>p_David_Beckham</SEO:objectAlias>
<SEO:objectId>P1</SEO:objectId>
<SEO:name>David Beckham</SEO:name>
<SEO:definition>interviewee</SEO:definition>
</SEO:person>
```

Fig. 2. Object David Beckham appears in a news program

```
<SEO:person type="soccerPlayer">
<SEO:objectAlias>p_David_Beckham</SEO:objectAlias>
<SEO:objectId>P1</SEO:objectId>
<SEO:name>David Beckham</SEO:name>
<SEO:definition>soccer player</SEO:definition>
</SEO:person>
```

Fig. 3. Object David Beckham appears in a soccer game

Based on our previous work [31], we can semi-automatically produce SEO metadata following the scheme. This scheme allows audiovisual clips from various genres to be described with common conventions of syntax and structure. Therefore, the metadata attached to these audiovisual clips can be automatically transformed into Topic Maps using one algorithm, SEO2TopicMaps.

SEO2TopicMaps Topiclization Algorithm. SEO2TopicMaps (read as SEO to Topic Maps) Topicalization Algorithm is the tool which utilizes xQuery to transform SEO metadata into Topic Maps. The algorithm constructs the Topic Maps based on SEO data instance by developing topics from segments, events and objects and building associations between them. The following is the algorithm which topicalizes objects to Topic Maps format based on SEO instance in Fig. 3.

The following is a segment of SEO2TopicMaps topiclization algorithm.

```
For $i in $vidLib/SEO:semanticObjectCollection/*
let $objectAlias:=$i/SEO:objectAlias(:get object
AliasName=P_David_Beckham:)
let $type:=string($i/@type)(:get object
Type=soccerPlayer:)
return
<topic id="{$objectId}">
<instanceOf>
<topicRef xlink:herf="#{$type}"/>
</instanceOf>
  <baseName>
    <baseNamesString>{objectAlias}</baseNameString>
  </baseName>
  .. ..
</topic>
```

4 Visualization: Browsing Beyond Boundary

In this section, we present novel strategies for multimedia data access, presentation and mining on mobile devices by visualizing XML- based Topic Maps into graph format. By associating relevant topics scattered in multiple documents in various genres, the line between documents and genres is erased, and viewers can browse audiovisual dataset beyond these boundaries. It also provides the possibility that inexperienced users can explore complex dataset with only mouse manipulation. As for mobile users, most of them are inexperienced user with a vague exploration goal and limited input ability when consuming multimedia contents. Therefore, we identified that information visualization can be well utilized. In the following sections, we will detail how Ontopia[4], a visualization engine, depicts the graph to facilitate browsing beyond the boundary between documents and genres.

Ontopia provides a complete set of tools for building, maintaining and deploying Topic Map based applications. Among the tools, Ontopia Vizigator uses java technology to provide a graphical navigation of Topic Maps, offering a visual alternative to text-only browsing. Fig 4A, B, C are rendered by Vizigator using the Topic Maps generated by the SEO2TopicMaps algorithm. As shown in these figures, Manchester united soccer club is the initial topic of a user's interest, Fig 4A shows the semantic object structure of Manchester united. User can keep on exploring David Beckham in Fig 4B and would like to know about David Beckham is a player of foul. Finally Fig4C denotes that this foul clip is in play2 of track2 which also contains one excitement clip and one text screen. Fig 5 depicts the Topic Map merged by two separate Topic Maps being a soccer game and sports news on the same day. Compared to Fig4B, we found that the topic David Beckham is the news of the day is also linked to the topic David Beckham. Therefore, browsing beyond genres (between soccer and news program) and documents (between file A and file B) is achieved.

---

[4] http://www.ontopia.net/

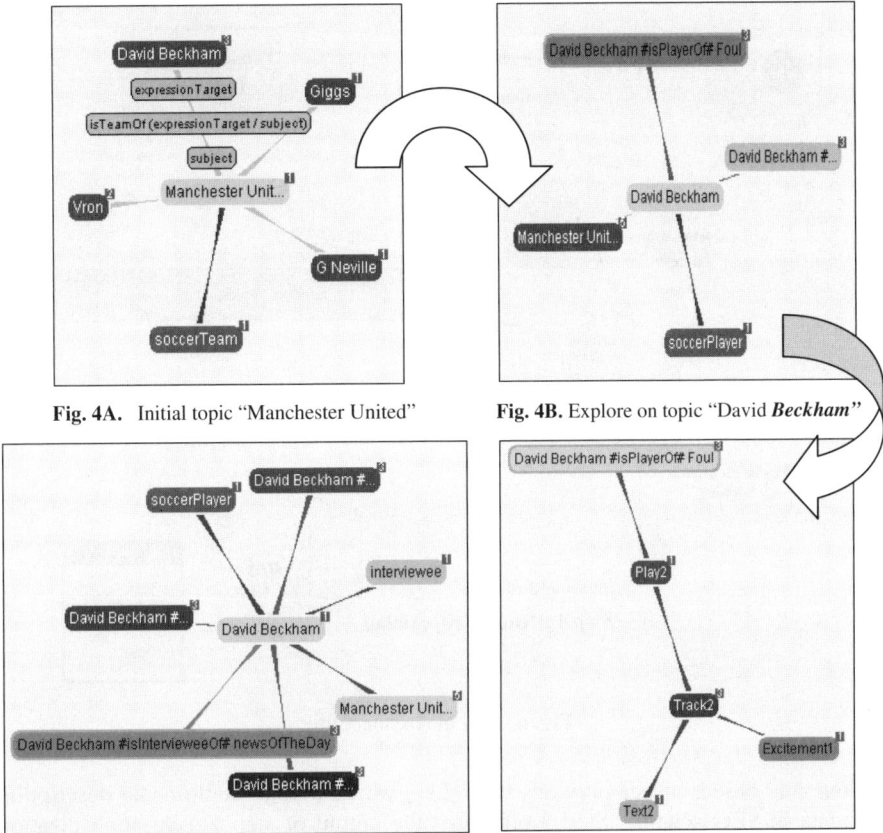

Fig. 4A. Initial topic "Manchester United"

Fig. 4B. Explore on topic "David *Beckham*"

Fig. 5. Topic Maps merged with soccer game and sports news

Fig. 4C. Explore on topic "David Beckham is a player of foul"

5 System Implementation

The system architecture is presented in Fig 6, which comprises two core components being topicalization processor and visualization processor. The system work flow is formulated in the following steps:

- Step 1: Raw video clips are described in various content description metadata languages, for example, MPEG 7 or SEO scheme.
- Step 2: Topicalization processor converts multimedia description metadata into the topic-based knowledge model.
- Step 3: Visualization processor converts the topic-based knowledge model into graphic navigation user interface according to user's request.
- Step 4: Users can request to stream down the intended clips based on the URL included in graphic navigation UI.

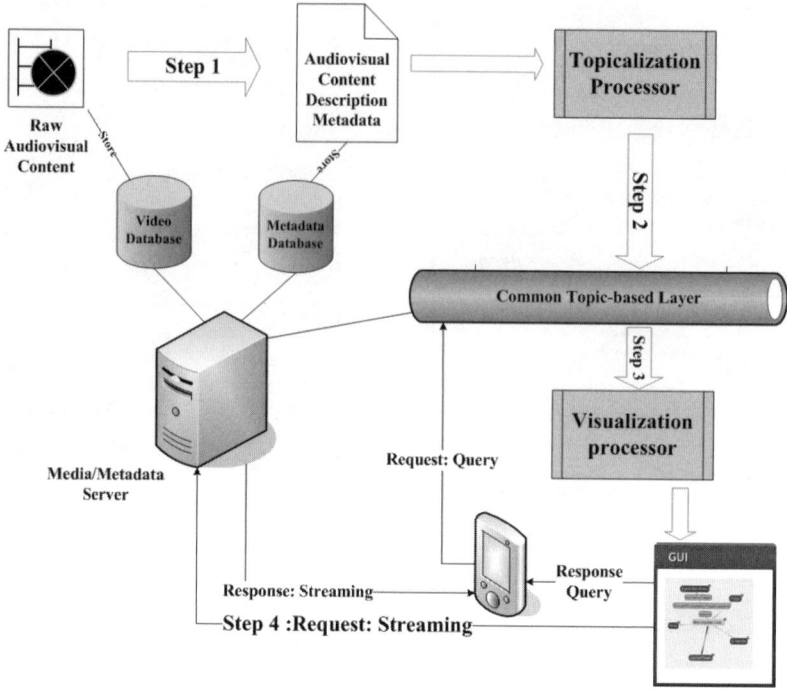

Fig. 6. System Architecture

Step 1 is based on our previous work [31] which outputs multimedia description metadata in SEO scheme. Step 2 processes the output of step 1 using topicalization processor. Currently, we are adopting SEO2TopicMaps conversion algorithm. After all multimedia description metadata are topicalized, they are stored centrally in eXist native XML database. When user send text-based queries, xQuery module will take the parameter (e.g. David Beckham) and search against eXist native XML database. A set of Topic Maps format result will return and feed into the visualization processor (currently using Ontopia), this graph processing engine will output topics, relationships as hyper-linkable graph. Users can then click on the graph and request to stream down the specific topic from video server. It should be acknowledged that both topicalization processor and visualization processor are designed pluggable, other conversion algorithms or visualization engines can be plugged in if necessary.

6 System Evaluation and Future Works

To examine the effectiveness of our proposed solution, we conducted a user evaluation on the prototype system with a group of 55 university students. The main objective of this survey is to gather users' feedback on the success of the two main aspects from the system, being: 1) Topic-Based Browsing module, 2) Overall functionality & features. Success was measured by three criteria, namely: effectiveness (to search personalized contents); intuitiveness; and enjoy-ability. On the overall functionalities, all aspects (locate interesting topics, mining videos across different genres, browsing between programs, and total control) scored average of 94.5% agreement (including strongly agree) ratio. Among the statistics, the most notable result is the facts that 53% participants indicated strongly agree and 43% indicated agree on "I like the idea of browsing between different genres of programs".

Currently, topicalization processor can only handle multimedia description metafile in SEO scheme. More other pluggable topicalization modules will be designed for other metafile formats such as Dublin[5]. An ontology which can narrow down semantic differences between metafiles will be deployed so that user can get more meaningful result. As for visualization, we will conduct an experiment on using DOT[6] format to represent graphs. In this perspective, users can manually associate relevant topics directly on graphs. We believe that collaborative annotation on media documents could be achieved more efficiently in this way.

7 Conclusions

In this paper, we have presented a users-oriented solution for efficient audiovisual content retrieval on mobile devices which is adaptive for mobile devices and mobile consumers' behavior. Given these two contexts, we identified a new method which is based on Topic Maps to organize and present multimedia data and metadata. We have used Topic Maps as the knowledge model to reorganize complex audiovisual dataset into a common topics-based layer. Therefore, interested topics scattered in multiple clips can be clustered, cross-related. SEO2TopicMaps algorithm supports automatic conversion from SEO to Topic Maps is proposed. By visualizing XML-based Topic Maps files to graphs, users are able to explore and locate interested topics intuitively and moreover across the boundary of media documents and genres.

References

1. Helin, J.: Developing mobile multimedia services to enable streaming to wireless devices. Presented at IIR 3rd Media Streaming Conference, Amsterdam, the Netherlands (2001)
2. Nordlof, P.: The Mobile triple play: bringing TV to mobiles. Available online at http://www.nsrc.se/The_Mobile_Triple_Play%20Per_Nordlof%20LM%20 Ericsson.pdf
3. Finnish mobile TV community. Available online at http://www.finnishmobiletv.com
4. O2: TV-to-mobile trial. Available online at http://www.o2.com/about/tv_to_mobile_trial.asp
5. Knoche, H., Sasse, M.A.: Can small be beautiful?: assessing image resolution requirement for mobile TV. In: Proceedings of the 13th annual ACM International Conference on Multimedia, Hilton, Singapore, pp. 829–838 (2005)
6. Knoche, H., McCarthy, J.D.: Design requirements for mobile TV. Presented at the 7th International conference on human computer interaction with mobile devices & services, Salzburg, Australia, pp. 69–76 (2005)
7. NOC: 3GSM mobile failing to get a good reception? Available online at http://www.noconline.org/newsdisplay.aspx?id=932

---

[5] http://dublincore.org/
[6] http://www.graphviz.org/

8. Seals, T.: Bandwidth variability, user interface problems may limit mobile TV. Available online at http://www.xchangemag.com/articles/07febonthetube01.html
9. Knoche, H.: Mobile TV users' needs and expectations of future services. Presented at WWRF (2004)
10. Jumisko-Pyykko, S., hakkinen, J.: Evaluation of subjective video quality of mobile devices. In: Proceedings of the 13th Annual ACM International Conference on Multimedia, Hilton, Singapore (2005)
11. Forum, M.T.: Completed pilots. Available online at http://www.mobiletv.nokia.com/pilots/finland/
12. instruments, T.: Mobile TV: about mobile TV. Available online at http://focus.ti.com/general/docs/wtbu/wtbugencontent.tsp?templateId=6123&navigationId=12499&contentId=4445
13. Motely, A.: Mobile TV – proven demand, but at what cost?. Presented at 3rd Annual Mobile TV and Video Forum 2007, Sofitel St. James London (2007)
14. Wiggin, N.: Providing a rich consumer experience. Presented at 3rd Annual Mobile TV and Video Forum 2007, Sofitel St. James London (2007)
15. Hollink, L., Nguyen, G.P., Koelma, D.C., Schreiber, A.T., Worring, M.: Assessing user behavior in news video retrieval. IEEE Proceedings of Vision, Image and Signal Processing 152, 911–918 (2005)
16. Yoo, H.Y., Cheon, S.H.: Visualization by information type on mobile device. Presented at Asia Pacific Symposium on Information Visualization (APVIS 2006), Tokyo, Japan (2006)
17. Chen, J.-Y., Taskiran, C., Delp, E.J., Bouman, C.A.: ViBE: a new paradigm for video database browsing and search. Presented at Proceedings of IEEE Workshop on Content-based Access of Image and Video Libraries, Santa Barbara, CA, USA (1998)
18. Guillemot, M., Wellner, P., Gatica-Perez, D., Odobez, J.-M.: A hierarchical keyframe user interface for browsing video over the internet. Presented at Human- Computer Interaction - INTERACT 2003 (2003)
19. Liu, Q., Zang, C., Tong, X.: Providing on-demand sports video to mobile devices. In: Presented at International Multimedia Conference. Hilton, Singapore (2005)
20. Zhang, W.Y.: An indexing and browsing system for home video. Presented at European Signal Processing conference, Finland (2000)
21. Carlsson, S.: Video browsing exploration and structuring (VIBES). Presented at Proceedings of the 4th European Workshop on Image Analysis for Multimedia Interactive Services, Queen Mary, University of London (2003)
22. Malaise, V., Aroyo, L., Brugman, H., Gazendam, L., Jong, A.: Evaluating a Thesaurus Browser for an Audio-Visual Archive. In: Staab, S., Svátek, V. (eds.) EKAW 2006. LNCS (LNAI), vol. 4248, Springer, Heidelberg (2006)
23. Keller, T., Tergan, S.-O.: Visualizing knowledge and information: an introduction. In: Tergan, S.-O., Keller, T. (eds.) Knowledge and Information Visualization. LNCS, vol. 3426, pp. 1–23. Springer, Heidelberg (2005)
24. Bertin, J.: Graphics and graphic information-processing. Walter de Gruyter, Berlin, New York, p. 273 (1981)
25. Herman, I.: Graph visualization and navigation in information visualization: A survey. IEEE Transaction on Visualization and Computer Graphics 6, 24–43 (2000)
26. Keim, D.A.: Information visualization and visual data mining. IEEE Transaction on Visualization and Computer Graphics 8, 1–8 (2002)

27. Feeney, A., Webber, L.: Analogical representation and graph comprehension. In: Butz, A., krüger, A., Olivier, P. (eds.) SG 2003. LNCS, vol. 2733, pp. 212–221. Springer, Heidelberg (2003)
28. Karstens, B., Kreuseler, M., Schumann, H.: Visualization of Complex Structures on Mobile Handhelds. Presented at International Workshop on Mobile Computing (2003)
29. Pepper, S.: The TAO of Topic Maps. Presented at XML Europe 2000, Paris, France (2000)
30. Garshol, L.M.: Living with topic maps and RDF, Available online at http://www.ontopia.net/topicmaps/materials/tmrdf.html
31. Tjondronegoro, D.: PhD Thesis: Content-based Video Indexing for Sports Applications using Multi-modal approach, Deakin University (2005)

Adaptive Video Presentation for Small Display While Maximize Visual Information

Yandong Guo, Xiaodong Gu, Zhibo Chen, Quqing Chen, and Charles Wang

Thomson Corporate Research, Beijing
eemars@gmail.com, xiao-dong.gu@thomson.net

Abstract. In this paper we focus our attention on solving the contradiction that it is more and more popular to watch videos through mobile devices and there is an explosive growth of mobile devices with multimedia applications but the display sizes of mobile devices are limited and heterogeneous. We present an intact and generic framework to adapt video presentation (AVP). A novel method for choosing the optimal cropped region is introduced to minimize the information loss over adapting video presentation. In order to ameliorate the output stream, we make use of a group of filters for tracking, smoothing and virtual camera controlling. Experiments indicate that our approach is able to achieve satisfactory results and has obvious superiority especially when the display size is pretty small.

Keywords: Adaptive video presentation, mobile device, optimal cropped region, maximize visual information, Kalman filter, and virtual camera control.

1 Introduction

It becomes more and more popular to watch videos through mobile devices and there is an explosive growth of mobile devices with multimedia applications. Unfortunately, there are two obstacles to browse videos on mobile devices: the limited bandwidth and the small display sizes of mobile devices. Thanks to the development of network, hardware and software, the bandwidth factor is expected to be less constraint while the limitation on display size remains unchangeable in the foreseeable future.

If sub-sampling each frame according to the resolution of the output device while preserving the intact video contexts, the excessive reduction ratio will lead to an ugly experience. It will be a good solution if cropping the most important part, called the region of interest (ROI), from the original video, discarding partial surroundings and then resizing the cropped region to the display size of the output device. How to get the optimal cropped region (OCR) is the key technique in the solution.

There are several approaches used for browsing videos on mobile devices by cropping region of interest [4-6]. In [4], the authors proposed a semi-automatic solution for this problem. The method proposed in [5] is focused on the technique of virtual camera control. And the approach in [6] is for special scenario of static panoramic capturing. However, the perceptual result is affected by the display resolution but none of these solutions has considered providing optimal cropped region (OCR) according to the display sizes, which means maximize the viewer received information.

In this paper, we presented an intact self-adaptive solution which can be used to all kinds of videos and can be adapted to various display sizes. Our main contribution is to propose a novel algorithm which can get OCR adaptively according to the display size while minimizing the information loss. Furthermore, we improved the visual camera control by adding zooming in/out operations when necessary.

The paper is organized as follows. In the following section, all the components of the system are explained briefly. The OCR choosing is described in Section 3 while the tracking and filtering is shown in Section 4. In Section 5, the virtual camera control is presented and the last section concludes this paper.

2 System Architecture

A complete framework of our approach is shown in Fig.1. First of all, we analyze the input sequence to extract the attention objects (AO), e.g., human faces, balls, texts or other saliency objects. These attention objects are divided into two groups according to the models by which we analyze the sequences: saliency attention objects by saliency models (bottom-up process) and semantic attention objects by semantic models (top-down process).

We adopt the Itti's model [1] as the saliency model to produce the saliency map and use the model mentioned in [10] as the semantic model to detect the saliency attention objects. Since "The purpose of the saliency map is to represent the conspicuity or the 'saliency'-at every location in the visual field by a scalar quantity and to guide the selection of attended locations, based on the spatial distribution of saliency" [1], we can get the conclusion that human attention is an effective and efficient mechanism for information prioritizing and filtering. Moreover, supposing that an object with larger magnitude will carry more information, we can calculate the information carried by the i-th saliency object as:

$$Infor_i^{saliency} = \sum_{x,y \in R} I_{x,y} . \tag{1}$$

Where $I_{x,y}$ denotes the scalar quantity of the pixel (x, y) in the saliency map, R denotes the region of the i-th object.

By employing the top-down model [2], we obtain the position, the region and the quantity of information of the semantic attention objects. In the following part of paper, we take football as an example of semantic attention objects, and use the trajectory-based ball detection and tracking algorithms mentioned in [3] to locate the ball. The information carried by semantic attention objects can be calculated as:

$$Infor_i^{semantic} = W_i \times area_i^{semantic} . \tag{2}$$

Where $area_i$ denotes the magnitude of the semantic attention degree and W_i is used to unify the attention models by giving a weight to each kind of semantic attention objects.

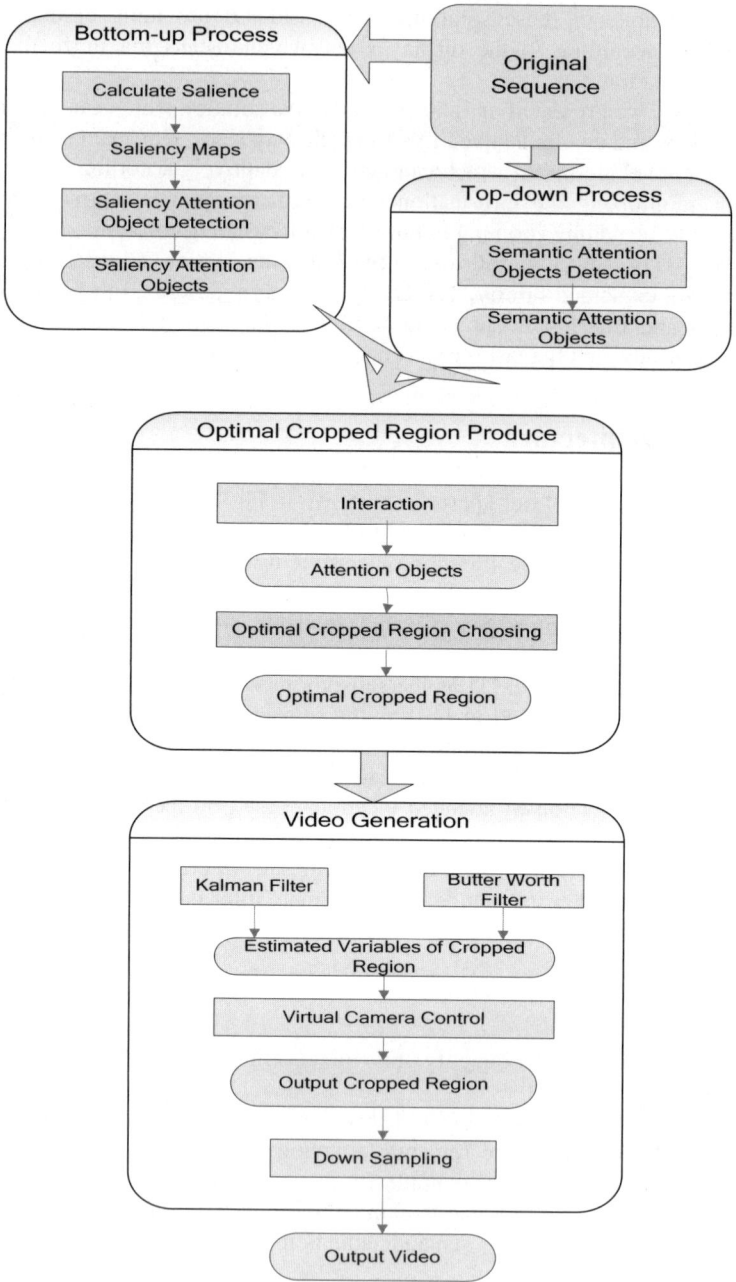

Fig. 1. The System Architecture

The saliency model by Itti determines the saliency objects by the nonlinear integration of low-level visual cues, mimicking processing in primate occipital and posterior parietal cortex. It makes no assumption on video content and thus is universally applicable. By making use of the saliency model our approach is able to deal with various videos. Moreover, we amend the results of the saliency model by adopting the semantic model, because the semantic model can provide more accurate location of attention objects. If a semantic object and a saliency object cover similar region (judged by the threshold) the region got by semantic model will be chosen for the attention object.

The saliency attention objects and the semantic objects are integrated to get a uniform attention model. An example frame is shown in Fig. 2, the saliency attention objects are marked by the black rectangles and the semantic attention objects are marked by the white rectangle.

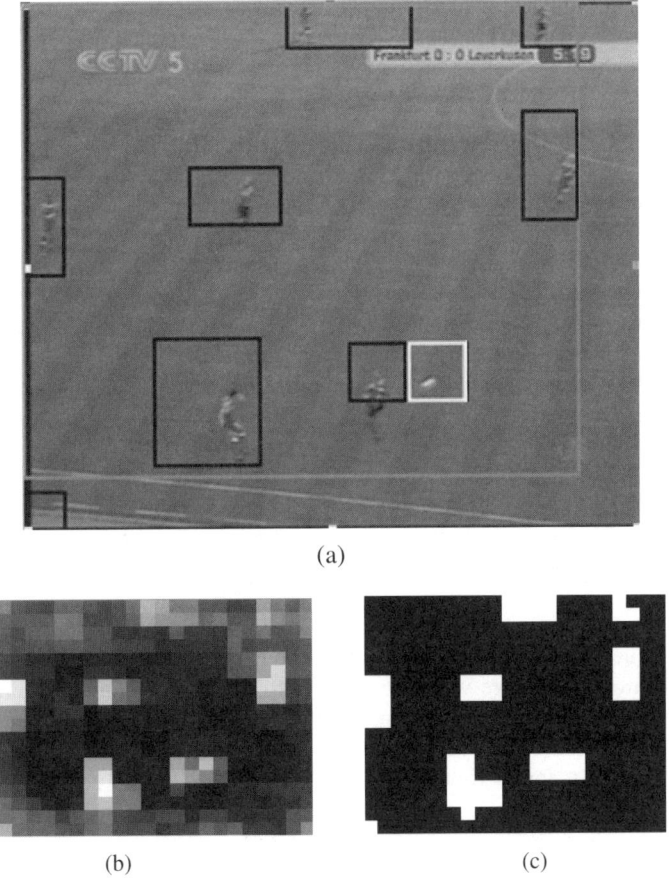

Fig. 2. (a) is the frame in which the attention objects were marked; (b) is the saliency map of the frame calculated by Itti's model and (c) is the map to show the saliency attention objects

As shown in Fig. 2, we usually get more than one AO. If we draw the cropped region including all AOs, we have to zoom out the cropped region at very large ratio when the display resolution is very low compared with the original resolution. However, when the display resolution is not very low compared with the original resolution, larger cropped region including more AOs may be the better choice. For the reason above, the cropped region should be adapted to different display sizes to preserve as much information as possible by trade-off between cropping and reduction.

The process of choosing OCR is a noisy process. We assume that the noise is Gaussian so we use the Kalman Filter to estimate the central coordinates of OCR and use IIR to smooth the areas of OCR temporally. Kalman Filter and IIR reduce most of the noise inherent in the OCR choosing, and produce a new region sequence (EOCR). If the EOCR is used to down-sample directly, the quality of the output video is often jittery. The movement and zoom in/out of EOCR is less smooth than that of a physical camera, which has inertia due to its mass. Therefore, we use an additional filtering step called virtual camera control (VCC) to produce a smoother and more pleasing video output.

3 Optimal Cropped Region (OCR) Choosing

When there are several AOs dispersed in one frame, if we draw the cropped region including all the AOs, we may get a large cropped region. Down sampling such a large cropped region into display size will lead to an excessive reduction ratio and the display sequence will be blurred badly when the screen used for display is pretty small. Because there is loss of details during the process of image resizing which cannot be recovered afterwards and when the resize ratio becomes larger the information loss increases rapidly [8], outputting all the AOs may not always be the best choice and we have to abandon some of the AOs to get the largest information output.

Which ones of the AOs should be involved in the cropped region and which ones should not? This is the problem we will solve in this section. In another words, that is to say, how to keep balance between information loss from down sampling and that from abandoning AOs?

Definition 1. The optimal cropped region (OCR) is the region which can generate display sequence with the minimum information loss.

In the following part, we will discuss how to measure the information loss along the process of cropping and down sampling.

When we choose some of the AOs to form a cropped region, the sum of information actually got by viewers is,

$$InforSum^{cropped\_region} = \sum_{i \in CR} infor_i^{Saliency} + \sum_{j \in CR} infor_j^{Sematic}. \qquad (3)$$

Where CR denotes the cropped region. Then, we will measure the information loss during the down sampling which is a fine to coarse image representation. In [9], Mario Ferraro considered the fine-to-coarse transformation of an image as an isolated irreversible thermodynamical system whose channels are dynamical subsystems. The information loss over the transformation is measured by P, the average of the density of entropy production across the image,

$$\sigma = \left(\frac{\nabla f(x,y,t)}{f(x,y,t)}\right)^2 \quad (4)$$

$$P = \iint_\Omega f(x,y,t)\sigma(x,y,t)dxdy$$

The density σ measures the local loss of information and t is a non-negative parameter that defines the scales of resolution at which the image is observed; small values of t correspond to fine scales (cropped region before down-sampling), while large values correspond to coarse scales (cropped region after down-sampling). (x, y) is used to denote the coordinates of a pixel in a discrete lattice.

In the case of image down-sampling, when the cropped region is scaled down from the original scale t_0 to the scale t_1, we use the operator T which takes the original image at one scale to another scale to give the transformation,

$$T_{downsample} : I(\Box,t_0) \rightarrow I(\Box,t_1). \quad (5)$$

In order to measure the information loss during the image down-scaling process, there must be a reference image to the original one. Therefore, we up-sample the down-scaled display image by the uniform algorithm to the resolution of cropped region at scale t_0 and choose it as the reference image:

$$T_{upsample} : I(\Box,t_1) \rightarrow I'(\Box,t_0). \quad (6)$$

Where $T_{downsample}$ and $T_{upsample}$ indicate the transformation of down-sampling and up-sampling, respectively.

We calculate the information loss ratio *InforLossRatio* over the transformation of $T_{downsample}$ and $T_{upsample}$ by (4).

With assumption that there is no information loss during the up-sampling process, we can get the remaining information of display image by,

$$InfoSum^{display} = (1-InforLossRatio) \times \sum_{i \in CR} infor_i^{Saliency} + \sum_{i \in CR} infor_i^{Sematic}. \quad (7)$$

Because the semantic attention objects do not lose its semantic meaning during the transformation, we do not multiply the information attenuation factor to them when they are larger than the minimal perceptible size which can be predefined according to the class of the objects.

We calculate the $InfoSum^{display}$ of display images got by cropped regions including different AOs and at last, we set the cropped region with the largest $InfoSum^{display}$ as the OCR and calculate the central coordinates (x_{OCR}, y_{OCR}) and area $Size_{OCR}$ of the optimal cropped region for the next use.

4 Tracking and Smoothing

After the OCR determination, we get a sequence of their central coordinates and a sequence of their size. The center of OCR tracking is generally a noisy process. We assume that the noise is Gaussian Noise, and estimate the coordinates and velocities of the centers by Kalman Filter [7]. We describe the model of the discrete-time system by the pair of equation below:

$$\text{System Equation: } s(k) = \Phi s(k-1) + \Gamma w(k-1). \tag{8}$$

$$\text{Measurement Equation: } z(k) = Hs(k) + n(k). \tag{9}$$

Where $s(k) = [x(k), y(k), v_x(k), v_y(k)]^T$ is the state vector, $x(k)$ and $y(k)$ are the horizontal and vertical coordinates of the center of OCR at time k, respectively. $v_x(k)$ is the velocity in the horizontal direction and $v_y(k)$ is the velocity in the vertical direction. $w(k-1)$ is the Gaussian noise caused by choosing, representing the center acceleration in the horizontal and vertical directions.

And $z(k) = [x_{OCR}(k), y_{OCR}(k)]^T$ is the measurement vector at time k. $n(k)$ denotes the noise caused by measure and supposed to be Gaussian white noise.

The state transition matrix and coupling matrix

$$\Phi = \begin{pmatrix} 1 & 0 & T & 0 \\ 0 & 1 & 0 & T \\ 0 & 0 & 1 & 0 \\ 0 & 0 & 0 & 1 \end{pmatrix}, \quad \Gamma = \begin{pmatrix} \frac{T^2}{2} & 0 \\ 0 & \frac{T^2}{2} \\ T & 0 \\ 0 & T \end{pmatrix}$$

And the measurement matrix, $H = \begin{pmatrix} 1 & 0 & 0 & 0 \\ 0 & 1 & 0 & 0 \end{pmatrix}$

The Kalman Filter estimates the coordinates and the velocities of the centers of OCRs from the results of OCR choosing. After the recursive procedure, we can get the estimated vector $\hat{s}(k) = [\hat{x}_{OCR}(k), \hat{y}_{OCR}(k), \hat{v}_x(k), \hat{v}_y(k)]$ with the minimum mean-squared error of estimation.

The area of OCR $Size_{OCR}$ may fluctuate intensively. We observed that the zooming of a physical camera is a more smooth process, so we use IIR filter to smooth the area of OCR temporally and the output of IIR is indicated by $\hat{Size}_{OCR}(k)$ [10].

5 Virtual Camera Control

If we use the tracking and smoothing results to move the cropped region directly, the quality of the output video is often jittery. The resulting motion is less smooth than that of a physical camera, which has inertia due to its mass. Virtual camera control (VCC) mentioned in [3] is used to solve the problem. The basic elements of VCC by

[3] are that the change of coordinates less than the threshold will be discarded and the coordinates are set to be constant, whereas the monotone continuous change of coordinates larger than the thresholod will be tracked and the coordinates are set to be changed smoothly. The VCC they mentioned is only used to control the centroid motion because the size of cropped region is constant in their application. However, the size of cropped regions can be changed in our algorithm and we ameliorate the VCC by adding zooming in/out function as one state of the state machine.

We use the $x_o(k), y_o(k), Size_o(k)$ to denote the central position and the size of region used to output at time k, respectively. When the inequalities

$$| Size_o(k-1) - \hat{Size}_{OCR}(k) | < \sigma_s$$

is satisfied for a certain constant σ_s, the sizes remain unchangeable.

$$Size_o(k) = Size_o(k-1). \tag{10}$$

Otherwise, we start the zoom operation of virtual camera,

$$Size_o(k) = \alpha_1 Size_o(k-1) + \alpha_2 \hat{Size}_{OCR}(k). \tag{11}$$

$$\alpha_1 + \alpha_2 = 1, \ \alpha_1, \alpha_2 > 0 \tag{12}$$

6 Experiment Results

We choose several video sequences, including sports match videos, news videos, home videos and surveillance videos. The normalized viewer received information

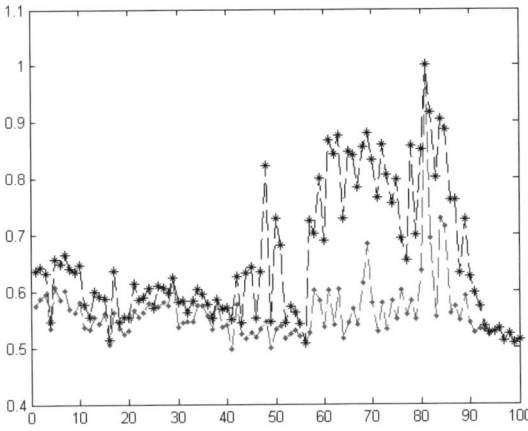

Fig. 3. The curve of the normalized viewer received information different cropped region
* With OCR choosing
•The cropped region with all attention objects

curve of a sequence about soccer match is shown in Fig. 3. The horizontal axis in the figure is the temporal axis while the vertical axis is used to indicate the normalized viewer received information. The figure indicates that the AVP with OCR choosing mechanism can output more visual information than AVP with the cropped region including all attention objects.

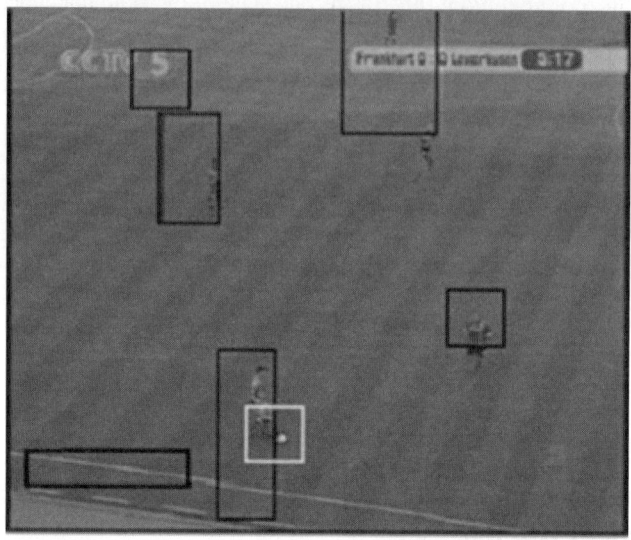

Fig. 4. The original frame with the attention objects marked. The saliency attention objects are marked in black rectangle while the semantic attention object is marked in white rectangle.

Since the evaluation of video presentation quantity is associated with the viewers' subjective feeling, we invited nine volunteers and showed the result sequences created by our approach, the sequences created by the framework without OCR choosing, and the sequences by down sampling the original sequences. We asked them to choose the sequence with the most satisfactory browsing experience. The feedback is encouraging and all sequences created by our approach got the highest evaluation.

There are the frames of the result sequences created by different approaches in different display sizes which are similar with the display sizes of real mobile devices as an example in figure 5. And their original frame with attention objects marked is shown in Fig. 4. The experiment results show that the larger the down-scaling ratio is, the larger the information loss ratio is, so the smaller display size usually leads to smaller cropped region. But there is no uniform function can describe the relationship between the down-scaling ratio and the information loss ratio precisely. We can see that the cropped regions in our approach are adjusted according to the display sizes of mobile devices to gain an optimal cropped region.

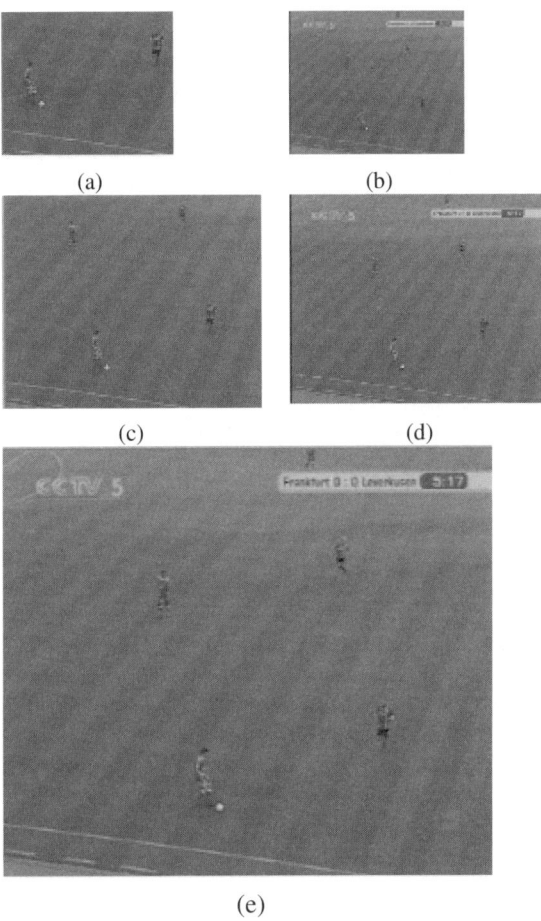

Fig. 5. (a), (c) are from the sequences created by our approach with the display size 88x72 and 132x108, respectively. (b), (d) are from the sequences created by the approach get cropped region including all the AOs, (without OCR choosing) with the display size 88x72 and 132x108, respectively. When the display size is 264x216, the approach with OCR choosing and the one without OCR choose gain the same cropped region, which is shown in (e).

7 Conclusion

In this paper we presented an intact framework for adapting video presentation. The framework integrates attention objects detection, optimal cropped region choosing, tracking and smoothing process and virtual camera control to output pleasant sequences. The whole process is automatic, robust and has obvious advantages when the display size is pretty small or when there are multi-attention-objects. This adaptive video presentation approach not only can be used for watch videos through mobile devices but also has potential use for transferring video bit streams through the network with limited bandwidth by transferring only saliency parts of the video. We will do experiments to demonstrate this potential use of the approach. Moreover, in our future work, we plan to accelerate the optimal cropped region choosing process and employ more rules from cinematography to ameliorate the mechanism of virtual camera control.

References

1. Itti, L., Koch, C., Niebur, E.: A Model of saliency-based visual attention for rapid scene analysis. IEEE Transactions on PAMI, 1254–1259 (1998)
2. Yu, X., Leong, H.W., Xu, C., Tian, Q.: Trajectory-Based Ball Detection and Tracking in Broadcast Soccer Video. IEEE Transactions on multimedia, 1164–1178 (2006)
3. Tovinkere, V., Qian, R.J.: Detecting semantic events in soccer games: Towards a complete solution. In: ICME, pp. 1040–1043 (2001)
4. Fan, X., Xie, X., Zhou, H.Q., Ma, W.Y.: Looking into video frames on small displays. In: MM 2003, November 2-8, pp. 247–250 (2003)
5. Aygun, R.S., Zhang, A.: Integrating virtual camera controls into digital video. In: Multimedia and Expo, ICME 2004, pp. 1503–1506 (2004)
6. Sun, X., Kimber, D., Manjunath, B.S.: Region of interest extraction and virtual camera control based on panoramic video capturing. IEEE Transactions on PAMI, 981–989 (2005)
7. Kay, S.M.: Fundamentals of statistical signal processing. Prentice-Hall, Englewood Cliffs (1993)
8. Cover, T.M., Thomas, J.A.: Elements of information theory. Rscience Publication (1991)
9. Ferraro, M., Boccigone, G.: Information properties in fine-to-coarse image transformations. In: Image Processing, ICIP 1998, pp. 757–761 (1998)
10. Mitra, S.K.: Digital signal processing. McGraw-Hill, New York (1993)
11. Ma, Y.F., Zhang, H.J.: Contrast-based image attention analysis by using fuzzy growing. In: MM 2003, Berkeley, California, USA, pp. 374–380 (2003)

An Efficient Compression Technique for a Multi-dimensional Index in Main Memory

Joung-Joon Kim, Hong-Koo Kang, Dong-Suk Hong, and Ki-Joon Han

School of Computer Science & Engineering, Konkuk University,
1, Hwayang-Dong, Gwangjin-Gu, Seoul 143-701, Korea
{jjkim9, hkkang, dshong, kjhan}@db.konkuk.ac.kr

Abstract. Recently, in order to retrieve data objects efficiently according to spatial locations in the spatial main memory DBMS, various multi-dimensional index structures for the main memory have been proposed, which minimize failures in cache access by reducing the entry size. However, because the reduction of entry size requires compression based on the MBR (Minimum Bounding Rectangle) of the parent node or the removal of redundant MBR, the cost of MBR reconstruction increases and the efficiency of search is lowered in index update and search. Thus, to reduce the cost of MBR reconstruction, this paper proposed a RSMBR (Relative-Sized MBR) compression technique, which applies the base point of compression differently in case of broad distribution and narrow distribution. In case of broad distribution, compression is made based on the left-bottom point of the extended MBR of the parent node, and in case of narrow distribution, the whole MBR is divided into cells of the same size and compression is made based on the left-bottom point of each cell. In addition, MBR was compressed using a relative coordinate and the MBR size to reduce the cost of search in index search. Lastly, we evaluated the performance of the proposed RSMBR compression technique using real data, and proved its superiority.

Keywords: MBR Compression Technique, Multi-Dimensional Index, Spatial Main Memory DBMS, Cache Access, R-tree.

1 Introduction

Recently with the development of geographic information systems (GISs), prompt and efficient processing of complex spatial data is keenly required in various GIS application areas including LBS (Location Based System), Telematics and ITS (Intelligent Transportation System). To meet the requirement, R&D is being made on the spatial main memory DBMS (Database Management System) that uploads the whole database onto the main memory and processes data there [6]. In addition, for the spatial main memory DBMS, multi-dimensional index structures have been proposed [1,11], which consider caches in order to optimize existing disk-based multi-dimensional indexes [5,8] in the main memory. The ultimate goal of the index structures is to increase fan-out by reducing the entry size and enhance the performance of the system by minimizing failures in cache access [2,3,4].

Sitzmann and Stuckey proposed pR-Tree (partial R-Tree) that removes child MBR coordinates overlapping with parent MBR coordinates in order to reduce cache failure of multi-dimensional index R-Tree [10]. pR-Tree removes pointers from child nodes and stores the nodes in the main memory in sequence, and identifies them by storing child MBR coordinates overlapping with parent MBR coordinates in four bits. However, with the increase of the number of entries, the percentage of child MBR coordinates overlapping with parent MBR coordinates drops considerably and, as a result, the performance is lowered. In addition, the performance of update is worsened by additional calculation for reconstructing removed child MBR coordinates.

Kim and Cha proposed CR-tree (Cache-conscious R-tree) that uses QRMBR (Quantized Relative MBR), which compresses MBR occupying the most part in R-Tree, as the key in order to reduce cache failure of R-Tree [7]. QRMBR is MBR that expresses the MBR of a child node as coordinates relative to the MBR of the parent node and compresses the relative coordinates through quantization. CR-tree shows higher performance in data insert and search than pR-tree because it can include more entries in a node by compressing the MBR of entries [9]. However, CR-tree may have errors occurring in the process of compressing MBR into QRMBR, and the errors may increase the number of objects involved in the refinement stage of search and, as a consequence, lower the performance of search. In addition, the performance of update is lowered by additional calculation for the reconstruction of QRMBR.

In this paper, we proposed a RSMBR (Relative-Sized MBR) compression technique that applies the base point of MBR compression differently in case of broad distribution and narrow distribution in order to reduce the additional cost of update for the reconstruction of MBR in index update. With the compression technique, in case of broad distribution compression is made based on the extended MBR of the parent node, and in case of narrow distribution the whole MBR is divided into cells of the same size and compression is made based on the left-bottom point of each cell. The application of compression base points reduces the change of base points in MBR compression and resultantly reduces the additional cost of update for the reconstruction of MBR. In addition, for reducing the cost of search in index search, the left-bottom point of MBR was expressed as relative coordinates and the right-top point as the size of MBR. In this way, by compressing MBR using relative coordinates and MBR size, we can reduce the number of objects involved in the refinement stage, which in turn reduces the cost of search.

The paper is organized as follows. Chapter 1 is introduction, and Chapter 2 analyzes existing MBR compression techniques. Chapter 3 suggests an efficient RSMBR compression technique for multi-dimensional indexes in the main memory. Chapter 4 proves the superiority of the proposed RSMBR compression technique by testing its performance in various ways. Lastly, Chapter 5 draws conclusions.

2 MBR Compression Techniques

Existing MBR compression techniques have problems such as the growing size of MBR and frequent repetition of MBR compression. Figure 1 shows the absolute coordinate MBR technique, the RMBR (Relative MBR) compression technique and the QRMBR (Quantized Relative MBR) compression technique proposed in CR-tree. R0 is a parent node with two child nodes, and R1 and R2 are the child nodes of R0.

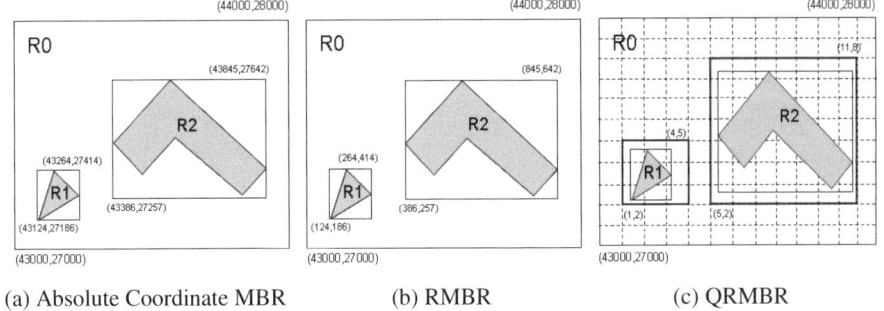

(a) Absolute Coordinate MBR   (b) RMBR   (c) QRMBR

Fig. 1. MBR Compression Technique

Figure 1(a) shows the MBR expression technique using absolute coordinates. In absolute coordinate MBR, the X and Y coordinates need 4 bytes each, so a total of 16 bytes of storage space is required. Figure 1(b) shows the RMBR compression technique that expresses the MBR of a child node based on the left-bottom point of the MBR of the parent node. In RMBR, the X and Y coordinates need 2 bytes each, so a total of 8 bytes of storage space is needed. In addition, Figure 1(c) shows the QRMBR compression technique, which applied the quantization process to RMBR coordinates.

QRMBR quantizes the RMBR of child nodes by dividing the size of the MBR of the parent node into a grid with N x M cells. For the quantization process, QRMBR needs 1 byte for each of the X and Y coordinates, so a total of 4 types of storage space are required. Therefore, QRMBR has 4 and 2 times, respectively, higher compression effect than absolute coordinate MBR and RMBR. However, QRMBR enlarges MBR, which increases the number of objects involved in the refinement stage and this, in turn, lowers search performance.

The compression of MBR needs a base point. Existing MBR compression techniques carry out compression based on the left-bottom point of the MBR of the parent node. Figure 2 shows the reconstruction of MBR in data insert and update.

As in Figure 2, when object R3 is inserted, the MBR of R0, the parent node of R1 and R2, is changed and accordingly MBR compression should be performed not only for newly inserted R3 but also R1 and R2. In addition, because the MBR of R0 is changed when object R1 is updated, MBR compression should be done again for R1 and R2. In this way, if MBR is compressed based on the MBR of the parent node, the compression is sensitive to the change of the MBR of the parent node. Therefore, if objects are inserted and updated frequently, the frequent reconstruction of MBR resulting from the change of the MBR of the parent node lowers update performance and increases overall execution time.

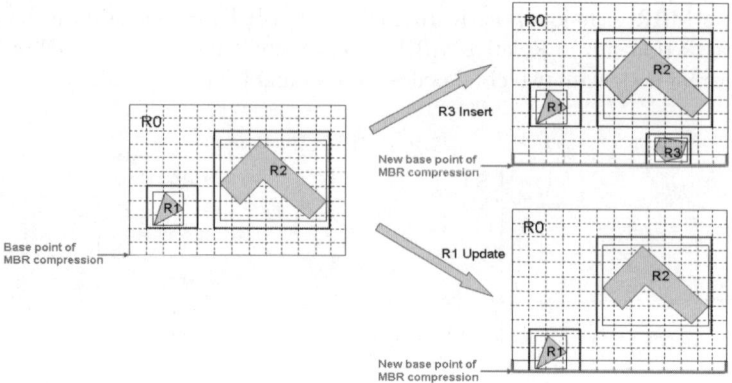

Fig. 2. MBR Reconstruction

3 RSMBR Compression Technique

This chapter explains the efficient RSMBR (Relative-Sized MBR) compression technique for multi-dimensional indexes in the main memory.

3.1 RSMBR Compression

In this paper, we propose the RSMBR compression technique that expresses the left-bottom point of MBR as relative coordinates and the right-top point as the size of MBR in order to prevent MBR from being enlarged in MBR compression. This technique can reduce the volume of data while maintaining the accuracy of MBR. Figure 3 shows the proposed RSMBR compression technique.

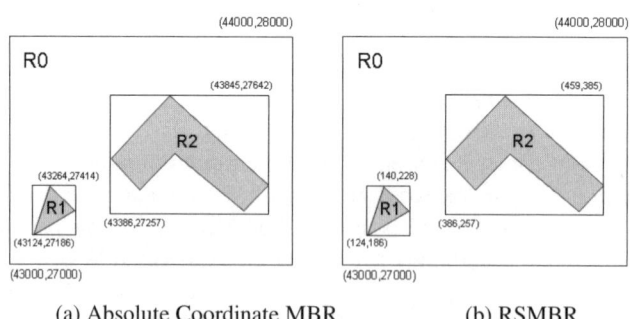

(a) Absolute Coordinate MBR   (b) RSMBR

Fig. 3. MBR Compression Technique

Figure 3(a) and Figure 3(b) show the absolute coordinate MBR technique and the proposed RSMBR compression technique, respectively. In Figure 3(b), the relative coordinates of the MBR of object R1 are (124, 186), and the size of the MBR is 140 in axis X and 228 in axis Y. In addition, the relative coordinates of the MBR of object R2 are (386, 257) and the size of the MBR is 459 in axis X and 385 in axis Y.

Accordingly, RSMBR of object R1 is (124, 186), (140, 228), and that of object R2 is (386, 257), (459, 385).

In the RSMBR compression technique, the coordinates of each right-top point are stored using the length flag and the actual value. Figure 4 shows the data structure for RSMBR coordinates.

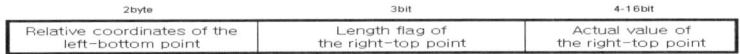

Fig. 4. Data Structure of RSMBR Coordinates

The length flag indicates the bit size occupied by the actual value. That is, 1, 2, 3 and 4 mean that the size of the actual value is 4, 8, 12 and 16 bits, respectively. Therefore, each size can express an actual value up to 15, 255, 4095 and 65535, respectively. For example, of the coordinates of the right-top point of RSMBR of object R1, 140 can be expressed as flag 010 and actual value 10001100, and of the coordinates of the right-top point of RSMBR of object R2, 459 can be expressed as flag 011 and actual value 000111001011. Consequently, X and Y coordinates of RSMBR can be expressed at least 6 bytes and up to 10 bytes, so the overall space needed for storing MBR can be reduced.

Because the size of MBR is below 255 in most spatial objects, it can be expressed with 1 byte. Thus, it is quite rare for RSMBR to exceed 6 bytes. In this way, when the RSMBR compression technique is used, some calculation may be necessary for reconstructing compressed RSMBR for spatial operations such as point queries and area queries. However, the RSMBR of objects compressed by the RSMBR compression technique does not need to be reconstructed, and spatial operation can be done by compressing the query coordinates of point queries or area queries in the same way as RSMBR compression of objects.

3.2 RSMBR Reconstruction

In this paper, we applied the base point of compression differently for broad distribution and narrow distribution in order to reduce the additional cost of update in the reconstruction of RSMBR. Broad distribution means the case that the size of the whole MBR is larger than a specific size (called, base size), and narrow distribution means the case that the size of the whole MBR is equal to or smaller than the base size. In this paper, we use 4 bytes as the base size that is used to distinguish between broad distribution and narrow distribution. Figure 5 shows how to reconstruct RSMBR in case of broad distribution.

As in Figure 5, when object R3 is inserted, even if the MBR of parent node R0 is changed, the MBR of R3 is included in the extended MBR of R0, the parent node of R1, R2 and R3, so RSMBR reconstruction is not necessary. In addition, when object R1 is updated as well, RSMBR reconstruction is not necessary because the MBR of R1 is included in the extended MBR of R0. In such broad distribution, if compression is made based on the left-bottom point of the extended MBR of the parent node, the compression is insensitive to the change of the MBR of the parent node, so the additional cost of update for RSMBR reconstruction can be reduced.

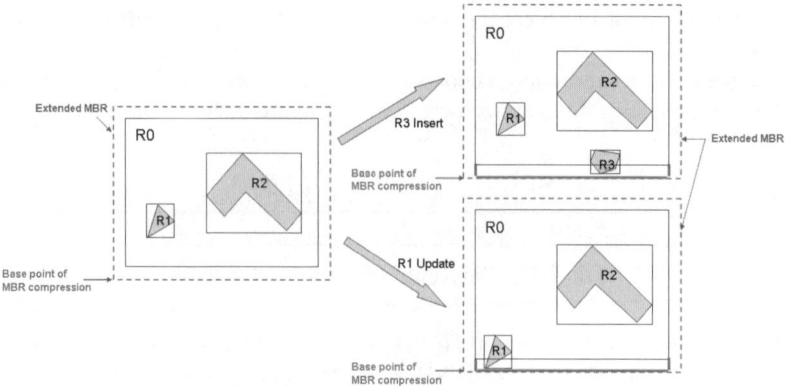

Fig. 5. RSMBR Reconstruction for Broad Distribution

Figure 6 shows how to reconstruct RSMBR in case of narrow distribution.

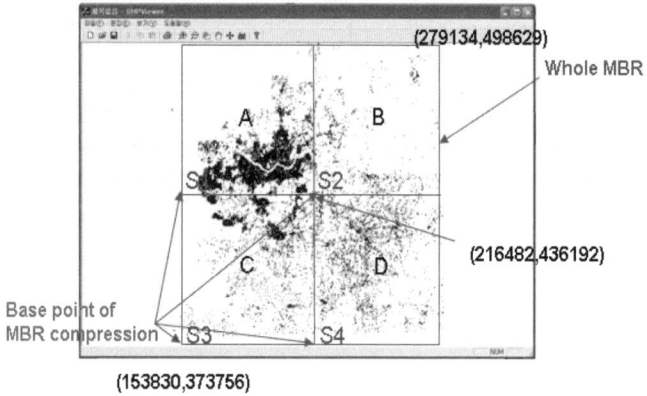

Fig. 6. RSMBR Reconstruction for Narrow Distribution

As shown in Figure 6, the size of the whole MBR is equal to or smaller than the base size (i.e., 4 bytes), the whole MBR is divided into 2-byte cells (called, A, B, C, D) and the base point of RSMBR compression is applied. That is, objects included in Cell A are compressed based on S1, the left-bottom point of Cell A, and objects in Cell B are compressed based on S2, the left-bottom point of Cell B. In the same way, objects in Cell C and Cell D are compressed based on S3 and S4, the left-bottom points of Cell C and Cell D, respectively. Like this, in case of narrow distribution, the whole MBR is divided into 2-byte cells and the left-bottom point of each cell is applied as the base point for RSMBR compression, and this can reduce the additional cost of update for RSMBR reconstruction.

3.3 RSMBR Algorithms

This chapter explains the RSMBR compression algorithm and the RSMBR reconstruction algorithm.

3.3.1 RSMBR Compression Algorithm

The RSMBR compression algorithm finds RSMBR, in which the left-bottom point has relative coordinates and the right-top point has the size of MBR. Figure 7 shows the RSMBR compression algorithm. In Figure 7, the difference between the left-bottom point of the object MBR and the left-bottom point of the base MBR is determined for obtaining the left-bottom point of RSMBR that has relative coordinates, and the difference between the right-top point of the object MBR and the left-bottom point of the object MBR is determined for obtaining the right-top point of RSMBR that has the size of MBR. In addition, the coordinates of each RSMBR are stored as the length flag and the actual value converted into a binary number.

3.3.2 RSMBR Reconstruction Algorithm

The RSMBR reconstruction algorithm reconstructs RSMBR in consideration of the size of the whole MBR. Figure 8 shows the RSMBR reconstruction algorithm. In Figure 8, in order to determine the base point of RSMBR, the whole MBR is obtained first and then in case the size of the whole MBR is larger than the base size, if the MBR of the updated object is included in the extended MBR of the parent node, RSMBR compression is done only for the MBR of the updated object based on the extended MBR of the parent node. And if not, RSMBR compression is done for all the objects included in the node based on the extended MBR of the changed parent node because the MBR of the parent node is changed by the updated node. In case the size of the whole MBR is equal to or smaller than the base size, RSMBR compression is performed based on the left-bottom point of the MBR of the cell containing the updated object.

4 Performance Evaluation and Analysis

This chapter evaluates and analyzes the performance of the absolute coordinate MBR technique, the QRMBR compression technique, and the RSMBR compression technique. Performance evaluation was made in the environment composed of 800 MHz Pentium III CPU, 512 MB memory and Redhat 9.0. Data on all buildings in Seoul (around 37MB) was used as test data. The test data is composed of 249,115 spatial objects, and each object has at least 8 and at most 8,174 coordinates.

In this chapter, we made comparative analysis of the performance of absolute coordinate MBR, QRMBR proposed in CR-tree, and RSMBR proposed in this paper by applying spatial index R-tree. That is, in the performance evaluation, we compared R-tree using absolute coordinate MBR, CR-tree which is R-tree using QRMBR, and RR-tree (RSMBR R-tree) which is R-tree adopting RSMBR. RR-tree is the same as R-tree except that in RR-tree each node uses RSMBR instead of absolute coordinate MBR in the entries.

4.1 Index Size

Figure 9 compares the index size of R-tree, CR-tree and RR-tree created when inserting the whole of 37MB test data. As in Figure 9, in terms of the size of spatial index, the performance of RR-tree is 5~10% lower than that of CR-tree but 45~50% higher than that of R-tree. This means that the index size of RR-tree is somewhat larger than that of CR-tree because the compression rate of RSMBR is slightly lower than that of QRMBR.

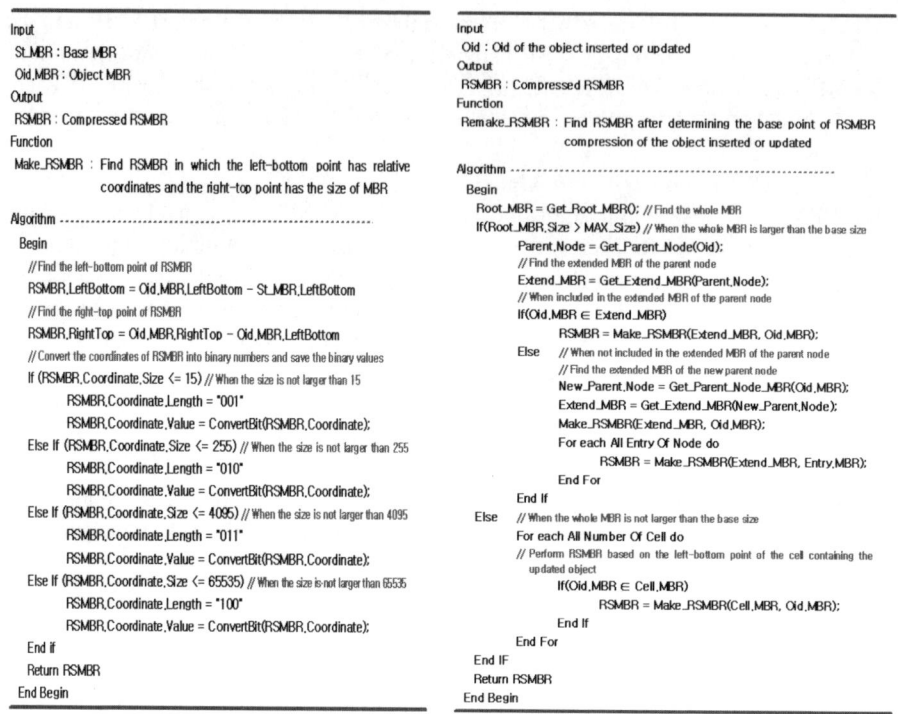

Fig. 7. RSMBR Compression Algorithm   **Fig. 8.** RSMBR Reconstruction Algorithm

4.2 Data Insert Time

Figure 10 compares data insert time for R-tree, CR-tree and RR-tree when inserting the whole of 37MB test data. As in Figure 10, in terms of insert time in the spatial index, the performance of RR-tree is 5~10% lower than that of R-tree but 10~15% higher than that of CR-tree. This means that data can be inserted faster in RR-tree because the process of RSMBR compression is simpler than that of QRMBR compression.

4.3 Data Update Time

Figure 11 compares data update time for R-tree, CR-tree and RR-tree when the number of objects to be updated is 10%, 20%, 30%, 40% and 50% of the total number of objects. As in Figure 11, in terms of data update time in the spatial index, the performance of RR-tree is 1~5% lower than that of R-tree but 20~25% higher than that of CR-tree. This means that because RSMBR is less sensitive than QRMBR to the change of the base point in MBR compression, MBR reconstruction of RR-tree is less frequent than that of CR-tree in data update during the MBR compression.

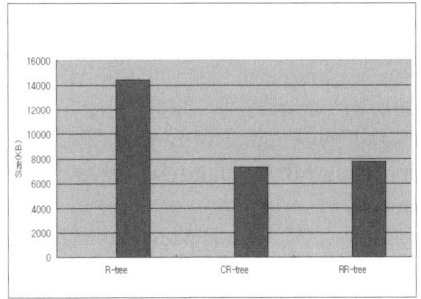
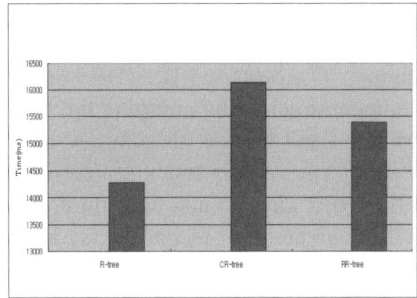

Fig. 9. Index Size **Fig. 10.** Data Insert Time

4.4 Data Search Time

Figure 12 compares data search time for R-tree, CR-tree and RR-tree when the query area is 10%, 20%, 30%, 40% and 50% of the whole area. As in Figure 12, in terms of data search time in the spatial index, the performance of RR-tree is 50~55% higher than that of R-tree and 10~15% higher than that of CR-tree. This means that because RSMBR prevents MBR from being enlarged in the process of MBR compression it can execute data search with less comparison operations than QRMBR.

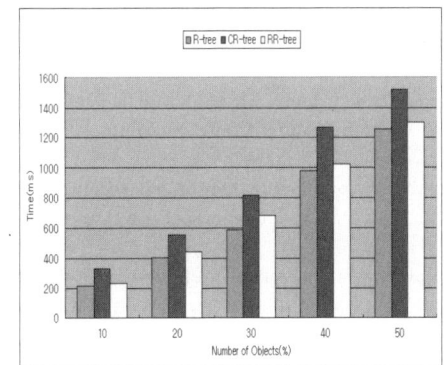
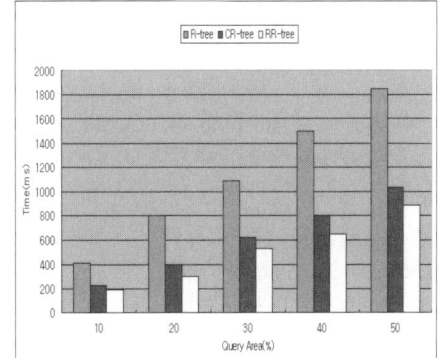

Fig. 11. Data Update Time **Fig. 12.** Data Search Time

5 Conclusions

Recently, many researchers are making different types of research on structure and algorithms for improving cache performance in order to optimize multi-dimensional indexes in the main memory. Representative methods are removing pointers from child nodes and compressing MBR. However, these methods lower update performance due to additional operations for reconstructing entry information in update, and also lower search performance because MBR is enlarged in the process of MBR compression and resultantly the number of objects involved in the refinement stage increases.

To solve these problems, this paper proposed the new RSMBR compression technique. The proposed RSMBR compression technique reduced additional cost of MBR reconstruction by applying the base point of compression differently for broad distribution and narrow distribution, and reduced the number of objects involved in the refinement stage in index search by compressing MBR using relative coordinates and the MBR size. In addition, through performance evaluation, we proved that the RSMBR compression technique is superior to other MBR compression techniques. Although the performance of the proposed technique was somewhat low in terms of index size, it was higher than those of other MBR compression techniques in data insert, update and search.

Acknowledgements

This research was supported by the Seoul Metropolitan Government, Korea, under the Seoul R&BD Program supervised by the Seoul Development Institute.

References

1. Berchtold, S., Bohm, C., Kriegel, H.P., Sander, J.: Independent Quantization: An Index Compression Technique for High-Dimensional Data Spaces. In: Proc. of the ICDE Conference, pp. 577–588 (2000)
2. Bohannon, P., McIlroy, P., Rastogi, R.: Main-Memory Index Structures with Fixed-Size Partial Keys. In: Proc. of Proceedings of the ACM SIGMOD Conference, pp. 163–174. ACM Press, New York (2001)
3. Boncz, P., Manegold, S., Kersten, M.: Database Architecture Optimized for the New Bottleneck: Memory Access. In: Proc. of the International Conference on VLDB, pp. 54–65 (1999)
4. Chen, S., Gibbons, P.B., Mowry, T.C., Valentin, G.: Fractal Prefetching B+-Trees: Optimizing Both Cache and Disk Performances. In: Proc. of the ACM SIGMOD Conference, pp. 157–168. ACM Press, New York (2002)
5. Guttman, A.: R-Trees: a Dynamic Index Structure for Spatial Searching. In: Proc. of the ACM SIGMOD Conference, pp. 47–54. ACM Press, New York (1984)
6. Kim, J.J., Hong, D.S., Kang, H.K., Han, K.J.: TMOM: A Moving Object Main Memory-Based DBMS for Telematics Services. In: Carswell, J.D., Tezuka, T. (eds.) Web and Wireless Geographical Information Systems. LNCS, vol. 4295, pp. 259–268. Springer, Heidelberg (2006)
7. Kim, K.H., Cha, S.K., Kwon, K.J.: Optimizing Multidimensional Index Tree for Main Memory Access. In: Proc. of the ACM SIGMOD Conference, pp. 139–150. ACM Press, New York (2001)

8. Mindaugas, P., Simonas, S., Christian, S.: Indexing the Past, Present, and Anticipated Future Positions of Moving Objects. ACM Transactions on Database Systems 31(1), 255–298 (2006)
9. Shim, J.M., Song, S.I., Min, Y.S., Yoo, J.S.: An Efficient Cache Conscious Multi-dimensional Index Structure. In: Proc. of the Computational Science and Its Applications, pp. 869–876 (2004)
10. Sitzmann, I., Stuckey, P.: Compacting Discriminator Information for Spatial Trees. In: Proc of the Australian Database Conference, pp. 167–176 (2002)
11. Zhou, J., Ross, K.A.: Buffering Accesses of Memory-Resident Index Structures. In: Proc. of the International Conference on VLDB, pp. 405–416 (2003)

RELT – Visualizing Trees on Mobile Devices

Jie Hao[1], Kang Zhang[1], and Mao Lin Huang[2]

[1] Dept. of Computer Science, Erik Jonsson School of Engineering and Computer Science,
The University of Texas at Dallas, Richardson, TX 75083-0688, USA
{jxh049000, kzhang}@utdallas.edu
[2] Department of Computer Systems, Faculty of Information Technology, University of
Technology, Sydney, PO Box 123 Broadway, NSW 2007, Australia
maolin@it.uts.edu.au

Abstract. The small screens on increasingly used mobile devices challenge the traditional visualization methods designed for desktops. This paper presents a method called "Radial Edgeless Tree" (RELT) for visualizing trees in a 2-dimensional space. It combines the existing connection tree drawing with the space-filling approach to achieve the efficient display of trees in a small geometrical area, such as the screen that are commonly used in mobile devices. We recursively calculate a set of non-overlapped polygonal nodes that are adjacent in the hierarchical manner. Thus, the display space is fully used for displaying nodes, while the hierarchical relationships among the nodes are presented by the adjacency (or boundary-sharing) of the nodes. It is different from the other traditional connection approaches that use a node-link diagram to present the parent-child relationships which waste the display space. The hierarchy spreads from north-west to south-east in a top-down manner which naturally follows the traditional way of human perception of hierarchies. We discuss the characteristics, advantages and limitations of this new technique and suggestions for future research.

Keywords: Tree visualization, mobile interface, screen estate, aesthetic layout.

1 Introduction

There is a dramatic increase in the population who use mobile computing devices. Although the hardware is becoming more powerful, online browsing and navigation tend to be not user-friendly. For example, when the user wishes to search for a favorite music on a mobile phone, many clicks or button-pushes are required. This is mostly due to the limited screen space where few music pieces can be presented on one screen.

Most current mobile online search is linear, possibly with scroll bars. Web browsing on mobile devices is also primarily based on the desktop browsing approach with scaled versions. Therefore, finding information on a mobile device has not been as fast as needed. There have been growing research activities in effective and efficient mobile user interfaces. Yet few hierarchical search methods that aim at minimizing the number of clicks and button-pushes have been developed for small screens.

This paper presents a new RELT method for visualizing hierarchical information on mobile devices. The remainder of this paper is organized as follows. Section 2 covers the related work by reviewing desktop-based tree visualization methods and mobile visualization methods. RELT is introduced in Section 3 and an application for music classification depicting this new algorithm in Section 4. Section 5 builds an estimate function for RELT and an optimized RELT algorithm. Section 6 concludes the paper and mentions the future work.

2 Related Work

2.1 Tree Visualization on Desktops

The current research on tree visualization can be generally classified into two categories:

- **Connection:** This method uses nodes to represent tree leaves, and edges to represent parent-child relationships. Much research has been done in this category, such as balloon view [4, 6], radial view [2, 6], and space-optimized tree visualization [7]. The connection-based approaches match the human perception of hierarchy. Their layouts are also easy to understand with clear structures.
- **Enclosure:** This method represents nodes as rectangles. The display area is recursively partitioned to place all the nodes inside their parent's regions. The tree map view [5, 9] is a typical enclosure-based approach. Enclosure-based approaches achieve economic screen usage, but do not provide a clear hierarchical view.

2.2 Visualization on Mobile Devices

Researchers have developed methods for mobile displays by deriving them from those for desktop displays.

Yoo and Cheon [12] introduced a preprocessor to classify the input information. It divides the input information into different types and each type maintains its corresponding visualization method. For hierarchical information, their approach applies the radial layout method [2, 3] with the mobile devices' restrictions [1, 11]. They also use the fisheye view algorithm to help the user to see highlighted regions.

Although, the above approach works for mobile devices, it fails to efficiently utilize the space, evidenced by the examples provided [12].

The main difference between the presented RELT approach and traditional radial approaches [11] is that the latter performs 360 degree circular partitioning while RELT uses 90 degree polygon partitioning, that is more appropriate to fit on mobile screens.

3 Radial Edgeless Tree Visualization

The RELT algorithm is designed to not only utilize the screen space but also maintain the tree layout. The following subsections first give an intuitive explanation of RELT, then describe the algorithm in detail, and finally discuss the complexity of the algorithm.

3.1 Basic Ideas

A tree is a connected graph $T=(V, E)$ without a cycle. A rooted tree $T=(V, E, r)$ consists of a tree T and a distinguished vertex r of T as the root. Each vertex v has an associated value $w(v)$, which we call the weight.

The entire rectangular display area is partitioned into a set of none-overlapping geometrical polygons (or nodes) $P(v_1), P(v_2), \ldots, P(v_n)$ that are used to visually represent vertexes $v_1, v_2, \ldots, v_n$. Each polygonal node $P(v)$ is defined by three or four cutting edges which may be shared with other nodes. These boundaries are defined as below:

1. A common boundary sharing with its parent represents the child-parent relationship.
2. A common boundary sharing with all its children represent the patent-child relationship.
3. One or two boundaries sharing with its siblings represent the sibling relationships.

The geometrical size of a node $P(v)$ is calculated based on its weight $w(v)$. We, therefore, use boundary-sharing to represent the parent-child relationships among nodes, rather than a node-link diagram. Thus, the display space utilization is maximized. The entire tree T is drawn hierarchically from the north-west at the root to the south-east in the top-down manner, which naturally follows the traditional way of human perception of hierarchies. Note the approach can be easily adapted to move the root to other screen locations.

An intuitive method combining the previous two approaches is constructed with three steps. First, a normal connection-based method, the classical hierarchical view [8] for example, is applied. Second, consider each node as a balloon and inflate all the balloons until they occupy the whole screen. Third, these anomalistic non-overlapping balloons are relocated to simulate the tree structure. Although the above step appears like a space-filling approach, the result is more like a radial display. As presented next, the difference from the typical radial approaches, such as InterRing [10], is that our approach computes area allocations based on the nodes' weights, rather than their angles.

3.2 Algorithm

For a given tree, the method recursively calculates the weight for each vertex. Vertexes are classified into four types and each type is assigned with a corresponding rule. The root is assumed to locate at the upper left corner. We employ depth-first search to traverse the tree. Whenever a new vertex is met, the corresponding rule is applied. Every rule considers two operations. One is the node area distribution operation. The other is how to recursively divide its area for its children. After completely traversing the tree, the entire display area is partitioned into a set of non-overlapping polygons which are used to represent vertexes $v_1, v_2 \ldots v_n$. In this case, the display area is fully utilized and a set of graphical links that are commonly used in traditional connection-based methods are avoided. The algorithm is given in peudocode as below:

```
procedure RELT (matrix ad_matrix)
begin
  Para_Creator (ad_matrix)
  // Calculate the necessary parameters for each node.
  DFS (ad_matrix)
  // Depth first search to traverse the tree
  if vertex  v is new then
       int L= Test (v)
       // Return rule L  to v
       Polygonal_Node (v, L)
       // Assign a region to v  with rule L
       Partition_Area(v)
       // Divide the area depending on the weights of v's
         children
       fi
end.
```

The RELT algorithm shown above consists of four major functions that are explained next.

Para_Creator(matrix *ad_matrix*) calculates the necessary parameters, including weight, depth, parent and children for each vertex. A vertex *v* is assigned with a weight $w(v)$, which is calculated in the following way:

- If vertex *v* is a leaf, $w(v) = 1$.
- Otherwise, if *v* is not a leaf and has *m* children
  $\{v_1, v_2, \ldots, v_n\}$ then

$$w_N = 1 + \sum_{i=1}^{m} w_i. \tag{1}$$

Test(vertex *v*) returns the rule that should apply to vertex *v*. We classify all vertexes into four types according to their characteristics in the tree. Specifically, a vertex *v* is of type:

1. If *v* is the only child of its parent.
2. If the parent of *v* has more than one child AND *v* is the first child of its parent.
3. If the parent of *v* has more than one child AND *v* is the last child of its parent (Note that child vertices are numbered from left to right. The left most child is the first and the right most child is the last).
4. If the parent of *v* must have more than one child AND *v* is neither the first nor the last child of its parent.

Before defining the function Polygonal_Node(Vertex *v*, int *L*), several notations and definitions should be introduced.

- L_{Ni}: is the i_{th} cutting edge of node $P(v)$. The cutting edges are numbered from left to right, as illustrated as Figure 1.

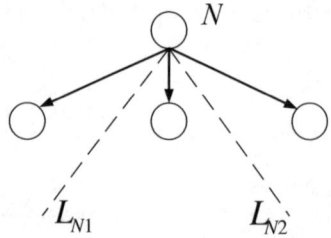

Fig. 1. Cutting edges

- N_{NAS}: N's nearest ancestor with sibling(s) as illustrated in Figure 2, is the nearest ancestor of N that has at least one sibling.
- N_{NAS_P}: The parent of N_{NAS}.
- N_{NASL}: N_{NAS} that is not the leftmost sibling.

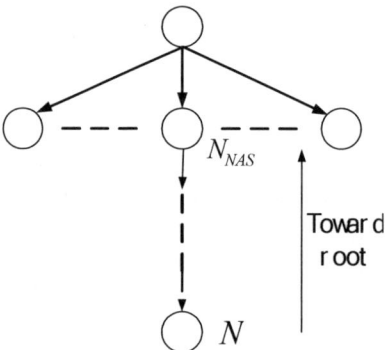

Fig. 2. N's nearest ancestor with siblings

- N_{NASL_P}: The parent of N_{NASL}.
- N_{NASR}: N_{NAS} that is not the rightmost sibling.
- N_{NASR_P}: The parent of N_{NASR}.

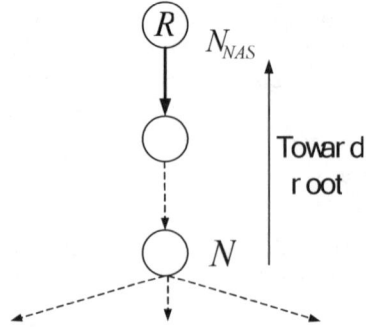

Fig. 3. Special case with the root being named NNAS

Consider such as case, as illustrated in the Figure 3, where a node N has no N_{NAS}. We treat it as a special case by making the root R as N_{NAS}.

Polygonal_Node(Vertex v, int L) represents a vertex in a polygonal shape, rather than a rectangular or circular shape that is commonly used in other connection-based visualizations. Each node is constructed by linking two division line, so this function essentially dictates how to choose these two cutting edges.

1. $L = 1$ (Node type 1)

The cutting edges used by node $P(v)$ are the same as those by N_{NAS}. Rule1 first finds cutting edges used by N_{NAS} and then link them together to form N's region.

2. $L = 2$ (Node type 2)

The first cutting edge chosen by Rule2 is $L_{N^1{}_i}$ where $N^1 = N_{NASR\_P}$ and N_{NAS_P} is the $(i+1)_{th}$ child of N_{NASR_P}. The second cutting edge is $L_{N^2{}_1}$ where N^2 is the parent of N.

3. $L = 3$ (Node type 3)

The first cutting edge selected by Rule3 is $L_{N^1{}_1}$ where N^1 is the parent of N. The second cutting edge is $L_{N^2{}_i}$ where $N^2 = N_{NASL\_P}$ and N_{NASL} is the $(i+1)_{th}$ child of N_{NASL_P}.

4. $L = 4$ (Node type 4)

The two cutting edges chosen by Rule4 are $L_{N^1(i-1)}$ and $L_{N^1{}_i}$ where N^1 is the parent of N and N is the i_{th} child of N^1.

Partition_Area (Vertex v), this function divides the remaining area of a vertex v based on the total weight of its children. The partitioned areas are allocated for the branches rooted at v's children.

Figure 4 shows an example tree. R, the tree's root, partitions the whole 90 degree angle to its children $P(v1)$, $P(v2)$ and $P(v3)$, the shade area in Figure 5 is the area given to the branch rooted at $N1$. $N1$ uses cutting edge $L_{N1,1}$ to recursively partition the area into $N11$ and $N12$ surrounded by a dashed line. Because $N11$ and $N12$ have the same weight by Equation <3>, they occupy the same sized areas.

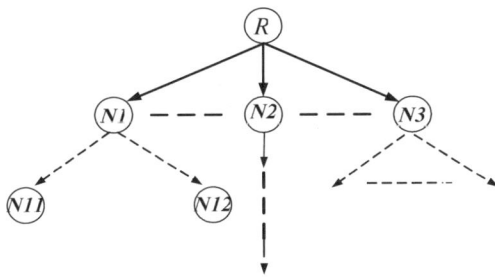

Fig. 4. An example tree

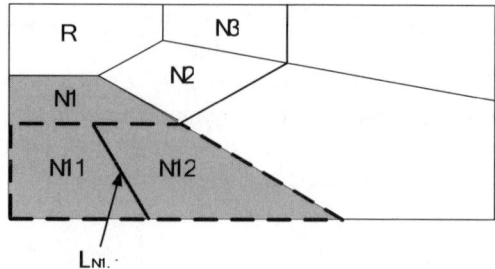

Fig. 5. Drawing of the branch for the tree in Figure 4

3.3 Complexity Analysis

For a n-node tree, the complexity of function Para_Creator is $O(n^2)$ because adjacent matrix is used for information structure storage. Th depth first search is used to traverse the tree, for each new node, functions Test(), Polygonal_Node() and Partition_Area() are applied. All these three functions are $O(1)$. So the complexity of this function is $O(n^2)$.

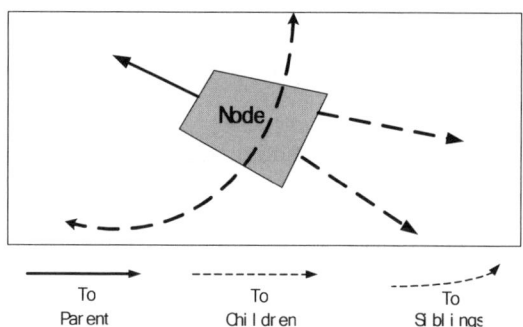

Fig. 6. General directions in parent-child relationships

Using the above RELT algorithm, the tree structure is clearly displayed, as shown in Figure 6. For any node, its parent is in the upper left direction, and its children are in the lower right direction. The nodes at the same level locate along the dashed curve. In addition, this method recursively divides the whole display area, maximizes the screen usage. The next section presents an application of the algorithm.

4 A Case Study: Music Selection

RELT works well for hierarchical information, especially with the overall structure revealed on a limited screen estate.

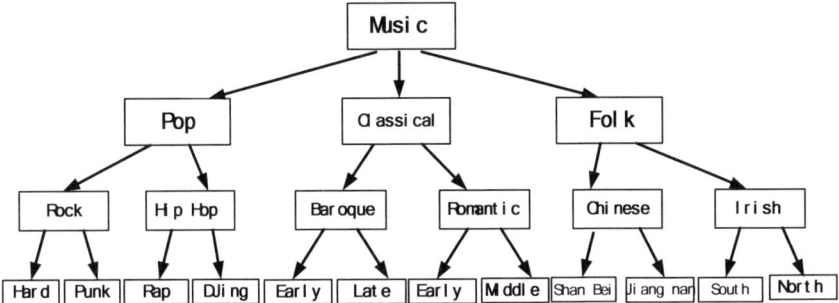

Fig. 7. A music classification example – a balanced tree

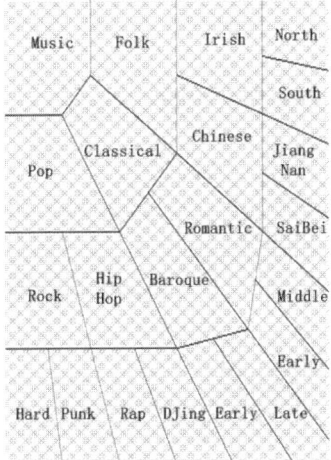

Fig. 8. RELT for the example music classification in Figure 7

One of the current trends is to combine mobile phones with MP3 player. With the available storage capacity and improved sound effect, consumers can download many music pieces from the Internet. With the increasing amount of music selections available on the Internet, there is an urgent need for a commonly accepted music classification system that can assist navigation and selection. The most common approach is using a menu bar. The structure of menu bar is very simple, like artist – album – track and may be defined by users (Ipod, for example). It however docs not show clearly the structure of the music categories.

Figure 7 shows an example music classification, whose. RELT display is shown in Figure 8.

This method works well when the input tree is almost balanced. The smaller difference between $w^{max}(l)$ and $w^{min}(l)$, the maximal and minimal weights at level l, the tree is more balanced and the layout is more aesthetic.

The example shown in Figure 7 is totally balanced because all the nodes on the same level have the same weight. Figure 9 gives an unbalanced tree since the node

Folk's (in a shaded ellipse) weight is 3 which is smaller than 5 of node Classical (in a white ellipse). This leads to an unaesthetic layout, as shown in Figure 10. Nodes Irish and Chinese, in shaded ellipses, are long, across several levels. Those long nodes make the hierarchical levels unclear. The next section discusses an optimization technique that reduces the number of such long nodes.

5 An Optimization

In an unbalanced tree, some leaf nodes may over-represent their areas. More specifically, such nodes take more than one level in the final RELT representation. Take the node "Chinese" in Figure 9 for example, it is a level 3 node but it covers across levels 3 to 5 as shown in Figure 10. The following subsections first introduce the notations and a layout estimate function, and then present the optimized algorithm in details.

5.1 Estimating Node Overrepresentation

The more number of levels over-represented by nodes in a RELT layout, the more misunderstanding of the tree structure may the layout lead to. The total number of over-represented levels is therefore a critical criterion in estimating the effectiveness of RELT for easy human perception.

Definition 1. Assume each leaf node is counted once, the *depth* of a node N, denoted $N.depth$, is the number of nodes from N (including N) to its nearest leaf.

Definition 2. A leaf node N is over-representing iff $N.depth < i.depth$, where i is another leaf node. We denote the set of over-representing leaf nodes as OR.

According to the RELT algorithm, non-leaf nodes will never over-represent. Every node in OR shares at least one of its boundaries with some other nodes at more than one level. For example, node "Chinese" in Figure 9 is in OR because it shares one of its boundaries with "Classical", "Baroque" and "Late" that are at different levels. Node "Irish" is not in OR because it shares each boundary with exactly one other node.

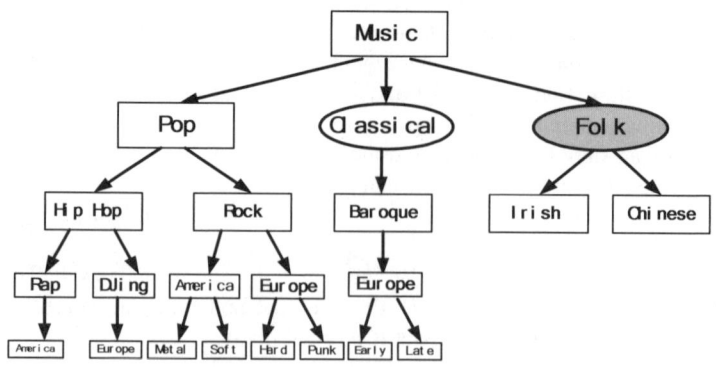

Fig. 9. An unbalanced tree

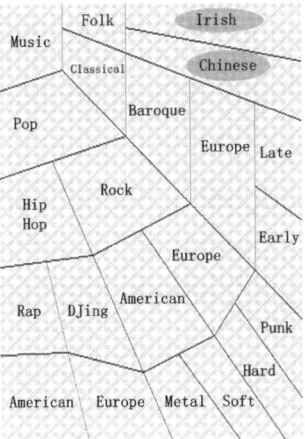

Fig. 10. RELT for the example in Figure 9

Definition 3. The number of *levels* over-represented by node N (in OR) is denoted L_OR_N. If both side boundaries of N are shared with N's sibling nodes, L_OK_N is the sum of the levels on these two boundaries.

The estimate function NOR for overall node over-representation can be constructed as follows:

$$NOR = \sum L\_OR_N \qquad (2)$$

Clearly, a small NOR is desirable.

5.2 Minimizing Overrepresentation

We consider the *state* of a tree as the one that uniquely determines the parent-child relationships and ordered (left to right) sibling relationships. For example, exchanging a node's left and right children will change the tree state. A change of relative positions of any two nodes will change the tree's state. A given state of a tree uniquely determines its layout by the RELT algorithm.

To obtain the best RELT layout, we need to investigate how to obtain the best state of a tree. A tree's nodes are initially divided into groups according to their levels as described in Section 3.2. At each level, the nodes are numbered from left to right. In the following, we will use $TN\_Depth(l_i)$ to represent a function that measures the maximum difference between the depths of any two leaf nodes at a given level i for a tree state.

Let

- N_{ij} be the node at level i numbered j.
- $N_{ij}.depth$ be the depth of node N_{ij}.
- l_{num} be the number of levels of T.
- l_{i_num} be the number of nodes at level i.

$$|N_{i(j+1)}.depth - N_{ij}.depth| = \begin{cases} |N_{i(j+1)}.depth - N_{ij}.depth| & (3) \\ 0 & (4) \end{cases}$$

$$TN\_Depth(l_i) = \sum_{j=1}^{l_{i\_num}-1} |N_{i(j+1)}.depth - N_{ij}.depth| \qquad (5)$$

(1) If one of N_{ij} and $N_{i(j+1)}$ is a leaf and the other is a non-lead node.
(2) If both N_{ij} and $N_{i(j+1)}$ are leaves or are non-leaf nodes.

The following function measures the maximum difference in depth between any two leaf nodes for a tree state:

$$TN\_Depth(State) = \sum_{i=1}^{l_{num}} TN\_Depth(l_i) \qquad (6)$$

Theorem 1. Let s_{old} and s_{new} be the old and new final states, and L_{old} and L_{new} be the corresponding old and new layouts of a tree, if $TN\_Depth(s_{new})$ is smaller than $TN\_Depth(s_{old})$, then $NOR(L_{new})$ is also smaller than $NOR(L_{old})$.

Proof. According to Equation <4>, the depth difference between two adjacent nodes contributes to the function only if one node is leaf and the other is a non-leaf node. Thus the leaf node is over-representing in the layout and the depth difference is exactly the number of levels over represented. For a given state of a tree and its layout, the values of TN_Depth and NOR are the same. The only difference is that TN_Depth explores estimates based on the tree's state and NOR estimates the outcome layout.

Now the strategy of minimizing overrepresentation becomes how to use function $TN\_Depth(State)$ to find the state that can result in the best layout of a tree. For a tree, the best state s_{BEST} satisfies Equation <5>:

$$TN\_Depth(s_{BEST}) = \min_{s_i \in S} TN\_Depth(s_i) \qquad (7)$$

where S is the set of all the tree's states.

Next finding the minimum number of $TN\_Depth(State)$ becomes the key. This can be done in two steps. The first step is expressed in Equation <3> and the second step in Equation 4. To simplify the problem, a basic assumption is initially constructed. Equation <4> will achieve its minimum point when all the results for each level in Equation <3> are the smallest. The process is then simplified to how to obtain the smallest number for each level in Equation <3> as described below.

Definition 4. A *unit* is the set of all the siblings who have the same parent.

A unit's position at its level is the number counted from left to right. Let U_{ij}, denotes the unit at level i and position j. For example, in Figure 10, "Hip Hop" and "Rock" make up a unit U_{31}, "Classical" is a unit U_{32} by itself, and "Chinese" and "Irish" make up the unit U_{33}.

Assume level i contains m units and each unit has exactly two nodes, a naive way to obtain the best layout is to compute overrepresentation by Equation <4> for the all the unit combinations and choose the state with the smallest depth. The complex is $O(2^m)$ because there are two possibilities in each unit. It is explicitly not a good choice especially when m is large.

Theorem 2. For a single unit, Equation <3> derives the minimum number if the nodes are sorted by their depths.

Proof. *If there is initially a sorted unit including m nodes and then two nodes D_i and D_j are exchanged.*

U_{SORT}:

$$(D_1, D_2 ... D_{i-1}, D_i, D_{i+1} ... D_{j-1}, D_j, D_{j+1} ... D_m)$$
$$1 \le h \le t \le m \quad D_h \le D_t$$

U_{UNSORT}:

$$(D_1, D_2 ... D_{i-1}, D_j, D_{i+1} ... D_{j-1}, D_i, D_{j+1} ... D_m)$$

$TN\_Depth(U_{SORT}) - TN\_Depth(U_{UNSORT})$
$= (|D_i - D_{i-1}| + |D_{i+1} - D_i| + |D_j - D_{j-1}| + |D_{j+1} - D_j|) -$
$\quad (|D_j - D_{i-1}| + |D_{i+1} - D_j| + |D_i - D_{j-1}| + |D_{j+1} - D_i|)$
$= (D_{i+1} - D_{i-1} + D_{j+1} - D_{j-1}) - (2D_j - D_{i-1} - D_{i+1} + D_{j-1} + D_{j+1} - 2D_i)$
$= 2(D_i + D_{i+1}) - 2(D_{j-1} + D_j) \ge 0$

$TN\_Depth(U_{SORT}) = TN\_Depth(U_{UNSORT})$ becomes true when $i + 1 = j$ or $D_i = D_j$.

Now we can derive an optimized RELT algorithm. Based on Theorem 2, function Sort (Node N) is inserted between Polygonal_Node(Node N, int L) and Partition_Area(Node N).

For every new non-leaf node N, function Sort (Node N) sorts its children by their depths. Assuming the j_{th} non-leaf node has n_i children, the complexity of function Sort (Node N) is $\sum_{j=1}^{n_{in}} n_j \lg n_j$.

Proof of complexity is given as follows:

$$\sum_{j=1}^{n_{in}} n_j \lg n_j \le \sum_{j=1}^{n_{in}} n_j \lg n_{in} \le \lg n_{in} \sum_{j=1}^{n_{in}} n_j < n \lg n_{in} < n \lg n$$

So, the complexity of optimized algorithm is $O(n^2) + O(n \log n)$.

6 Conclusions and Future Work

The small screen on mobile devices severely challenges the traditional hierarchy information visualization methods for desktop screens. Based on the analysis of the traditional tree visualization method, we observe the following:

- The two major approaches, i.e. connection and enclosure, for tree visualization are no longer suitable for mobile devices without proper adaptation.
- To achieve both a clear hierarchical structure and the maximum use of the display area, the current algorithms for desktop displays do not work well.

This paper has presented the RELT approach to tree visualization on small screens. The RELT algorithm traverses the tree with depth first, recursively partitions the remaining area, and allocates each partitioned area for a node. In this way, the entire display area is fully used while the hierarchical structure is clearly visualized.

The algorithm has been tested on several tree applications. Although our evaluation work is still underway, this approach has demonstrated to be feasible for visualizing tree-based hierarchical information on mobile devices. We plan to further optimize the RELT algorithm for more aesthetic layout while maintaining the clear tree structure. Adaptive layout features in response to browsing and labeling techniques will also be investigated.

References

1. di Battista, G., Eades, P., Tamassia, R., Tollis, I.G.: Graph Drawing: Algorithms for the Visualization of Graphs. Prentice Hall, Englewood Cliffs (1999)
2. Eades, P.: Drawing Free Trees, Bulletin of the Institute for Combinatorics and Its Applications, pp.10-36 (1992)
3. Herman, I., Melançon, G., Marshall, M.S.: Graph Visualization in Information Visualization: a Survey. IEEE Transactions on Visualization and Computer Graphics, 24–44 (2000)
4. Jeong, C.S, Pang, A.: Reconfigurable Disc Trees for Visualizing Large Hierarchical Information Space. In: InfoVis 1998. Proc. 1998 IEEE Symposium on Information Visualization, pp. 19–25. IEEE CS Press, Los Alamitos (1998)
5. Johnson, B., Shneiderman, B.: Tree-maps: A Space-filling Approach to the Visualization of Hierarchical Information Structures. In: Proc. 1991 IEEE Symposium on Visualization, pp. 284–291. IEEE, Piscataway, NJ (1991)
6. Lin, C.C., Yen, H.C.: On Balloon Drawings of Rooted Trees. In: Healy, P., Nikolov, N.S. (eds.) GD 2005. LNCS, vol. 3843, pp. 285–296. Springer, Heidelberg (2006)
7. Nguyen, Q.V., Huang, M.L.: A Space-Optimized Tree Visualization. In: InfoVis 2002. Proc. 2002 IEEE Symposium on Information Visualization, pp. 85–92. IEEE Computer Society Press, Los Alamitos (2002)
8. Reingold, E.M., Tilford, J.S.: Tidier Drawing of Trees. IEEE Trans. Software Eng. 7(2), 223–228 (1981)
9. Shneiderman, B.: Treemaps for Space-Constrained Visualization of Hierarchies (December 26, 1998) (last updated April 26, 2006), http://www.cs.umd.edu/hcil/treemap-history/

10. Yang, J., Ward, M.O., Rundensteiner, E.A.: InterRing: An Interactive Tool for Visually Navigating and Manipulating Hierarchical Structures. In: InfoVis 2002. Proc. 2002 IEEE Symposium on Information Visualization, pp. 77–84. IEEE Computer Society Press, Los Alamitos (2002)
11. Yee, K.P., Fisher, D., Dhamija, R., Hearst, M.: Animated Exploration of Dynamic Graphs with Radial Layout. In: InfoVis 2001. Proc. 2001 IEEE Symposium on Information Visualization, San Diego, CA, USA, pp. 43–50. IEEE Computer Society Press, Los Alamitos (2001)
12. Yoo, H.Y., Cheon, S.H.: Visualization by Information Type on Mobile Devices. In: Proc. 2006 Asia-Pacific Symposium on Information Visualization, vol. 60, pp. 143–146 (2006)

Auto-generation of Geographic Cognitive Maps for Browsing Personal Multimedia

Hyungeun Jo, Jung-hee Ryu, and Chang-young Lim

Graduate School of Culture Technology,
Korea Advanced Institute of Science and Technology (KAIST),
373-1 Guseong-dong, Yuseong-ku, Daejeon, Korea
acid@kaist.ac.kr, ryu@business.kaist.ac.kr, cylim@kaist.ac.kr

Abstract. A geographic map is an important browsing tool for multimedia data that can include personal photos, but geographically correct maps are not always easy to use for that purpose due to the frequent zooming and panning, as well as the existence of extraneous information. This paper proposes a new user-interface concept for geo-tagged personal multimedia browsing in the form of a cognitive map. In addition, design criteria are defined and an auto-generation method is presented for this map. The proposed method produces a map represented as a clustered graph with vertices and edges in real time. It is visually compact, preserves geographical relationships among locations and is designed for both PCs and mobile devices. An experiment was conducted to test the proposed method with real-life data sets.

1 Introduction

Time and location are two main factors that make up personal episodic memory [3]. For this reason, many commercial services such as Flickr and Google use geographic mapping as a main user interface for personal photos and text data browsing.

Most services in this area use vector-based graphic maps, while some also use satellite photo maps. A common point among them is their realistic scale and the inclusion of a large amount of information. As a map for ordinary use, either of these approaches may be satisfactory, but there are issues in terms of usability when this map is used to browse geo-tagged personal multimedia data. Essentially, two factors limit the use. First, the number of important places for one person is not high, but the density can vary greatly. This can include ten restaurants on one street which the user often visits, and three sightseeing places in different parts of a nation. In such a case, in a nation-wide view the restaurants cannot be depicted or made clickable due to a lack of space for separate vertices and labels, while other areas in the nation map except for a small number of sightseeing places are cognitively empty, and the density of information radically decreases, finally leading to frequent zooming and panning. The second factor arises while browsing. A user must see all unnecessary general information drawn on the map, thus resulting in reduced space for information regarding desired places, which hinders quick perception.

Given that no one perfect map exists that satisfies total functionality [12], [16], this draw-back can be understood as a mismatch between the purpose and the tool. For the same reason that the schematized map in Fig. 1.(b) is used for a subway map as opposed to that in Fig. 1.(a), another style of map can be more suited to personal multimedia browsing; here the concept of a cognitive map is suggested.

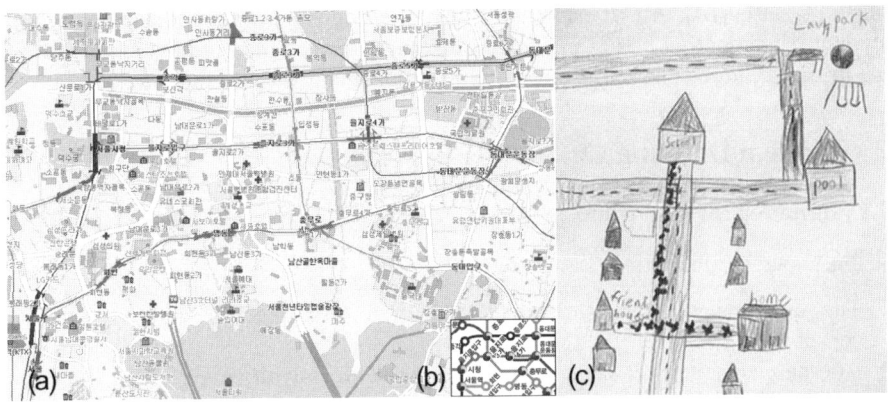

Fig. 1. (a) A map of the downtown in Seoul, at the most condensed zoom level as far as the map software can render each subway station separately. (b) A part of the Seoul subway map covering the same downtown area with the same font size. The display area is greatly reduced because the map is dedicated to depicting the subway system. (c) A hand-drawn map by a child.

After Lynch showed that a subject's understanding and bearing regarding a city can differ from the actual structure of the city, and that this can differ according to the individual [10], Downs and Stea defined a cognitive map as "a person's organized representation of part of the spatial environment"[7]. In a cognitive map, locations of strong interest are exaggerated while those of little interest or associated with bad feelings are eliminated. In addition, distances as well as the importance between/of locations can vary largely according to the knowledge, preference, social class, time/cost to move between, and other diverse factors related to different individuals [7], [8]. A cognitive map is not restricted in one fixed format, but can reflect some or all of the above characteristics. The result can be acquired using a drawing by a subject or via analysis from a verbal interview. Fig. 1(3) shows an example of a hand-drawn cognitive map of the local area of a subject.

In terms of the aforementioned multimedia browsing problem, the human cognitive map contains features that are needed for the geographical browsing referred to above but that are lacking in realistic maps. In addition, considering that a location at which a person has taken a photo is more likely to have a special meaning to the person compared to other locations, it is true that a geo-tagged photo itself can be a good source for the cognitive map of the person.

In this study, the goals are to suggest criteria for a good cognitive map for multimedia browsing that can be represented as a graph G and sets of vertices and edges of G, and to suggest algorithms that automatically generate and implement a map from a set of vertices with geo-tagged photos.

Thus far, the most similar work in this area is Agrawala and Stolte's LineDrive system [1], which draws "cognitively right" route maps for car navigation as an alternative for a realistic map. Although the work is very similar to the goals of this study, it cannot be adapted to an algorithm, as it represents a matter of path simplification which is contrary to that in the present study, in which vertices only, and no edges, are received as input. Other works related to geography-based graph drawing include a schematization of road networks [2], [6], and auto-mated drawing of metro maps [4], [9], [14]. All of these, however, use edges as input and require a large amount of computing power in order to determine the optimal orthogonal or octagonal layout solutions that preserve edge embedding and avoid line crossing.

2 Problem Defining

2.1 Input for the Algorithm

In this paper, the following is assumed:

1. The user generates multimedia data with a GPS-attached digital camera or cell-phone.
2. Alternatively, the user generates data using devices without GPS, and does geo-tagging with another application using a user interface such as a map on which the user clicks on a location and relates photos to the location.

In either case, the user would be encouraged to name each location, but this is not mandatory in all cases, regarding a random shot and other diverse situations in which the naming task is difficult.

Input: A set of location records that have the latitude, longitude, name, date, number of photos that belong there.

2.2 Criteria for Output Design

C1: Compactness. The most important factor, as explained in the introduction.

C2: Preserving the orthogonal order. If vertex u was left / up from vertex v in the input, it should also be preserved in the output. It is known that people seldom miss qualitative directions such as right/left. This is in contrast to quantitative attributes such as distance [10].

C3: Preserving the proximity. If vertex u was closest to vertex v in the input, it should also be preserved in the output. However, as shown in studies of layout adjustments, such as [15], this represents a trade-off relationship with C2. C2 receives priority in this paper.

C4: Multiple scales in one map, reflecting the density of each area. In the cognitive map, different scales can be used simultaneously, and human perception of distances can change according to the number of landmarks between two locations [5]. It is also possible to free the user from frequent zooming by drastically adapting this concept.

C5: Octagonal drawing of the edges. It is true that with this type of input, there is no firm reason to generate edges. However, given that when people see maps they perceive locations connected via a short route as closer to each other compared to locations with a longer route, even when the Euclidean distances are identical [13], arbitrary edges are also useful. If edges are drawn according to closeness in the original coordinates, this will be perceived as close though the coordinates may become distant in the resulting graph. Thus, in this case the edges act as complements to the lack of the proximity. Additionally, according to Gestalt theory, visual flow in the map will aid with the reading.

For this purpose, octagonal straight line drawing is preferable regarding human behavior that tends to understand and explain road direction as multiples of 45°[5]. But in this paper, the orthogonal drawing that *looks octagonal* is proposed instead of actual octagonal drawing. The orthogonal drawing that looks octagonal means that the drawing has 4 ports in maximum for each location but the connection edges containing an orthogonal bend are smoothen to the diagonal form in the visualization stage. This can be better approach because the edges in this method have no fixed meaning as aforementioned and the purely octagonal drawing can be visually more cluttering when the locations with the connection degree of 5~8 occur. Also by this approach, the proposed method maintains the same connection structure with normal orthogonal drawing for mobile devices which will be explained in the next criterion. This can be the essential feature further regarding the data exchange between PCs and mobile devices.

C5-1: Orthogonal drawing of the edges. This is a modified criterion for use in mobile devices with four directional arrow keys. The cognitive map proposed in this paper, with orthogonal edges and a grid layout, enables natural mapping between the movement and the arrow keys so that a user can easily predict the point to move by each arrow key and browse the whole map by simply moving from point to point.

3 Proposed Method

The proposed method in this paper is consisted of 6 consecutive stages: preprocessing, location layout, connecting locations, weighting locations, administrative district layout, and label layout. Among these, location layout described in section 3.2 and connecting locations in section 3.3 are the key stages of the proposed method.

3.1 Preprocessing

Preprocessing is performed firstly as listed below to modify the data set for the use from the next stage.

– Acquisition of city, state and country name: by reverse geo-coding service.
– Filling in empty location names: Street names also can be obtained from reverse geo-coding service. If there is no street, the location name field is just filled with the city name.
– Rounding off coordinates: Inputted latitudes and longitudes are rounded off to four decimal points to get rid of the GPS error effect and simplify the calculation.
– Merging the same locations: In this process, if locations $v_1, v_2, ... v_k$ have same name and identified as same by input, they are merged into one average location v_{ave}. The coordinate of v_{ave} is the average of the coordinates of $v_1, v_2, ... v_k$.

3.2 Location Layout

This stage makes condensed, grid aligned layout of locations which preserves geographical relation to the minimal degree. About this issue, [11] describes a simple algorithm, which firstly sorts n given vertices both in x-order and y-order, and if vertex v is i^{th}/j^{th} in x/y-order, allocates it at the grid location (i, j). The algorithm to be proposed in this stage is a modified version of that which obtains more condensed layout.

Supposed that there are real-world locations v_i, v_j of which x-coordinates $x_i < x_j$, and they are distant as d_x in x-coordinate and d_y in y-coordinate. If d_y is larger enough than d_x, it's difficult to recognize the difference in x-coordinate and more difficult if $|d|$ increases. By this heuristic, in this case the proposed algorithm moves v_j so that v_i and v_j are on same x-coordinate as shown in Fig. 2(a).

For convenience, this transformation will be called as *line-overlay* from the following description, because supposed that more than one vertex have the same x / y coordinate and there exists virtual vertical / horizontal line which crosses the vertices, *line-overlay* should be executed line-by-line.

For example, Fig. 2(b) shows the situation which has 2 virtual vertical lines l_1, l_2 consisted of n, m vertices on x-coordinate x_1, x_2, and θ_k^{high}, θ_k^{low} defined as the angles between vertex v_k on l_1 and the two closest vertices u_i, u_j each from upper and lower part of l_2. Then *line-overlay* only can be executed when:

$$\min\{\theta_k^{high}, \theta_k^{low} \mid k = 1, 2, \ldots n\} \geq \Theta \quad (1)$$

where Θ is a constant that can be set as any number under $\pi/2$. In the test, when $\pi/4 \leq \Theta \leq 3\pi/8$, fairly reasonable results were be able to obtained.

Note that the lines for the *line-overlay* are based on the positions during the transition, but θ_k^{high} and θ_k^{low} should be based on their original coordinates. Moreover, if there is no vertex on the upper part of l_2, θ_k^{high} is defined as $\pi/2$. The same is true for the lower part and θ_k^{low}.

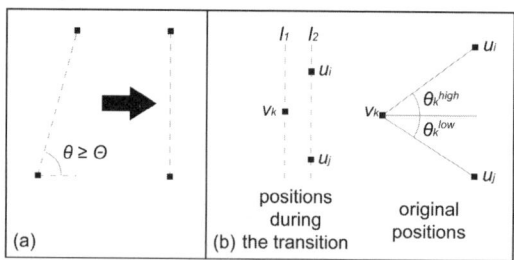

Fig. 2. (a) *Line-overlay*: the process to make the map more condensed. (b) Two angles to be compared to Θ. In most cases with real-world data sets, there exists either of θ_k^{high} or θ_k^{low} because the right side of l_1 is not yet condensed in x-coordinate; it seldom happens that two locations have the same longitude to four decimal points.

The problem that occurs when the *line-overlay* is executed vertex-by-vertex is shown in Fig. 2(c). In addition, Fig. 2(d) and Fig. 2(e) show cases in which the *line-overlay* fails.

Here is described the location layout algorithm that applied the *line-overlay*. In this algorithm, two Θ values, Θ_s and Θ_d are used to generate a more elaborate result. Θ_s is used in the regions where the distribution of the vertices is *sparse*, and Θ_d is used where they are *dense*. The density is measured by comparing the distance between the vertices with D_{den}, which can be computed from:

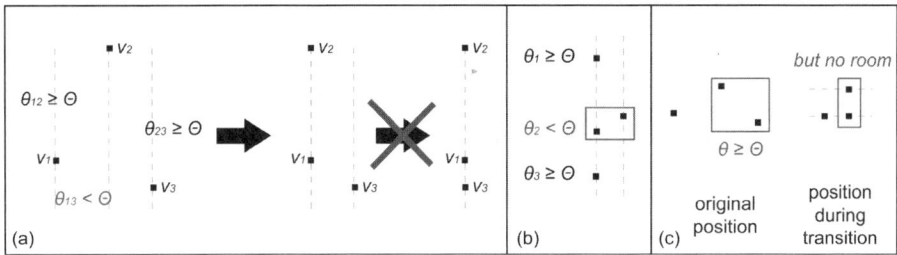

Fig. 3. (a) The problem that occurs when the *line-overlay* is executed vertex-by-vertex. (b) The case that fails because one angle is smaller than Θ. (c) The case in which θ is larger than Θ but fails because there is no room for the vertex to move in the next line.

$$D_{ave} = \sqrt{(x_{max} - x_{min})(y_{max} - y_{min})/n}$$
$$D_{den} = D_{ave}/K \qquad (2)$$

where x_{max} / y_{max} denote the maximum x / y values, and x_{min} / y_{min} denote the minimum x / y values in the original coordinates of $v_1, v_2, \ldots v_n$. If the entire map area is virtually divided into n square cells of even sizes, D_{ave} becomes the side length of one cell. K is a constant, and 40 were used in the implementation.

The proposed algorithm is described below, where x^{orig} denotes the x-coordinate of the original position of vertex v.

Algorithm CondenseLocations

```
Sort all the vertices in x-order and name v₁, v₂, ⋯vₙ;
for i ← 1 to n do
    Supposed that vᵢ, vᵢ₊₁⋯vₖ have the same x value;
    //in other word, aligned in the same vertical line
    Δ ← x_{k+1} - xᵢ; Δ^{orig} ← x^{orig}_{k+1} - x^{orig}_k; ∂ ← 0;
    for j ← i to k do
        if (x_{k+1},yⱼ) is not empty then {∂ ← 1; break;}
        if (Δ^{orig} ≤ D_{den}) then Θ ← Θ_d else Θ ← Θ_s;
        if (θⱼ^{low} or θⱼ^{high} < Θ) then {∂ ← 1; break;}
        for j ← k + 1 to n do xⱼ ← xⱼ - (Δ - ∂);
    if ∂ = 1 then i ← k + 1;
    //if ∂ = 0, it indicates now vᵢ, vᵢ, ⋯vₖ and v_{k+1} are
    //on the same line, thus, this loop must be executed
    //from i again.
(Repeat the same with respect to y-order)
```

3.3 Connecting Locations

The algorithm used in this stage firstly selects the smallest cluster – naturally, at the beginning all clusters have only one location and thus are the same size – and finds the closest location from the cluster and links them, making them into one relatively large cluster. By simply repeating this until there remains only one cluster, all of the locations are connected quickly with nearly no edge-crossing.

To find the closest location from v more rapidly, all of the locations are sorted in x-order and an examination is executed from vertices that are closer in the x-order. If the difference in the x-coordinate becomes greater than the closest distance up to that point, the searching stops.

The proposed algorithm for connecting lines are described below, where C_{min} indicates the minimal cluster, $|C_{min}|$ is the number of locations in C_{min}, $v.parent$ is the cluster that vertex v belongs to, P is a pair of two empty ports of vertices, and d is the Euclidean distance between the original positions of the two vertices.

Algorithm ConnectLocations

```
Sort all the vertices in x-order and name v₁, v₂, ... vₙ;
do
    for i ← 1 to |Cmin| do
        Supposed that vk indicates ith vertex in Cmin;
        for j ← k-1 to 1 do
            if |xkorig - xjorig| > dres then break;
            if vk.parent = vj.parent then continue;
            Ptemp ← Empty ports of vk, vj which makes the
                    connection of minimal bend;
            if Ptemp do not exist then continue;
            if Bend(Ptemp) ≥ 2 then continue;
            dtemp ← Euclidian distance(vkorig, vjorig);
            if (dres = NULL or dres > dtemp) then {dres =
                dtemp; Pres = Ptemp;}
        for j ← k+1 to n do the same process above;
    Connect(Pres);
while the number of cluster is larger than 1
```

3.4 Weighting Locations

In this stage, N_k, the number of photos taken in location v_k and F_k which is the number counting how many visits was made at v_k are used to compute the weight of v_k. By this weight the size of location vertex on rendered map can be decided. If there are n locations the weight function is obtained by the following equation:

$$w(v_k) = n\left(c_1 \frac{\log_{10} N_k}{\sum_{i \in V} \log_{10} N_i} + c_2 \frac{F_k}{\sum_{i \in V} F_i}\right), \quad c_1 + c_2 = 1 \quad (3)$$

The average of the weights $\{w(v_k) \mid k=1,2,\ldots n\}$ is 1. So given S, the normal size of the location, the size of location v_k can be calculated as following:

$$size(v_k) = S \times w(v_k) \quad (4)$$

3.5 Administrative Districts Layout

Cities, states, and countries are represented as rectangle boxes in this method, bounding the locations that belong to the districts. The drawing order among these boxes is important be-cause a box drawn later hides a box drawn earlier, and this can cause misconceptions about the coverage of the district (See Fig. 4(a)).

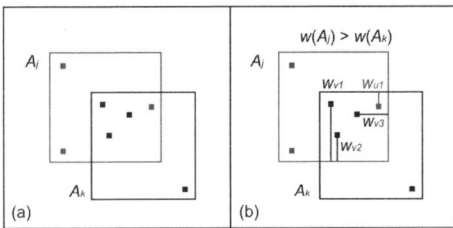

Fig. 4. (a) It is clear that the overlapped area belongs to A_k. Thus, in this case A_k should be drawn later than A_j. (b) Illustrates how w_v is measured.

To solve this problem, the weight of each box is calculated when the two boxes A_j and A_k at the same level overlap:

$$w(A_k) = \sum_{v \in A_k} w_v \quad (5)$$

where w_v denotes how much the vertex v intruded into the opposite box. This can be measured by selecting the smaller between the distances from v to the two sides which were divided by A_k. If v did not intruded into opposite box, then $w_v=0$. If $w(A_k)$ is larger than $w(A_j)$, A_k is drawn later than A_j.

3.6 Label Layout

There are four types of labels in this paper – location names, city names, state names, and country names. Location names are placed inside of the grid cells, and if the right side of the location is empty, the label expands to the right. City and state names have two candidates for label position: above and left of the upper-left corner of the bounding box. They are placed at the location where less overlapping occurs. Country names are placed above the upper-left corner.

4 Implementation and Experimental Result

This method was implemented in Visual C++ using a GDI-graphic API. The test was executed on a single processor 3.0GHz Pentium 4 machine with 1GB of RAM.

All the data sets used in this test are real-life sets consisting of photos that were collected from Flickr, which provides coordinate data of each geo-tagged photo. Each set consisted of one person's data. As a reverse geo-coding service, the one serviced by ESRI[1] was used first. This provides results at the street level. For the coordinates that cannot be transformed in ESRI because no streets or town were given, area names displayed in Flickr were used as a supplement.

One of the results is shown in Fig. 5(a). It consists of 23 locations, the map size (in grid) is 11x10, the map size in pixels is 676x450 pixels, and the rendering time excluding pre-processing was 0.078 seconds. The rendering time is sufficiently rapid, so that this method can be expected to show reasonable performance with even a simple mobile device. Moreover, the map size is feasible for browsing on both a PC and a mobile device. In a recent cellular phone equipped with a 640x480 display, it can be shown with nearly no scrolling. Fig. 6 shows an orthogonal version that can be easily browsed using only arrow keys.

In the layout, the locations aligned provide a geographically precise sense, and drastically different scales applied to different areas do no harm to a quick perception of geographic relationships. Note that in the Flickr map, zooming should be done from the 4th level to the 14th level in order to identify all locations separately, as seen in Fig. 5(b).

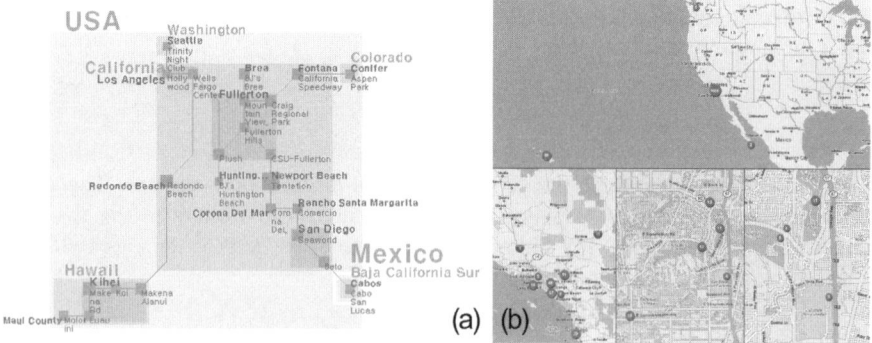

Fig. 5. (a) Map of user A drawn by the proposed method (b) The same data set on a Flickr map

---

[1] www.esri.com

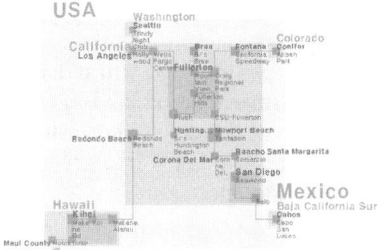

Fig. 6. Map of user A drawn by the proposed method as a normal orthogonal drawing

Fig. 7(a) is one of the most difficult cases. It consisted of 194 locations with a map size (in grid) of 84x60. The map size in pixels is 3186x2170 pixels and the rendering time excluding pre-processing was 0.312 sec. The rendering time is rapid but the map size in screen requires several instances of panning to see the entire map. However, when compared to the realistic map in Fig. 7(c), in which zooming should be done from the 3rd level to the 15th level, it is still compact. The proximity problem occurs in the upper side of the map due to the preserving of the orthogonal order around the dense area.

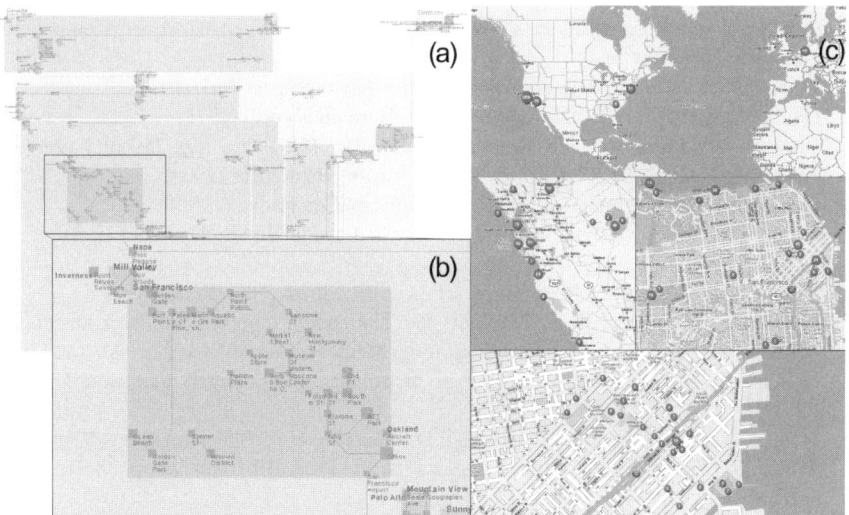

Fig. 7. (a) Map of user B drawn by the proposed method (b) Magnified view around San Francisco in the same map (c) The same data set in Flickr map

5 Conclusion and Future Work

In this paper, a new method that automatically generates a cognitive map to browse geo-tagged personal multimedia is proposed. Examples based on a real-life data set showed that the proposed method drastically reduces the overall map size and zoom level, while it pre-serves well-organized visual information as well as the geographical relationships among the locations.

In a further study, a modified algorithm that will reduce the distortion of the proximity around dense areas is under investigation, as is a field test with actual users on PCs and mo-bile devices. In addition, a more sophisticated system that can reflect actual road data or day-long GPS data that takes into account more than merely photo-taking moments is worthy of exploration.

References

1. Agrawala, M., Stolte, C.: Rendering Effective Route Maps: Improving Usability Through Generalization. In: Proc. SIGGRAPH, pp. 241–249 (2001)
2. Avelar, S., Muller, M.: Generating topologically correct schematic maps. In: Proc. International Symposium on Spatial Data Handling, pp. 4a.28-4a.35 (2000)
3. Baddeley, A.: Human Memory, Theory and Practice. LEA Publishers (1990)
4. Barbowsky, T., Latecki, L.J., Richter, K.: Schematizing maps: Simplification of geographic shape by discrete curve evolution. In: Habel, C., Brauer, W., Freksa, C., Wender, K.F. (eds.) Spatial Cognition II. LNCS (LNAI), vol. 1849, pp. 41–48. Springer, Heidelberg (2000)
5. Byrne, R.W.: Memory for urban geography. Quarterly Journal of Experimental Psychology 31(1), 147–154 (1979)
6. Cabello, S., de Berg, M., van Kreveld, M.: Schematization of networks. Computational Geometry 30(3) (2005)
7. Downs, R.M., Stea, D.: Maps in Minds. Harper & Row, New York (1977)
8. Gould, P., White, R.: Mental Maps. Penguin, Harmondsworth (1974)
9. Hong, S.-H., Merric, D., do Nascimento, H.A.D.: The metro map layout problem. In: Pach, J. (ed.) GD 2004. LNCS, vol. 3383, Springer, Heidelberg (2005)
10. Lynch, K. (ed.): The image of the City. MIT Press, Cambridge (1960)
11. Lyons, K.: Cluster busting in anchored graph drawing. In: Proc. CAS Conference, pp. 7–16 (1992)
12. MacEachren, A.M.: How Maps Work. The Guilford Press, New York (1995)
13. McNamara, T.P., Ratcliff, R., McKoon, G.: The mental representation of knowledge acquired from maps. J. Exp. Psychol., Learning, Memory, Cognition 10(4), 723–732 (1984)
14. Nollenburg, M.: Automated drawing of metro maps. PhD paper, Karlsruhe University (2005)
15. Storey, M.A.D., Wong, K., Fracchia, F.D., Mueller, H.A.: On Integrating Visualization Techniques for Effective Software Exploration. In: Proc. IEEE InfoVis, p. 38. IEEE Computer Society Press, Los Alamitos (1997)
16. Tufte, E.: Envisioning Information. Graphics Press, Conneticut (1990)

Automatic Image Annotation for Semantic Image Retrieval

Wenbin Shao, Golshah Naghdy, and Son Lam Phung

SECTE, University of Wollongong,
Wollongong NSW, 2522 Australia
{ws909,golshah,phung}@uow.edu.au

Abstract. This paper addresses the challenge of automatic annotation of images for semantic image retrieval. In this research, we aim to identify visual features that are suitable for semantic annotation tasks. We propose an image classification system that combines MPEG-7 visual descriptors and support vector machines. The system is applied to annotate cityscape and landscape images. For this task, our analysis shows that the colour structure and edge histogram descriptors perform best, compared to a wide range of MPEG-7 visual descriptors. On a dataset of 7200 landscape and cityscape images representing real-life varied quality and resolution, the MPEG-7 colour structure descriptor and edge histogram descriptor achieve a classification rate of 82.8% and 84.6%, respectively. By combining these two features, we are able to achieve a classification rate of 89.7%. Our results demonstrate that combining salient features can significantly improve classification of images.

Keywords: image annotation, MPEG-7 visual descriptors, support vector machines, pattern classification.

1 Introduction

Traditional image retrieval techniques are mainly based on manual text annotation [1]. Given the rapid increase in the number of digital images, manual image annotation is extremely time-consuming. Furthermore, it is annotator dependent. Content-based image retrieval (CBIR) promises to address some of the shortcomings of manual annotation [1,2,3]. Many existing CBIR systems rely on queries that are based on low-level features. One of the main challenges in CBIR is to bridge the *semantic gap* between low-level features and high-level contents [2,4,1]. For example, consider an image of a mountain: in low-level terms, it is a composition of colours, lines of different length, and different shapes; in high-level terms, it is a mountain. If users want to search for mountain images, they need to specify the low-level features such as green texture or they could enter the keyword *mountain*. Automatic annotation at semantic level employs keywords to represent images. It is a powerful approach, because people are better at describing an image with keywords than with low-level features.

In this paper, we aim to identify visual features that are suitable for automatic semantic annotation tasks. We propose an image classification system that combines MPEG-7 visual descriptors and support vector machines (SVMs), and apply the system to annotate cityscape and landscape images. Note that the system can be also extended to process other image categories. This paper is organised as follows. In Section 2, we review existing techniques for classifying images, especially cityscape versus landscape images. In Section 3, we describe the proposed system that combines MPEG-7 visual descriptors and support vector machines. In Section 4, we present and analyse the experiment results. Finally in Section 5, we give the concluding remarks.

2 Background

Many attempts at bridging the *semantic gap* have been made [2,4,1]. Yiu [5] uses colour histogram and dominant texture orientation to classify indoor and outdoor scenes. When the k-nearest neighbour classifier is used, Yiu finds that colour features outperform texture features. However, when support vector machine classifier is used, the texture features perform better than the colour features. The SVM classifier combining colour and texture features has a classification rate of 92% on a test dataset of 100 images.

Szummer and Rosalind [6] study indoor and outdoor image classification with four features: colour histogram in the Ohta colour space, texture feature based on a multi-resolution, simultaneous autoregressive model, and frequency features based on the two-dimensional DFT and DCT. They report a classification rate of 90.3% on a dataset of 1343 Kodak consumer images.

Vailaya et al. [7] propose a system on classification of cityscape versus landscape images that is based on the k-nearest neighbour classifier. They evaluate five features, namely colour histograms, colour coherence vectors, moments of image DCT coefficients, edge direction histograms, and edge direction coherence vectors. For features based on edge direction histograms and edge direction coherence vectors, Vailaya et al. report a classification rate of more than 93%, on a dataset of 2716 images (mainly Corel stock photos).

Vailaya et al. [8] later use a hierarchical architecture to first separate indoor from outdoor images and then divide outdoor images into subcategories such as *sunset*, *forest* and *mountain*. Their approach is based on Bayesian classifiers. For classifying indoor versus outdoor images, Vailaya et al. report an overall classification rate of 90.5% and find that features based on spatial colour distribution are better than colour and texture features. For classifying sunset versus forest and mountain images, they observe that colour histogram is better than the edge direction features. Vailaya et al. investigate the incremental learning technique and show that, as the training set increases, this technique improves classification accuracy. They also explore the affects of feature subset selection.

Lienhart and Hartmann [9] propose an image classification approach based on the AdaBoost learning algorithm. To classify graphical versus photo-realistic images, they use low-level features based on colour and pixel proportion. To classify real photos versus computer-generated images, they use texture features. In the classification of comics versus presentation slides, they use heuristic features based on text width and text position with respect to the entire image. They use about 5300 images for training and 2250 images for testing, and report classification rates between 93.7% and 98.2% on the test set.

Hu et al. [10] propose a Bayesian method with relevance feedback for indoor and outdoor image classification. The features they use include colour histograms, colour coherence vectors, edge direction histograms, edge direction coherence vectors and texture orientations. Hu et al. show that, on the same dataset, their method achieves a higher classification rate, compared to the method proposed by Szummer and Rosalind [6].

3 Methodology

We propose an image annotation system that combines MPEG-7 visual descriptors and support vector machines. The block diagram of the system is shown in Fig. 1. First, the input image is segmented into regions; this stage is only required for visual descriptors that are based on object shapes. Next, MPEG-7 visual descriptors are extracted from the image. Finally, support vector machines are used to classify the visual descriptors into different image categories such as *landscape*, *cityscape*, *vehicle* or *portrait*.

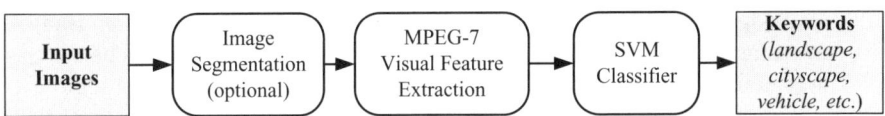

Fig. 1. Proposed image annotation system

3.1 Image Segmentation

While most low-level features are extracted from the entire image, some features such as region-based shape and contour-based shape require the image to be segmented into regions or objects. Therefore, our system includes an optional image segmentation stage that relies on multi-resolution watershed segmentation and image morphology [11, 12].

In watershed segmentation, a grey image is considered as a topographic surface. Suppose the surface is flooded by water from its minima. When water from different sources is about to merge, dams are built to prevent the water from merging. The rising water finally partitions the image into two different sets: the catchment basins and the watershed ridge lines. We realise that the watershed method may cause over segmentation if it is applied directly to the gradient image. This problem can be alleviated if the topographic surface is flooded from a set of predefined markers. We propose extra processing steps that are based on two morphological operations: *opening by reconstruction* and *closing by reconstruction*. Image opening operation removes small regions (caused by noise) while preserving the shape of foreground objects. Image closing fills in holes in the objects while keeping their shapes.

3.2 MPEG-7 Visual Descriptors

Multimedia Content Description Interface (MPEG-7) is a standard for multimedia content description for a wide range of applications involving image, video and audio search. MPEG-7 defines three major categories of visual descriptors for still images: colour, texture and shape descriptors [13,14,15,16].

- **Colour descriptors.** All colour descriptors can be extracted from an image or an image region. There are four main descriptors: dominant colour, scalable colour, colour structure and colour layout. *Dominant colour* is a compacted description that consists of the representative colours in an image or an image region. *Scalable colour* is a colour histogram in the HSV colour space. It is encoded by a Haar transform. *Colour structure* is a colour structure histogram that consists of the information of colour content and the corresponding structure. *Colour layout* represents images by spatial colour structure. It is resolution-invariant and extracted in YCbCr colour space.
- **Texture descriptors.** All texture descriptors can be extracted from an image or an image region in MPEG-7 monochrome colour space. Three common texture descriptors are edge histogram, homogeneous texture and texture browsing. *Edge histogram* describes the local spatial distribution of edges in an image. It is extracted from 16 subimages of an image. There are four directional edges and one non-directional edge defined in MPEG-7. *Homogeneous texture* employs the mean energy and the energy deviation to characterise the region texture. *Texture browsing* specifies a texture in terms of regularity, coarseness and directionality.
- **Shape descriptors.** For each two dimensional region, there are two types of shape descriptors: a *region-based shape* descriptor represents the shape of a region whereas a *contour-based shape* descriptor reveals the properties of the object contour.

3.3 Support Vector Machines

In machine learning and pattern classification, support vector machines are a supervised learning approach that has been demonstrated to perform well in numerous practical applications [17, 18, 19]. In two-class classification problem, the SVM's decision boundary is constructed from the training data by finding a separating hyperplane that maximizes the margins between the two classes; this is essentially a quadratic optimization problem. This learning strategy is shown to increase the generalization capability of the classifier. We can apply SVMs to complex nonlinear problems by using kernel methods and projecting the data onto a high-dimensional space. Apart from its good generalisation capability, the SVM approach works well when the number of training samples is small. The main challenge in applying SVMs is to find the appropriate features, the kernel function, and the training parameters. In this work, we will use SVM as the basic tool for classification of image features. Our main aim is to identify salient image features for the task of semantic image annotation.

4 Results and Analysis

This section describes an application of the proposed image annotation system in classifying landscape and cityscape images. First, we describe the data collection and experimental procedure. Next, we present the classification results of different MPEG-7 visual descriptors and compare our system with an existing image classification system. Finally, we discuss techniques to improve the classification performance of the system.

4.1 Data Preparation and Experimental Steps

Our work is based on a large-scale dataset of images. To provide classification results on real-world images, we have collected about 14,000 images from online repositories of digital photos and manually annotated these images into four categories: landscape, cityscape, portrait and vehicle. For the task of classifying landscape versus cityscape images, we use a dataset of 3600 landscape images and 3600 cityscape images. These images vary widely in size, quality, and contents; some images even have blurred or monochrome (red=green=blue) appearance. Examples of the images are shown in Fig. 2. We use 4200 images for training and 3000 images for testing; the number of landscape images and cityscape images are equal.

In the experiments, the MPEG-7 reference software called *eXperimentation Model* (XM) [20] was used to extract most MPEG-7 visual descriptors. We used MATLAB to extract the dominant colour descriptor, because there was a bug in the XM software. To train and evaluate SVM classifiers, we chose an SVM library called *LIBSVM* [21], developed by Chang et al. at National Taiwan University. After trying different kernel functions, we selected the radial basis function kernel. We experimented with different SVM parameters: training cost c and kernel radius γ.

Because of the large number of MPEG-7 visual descriptors, we conducted a preliminary experiment on a small set of about 900 images to short-list the descriptors. We excluded from further analysis any descriptor that is computationally intensive or does not perform well. For example, the texture browsing descriptor for each image of size 600×800 pixels requires over 60 seconds to compute (using XM software on a P4 3GHz computer). This descriptor is an extension of the homogeneous texture descriptor, which can be computed more efficiently.

Furthermore, both region-based and contour-based shape descriptors performed poorly: after training, the system based on these descriptors misclassified most test images. There are two possible explanations for this result. First, robust image segmentation is still a challenging task; in the shape-based approach, failure to segment objects will impede the shape descriptors from representing the objects accurately. Second, landscape and cityscape images may contain many common shapes.

(a) landscape

(b) cityscape

Fig. 2. Example images in the dataset of 7200 images

4.2 Comparison of MPEG-7 Visual Descriptors

In this experiment, we aim to identify the visual descriptors that perform well in the task of landscape and cityscape image classification. We constructed SVM classifiers that use each of the following MPEG-7 visual descriptors:

- dominant colour,
- colour layout
- scalable colour
- colour structure,
- homogeneous texture and
- edge histogram.

On the training set of 4200 images, the classification rates for the above descriptors are 84.5%, 80.9%, 89.3%, 99.0%, 85.5% and 93.1%, respectively. Note that the classification rates on the training set can change depending on two parameters: cost c and kernel radius γ. The training performance is reported for the parameter combination that gives the best performance on the test set.

The classification rates on the training set and test set for the scalable colour, colour structure and edge histogram descriptors are shown in Tables 1, 2, and 3, respectively. We have experimented with c values from 2^{-5} to 2^{15} and γ values from 2^{-15} to 8.

Classification performance, on the test set of 3000 images, of the six MPEG-7 visual descriptors is shown in Fig. 3. The dominant colour descriptor is the only one with a classification rate below 70%. There are three MPEG-7 visual

Table 1. Classification rates of the scalable colour descriptor on (*training set*, *test set*) for different SVM parameters

γ \ c	2^{-7}	2^{-5}	2^{-3}	0.25	0.5	4
0.5	76.2/73.4	81.0/78.5	89.4/80.4	94.3/78.4	99.1/74.2	99.7/49.1
4	80.8/78.2	89.3/80.5	99.3/78.4	99.5/78.8	99.7/78.6	99.7/49.1
16	84.9/80.1	96.5/78.0	99.6/77.9	99.7/79.2	99.7/78.7	99.8/49.2
64	89.8/80.0	99.4/74.7	99.7/78.7	99.7/79.3	99.8/79.0	99.8/49.4
256	96.3/76.9	99.6/74.8	99.7/78.5	99.8/79.7	99.8/79.3	99.9/49.4
1024	99.4/73.4	99.7/75.9	99.8/77.8	99.7/79.3	99.8/79.0	99.9/49.4

Table 2. Classification rates of the colour structure descriptor (*training set*, *test set*) for different SVM parameters

γ \ c	2^{-7}	2^{-5}	2^{-3}	0.25	0.5	4
0.5	76.3/72.8	79.7/77.7	84.6/80.0	86.9/80.8	90.7/82.6	100.0/55.6
4	79.3/76.3	84.0/79.4	90.9/81.7	95.6/82.3	99.0/82.8	100.0/72.6
16	82.0/78.1	86.9/80.7	95.9/81.7	99.1/81.7	99.9/82.3	100.0/72.6
64	84.3/79.7	90.9/80.9	99.0/80.7	99.9/81.4	100.0/82.5	100.0/72.6
256	87.1/80.4	95.8/81.0	99.9/80.2	100.0/81.4	100.0/82.2	100.0/72.6
1024	90.9/80.2	98.9/80.2	100.0/80.3	100.0/81.2	100.0/82.3	100.0/72.6

Table 3. Classification rates of the edge histogram descriptor on (*training set*, *test set*) for different SVM parameters

γ \ c	2^{-7}	2^{-5}	2^{-3}	0.25	0.5	4
0.5	84.3/81.6	86.7/83.2	90.1/84.4	92.8/84.5	95.3/83.5	100.0/50.4
4	86.4/83.2	89.5/84.3	97.7/84.5	99.9/84.4	100.0/84.9	100.0/50.4
16	87.8/83.7	93.1/84.6	100.0/83.8	100.0/84.2	100.0/84.9	100.0/50.4
64	89.7/84.4	97.4/83.8	100.0/83.5	100.0/84.2	100.0/84.9	100.0/50.4
256	93.3/84.2	99.9/81.9	100.0/83.5	100.0/84.2	100.0/84.9	100.0/50.4
1024	96.9/83.2	100.0/81.8	100.0/83.5	100.0/84.2	100.0/84.9	100.0/50.4

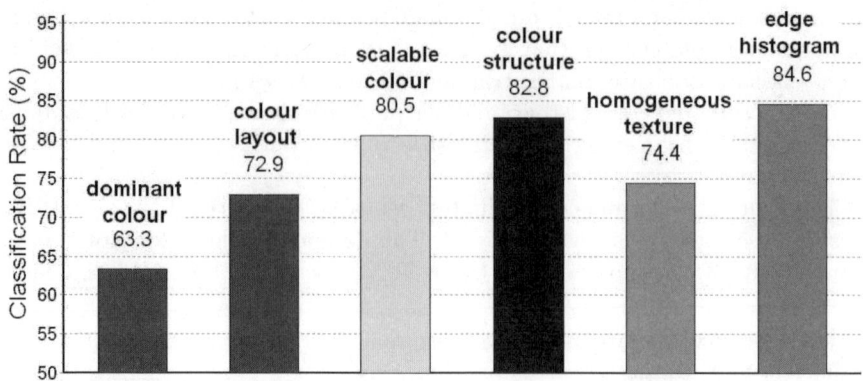

Fig. 3. Comparison of MPEG-7 visual descriptors in landscape versus cityscape image classification task, on a test set of 3000 images

descriptors that have a classification rate above 80%: scalable colour, colour structure and edge histogram. The edge histogram descriptor has the highest classification rate of 84.6%; this result shows that edge histogram is a salient visual feature in differentiating landscape and cityscape images.

4.3 Comparison with Other Techniques

For comparison purposes, we study the performance, on the same dataset described in Section 4.1, of an image classification approach proposed in [7]. Vailaya et al. [7] use a weighted k-nearest neighbour classifier to differentiate cityscape versus landscape images and study several feature vectors. They find that the edge direction histogram (EDH) feature vector performs better compared to the others. An EDH feature vector has 73 elements. The first 72 elements are the normalized histogram of edge directions (72-bin). The last element is the normalized count of non-edge pixels.

We experimented with the k-NN classifier where k varies from 1 to 15; note that in [7], k goes from 1 to 9. Table 4 shows the classification results of Vailaya et al.'s method on our image dataset: the best classification rate achieved is 82.8% when k is equal to 7. This classification rate is similar to that of our colour structure descriptor (82.8%), and is lower compared to the CR of our edge histogram descriptor (84.6%).

Table 4. Classification rates of the k-NN classifier and the EDH feature

Number of nearest neighbours k	1	3	5	7	9	11	13	15
Classification Rate (CR) (%)	80.4	81.9	82.7	82.8	82.6	82.7	82.7	82.6

4.4 Improving the System

So far, we have identified a number of MPEG-7 visual descriptors that are suitable for the task of classifying landscape and cityscape images. The system performance can be improved by combining these salient features and there are different approaches in doing so.

- **Using a single SVM:** We assemble all salient descriptors into a single feature vector and use only one SVM to classify the feature. When we combine the edge histogram and the colour structure in this way, the classification rate on the test set is 88.5%.
- **Using multiple SVMs:** We build individual SVMs that use separate visual descriptors and a final SVM to process the ensemble of confidence scores produced by the individual SVMs. The system implementing this approach has a CR of 89.7% on the test set. However, when we combine the best three salient descriptors, the classification rate increases to only 88.6%.

Our results show that combining salient features has a clear advantage to classification accuracy and is a promising research direction.

5 Conclusion

In this paper, we have presented an image classification system that combines MPEG-7 visual descriptors and support vector machines. We analysed a wide range of MPEG-7 visual descriptors including colour, edge, texture and shape. On a large dataset of 7200 landscape and cityscape photos, our system achieves a classification rate of 89.7%. We find that for landscape versus cityscape classification, the edge histogram and colour structure descriptors outperform other MPEG-7 visual descriptors and classification rate is improved by combining these two features.

References

1. Long, F., Zhang, H., Feng, D.: Fundamentals of content-based image retrieval. In: Feng, D., Siu, W.C., Zhang, H.J. (eds.) Multimedia Information Retrieval and Management - Technological Fundamentals and Applications, Springer, Heidelberg (2002)
2. Smeulders, A.W.M., Worring, M., Santini, S., Gupta, A., Jain, R.: Content-based image retrieval at the end of the early years. IEEE Transactions on Pattern Analysis and Machine Intelligence 22(12), 1349–1380 (2000)
3. Datta, R., Li, J., Wang, J.Z.: Content-based image retrieval: approaches and trends of the new age. In: The 7th ACM SIGMM International Workshop on Multimedia Information Retrieval, pp. 253–262. ACM Press, New York (2005)
4. Eakins, J.P.: Retrieval of still images by content. In: Agosti, M., Crestani, F., Pasi, G. (eds.) ESSIR 2000. LNCS, vol. 1980, pp. 111–138. Springer, Heidelberg (2001)
5. Yiu, E.C.: Image classification using color cues and texture orientation. PhD thesis, Massachusetts Institute of Technology (1996)

6. Szummer, M., Picard, R.W.: Indoor-outdoor image classification. In: IEEE International Workshop on Content-Based Access of Image and Video Database, pp. 42–51. IEEE Computer Society Press, Los Alamitos (1998)
7. Vailaya, A., Jain, A., Zhang, H.J.: On image classification: city vs. landscape. In: IEEE Workshop on Content-Based Access of Image and Video Libraries, pp. 3–8. IEEE Computer Society Press, Los Alamitos (1998)
8. Vailaya, A., Figueiredo, M., Jain, A., Zhang, H.J.: Content-based hierarchical classification of vacation images. IEEE International Conference on Multimedia Computing and Systems 1, 518–523 (1999)
9. Lienhart, R., Hartmann, A.: Classifying images on the web automatically. Journal of Electronic Imaging 11(4), 445–454 (2002)
10. Hu, G.H., Bu, J.J., Chen, C.: A novel bayesian framework for indoor-outdoor image classification. International Conference on Machine Learning and Cybernetics 5, 3028–3032 (2003)
11. Gonzalez, R.C., Woods, R.E.: Digital image processing. Prentice Hall, New York (2002)
12. Gonzalez, R.C., Woods, R.E., Eddins, S.L.: Digital image processing using MATLAB. Prentice Hall, New York (2004)
13. Manjunath, B.S., Salembier, P., Sikora, T. (eds.): Introduction to MPEG-7: Multimedia content description interface. Wiley, Milton (2002)
14. MPEG-7 Video Group. Text of ISO/IEC 15938-3/FDIS information technology - Multimedia Content Description Interface - Part 3 Visual. In: ISO/IEC JTC1/SC29/WG11/N4358, Sydney (2001)
15. Nack, F., Lindsay, A.T.: Everything you wanted to know about MPEG-7, part 1. IEEE Multimedia 6(3), 65–77 (1999)
16. Nack, F., Lindsay, A.T.: Everything you wanted to know about MPEG-7part 2. IEEE Multimedia 6(4), 64–73 (1999)
17. Burges, C.J.C.: A tutorial on support vector machines for pattern recognition. Data Mining and Knowledge Discovery 2(2), 121–167 (1998)
18. Cristianini, N., Shawe-Taylor, J.: An introduction to support vector machines and other kernel-based learning methods. Cambridge University Press, Cambridge (2001)
19. Abe, S.: Support vector machines for pattern classification. Springer, New York (2005)
20. Institute for Integrated Systems. MPEG-7 eXperimentation Model (XM), Software (2005), available at http://www.lis.e-technik.tu-muenchen.de/research/bv/topics/mmdb/mpeg7.html
21. Chang, C.C., Lin, C.J.: LIBSVM: a library for support vector machines, Software (2007), available at http://www.csie.ntu.edu.tw/~cjlin/libsvm

Collaterally Cued Labelling Framework Underpinning Semantic-Level Visual Content Descriptor

Meng Zhu and Atta Badii

IMSS Research Centre, Department of Computer Science,
School of Systems Engineering, University of Reading, UK
{meng.zhu, atta.badii}@reading.ac.uk

Abstract. In this paper, we introduce a novel high-level visual content descriptor devised for performing semantic-based image classification and retrieval. The work can be treated as an attempt for bridging the so called "semantic gap". The proposed image feature vector model is fundamentally underpinned by an automatic image labelling framework, called Collaterally Cued Labelling (CCL), which incorporates the collateral knowledge extracted from the collateral texts accompanying the images with the state-of-the-art low-level visual feature extraction techniques for automatically assigning textual keywords to image regions. A subset of the Corel image collection was used for evaluating the proposed method. The experimental results indicate that our semantic-level visual content descriptors outperform both conventional visual and textual image feature models.

Keywords: Automatic image annotation; fusion of visual and non-visual features; collateral knowledge; semantic-level visual content descriptor; multimodal data modelling; image indexing and retrieval.

1 Introduction

Text and image are two distinct types of information from different modalities, as they represent the world in quite different ways. However, there are still some unbreakable, implicit connections between them somehow. Text can be used to describe the content of image, and image may convey some semantics within the context of text. The reason that we call them implicit connections is due to a gap between text and image information, which is usually referred to as the "semantic gap". "The semantic gap is the lack of coincidence between the information that one can extract from the visual data and the interpretation that the same data may have for a user in a given situation." [21]. Different people may describe the same image in various ways using different words for different purposes. The textual description is almost always contextual, whereas the images may live by themselves. Moreover, the textual descriptions of images often remain at a higher level to depict the properties that are very difficult to infer purely by vision, like the name of the person, disregarding the low-level visual features, such as the colour of the person's clothing.

As a result of the rapid advances in digital imaging technology, massive amount of annotated digital images are widespread all over the world. Typical examples include the online galleries (e.g. the National Gallery, London[1]), online news photo archives, as well as the commercial digital image libraries, such as Corel Stock Photo Library which contains 60,000 annotated images. The most straightforward use of such collections is for the users to browse and search the images that match their needs. Therefore, there is a growing interest in user's search requirements which motivates research to increase the efficacy of indexing and retrieval of images. The annotated image collections can either be indexed and searched by text (i.e. annotation) [12], or by image content (i.e. visual features). The annotations, typically but not always, refer to the content of the image, and are usually neither fully specific nor comprehensive. For example, the annotations in the digital image gallery often include some content-independent information [4], such as artist's name, date, ownership, etc. For most annotated image collections, such as Corel Stock Photo Library and news photo archives, the annotations focus on describing specific content of the images, but cannot cover all of the information conveyed by the images. Content-Based Image Retrieval (CBIR) concentrates on exploiting computer vision and image processing methods to facilitate image indexing and retrieval. It was proposed mainly due to the inadequacy of text in representing and modelling the perceptual information conveyed by the rich content of images. A lot of research has been done by many researchers in this area e.g. [20], [21], [17] and [15]. Despite the significant achievements these researchers have made, it is widely acknowledged that strong retrieval performance can only be achieved on low-level visual features matching such as colour, texture, edge, etc. The research progress in CBIR is limited when matching the images content towards object-level semantics. This is to be expected, because it is always very difficult for computer vision approaches alone to identify visually similar but conceptually different objects.

In this paper, we propose a high-level image feature vector model which combines multi-modal cues i.e. both visual and textual cues for semantic-based image retrieval in order to bridge the so called "semantic gap". The construction of this novel multi-modally sourced image content descriptor is based on the automated image labelling framework that we refer to as the Collaterally Cued Labelling (CCL). Both the image content (i.e. visual features) and the collateral texts are used as the input to the labelling system. Content and context-associative knowledge is extracted from the collateral textual information to enhance the performance of the labelling framework. Two different feature vector models are devised based on the CCL framework for clustering and retrieval purposes respectively. The effectiveness of the clustering and retrieval as supported by our proposed feature vector models is examined using Self-Organizing-Maps (SOMs).

In section 2, we state the motivation of our research. Section 3 reports the design of the CCL framework. In section 4, we describe how our proposed high-level image feature vector models are built up. Section 5 reports the experiments conducted for evaluating our proposed method. Section 6 presents our conclusions.

---

[1] http://www.nationalgallery.org.uk/

2 Motivation

According to the studies performed by many researchers (e.g. [8], [11], [14], [18]), who have focused on analysis of users' needs for visual information retrieval systems, we can conclude that: **i)** users search for visual information both by types and identities of the entities in images, **ii)** users tend to request images both by the innate visual features and the concepts conveyed by the picture. With this conclusion, clearly we can see that using text or image features alone cannot adequately support users' requests. It has been argued that the low-level visual primitives exploited by CBIR systems are not sufficient for depicting the semantic (object) level meaning of the images. On the other hand, dominant commercial solutions like Google and Yahoo image search engines which take advantage of the collateral textual information co-occurring with the images, such as title, caption, URL, etc., cannot always meet the user's perceptual needs and sometimes the retrieval results may be unpredictable due to the inconsistency between the collateral text and image content.

Therefore, any attempt to combine both visual and textual cues for image representation and indexation should take the priority in realising the next generation of Semantic-Based Image Retrieval (SBIR). Srihari has reported some pioneering research in this area of integrating visual and linguistic modalities for image understanding and object recognition [22], [23]. Paek et al [19] presented an integration framework for combing the image classifiers which are based on the image content and the associated textual descriptions. Recently researchers started to develop methods for automatically annotating images at object level. Mori et al [16] proposed a co-occurrence model which formulated the cooccurrency relationships between keywords and sub-blocks of images. Duygulu et al [7] improved Mori et al's co-occurrence model with Brown et al's machine translational model [5] which assumes image annotation can be considered as a task of translating blobs into a vocabulary of keywords. Barnard et al ([1], [2], [3]), proposed a statistical model for organising image collections into a hierarchical tree structure which integrates semantic information provided by associated text and perceptual information provided by image features. Li and Wang [13] created a system for automatic linguistic indexing of pictures using a statistical modelling approach. The two-dimensional multi-resolution Hidden Markov Model is used for profiling categories of images, each corresponding to a concept. Zhou and Huang [26] explored how keywords and low-level content features can be unified for image retrieval. Westerveld [24] and Cascia et al [6] both chose on-line newspaper archives with photos as their test dataset to combine the visual cues of low-level visual features and textural cues of the collateral texts contained in the HTML documents. Promising experimental results have been reported throughout the literature in this novel research area. These studies have largely motivated our research in combining both visual and textual cues in high-level image feature vector model construction.

3 Collaterally Cued, Primarily Confirmed Image Labelling

According to the information need, there is always a distinction between the primary and collateral information modalities. For instance, the primary modality of an image retrieval system is of course the images, while all the other modalities that explicitly or implicitly related to the image content, like collateral texts, could be considered as the collateral modality. The proposed Collaterally Cued Labelling (CCL) framework constituting the fundamental part of the semantic-level visual content descriptor construction is designed with the expectation of taking fully advantage of both the primary (visual) and collateral (textual) cues in order to perform automatic blob-based semantic-level image labelling.

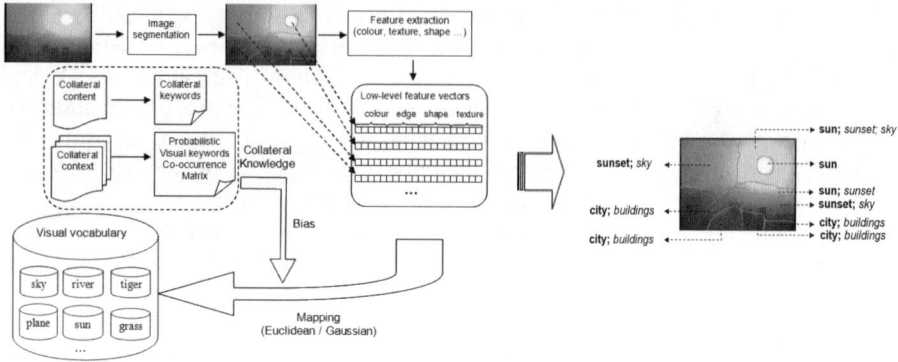

Fig. 1. Architecture and workflow of the CCL framework

Fig. 1 illustrates the architecture and workflow of the CCL framework. The raw image data is firstly segmented into a number of regions and then low-level visual features, like colour, edge, shape, texture, are extracted from each segment. We then map those blob-based low-level visual features to the visual concepts defined in the visual vocabulary with the help of the domain knowledge extracted from the collateral texts. In the following sections, we will discuss each of the above components in detail.

3.1 Blob-Based Low-Level Visual Feature Extraction

We use the Normalised Cuts [10] method to segment the images into a number of regions. A 54-dimesional feature vector, comprising colour, edge, shape and texture features, is created for each blob. 21 colour features are extracted using the intensity histogram. 19 edge features are extracted by applying an edge filter and the water-filling algorithm [25] on the binary edge image. A statistical texture analysis method proposed by Haralick et al [9], the Grey Level Co-occurrence Matrices, was used to extract 6 texture features from each segment. Finally, 8 features related to shape were extracted using Matlab-based functions for extracting the statistics of image regions, e.g. eccentricity, Euler number, rectangularity, etc. We constructed the final feature vector by sequentially assembling all the four kinds of features together.

3.2 A Visual Vocabulary

The construction of the visual vocabulary is based on a dataset provided by Barnard [2], which consists of 3478 manually annotated regions of interest segmented from 454 Corel images. The image segments were annotated by Duygulu [2] with a controlled vocabulary which is a subset of the terms that had been used for annotating the Corel images. An image segment may be annotated using single or multiple keywords. The visual vocabulary developed in our system is a set of clusters of blob-based visual feature vectors labelled by textual keywords. With the manual blob-based annotations, we simply group the blobs labelled by the same textual keyword together as the clusters which finally constitute a bank of 236 "visual keywords".

3.3 Constructing a Collateral Knowledgebase

We constructed our collateral knowledgebase by extracting and representing both the collateral content and context related knowledge from the collateral textual descriptions of the images.

Collateral Content

The collateral content refers to the knowledge that can be extracted directly from the collateral texts. Such kind of collateral information becomes easier and easier to acquire due to the multimodal nature of the modern digitised information dissemination. There are many different sources where the collateral keywords can be extracted, such as captions, titles, URL, etc. Such keywords are expected to depict the subject or concepts conveyed by the image content.

Collateral Context

Collateral context is another new concept we introduced in this research, which refers to the contextual knowledge, representing the relationships among the visual concepts. We use a conditional probabilistic co-occurrence matrix which represents the cooccurrency relationships between the visual concepts defined in the visual vocabulary. Hence, a 236 × 236 matrix is created based on the manual annotations of the image segments. Each element of the matrix can be formally defined as follows:

$$P(K_j|K_i) = \frac{P(K_i, K_j)}{P(K_i)} \quad (i \neq j) \tag{1}$$

where $P(K_i, K_j)$ can be considered as the co-occurrence of each pair of keywords i and j appeared in the manual annotation and $P(K_i)$ as the occurrence of the keyword i.

The relationships could be considered as bidirectional, because the value of each element is calculated based on the conditional probability of the cooccurrency between two visual keywords. The relationships appear to be reasonable in many cases (see Fig. 2). For instance, the relationship between sky and clouds shows that clouds has bigger cooccurrency probability against sky than the other way round. And this appears to be reasonable because clouds must appear in the sky while there may be a lot of other things appearing in the sky.

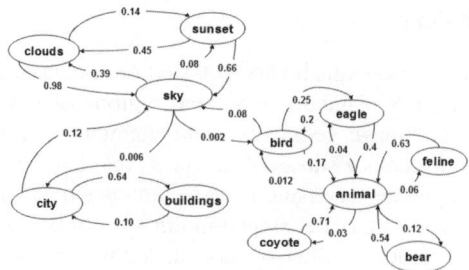

Fig. 2. A graph representation of the visual keyword co-occurrence matrix

The exploitation of this co-occurrence matrix is a novel way of bridging the gap between the computer generated image segments and the semantic visual object. So far, even the most state-of-the-art image segmentation techniques cannot generate perfect segments which separate the whole objects from each other. Normally, an image segment either contains a part of an object or several parts of different objects which are occluded by each other. For instance, it is very likely to get segments containing a wing of an airplane with the background of sky. With our co-occurrence matrix, if the segments can be firstly recognized as a wing then system will probably provide other labels like plane, jet, sky, etc. according to the conditional probability between the visual keyword 'wing' and the others.

3.4 Mapping Towards Visual Concepts

Having constructed the visual vocabulary and the collateral knowledgebase, we can proceed to map the low-level blobs to the visual keywords. We used two methods to do the mapping, namely Euclidean distance and Gaussian distribution. For the Euclidean distance method, we calculate the distance between the blob features and the centroids of each cluster within the visual vocabulary. For the Gaussian distribution method, we calculate the Gaussian probabilities between the blob feature vectors and the visual feature clusters. However, labelling assignments based only on the shortest Euclidean distance or highest Gaussian probability cannot always provide accurate labels. Therefore, we introduce the collateral content knowledge as a cue to bias the mapping procedure and use the primary (i.e. visual) feature matching as a reconfirmation to improve the labelling quality. The process is quite straightforward. Instead of finding the shortest Euclidean distance or highest Gaussian probability against all the clusters within the visual vocabulary, we seek to determine the best matching visual keywords which appear as collateral keywords, which can be formally described as follows:

Euclidean Distance Approach:
$$Min\left(\sqrt{\left\|\vec{f}-\vec{u}\right\|^2}\right) \quad (2)$$

Gaussian Distribution Approach:
$$Max\left(\frac{1}{\sqrt{2\pi\sigma^2}}\exp\left(\frac{-\left(\vec{f}-\vec{u}\right)^2}{2\sigma^2}\right)\right) \quad (3)$$

where f is the blob feature vector; $\vec{u}$ and σ are the mean vector and standard deviation of selected *clusters* which can be defined as:

$$clusters \in \{k \mid k \in collateral\_keyword\} \cap \{k \mid k \in visual\_vocabulary\}$$

We then exploit the collateral context knowledge (i.e. the co-occurrence matrix) to expand the labelling results with some more context related labels. A Thresholding mechanism is devised to combine both visual similarity between the blob features and visual keywords, and, co-occurrence probability between the collateral content-based labels and the rest of the visual keywords within the visual vocabulary. See formula 4 and 5.

Euclidean Distance Approach:
$$\left(Min\left(\sqrt{\|\vec{f} - \vec{u}_{kn}\|^2} \right) \Big/ \sqrt{\|\vec{f} - \vec{u}_{km}\|^2} \right) \times P(K_m | K_n) \geq Threshold \quad (4)$$

Gaussian Distribution Approach:
$$\left(\frac{1}{\sqrt{2\pi\sigma^2}} \exp\left(\frac{-(\vec{f} - \vec{u}_{km})^2}{2\sigma^2} \right) \right) \times P(K_m | K_n) \geq Threshold \quad (5)$$

where
$$kn \in \{k \mid k \in collateral\_keyword\} \cap \{k \mid k \in visual\_vocabulary\}$$
$$km \in \{k \mid k \in visual\_vocabulary\}$$

The determination of *Threshold* is based on experiment and can be adjusted according to the different needs. For example, the lower the value is the more context related labels you get, however the relevance of the label may fall down accordingly. In contrast, greater value leads to more relevant but less labels. Fig. 1 shows an example labelling results produced by the CCL using Euclidean distance approach. The keywords in Bold are the labels assigned to the region cued by the collateral content (i.e. collateral keywords). The keywords in Italic are the labels given based on the visual keywords co-occurrence matrix by following formula 4.

4 Semantic-Level Visual Content Descriptor

Based on the region-based labels generated using the CCL framework, we constructed a novel homogenous semantic-level image feature vector model. Thus the new feature vector model combines both the visual features of the image content and the textual features of its collateral text. Two different kinds of feature vectors were created for retrieval and classification purposes respectively. Based on the blob-based linguistic labels, we can calculate the weight of each term on the basis of the frequency of its appearance in the labels of all the segments. Therefore, it is expected that the weights should reflect the proportion of the occurrence of the visual object within the whole scene. For the example shown in Fig. 1, we can generate the proportions of the key terms as: Sunset: 4/18; City: 4/18; buildings; 4/18; Sun 3/18; Sky: 3/18.

A 236-dimensional feature vector, where each dimension represents a visual keyword in the vocabulary, can be created with each element be assigned with the proportion value described above. Another feature vector model can be applied in a similar way. However instead of defining each visual keyword as a dimension, this time we group the keywords into exponential partitions using a function of $f(x)=a^x$ where $x \in \{0, 1, 2, 3, \ldots\}$, according to the *tf×idf* values for the key terms appearing in each pre-classified category in the training dataset. a is a statistic parameter which can be specified accordingly. The aim of developing such kind of feature vector model is to give more weight to the most important keywords for each category. Accordingly the value of the element of the feature vector will be the quotient of the keyword proportion divided by the number of keywords with in the exponential partition. Table 1 shows an example of an exponential partition based high-level image feature vector, where the value of each element of the vector can be defined as p/a^x where p is the sum of proportion values of labels appeared in the corresponding group and a^x is number of keywords in the exponential partition of each pre-defined category. The number of the category depends on how the training dataset are organized and classified. And the range of x should be adjusted according to how many key terms appeared in each category.

Table 1. An example of exponential partition based feature vector

	Category 1				Category 2				
	Part 1	Part 2	Part 3	Part 4	Part. 1	Part. 2	Part. 3	Part. 4	...
No. of Words	$3^0=1$	$3^1=3$	$3^2=9$	$3^3=27$	$3^0=1$	$3^1=3$	$3^2=9$	$3^3=27$	...
Proportion of Words in Partition	3/5	1/5	0	0	0	0	1/5	0	...
Weight value	0.6	0.06	0	0	0	0	0.02	0	...

We believe that this feature vector model can facilitate the classification, because more weights are given to the most significant and representative key terms for each category which are encoded at the beginning while assigning less weights to the keywords which are located towards the end of each partition.

5 Experiments

We conducted all the experiments based on the Corel image dataset. An interesting aspect of the collection is that each image has been annotated with a set of keywords which depict the key objects appeared in the image. In our experiments, we took advantage of the given keywords as collateral keywords for the images. Those keywords can be considered to be extracted from the collateral textual information of the image using natural language processing or keyword extraction techniques. We selected 30 categories of Corel images as our experimental dataset (see Table 2).

Table 2. 30 categories of Corel images

Class no.	Category Name	Class no.	Category Name	Class no.	Category Name
1	Sunsets & Sunrises	11	Wildlife of Antarctica	21	Divers & Diving
2	Wild Animals	12	Elephants	22	Grapes & Wine
3	The Arctic	13	Foxes & Coyotes	23	Land of the Pyramids
4	Arizona Desert	14	Rhinos & Hippos	24	Nesting Birds
5	Bridges	15	Arabian Horses	25	Helicopters
6	Aviation Photography	16	Greek Isles	26	Models
7	Bears	17	Coins & Currency	27	Virgin Islands
8	Fiji	18	English Country Gardens	28	Tulips
9	North American Deer	19	Bald Eagles	29	Wild Cats
10	Lions	20	Cougars	30	Beaches

5.1 Extracting Blob-Based Visual Features

The 3000 Corel images were firstly segmented into 27953 regions of interest. The segments for each image were sorted in the descending order according to the segment's size (or area), and we selected the top 10 segments, if the image was segmented into more than 10 regions, as the most significant visual objects in the image to label. We used different mapping methods, namely Euclidean distance and Gaussian distribution, to label the 27953 blobs. Two different kinds of feature vectors as proposed in section 4 were generated for all the 3000 Corel images. One was the 236-dimensional feature vector, whereby the value of each dimension is determined by the weights of the visual keywords defined in the visual vocabulary, which is the ratio between the occurrence of the label and the total number of the labels assigned to the image. The other kind of feature vector is based on the exponential partitioning on the keywords appearing in each category. In our case, we divided the keywords into 4 groups by applying the function $f(x) = 3^x$ where $x \in \{0, 1, 2, 3\}$ according to the $tf \times idf$ values of the keywords in the category. Therefore, there were 120 (4×30) dimensions in this feature vector. Because we also used different methods for the mapping from the low-level blobs to the visual keywords, i.e. Euclidean distance and Gaussian distribution, we finally generated 4 sets of feature vectors for each of the 3000 Corel images, namely Euclidean 236D and 120D, Gaussian 236D and 120D.

In order to compare the performance of the proposed feature vector models with the conventional image feature extraction methods, we also extracted 3000 image feature vectors based on the visual content of the image and text feature vectors based on the Corel keywords. For the visual feature vector, we constructed a 46-dimensional feature vector which consisted of colour, edge and texture features for each image. For the textual feature vector, we selected the top 15 keywords with the highest $tf \times idf$ values for each category. Then we merged the duplicated keywords, and finally built up a 333-dimesional text feature vector model. The value of each element is the occurrence of the keywords.

5.2 Semantic-Based Image Classification and Retrieval

We created four 30 × 30 SOMs (i.e. Self Organising Maps) for learning to classify the four kinds of feature vectors, i.e. Euclidean 120D, Gaussian 120D, Visual 46D, and Textual 333D. We divided the 3000 Corel images into training and test set with a ratio of 1:9. 10 images were randomly selected from each category making a total number of 300 test images, and the rest 2700 images as training set. Each one of the four SOMs was trained with the 2700 feature vectors extracted from the training images using each method for 100 epochs.

The trained SOMs were firstly used to perform the image retrieval on the four different kinds of feature vectors. The SOM-based retrieval was performed on the basis of the principle that we would retrieve the training items with respect to their activation values to the Best Matching Unit (BMU) of the test input. The Euclidean 120D and Gaussian 120D outperform both the textual and visual feature vectors for the top 5 retrieved items, and significantly increased the precision as compared to the results achieved by purely using content-based image feature vectors (see Fig. 3).

Thereafter we examined the classification effectiveness of our proposed high-level image content descriptor by testing the trained network using the 300 test feature

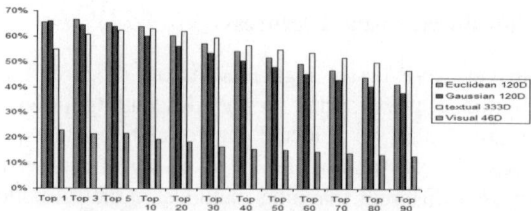

Fig. 3. Precisions of the SOM-based image retrieval using 4 different feature vector sets

vectors. 4 confusion matrices were created for the test results based on which we calculated the classification accuracy using the four different kinds of feature vectors, (see Fig. 4). The Euclidean 120D showed the best average accuracy at 71% followed by the Gaussian 120D at 70%. Also, the performance of the Euclidean 120D and Gaussian 120D appeared to be more stable than that of Textual 333D and Visual 46D.

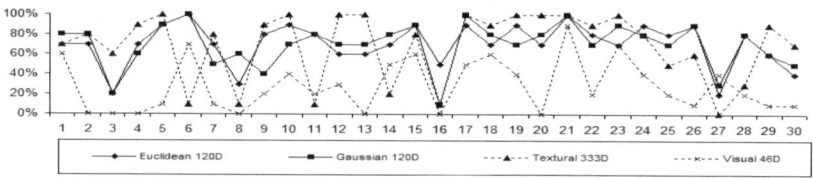

Fig. 4. SOM classification accuracy using four kinds of feature vectors

	Query Image	Corel keywords
		Animal bear fish water
Textual 333D		
Euclidean 236D		
Gaussian 236D		

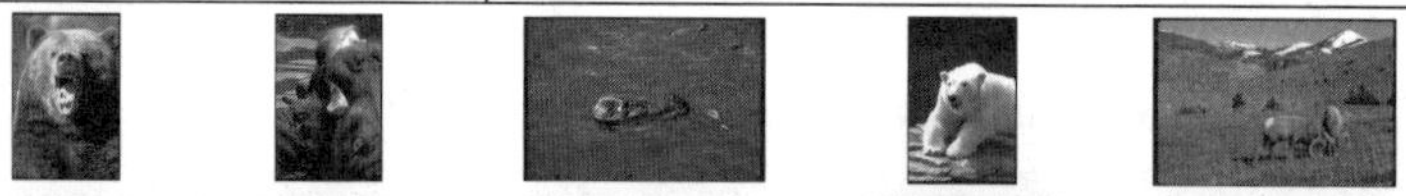

Fig. 5. Example retrieval results using different feature vector models

However, the statistical calculations alone sometimes are not enough for such evaluation. We believe that one of the advantages of our proposed semantically-cued high-level visual content descriptors is that they can combine both the visual and semantic similarity of the image. Take the retrieval results shown in Fig. 5 as an example. Although Textual 333D got the best statistical results in terms of precision, it cannot meet the user's perceptual and conceptual needs on the objects of the interest. However, by using our proposed high-level image feature vector models (i.e. Euclidean 236D and Gaussian 236D) the objects of interest can be successfully identified, e.g. bear and water in this case. Also, we can retrieve images that are both visually and conceptually similar to the query images, even though the retrieved images were categorised into the different classes by the annotator with that of the query image. It is also this reason that sometimes the statistical calculations, e.g. precision, appears to be lower for the proposed features than the textual features.

We also believe that in realistic scenarios, where the collateral text may contain a lot of noises or may prove to be too subjective, the advantage of the proposed high-level image feature vector model will become more obvious due to its ability of visually re-confirming that the concept does indeed exist in the images.

6 Conclusions

We have proposed a semantically cued intermediate image content descriptor model underpinned by a Collaterally Cued Labelling (CCL) framework. The CCL image labelling framework can automatically associate the regions of interest of given images with linguistic labels. The test results using 4 different kinds of high-level image feature vector models were compared with the results obtainable using traditional text-based and content-based image feature models. This showed that the former consistently yield better performance than the latter in terms of satisfying both the perceptual and conceptual needs of the user.

Future work is planned to further exploit the collateral knowledge with more sophisticated knowledge representation schemes such as ontological networks. We fully expect to explore the potential of CCL in a range of realistic retrieval scenarios which could be another challenging proving ground.

Acknowledgements

The authors wish to thank K. Barnard and P. Duygulu for providing us with their manually labelled Corel image segments. We would also like to thank for the support of the UK Engineering and Physical Sciences Research Council Grant (DREAM Project, DT/E006140/1).

References

1. Barnard, K., Duygulu, P., Forsyth, D.: Clustering Art. In: Proc. IEEE Conf. on Computer Vision and Pattern Recognition, vol. 2, pp. 434–441 (2001)
2. Barnard, K., Duygulu, P., de Freitas, N., Forsyth, D., Blei, D., Jordan, M.I.: Matching Words and Pictures. Machine Learning Research 3, 1107–1135 (2003)
3. Barnard, K., Forsyth, D.: Learning the Semantics of Words and Pictures. In: Proc. Int. Conf. on Computer Vision, II, pp. 408–415 (2001)
4. Bimbo, A.D.: Visual Information Retrieval. Morgan Kaufmann Publishers, Inc, San Francisco, California, US (1999)
5. Brown, P., Pietra, S.D., Pietra, V.D., Mercer, R.: The mathematics of statistical machine translation: Parameter estimation. Computational Linguistics 19(2), 263–311 (1993)
6. Cascia, M.L., Sethi, S., Sclaroff, S.: Combining Textual and Visual Cues for Content-Based Image Retrieval on the World Wide Web. In: Proc. of IEEE Workshop on Content-Based Access of Image and Video Libraries, IEEE Computer Society Press, Los Alamitos (1998)

7. Duygulu, P., Barnard, K., Freitas, N., Forsyth, D.: Object recognition as machine translation: Learning a lexicon for a fixed image vocabulary. In: Heyden, A., Sparr, G., Nielsen, M., Johansen, P. (eds.) ECCV 2002. LNCS, vol. 2351, pp. 97–112. Springer, Heidelberg (2002)
8. Enser, P.G.: Query analysis in a visual information retrieval context. Document and Text Management 1, 25–52 (1993)
9. Haralick, R.M., Shanmugam, K., Dinstein, I.: Texture features for image classification. IEEE Trans. On Sys. Man, and Cyb. SMC 3(6), 610–621 (1973)
10. Jianbo, S., Jitendra, M.: Normalized Cuts and Image Segmentation. IEEE Trans. on Pattern Analysis and Machine Intelligence 22(8) (2000)
11. Keister, L.H.: User types and queries: impact on image access systems. Challenges in indexing electronic text and images. Learned Information (1994)
12. Lew, M.S.: Next-generation web searches for visual content. IEEE Computer 33, 46–53 (2000)
13. Li, J., Wang, J.Z.: Automatic linguistic indexing of pictures by a statistical modelling approach. IEEE Trans. Pattern Analysis and Machine Intelligence 25(9), 1075–1088 (2003)
14. Markkula, M., Sormunen, E.: End-user searching challenges indexing practices in the digital newspaper photo archive. Information retrieval 1, 259–285 (2000)
15. Marques, O., Furht, B.: Content-Based Image and Video Retrieval. Kluwer Academic Publishers, Norwell, Massachusetts, US (2002)
16. Mori, Y., Takahashi, H., Oka, R.: Image-to-word transformation based on dividing and vector quantizing images with words. In: 1st Int. Workshop on Multimedia Intelligent Storage and Retrieval Management (1999)
17. Morris, T.: Computer Vision and Image Processing. Palgrave Macmillan Publishers, Ltd, New York, US (2004)
18. Ornager, S.: View a picture: Theoretical image analysis and empirical user studies on indexing and retrieval. Swedis Library Research 2(3), 31–41 (1996)
19. Paek, S., Sable, C.L., Hatzivassiloglou, V., Jaimes, A., Schiffman, B.H., Chang, S.F., McKeown, K.R.: Integration of visual and text based approaches for the content labelling and classification of Photographs. In: ACM Workshop on Multimedia Indexing and Retrieval. ACM Press, New York (1999)
20. Rui, Y., Huang, T.S., Chang, S.F: Image Retrieval: current techniques, promising directions and open issues. Visual Communication and Image Representation (1999)
21. Smeulder, A.W.M., Worring, M., Anntini, S., Gupta, A., Jain, R.: Content-Based Image Retrieval at the End of the Early Years. IEEE Trans. on Pattern Analysis and Machine Intelligence 22(12) (2000)
22. Srihari, R.K.: Use of Collateral Text in Understanding Photos. Artificial Intelligence Review. Special Issue on Integrating Language and Vision 8, 409–430 (1995)
23. Srihari, R.K.: Computational Models for Integrating Linguistic and Visual Information: A Survey. Artificial Intelligence Review, Special Issue on Integrating Language and Vision 8, 349–369 (1995)
24. Westerveld, T.: Image Retrieval: Content Versus Context. In: Proc. of Content-Based Multimedia Information Access, pp. 276–284 (2000)
25. Zhou, X.S., Huang, S.T.: Image Retrieval: Feature Primitives, Feature Representation, and Relevance Feedback. In: IEEE Workshop on Content-based Access of Image and Video Libraries (2000)
26. Zhou, X.S., Huang, S.T.: Unifying Keywords and Visual Contents in Image Retrieval. IEEE Trans. Multimedia 9(2), 23–33 (2002)

Investigating Automatic Semantic Processing Effects in Selective Attention for Just-in-Time Information Retrieval Systems

John Meade and Fintan Costello

School of Computer Science & Informatics
University College Dublin
Ireland
{john.meade,fintan.costello}@ucd.ie

Abstract. Just-in-Time Information Retrieval (JITIR) systems aim to automatically retrieve useful information on the basis of the user's current task and to present this information to the user without disrupting that task. We ask whether the cognitive mechanism of 'selective semantic processing' can minimise the disruptive nature of presenting JITIR information to the user. This mechanism may allow users to subconsciously filter out irrelevant information presented in the periphery of the visual field, while maintaining awareness of relevant information. We report an experiment assessing both attention to peripherally presented information (measured via recall) and level of distraction (measured via typing keystroke rate) in a JITIR system used to write reports on various topics. The experimental results showed that peripherally presented information that was relevant to a user's writing topic reliably entered their attention significantly more often than irrelevant information, and was significantly less distracting than similar but irrelevant information.

1 Introduction

Just-in-Time Information Retrieval (JITIR) systems are designed to automatically retrieve relevant information for users on the basis of the user's current activity. If the user is currently writing a document about genetic engineering, for example, a JITIR system will attempt to automatically retrieve and present the user with information that they may need for that document (without the user having to carry out an explicit search for that information). There has been a notable increase in interest in such systems in recent years [3,5,6], motivated by the fact that JITIR systems have the potential to increase user productivity and enhance the user experience of information retrieval tasks. However, the information retrieved by a JITIR system is presented to the user at the cost of potentially distracting the user from their work. Currently the focus of JITIR research has been on improving the quality of information returned [2,3,4,5,6] which ensures that the benefit of quality information outweighs the cost of distracting the user. In this paper, we report an experiment investigating the cost of presenting potentially distracting information to a user.

Recent research on selective attention and automatic semantic processing [8] has shown that people are capable of subconsciously processing words presented outside their focus of attention, and can subconsciously register words of relative significance to them, such as personal surnames. A similar ability may come into play within a user's work, allowing the user to use selective attention to subconsciously register words that are relevant to the task at hand, while not registering irrelevant words (and so not being distracted by those words). Our experiment is designed to ask whether these selective attention effects do occur in work tasks, and if they do occur, whether these effects could be exploited by a JITIR system to decrease the distracting nature of information presentation.

The primary task for a JITIR system is to retrieve information that is relevant to the user's current task. Current systems take a variety of approaches to this problem. We base our experiment in this paper on the 'topic probability' JITIR system described by [6]. This system is designed to analyse the document that the user is currently writing and to identify the topics that the user is working on. These topics are then used to construct queries that are fed to a standard web-based search engine, with the results returned to the user. The results are presented to the user outside their focus of attention (in a 'ticker' type stream flowing across the bottom of the word-processing window in which they are working). We use this 'topic probability' JITIR design in our experiment because it allows us to examine the degree to which information from different sources (on topic, close to topic, distant from topic) can potentially distract and interrupt the user. Our experiment requires a user to write a short essay on a particular subject while information is presented to them peripherally (flowing across the bottom of the edit window). This information is designed to be either directly on the topic of their essay, from a slightly related topic, or from a distant topic. We treat the user's ability to recall words from the information presented to them by the JITIR system as a measure of how much attention is paid to the information outside the focus of attention.

The remainder of this paper is organised as follows. Section 2 describes the 'topic probability' JITIR system and also describes two alternative systems ('Watson' [3] and 'Remembrance Agent' [2]). Section 3 gives an overview of the phenomenon of automatic semantic processing and details the effects of task-related stimuli on such processing. Our experiment is explained in Section 4, with an analysis of the users recall in Section 5 and the analysis of the users typing performance in Section 5.2. In Section 6 we draw a number of conclusions which may be useful for designers of JITIR systems.

2 Just-in-Time Information Retrieval

One of the earliest JITIR systems is 'Remembrance Agent' [2]. Remembrance Agent focuses on retrieving local documents related to the document a user is manipulating. The agent uses the information retrieval system Savant, which indexes a user's local documents, to find similar documents. The agent submits the document that the user was viewing as a query to Savant, this in turn uses document similarity between the query document and local documents, to find similar documents. These similar documents are then presented to the user to assist them in completing the current document.

A second well known JITIR system is 'Watson' [3]. Watson gathers contextual text information from the document a user is manipulating, and uses that information to find related documents from varying information repositories such as the web and the users local documents. It uses $TFIDF$ to weight terms in order to construct queries. However, it does not solely rely on $TFIDF$: it also uses a fixed set of heuristics to adjust the weight of words according to their position and emphasis in the document (e.g. font size). For example, words that appear at the beginning of a document are assigned more value than words at the end of a document. Watson removes redundant information from the results retrieved by removing duplicate information from multiple sources. Additionally Watson checks that the information it returns is valid by detecting broken links removing any result that does not provide live information. By doing so the user is assured that anytime spent processing the information will be of benefit to them. Watson has been under development since 1999 [1] and was released commercially in 2005.

The previous two systems treat a user's document as a bag of words and do not investigate any underlying structure of meaning in the documents. In our experiments we focus on the 'topic probability' JITIR design taken by [6]. This JITIR system finds information based on the topical content of the user's work. The hierarchical structure of the lexical database WordNet-2.1 [13] is used to define the set of candidate topics. Topics are defined as any noun sense in WordNet that subsumes one or more words in the user's document. The probability of a candidate topic is calculated as the probability that the words, from the document that fall under the topic, do not occur together at random. The probability for words occurring under a topic at random is given by Equation 1, which is the hypergeometric distribution for selection without replacement.

$$P_{random}(T|S,W) = \frac{\binom{T}{S}\binom{N-T}{W-S}}{\binom{N}{W}}. \quad (1)$$

Where T is the number of words in WordNet that occur under the topic, W is the complete number of unique words within the document. S is the number of words within the document that fall under the topic and N is the total number of words in WordNet. Therefore the probability that the words did not occur at random is:

$$P_{correct}(T|S,W) = 1 - P_{random}(T|S,W). \quad (2)$$

The most probable topics are selected as the topics in the user's document. These topics are then used to construct queries whose results are presented to the user. This ability to determine the topics of a document allows information to be retrieved on the basis of topic coverage and can be used to filter out information that does not discuss or is not related to the topics within the user's document.

3 Selective Attention and Semantic Processing

The extent to which word meanings are automatically processed both inside and outside a persons focus of attention has been the subject of continuing interest. The review provided by [8] highlights current work in automatic semantic access. To summarise it has been found that two paradigms may hold for selective attention. The first, 'early selection', predicts that unattended stimuli (stimuli outside the centre of focus) will be seen as irrelevant at an early stage. For example, words that appear at the periphery of attention would be ignored without processing. The second theory 'late selection' states that all stimuli are processed without attention or conscious thought, and are excluded from consciousness if irrelevant. Therefore, words, no matter where they appear, are processed up to and including the level of meaning but are masked from consciousness if irrelevant. It has been demonstrated that stimuli that are personally significant, such as a person's name, are more likely to be processed for meaning regardless of being centrally or peripherally located [11]. These observations support the late selection theory.

3.1 Task-Relevant Stimuli

Experiments carried out by [7] asked whether task-relevant stimuli affect automatic semantic processing. In these experiments a colour appears on screen at the same time a word appears. The task in these experiments is to correctly call out a predefined personally significant word associated with the colour while ignoring the on screen word. For example, if the colour red appears a user must utter her first name ('Michelle') in response. The location of both word and colours are at times spatially separated to examine unattended stimuli effects. Some of the words consisted of task related terms i.e. colour names. It was found that words appearing in the centre of someone's focus of attention led to slower reaction times if they were of personal significance i.e. a person's surname or mothers name, compared to neutral items. Words placed at the peripheral of someone's attention had no effect on reaction times and that performance was reduced only when words were relevant to the predefined task, this supports the 'late selection' theory. This work highlights that it is possible to present information to a user outside the focus of attention and yet not distract them if it is unrelated to a set task. Similarly, information that is task-related will attract the users attention.

Following these results, we wish to investigate the effect of automatic semantic processing when a user is carrying out real world tasks while they are presented with information peripherally. Our hypothesis is, information that is related to or somewhat related to a user's work will be more likely to attract their attention. However, information that is unrelated will not affect their attention. We expect, as the information moves from specific to general the users ability to ignore it increases.

4 Experiment

To test our hypothesis we set the task of writing an essay, to focus the user's attention on a particular area of the screen and on a particular task. We then present words below the focus of attention that can be classed as on-task, near to the task and distant from the task. The user then must try to remember the words shown to them, in order to gauge the extent a user processed the words presented. In the experiments, a participant was asked to write an essay on three different topics. As they type their essay, words are scrolled under the typing area, in a ticker tape window. These words are taken from three associated topic categories *Topic*, the essay topic (on-task), *Near*, a topic slightly related to the essay (near-task) and *Distant*, a topic unrelated to the essay (distant from task). We recorded when each word appeared on screen and when a word went off screen. In addition, the user's keystrokes with a corresponding timestamp for when they occurred were recorded.

In total 27 words are shown to the participant, 9 from each category, these are randomly selected and presented randomly. Words start scrolling by 20 seconds after the participant begins typing. A new word appears when the current word on screen has passed halfway across the ticker tape. The interface prevented the user from typing after the last word went off screen. The 9 words from each topic category were divided into three equally sized subcategories by frequency, Low Frequency (*LF*), Medium Frequency (*MF*) and High Frequency (*HF*). The frequency of a word was defined as the number of pages returned by the Google search engine containing that word. The frequency levels are: $LF \leq 11 \times 10^5$, $11 \times 10^5 < MF < 11 \times 10^6$ and $HF \geq 11 \times 10^6$.

After the essay writing task the user was then given a recall task. In this recall task they were shown 54 words, 27 of which had been presented in the scrolling area. The other 27 were foils, which the user had not seen but were from the same topic categories. Participants were asked to indicate which words they recalled seeing scrolled across the bottom of the screen.

4.1 Preparation

We selected 6 unique essay topics such that 3 pairs existed and each pair of topics fell under a unique parent topic. Two topics that fall under a common parent are defined as slightly related to each other. For example, consider the topics 'Stem Cell Research' and 'Genetically Modified Food'. Both topics discuss distinct areas of genetic engineering, they are slightly related to each other yet distinct and fall under the common parent topic of 'Genetic Engineering'. Table 1 shows the topics and their associated parent topic. For each topic two sets of words were found, each set consisted of 6 *HF*, 6 *MF* and 6 *LF* words. The two sets will be denoted by *word_set_1* and *word_set_2*.

A check verified that each word under a topic had a higher co-occurrence with that topic than with any other topic (using Google hits as a measure of co-occurrence). Another check ensured that the average co-occurrence between words under a parent topic and that parent topic was higher than their average

Table 1. Essay Topics and Parent Topic

Parent Topic:	Genetic Engineering	
Topics:	Stem Cell Research	Genetically Modified Food

Parent Topic:	Conservation	
Topics:	Renewable Energy	Waste Management

Parent Topic:	Expedition	
Topics:	Space Exploration	Tourism

co-occurrence with any other parent topic (again using Google hits as a measure of co-occurrences). Therefore, the two topics under a parent were closely related.

4.2 Design

Twelve students from the University College Dublin voluntarily participated in the experiment. There were two possible sets of essays that a student could be asked to write about, Group A and Group B. Within each group the students were split into two subgroups. These subgroups are based on the sets from which the words shown to the user are taken. As discussed in Section 4.1 there are two sets of words for each topic, given by $word\_set\_1$ and $word\_set\_2$. This was to minimise the possibility of any effects being attributed to the words shown to participants. The complete groups are defined in Tables 2-3, where 'Essay Topic' is the essay a user was asked to write about and from what set the *Topic* words are taken. Similarly 'Near Topic' and 'Distant Topic' are the sets the *Near* and *Distant* words are taken from.

Table 2. Essay Group A

(a) Group A

Essay Topic	Near Topic	Distant Topic
GM Food 1	Stem Cell 2	Tourism 2
Space 1	Tourism 1	Renewable 2
Waste 1	Renewable 1	Stem Cell 1

(b) Group A Complement

Essay Topic	Near Topic	Distant Topic
GM Food 2	Stem Cell 1	Renewable 1
Space 2	Tourism 2	Stem Cell 2
Waste 2	Renewable 2	Tourism 1

Table 3. Essay Group B

(a) Group B

Essay Topic	Near Topic	Distant Topic
Stem Cell 1	GM Food 2	Waste 2
Tourism 1	Space 2	GM Food 1
Renewable 1	Waste 1	Space 1

(b) Group B Complement

Essay Topic	Near Topic	Distant Topic
Stem Cell 2	GM Food 1	Space 2
Tourism 2	Space 1	Waste 1
Renewable 2	Waste 2	GM Food 2

4.3 System

Our test system is built as a component for KOffice-1.4-2 [9] an office suite which is distributed under the Free Software/Open Source license. We modified

KWord, the text editor application in the suite, so that words would scroll across a ticker tape window within the application. Preliminary testing showed that words scrolling beneath the area where text is entered was the least distracting, with text scrolling above the most distracting. The ticker tape is fixed below the area of text for the purpose of our experiments as we did not wish the location of where the words may appear to be a factor in our analysis. Figure 1 is a screenshot of the test system in use.

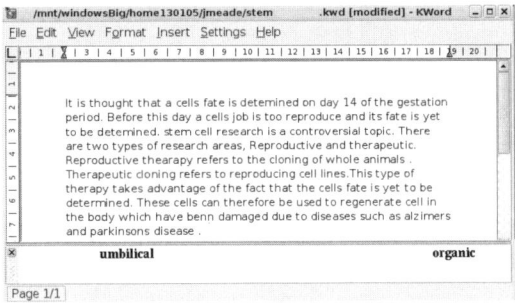

Fig. 1. Test System Screenshot

5 Analysis

In our analysis we focused on two aspects of users' performance: their ability to recall the words presented to them and their average number of keystrokes while words presented were on screen. We use the participants recall as a measure of their awareness of the words presented to them and their typing performance as measure of how distracted they are by these words.

To determine what factors influenced the participants' performance we carried out a Multivariate Repeated Measures analysis on the number of words recalled by a participant and on the average number of keystrokes when a word is on screen. The analysis design for both recall and typing performance is given by Table 4.

Table 4. Analysis of Recall Data Design

Subject Type	Factor	No. of Levels
Within-Subject	Essay Order	3
Within-Subject	Topic Category	3
Within-Subject	Word Frequency	3
Between-Subject	Group	2
Between-Subject	Word Set	2

In this table, 'Essay Order' is the order that the essay was written i.e. first, second or third. 'Topic Category' consists of *Topic*, *Near* and *Distant*, 'Word frequency' is the frequency of the word. The 'Between-Subject' factor 'Group' specifies which set of essays, Group A or B, and 'Word Set' is the set of words i.e. *word_set_1* or *word_set_2*.

5.1 Recall Results

Our analysis of word recall showed that only the 'Topic Category' factor had a significant affect (F(2,16)=6.459, p=0.009, MSS=6.040) on how words were recalled. In particular, the recall of category *Topic* words was significantly greater than that of *Distant* words. Figure 2 shows the average recall between topic categories with Standard Error of the Mean (SEM).

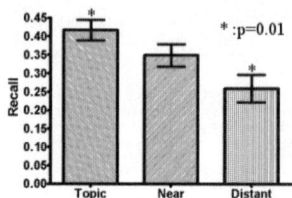

Fig. 2. Average recall between topic categories

This confirms that participants were more aware of information that was related or near to their essay topic. It is also worth noting that as the participants wrote the three essays they saw topics which were alternated between *Near* and *Distant* categories i.e. Group A saw 'Stem Cell' as *Near* for the first essay and as *Distant* for the third. When the words are seen as *Near* there was no significant difference in recall to the *Topic*, but once they fall into the category of *Distant* the recall difference is significant. This confirms that irrespective of the topic words fall under or the words themselves, once they are classified as unrelated to the current essay topic, the user becomes less aware of them. Next, we analyse the typing performance of the participants and determine what factors affected their typing speed.

5.2 Distraction Results

Initially we believed the location of the words within the ticker tape would be a factor i.e. if the words were in the first half (just appeared) or in the second half (about to go off screen). We performed an initial analysis of the average number of keystrokes when a word was in the first half or second half of the ticker tape's screen area. However, an analysis with this factor showed that where the words were on the ticker tape was not a significant factor so we did not include it in our final analysis. Table 4 gives our analysis design. From the analysis we found that only the 'Topic Category' of the word significantly affected the number of keystrokes (F(2,16)=6.497, p=0.009, MSS=447.133). Figure 3 shows the average number of keystrokes for each topic category, with SEM and the p values obtained from our analysis.

On average a user typed faster for *Distant* words and slowest for *Near* words. The number of keystrokes was significantly lower for *Near* words compared with *Topic* and *Distant* words. The *Near* words caused the user to slow down the

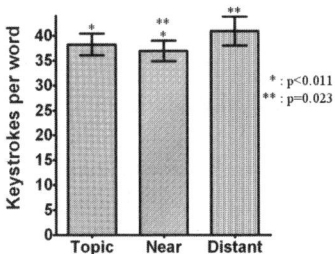

Fig. 3. Average Keystrokes per Topic Category Word

most. This would indicate that *Near* words were the most 'distracting' as the *Distant* words are ignored and the *Topic* words are related to the user's task.

Next we examined if presenting extra information to the user caused a decrease in typing performance and what factors influenced the decrease. The ticker-tape we used only started presenting information to the user 20 seconds after they had started typing. We therefore took the typing data from the first 20 seconds of each essay as a control measure of the participants' average number of keystrokes per second.[1] The number of keystrokes per second was also calculated when words were presented on screen to the participants. As before, we adhered to the analysis design outlined by Table 4. Presenting extra information to the user did decrease performance compared to the performance when no information was presented. However, the 'Topic Category' of the words on screen was again the only significant factor affecting how much the performance decreased (F(2,16)=3.636, p=0.031, M=4.351). Figure 4 shows the decrease in performance by topic category and gives the p values of significant difference.

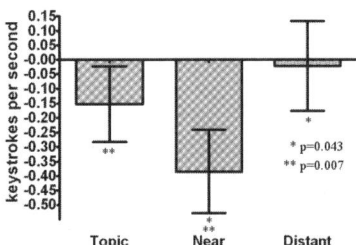

Fig. 4. Number of Keystrokes Per Second by Topic Category

Near words produced the greatest decrease and were significantly greater than *Topic* and *Distant* words. Its worth noting that *Distant* words had almost no effect on typing speed. This suggests that the presence of such words are not distracting to the user, this correlates with the recall results for *Near* words.

---

[1] The use of words from the first 20 seconds is not a substitute for a comprehensive control measure. We hope to perform a comprehensive control test in the future.

5.3 Word Frequency Effects

Although word frequency was not deemed a significant factor by our analysis of recall there were some interesting trends apparent within the data, in particular for low and high frequency words. Work carried out by [10], found when participants are shown a list of words containing both high frequency and low frequency the recall for low frequency words is significantly greater than for high frequency words. We had expected to find similar behaviour in our results and we did find that recall of low frequency words was consistently high. However, when we compared low frequency and high frequency word recall by topic category, we found that there was no significant difference for low and high frequency words in *Topic* and *Near*. There was however, a significant difference for high and low frequency words in the topic category *Distant*, see Figure 5.

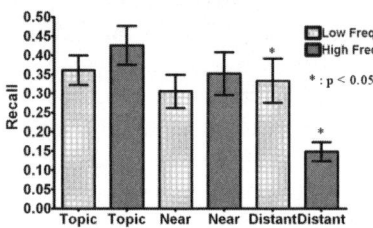

Fig. 5. Comparison of Topic Category Low and High Frequency recall

From this observation we can infer that the topic categories *Topic* and *Near* altered the effects of word frequency on recall, these categories boosted the participant's ability to recall high frequency words which is in contrast with [10]. The category *Distant* did not alter the recall from the behaviour observed by [10]

For typing performance we had expected that the frequency of the words appearing on screen would have an effect on typing performance, with low frequency words deteriorating performance. However, from our analysis in Section 5.2 we found that the performance across the frequency categories was consistent with no difference in the number of keystrokes for low, medium or high frequency words.

5.4 Discussion

To summarise, from our experiment we found that words classified as related or slightly related (*Topic* or *Near*) to the topic of a user's work are more likely to be recalled by a user. This suggests that users are more aware of words from these categories. The boosted recall of high frequency words for *Topic* and *Near* reinforces this supposition. In conjunction with these results we found that *Near* words are the most distracting to the user and reduce their typing performance when on screen. This emphasises that the words most detrimental to the user's work performance are *Near* words as the user is very aware of them and distracted by them. Therefore, JITIR systems should focus on removing from presented information such 'near topic' words. Our results show by contrast that 'distant' topic words are not distracting and that the user is relatively unaware of them. The fact that 'on topic' words do not diminish the user's performance and are accessible by the user, i.e. the user is aware of them, is encouraging. This confirms that presenting information in the periphery would be beneficial for JITIR systems.

6 Conclusions

We found that a user is more aware of information that is related or slightly related to the topic of their own work. JITIR systems could take advantage of this phenomenon to minimise their impact on a user, by filtering the information retrieved according to topic category. For example if the information retrieved is slightly related (*Near*) to the user's current work the JITIR should ensure that the information is beneficial to the user as they are more likely to pay attention to it. Similarly if the information is unrelated to the topic of the current task, a JITIR can readily discard it if the system is certain that the information is of no benefit. Otherwise, there is a low cost in presenting it to the user as they are unlikely to be aware of it and there is a potential gain from the user using it.

Low frequency words, irrespective of the topic category they fall under, will attract a user's attention. Therefore, if the JITIR system is confident the information is related to the topic of the user's work then and only then it should allow low frequency words. JITIR systems should avoid presenting low frequency words within information that it is unsure will be of benefit to the user i.e. distant topic words. This could be done by replacing low frequency words with high frequency synonyms, as we have shown high frequency distant topic words are less likely to attract attention.

It's worth noting that these results from our experiments can also be applied to areas other than JITIR. For example, Google AdSense [12] displays adverts on web pages, by modifying the wording of these adverts to better reflect the topics within web pages they would attract a user's attention. Similarly to boost the attention of unrelated ads the use of low frequency words can attract attention.

Finally we showed from our analysis of typing performance that words which are near the user's work topic had the greatest impact on a users performance. This would suggest that a JITIR system must ensure that the information presented to the user must be either about the topic of the user's work as it will be benefit to the user, or distant from the topic as it will not affect the user's performance and the user will, from our results in Section 5, not attend to those words. Words from topics that are near to the user's topic of work should be avoided as these words will decrease the user's performance.

7 Future Work

From Section 5.2 we showed our preliminary analysis into how words from different topic categories affect the user's typing performance. We hope to perform a more in-depth analysis of this data to ascertain what further affects topic words have on a user's performance. We also hope to investigate if the effects observed for single words will transfer to sentences of information. This would further benefit a JITIR system and allow such systems to be optimised.

Acknowledgements

We would like to thank all the people who took part in the experiments.

References

1. Budzik, J., Hammond, K.: Watson: Anticipating and Contextualizing Information Needs. In: 62nd Annual Meeting of the American Society for Information Science, Medford, NJ (1999)
2. Rhodes, B., Maes, P.: Just-in-Time Information Retrieval Agents. IBM Systems Journal 39(3-4), 685–704 (2000)
3. Budzik, J., Hammond, J., Birnbaum, L.: Information Access in Context. Knowledge Based Systems 14(1-2), 37–53 (2001)
4. Henzinger, M., Chang, B., Milch, B., Brin, S.: Queryfree news search. 12th International Word Wide Web Conference (WWW) (1-10 May, 2003)
5. Goodman, J., Varvalho, V.R.: Implicit Queries for Email. In: The Conference on Email and Anti-Spam (CEAS) (2005)
6. Meade, J., Costello, F., Kushmerick, N.: Inferring Topic Probability for Just-in-Time Information Retrieval. In: 17th Irish Conference on Artificial Intelligence and Cognitive Science (AICS 2006), Belfast, Northern Ireland (2006)
7. Gronau, N., Cohen, A., Ben-Shakhar, G.: Dissociations of Personally Significant and Task-Relevant Distractors Inside and Outside the Focus of Attention: A Combined Behavioural and Psychophysiological Study. Journal of Experimental Psychology: General 132, 512–529 (2003)
8. Deacon, D., Shelley-Tremblay, J.: How automatically is meaning accessed: A review of the effects of attention on semantic processing. Frontiers in Bioscience 5(Part E), 82–94 (2000)
9. KOffice Office Suite: http://www.koffice.org
10. Dewhurst, S.A., Hitch, G.J., Barry, C.: Separate effects of word frequency and age of acquisition in recognition and recall. Journal of Experimental Psychology: Learning, Memory, and Cognition 24(2), 284–298 (1998)
11. Wolford, G., Morrison, F.: Processing of unattended visual information. Memory & Cognition 8(6), 521–527 (1980)
12. Google AdSense: https://www.google.com/adsense/
13. Miller, G.A.: WordNet: A lexical database for English. Communication of the ACM 38(11), 39–41 (1995)

News Video Retrieval by Learning Multimodal Semantic Information

Hui Yu, Bolan Su, Hong Lu, and Xiangyang Xue

Shanghai Key Laboratory of Intelligent Information Processing
Department of Computer Science & Engineering
Fudan University, Shanghai, China
{052021193, 0472422, honglu, xyxue}@fudan.edu.cn

Abstract. With the explosion of multimedia data especially that of video data, requirement of efficient video retrieval has becoming more and more important. Years of TREC Video Retrieval Evaluation (TRECVID) research gives benchmark for video search task. The video data in TRECVID are mainly news video. In this paper a compound model consisting of several atom search modules, i.e., textual and visual, for news video retrieval is introduced. First, the analysis on query topics helps to improve the performance of video retrieval. Furthermore, the multimodal fusion of all atom search modules ensures to get good performance. Experimental results on TRECVID 2005 and TRECVID 2006 search tasks demonstrate the effectiveness of the proposed method.

Keywords: Video Retrieval, Rich Semantic Information, TRECVID, Manual Search Task.

1 Introduction

With the development of multimedia technologies and broadband internet, large amount of video data has been generated and made accessible in digital form throughout the world. More and more people are used to accessing video data for entertainment. However, finding the desired video data from a large amount of video database remains a difficult and time-consuming task. As a result, efficient video retrieval becomes more challenging. TRECVID [1], which is sponsored by the National Institute of Standards and Technology (NIST), provides us a benchmark for video retrieval. The video date provided by TRECVID is normally 100-hours long news videos. And the query topics include querying specific person, specific object, specific event or scene.

In this paper we propose a novel compact searching model for video retrieval based on learning rich semantic information from video. In extracting semantic information, we explore data mining methods and extraction of semantic ontology. Experimental results and evaluation on TRECVID2005 and TRECVID2006 illustrate the effectiveness of the proposed video retrieval method.

The remainder of this paper is organized as follows. In Section 2 we briefly review some video retrieval methods. Section 3 introduces our searching model for news video retrieval in TRECVID. Experimental results and analysis are given in Section 4 and the conclusion and introduction of our future work are given in Section 5.

2 Related Work in Video Retrieval

In content-based video retrieval, there are two main categories of methods: text-driven method and visual/aural-driven method. TRECVID research work has shown that text-driven method can achieve promising performance. For example, Yang *et al.* adopt text feature from transcript clues, video OCR text, and speech text to retrieve relevant persons in the video [2]. There are some situations that the text is absent. Under this case, the text feature can not be extracted. Christel *et al.* solved this problem by finding a class of topics for which the visual/aural-driven method would perform better than text-driven method [3]. Some work considered the combination of text, visual and aural information. After the experiment of analyzing the performance by adopting visual, concept and text features in content-based retrieval, Rautiainen *et al.* found weighted fusion of all kinds of features improved the performance over text search baseline [4]. Hauptmann *et al.* adopted multi-model fusion methods to retrieve video clips/shots, which re-ranked the results of text (transcript)-based retrieval results based on multimodal evidences using a linear model [5] [6]. They formulated video retrieval in multi-modalities including free text description, example images, and example video clips. Then multi-models including text retrieval, content-based image retrieval, specific shot detection and face similarity query were used to retrieve relevant video shots. Their results showed that the query results of text were the most reliable; however the ranks were not good. Hence linear fusion was proposed to re-rank the query results of text queried by multimodal features. Chua *et al.* adopted the similar video retrieval method [7][8], and learned the weight of linear fusion by query-type-specific methods [5]. In addition, to improve retrieval recall, they added the weight of text retrieval by expanding the query words using WordNet [9], and Rautiainen *et al.* also found expanded query term list could give notable increase in performance [4].

Nowadays in the research of video retrieval, more and more attention has paid to making better using of multimodal information including text, image, and audio. Most of these methods' emphasis is the fusion of the multimodal features, and they adopt different machine learning methods such as logistic regression, MGR [10] for fusion or boost [6] to train the weights of multimodal features in their experiments. But there is bottleneck in video retrieval research nowadays. Although different kinds of machine learning methods are used to help retrieval, the performance of video retrieval promotes little. Our works' focus is to improve the retrieval performance by mining more useful features from the original information and learning the relation among rich semantic information. Specifically, compared with the traditional text-driven retrieval method, our method mined more textual information such as query keywords selection and so on. Rich visual-driven features are also extracted to help promote performance. The model we propose performed well in TRECVID2005 and TRECVID2006 manual search task.

3 Compound Video Retrieval Model by Learning Multimodal Information

Video is a set of continuous frames in temporal, which combines multimodal information including images, text, audio, etc. Our model makes full use of these information.

3.1 Multimodal Information in Video

In our study, among the main categories of multimodal information, such as textual information, visual information, and aural information, we use visual and textual information. The details will be introduced as follows.

3.1.1 Textual Information
Early year's research has shown that text information plays an important role in video retrieval. Compared to visual information and aural information, text information is more reliable in video retrieval. In our retrieval model, we extract textual information by these ways. For example, by using automatic speech recognition (ASR) and video optical character recognition. Also, NIST provided close caption information in TRECVID. In our opinion, the textual information corresponding to the shots provides the evidence on the relevance of the shot. After the extraction of the textual information, we prune the stop words and index the remained words in the form of reverse table in order to improve retrieval efficiency.

3.1.2 Visual Information
Compared to textual information, visual information seems more intuitionistic for person. In our model, there are two main categories of visual information:

Low-level visual information: there are two kinds of low-level visual information extracted in our system. One kind of feature is image low-level feature. There are four global features are used in our model including HSV color histogram, Lab color histogram, Gabor texture feature, and edge histogram descriptor [11]. Another kind of feature is camera motion feature. We used a feature-based camera motion detection approach [12] to analyze the camera motion in a shot. This approach utilizes the motion vectors in P-frames from compressed video stream to detect human perceptive camera motions such as pan, till, zoom, and still in an individual shot.

High-level visual information: there exists gaps between high-level semantic information and low-level image features. In TRECVID 2005 [13], we extracted high-level feature by means of computing global low-level features. These low-level features are color layout descriptor (CLD), scalable color descriptor (SCD), Lab color histogram (LAB), edge histogram descriptor (EHD), and Gabor texture feature (GAB). For LAB and GAB we split the image into a 5 by 5 grid in order to reflect spatial information in some degree. We adopt (PCA) for feature dimensionality reduction and (SVM) for classification in the process of training. Finally classifier combination method such as Borda Count (BC) is used to fuse the multiple classification results. In TRECVID 2006 [14], we used JSEG [15] to segment key frames of each shot into several regions and label these regions as basic concepts. We extract low-level features from these regions. In the process of training, we used different methods such as SVM and GMM for classification. Since news video has its own characteristic, we designed a detector to detect anchor shots in new video. The detector first selects shots which contain faces as candidates by face detection algorithm, then extracts facial features such as size, location, orientation, etc. Clustering is then used to find anchor shot according to the particular temporal repeat property in a phase of news video.

3.2 Compound Retrieval Model

Based on the information extracted from videos, we construct our compound retrieval model. Our model consists of several feature retrieval modules. Each module plays a different role in video retrieval. This model has been experimented in TRECVID2005 and TRECVID2006 manual search task.

3.2.1 Feature Retrieval Module

Our compound model consists of five main classes of feature query module. We introduce these modules in detail in the following.

Text Query Module (TQ): Text query module is a typical Information Retrieval (IR) search engine based on the match between the textual query and the words in textual index constructed by extracted textual information. For the text query results in our study, the TF*IDF weighting scheme is adopted to generate the textual retrieval scores of shots. Considering a relevant shot does not always have keyword hit on itself, we use a window to overcome this temporal mismatch. Specifically, we propagate IR score of a "hit" shot S_0 to its neighboring shots S_i in a window by an exponential decay function, i.e., $r(S_i) = r(S_0) \cdot \alpha^i$, where α is within [0, 1]. From this equation, the closer the shot is to the position of keyword hit, the larger score it gets. The parameter α in exponential decay function is set as 0.5 in our experiment. Furthermore in TRECVID2006, We focus three points in the process of textual query module. One point is to adaptively determine the neighboring shots similar to shot S_0. We adopt the algorithm introduced in [16] to compute the similarity between neighboring shots and set a threshold T1 on the similarity. If the similarity is smaller than T1, we regard the two shots are irrelevant. If there are two continuous shots irrelevant, we regard the compared neighboring shot as the boundary of the related window. The second point we focus is the selection of query text. It is not an easy job to select suitable query text. There exist some reasons: Domain knowledge helps person to extract query keywords, but he can hardly establish exact query keywords of all topics which influence the final retrieval results greatly. For example, query topic 0190 in TRECVID2006 is to "Find shots of at least one person and at least 10 books". We construct a statistical method: Keywords Selection by Feedback (KSF) to obtain each topic's query keywords. First we select basic query keywords of each topic manually and expand them by means of WordNet [9]. We use expanded keywords to do retrieval for each topic in develop dataset. Then for each topic those words exist in positive result shots are keyword candidates. We compute keyword candidates' frequency and rank them. Finally we eliminate those candidates which have too high or too low frequency because it means they are too popular or too rare in textual information and the left are as query keywords in test dataset. The process of KSF loops several times in our system. Figure 1 shows the framework of KSF. The last point we find that due to some text transcript operation like ASR and OCR doesn't provide promising text detection precision, some textual information we extracted contains a lot of mistakes, e.g. word spelling error, etc.

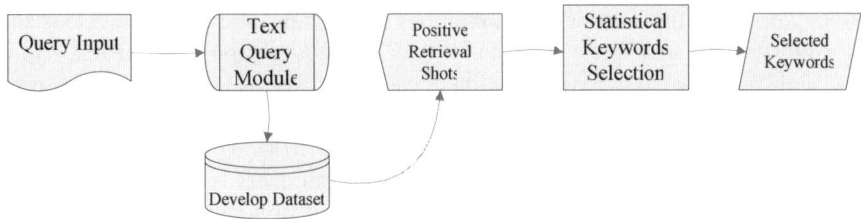

Fig. 1. Framework of KSF

These mistakes often occur in personal names and place names. So when we match two words in textual query module, we calculate edit distance [17] of two strings, if the distance is smaller than a threshold, that means maybe these two words are similar even same.

Image Query Module (IQ): Image query module computes the similarity of the key-frame of the shot to image samples or key-frame of video samples. We use the features extracted from low-level visual information, including HSV color histogram, Lab color histogram, Gabor texture feature and edge histogram descriptor. For data consistency, we only select the images of key-frames of the provided video examples for querying, while the provided image examples will be discarded if they are not from videos because there exist great diversity among images from different datasets.

Semantic Feature Extraction Module (SFE): We use the features extracted from high-level visual information in this query module. These semantic concepts we used include (1) person, e.g. leader, prisoner, etc; (2) object, e.g. car, building, flag, etc; (3) nature scene, e.g. water, sky, etc; (4) event, e.g. people-marching, meeting, etc. We extract 39 semantic concepts defined by TRECID 2006 by means of our method. Another 101 semantic concepts provided by MEDIAMILL [18] are also used in our module.

Motion Analysis Module (MA): Motion analysis module helps mining the camera motion in the video. In some videos such as sports video, motion analysis play an important role in video retrieval. There are four kinds of camera motion detected in our system: pan, tilt, zoom, and still.

Filter Module (FM): In TRECVID search task, we particularly design a filter module to filter those unsure resulted similar shots. The module consists of two parts. One part is the stable filter which uses the feature of anchor shots in high-level visual information. Each topic's query results should be filtered by it. The other part is the variable filter which consists of semantic concept features. And for different query topics, this filter may not be necessary. The second part is only used in TRECVID2006.

For example, the query topic 0187 in TRECVID2006 "Find shots of one or more helicopters in flight", we should filter those resulted shots contain semantic concepts e.g. building, water, etc. For another query topic 0192 in TRECVID2006 "Find shots of a greeting by at least one kiss on the cheek", we may filter those shots contain semantic concepts e.g. plane, ship, etc. So we construct concepts relation ontology manually to help we establish the ingredient of each topic's variable filter. In our ontology network there are two main relations among semantic concepts: coexistence and exclusion. For example, "ship" and "water" have a relation of coexistence while "sky" and "building" may have a relation of exclusion. We just want filter those result shots which contained excluded semantic concepts. If concept A can coexist with concept B, then $P(A|B)=P(B|A)=1$, otherwise $P(A|B)=P(B|A)=0$. In the process of query, we will first extract semantic concept set T from query text. Then construct its variable filter with the semantic concept a following the principle $\{P(a|t)=0;\ t \in T\}$.

Figure 2 shows the work flow of five query modules in our model.

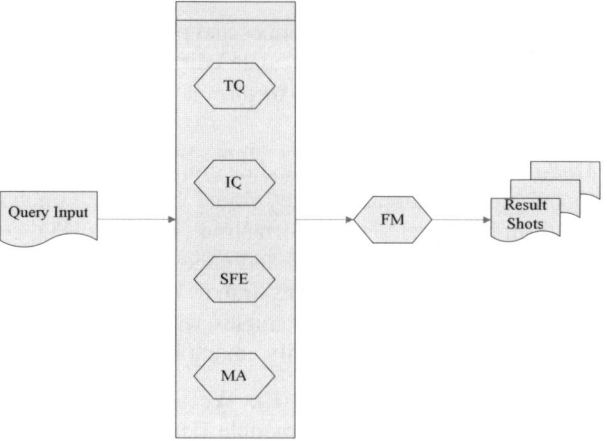

Fig. 2. Work flow of query module in our framework

3.2.2 Query Topic Style Analysis in Our Model

In traditional video retrieval method, an important problem has been ignored. People often prefer to select video features as many as possible which they would like to apply for all styles of query topics. Actually not all features are suitable for all styles of query topics, because for some topics we could extract little information of some feature. For example, textual feature is available for topics such as searching person, but it is not suitable for searching object sometimes. We may have no idea of query textual input for the topic 0166 "Find shots of one or more palm trees" in TRECVID2005. Even using our KSF model, we find little textual information can be extracted for this topic. The same problems may occur in other feature information. Invalid feature information may bring noise to the retrieval system. So it is necessary to do query topic style analysis in the framework of video retrieval. In our study we define four categories of style:

T Style: Mainly rely on textual information in the process of retrieval. Query topics relate to persons often belong to this category. Because face detection helps little for person detection in real news video retrieval due to complex camera angles and camera lights, while we can extract rich textual information from transcript operations. Textual information acts a key role here.

V Style: Mainly rely on visual information (including both low-level features and high-level features) in the process of retrieval. Query topics relate to concrete objects such as palm tree, building, etc. often belong to this category. We can extract little textual information for this category's query topics. Visual information is mainly used here.

M Style: Both textual information and visual information play important parts in the process of retrieval. Query topics relate to scene and event often belong to this category. We can extract rich textual and visual information for this category's topics.

N Style: We can extract neither rich textual information nor visual information for this category's topic. So our system often provides a disappointing performance for this category's topic which remains the challenge in our future's research.

Table 1 shows the style of query topics we defined in TRECVID2006.

Table 1. Styles of Query Topic

Topic Number	Topic Content	Topic Style
0173	one or more emergency in the motion	M
0174	tall buildings (more than 4 stories)	V
0175	people leaving or entering a vehicle	V
0176	soldiers/police escorting a prisoner	N
0177	daytime demonstration or protest	M
0178	US Vice President Dick Cheney	T
0179	Saddam Hussein	T
0180	people in uniform and in formation	N
0181	President Bush, Jr. walking	T
0182	soldiers or police with weapons	M
0183	boats or ships	V
0184	people seated at a computer	V
0185	People reading newspaper	N
0186	Nature scene	V
0187	helicopters in flight	V
0188	something burning with flames	M
0189	a group including least four people	M

Table 1. (*continued*)

0190	one person and at least 10 books	N
0191	one adult person and at least one child	M
0192	a greeting by at least one kiss on the cheek	N
0193	Mokestacks, chimneys, cooling towers	V
0194	Condoleeza Rice	T
0195	soccer goalposts	M

3.2.3 Module Fusion and Rank

According the results of the analysis of query topics' styles, we establish each query module's weight in that topic's retrieval. For ***T Style*** topic, TQ has higher weight while IQ, SFE, MA have lower weights. To the opposite, for ***V Style*** IQ, SFE and MA should have higher weight than TQ. For ***M Style*** and ***N Style***, which module has higher weight depends on each module's performance in that topic's retrieval. From our early research and TRECVID annual report [19], we obtain some statistical information about query module's performance which is shown in Table 2. In our retrieval module, relation algebra expression is used to fuse multimodal information from various query modules. Since relation algebra is a set of operations which aim to the query operation of relation database, it can effectively express all kinds of complex query. The multimodal information that comes from various retrieval modules is independent with each other, which are similar to the "table" in relation algebra. Hence relation algebra expression can be used to fuse multimodal information and achieve good fusion performance in complex query topics. Final output results would be ranked by the strategy: Borda count. In our experiment, we ranked the retrieval precision of 24 topics in TRECVID. We think retrieval modules show good performance on the topics which ranked from 1st to 8th places and show common performance on the topics ranked fro 9th to 16th places. For the remaining topics, retrieval module show bad performance.

Table 2. Statistical Information of Query Module's Performance for Some Topics

Retrieval Topic	TQ's Performance	SFE's Performace
person	good	bad
sports game	good	good
sky	bad	good
building	bad	common
car	bad	common
battle	common	common
map	common	common
garden	bad	good
palm tree	bad	bad

4 Experiments

4.1 Experimental Data and Evaluation Metrics

We participate in TRECVID2005 and TRECVID2006 manual search task. Each year, NIST provides 277 news videos (about 160 hours) for test. All the videos are from the programs of six different TV stations in the world, which include three kinds of language: Chinese, English, and Arabic. There are 24 query topics each year for evaluation, which include searching specific person, specific object, and specific scene and event.

There are two evaluation metrics involved in all retrieval evaluation, i.e., recall and precision. In TRECVID Evaluation, since there are few relevant items in test data, we will achieve low precision even we have searched all the relevant items. To resolve this problem, average precision (AP) is proposed to evaluate the query results, which is defined as follows: For a given query, it assumed that we only consider the performance of N_r items of the results. Assume that the system only retrieves k relevant items in N_r items and they are ranked as $r_1, r_2, ..., r_k$. The average precision (AP) is computed as

$$AP = \left\{ \sum_{i=1}^{k} i / r_i \right\} / N_r \tag{1}$$

As shown in Equation (1), this metric reward the systems that put the relevant items near the top of the retrieval list and punish those that add relevant items near the bottom of the list. Another metric mean average precision (MAP) is defined as the mean of average precision of all topics.

4.2 Experimental Results and Analysis

In TRECVID 2005 manual search task evaluation, the MAP (Mean Average Precision) of all the submitted runs are illustrated in Fig. 3. Among them the one which is labeled as "M_A_2_D_MM_BC_1" is the run we submitted, which is the search result of our compound video retrieval model. The MAP of this result is 0.105, which ranks 4 among 26 submitted results.

In TRECVID 2006 manual search task evaluation, we refined our model based on that on TRECVID2005. Update some algorithms including adaptive neighboring shots selection, edit distance's calculation, and SKF method in TQ, query topic style's analysis, etc. MAP (Mean Average Precision) of all the submitted runs of TRECVID 2006 is illustrated in Fig. 4. Among them the one which is labeled as "M_A_2_FD_M_TEXT_1" is the run we submitted, which is the search result of the retrieval only used textual information. The MAP of this result is 0.036, which ranks 1 among 11 submitted results. Another submitted run which is labeled as "M_A_2_FD_MM_BC_3" which is the result of our compound model ranks 7 with its MAP 0.026.

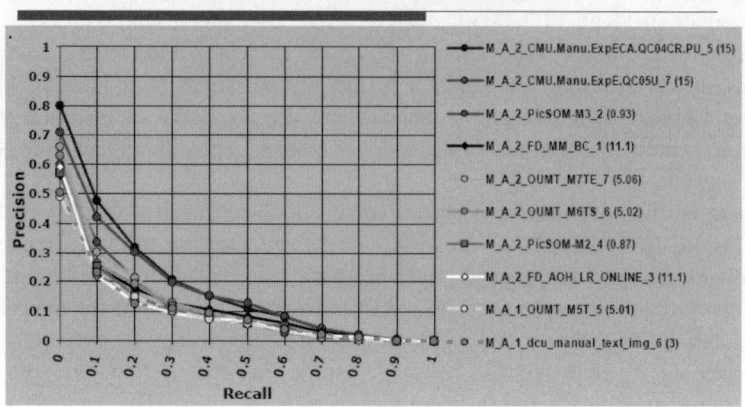

Fig. 3. Query results of TRECVID 2005 manual search

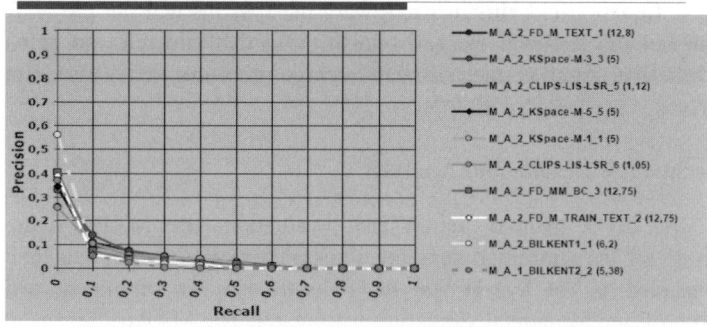

Fig. 4. Query results of TRECVID 2005 manual search

Our textual query module achieves highest MAP in TRECVID2006 manual search task that shows updated methods in TQ have played key roles. But our compound model didn't perform well this time as expected. There exists some reasons. The highest MAP in TRECVID2006 is just 0.036 which is much less than TRECVID2005's 0.105. One main factor is TRECVID2006's query topic is much difficult than TRECVID2005's. Statistical data in Table 1 and Tables 2 shows most topics in TRECVID2006 are non-T Style topics, and for most of those non-T Style topics, non-TQ modules (SFE module, IQ module, MA module) couldn't achieve a promising performance. So after the fusion of all query modules, the final MAP is greatly declined. Fig 5 shows each topic's MAP in TRECVID2006. It can be observed that most of the topics that have high MAP are *T Style* topics which may confirms that textual information is more reliable in video retrieval.

5 Conclusion

In this paper we proposed a compound model for video retrieval. We depicted each query module in our model in details which include adaptive neighboring shots selection, edit distance's calculation, SKF method in TQ and so on. We also introduced the concept of analysis of query topical style which could help us understand the essence of video retrieval. The results of the evaluation of TRECVID2005 and TRECVID2006 manual search task showed that our model performed well. The advantage of our model is the special algorithm designed in each retrieval module which helps to improve each module's performance in retrieval. Besides that, statistical information could also help us in video retrieval analysis. In the future work, we will explore more in visual-based features and use some machine learning method in our study, e.g. the establishment of each module's weight.

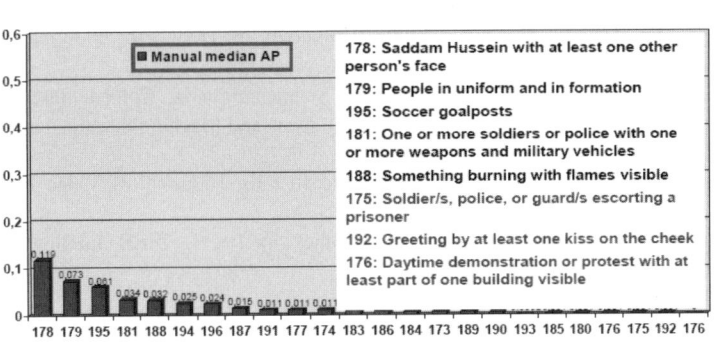

Fig. 5. Manual Runs' MAP by Topic

Acknowledgments. This work was supported in part by Natural Science Foundation of China under contracts 60533100 and 60402007, and Shanghai Municipal R&D Foundation under contracts 05QMH1403, 065115017 and 06DZ15008.

References

1. TRECVID Video Retrieval Evaluation: http://www-nlpir.nist.gov/projects/trecvid/
2. Yang, J., Hauptmann, A.: Naming Every Individual in News Video Monologues. Proceedings of 12th ACM International Conference on Multimedia, October 10-16, pp. 580-587.(2004)

3. Christel, M., Conescu, R.: Addressing the Challenge of Visual Information Access from Digital Image and Video Libraries. JCDL'05, June 7-11, pp. 69-78, (2005)
4. Rautiainen, M., Ojala, T., et al: Analyzing the Performance of Visual, Concept, Text Features in Content-Based Video Retrieval. MIR'04, October 15-16, pp. 197-204, (2004)
5. Hauptmannn, A., Chen, M.Y., Christel, M.C., et al.: Confounded Expectations: Informedia at TRECVID 2004. In: TRECVID Workshop, USA (2004)
6. Yan, R., Hauptmann, A.: Co-Retrieval: A Boosted Reranking Approach for Video Retrieval. In: IEE Proc. -Vis. Image Signal Process 152(6), 888–895
7. Chua, T.S., Neo, S.Y., Li, K.Y., et al.: TRECVID 2004 Search and Feature Extraction Task by NUS PRIS. In: TRECVID Workshop, USA (2004)
8. Yang, H., Chaisorn, L., Zhao, Y.L., Neo, S.Y., et al.: VideoQA: Question Answering on News Video. In: Proceedings of the ACM conference on Multimedia, pp. 632–641. ACM Press, New York (2003)
9. WordNet, a lexical database for the English language: http://wordnet.princeton.edu/
10. Melnik, O., Vardi, Y., Zhang, C.H.: Mixed Group Ranks: Preference and Confidence in Classifier Combination. IEEE Trans on Pattern Analysis and Machine Intelligence 26, 973–981 (2004)
11. MPEG-7 Overview, http://www.chiariglione.org/mpeg/standards/mpeg-7/mpeg-7.htm
12. Zhu, X.Q., Elmagarmid, A.K., Xue, X.Y., Wu, L.D., Catlin, A.C.: InsightVideo: Toward Hierarchical Video Content Organization for Efficient Browsing, Summarization and Retrieval. IEEE Trans. on Multimedia 7, 648–666 (2005)
13. Xue, X., Lu, H., Wu, L., et al.: Fudan University at TRECVID 2005. In: TRECVID workshop, Gaithersburg, MD (November 2005)
14. Xue, X., Lu, H., Yu, H., et al.: Fudan University at TRECVID 2006. In: TRECVID workshop (November 2006)
15. Deng, Y., Manjunath, B.S: Unsupervised Segmentation of Color-texture Regions in Images and Video. IEEE Trans. on Pattern Analysis and Machine Intelligence 23(8), 800–810 (2001)
16. Peng, Y., Ngo, C.: Clip-based Similarity Measure for Hierarchical Video Retrieval. In: MIR 2004, October 15-16, pp. 53–60 (2004)
17. Houghton, R.: Named Faces: Putting Names to Faces. IEEE Intelligence System Magazine 14(5), 45–50 (1999)
18. MediaMill, http://www.mediamill.nl/
19. Over, P., et al.: TREC 2005 Video Retrieval Evaluation. In: TRECVID workshop, Gaithersburg, MD (November 2005)

Visualization of Relational Structure Among Scientific Articles

Quang Vinh Nguyen, Mao Lin Huang, and Simeon Simoff

Faculty of Information Technology,
University of Technology, Sydney, Australia
{quvnguye,maolin,simeon}@it.uts.edu.au

Abstract. This paper describes an ongoing technique to collecting, mining, clustering and visualizing scientific articles and their relations in information science. We aim to provide a valuable tool for researchers in quick analyzing the relationship and retrieving the relevant documents. Our system, called *CAVis*, first automatically searches and retrieves articles from the Internet using given keywords. These articles are next converted into readable text documents. The system next analyzes these documents and it creates similarity matrix. A clustering algorithm is then applied to group the relevant papers into corresponding clusters. Finally, we provide a visual interface so that users can easily view the structure and the citing relations among articles. From the view, they can navigate through the collection as well as retrieve a particular article.

Keywords: Information Visualization, Citation, Clustering, Article Retrieval, Text-Mining.

1 Introduction

Scientific article's analysis usually shows the coherence in literatures as well as their change in intelligible ways over time. One of the significant areas of the scientific analysis is citation and co-citation networks. This area has been long studied in information science and other disciplines. Most of the recent techniques cooperate with an advanced visualization technique for assisting the analysis. Within the scope of this paper, we briefly review a few typical techniques in citation and co-citation analysis.

CiteSpace [4] and its new version *CiteSpace II* [5] are the one of the most recent examples along this line of research. This system can detect, conceptualize and visualize emerging trends and transient patterns of citation and co-citation footprints in scientific literature. Although it provides a good visualization which shows both of cluster views and time-zone views, its 2D spring-force layout algorithm is slow and could create a crowded view for a large scientific citation networks.

CiteWiz [8] is another framework for bibliographic visualization of scientific network. This technique can graphically present the hierarchies of articles with potentially very long citation chains. Besides the visualization of citation hierarchies, the authors also provide an interaction technique and an attributed property for enhancing the overview of important authors and articles which make it more useful for a wide range of scientific activity. Modjeska et al. [14] described a similar technique that can capture the user-required relationships between bibliographic entries or articles. This system also supports chronological structure with multi-articles overview, single-article view and spatial attribute-relevance view. Other visualization techniques for citation and co-citation analysis can be found at [2], [6], [13], [17], [19] and [22].

Although citation and co-citation analysis has been received some attention from researchers, most of the current systems do not analyze the real article in order to show the coherence and property of literatures. This paper proposes a new technique that can collect automatically real articles from the Internet from given keywords. These actual documents are then converted into a readable text format for data mining and clustering based on their content similarity. A visualization and interaction interface is provided in which user can view the pattern of collected articles as well as access to the actual document via an interactive navigation. Our initial prototype aims to show the feasibility of a system that can automatically collect, analyze and visualize the scientific articles so that the researchers can understand the trend and pattern of the research in information science. Section 2 shows a framework of the *CAVis* system. Section 3 describes further detail of components in our prototype including text mining, clustering and visualization. The following section presents a technique for visualizing article citations. Final section is the conclusion and future work.

2 The Framework of CAVis

The *CAVis* technique includes four main modules including article collection, file conversion, text-mining and clustering, and visualization (see Figure 1). These first three modules involve heavily computational resource and Internet bandwidth, thus they are operated independently from the last module, i.e. the visualization and interaction. These main modules from CAVis's framework are now further described.

Article collection - first, the raw data is collected from the Internet based on given search topic keywords. These raw data theoretically includes scientific articles which can be formatted as Portable Document Format (PDF), Post Script (PS), Microsoft Word documents and others. The searching of these articles operates automatically via search engines, IEEE's and ACM's websites. The process usually requires a huge Internet bandwidth and computer resources. Within the scope of the project, this prototype only collects scientific articles in the PDF format from the Google Web APIs [23]. The APIs service includes a rich library that allows users to query billions of web pages directly from their own computer programs. From this APIs, we implemented a program that collect all PDF files (possibly other types such as PS and Doc files in the future) which are returned from the Google's search for further processing.

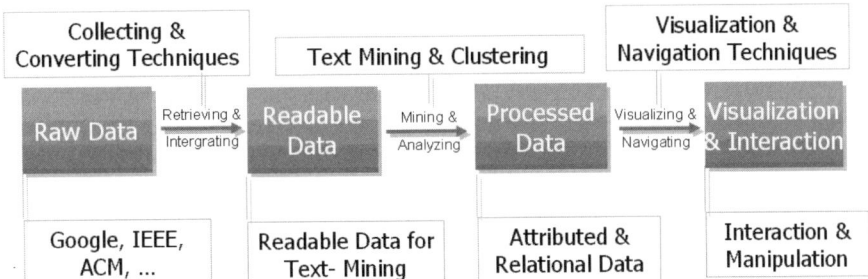

Fig. 1. The framework of the CAVis system

These collected articles are then stored locally at associated folders. We provide a mechanism to filter out those broken links, duplicated links as well as irrelevant and bad files. Names of those files are also modified to ensure there are no duplications. Finally, all associated information from the retrieval process is recorded for later management and operation. Figure 2 shows an example of the article collection interface and the recorded information.

Format conversion - this process is responsible for converting formats into unified readable content for further text-mining, i.e. from PDF to TEXT. We applied the PDFBox library [24] for converting those PDF documents into text documents. Similar to the previous process, this step also requires a large computer resources and it runs independently from a server to ensure that it does not affect with user interaction.

Text-mining and clustering - this process analyzes those text files in order to find the clusters of similarity and possible inter-relationship or inter-citations among the articles. In this prototype, we only apply a text mining algorithm for discovering the article relationships. We then use the *FACADE* clustering algorithm [18] to identify the groups of relevant articles.

Visualization and Interaction - the visualization interface allows users to view, learn, interact and retrieve the corresponding articles. This interactive interface is able to show not only the abstract view of the entire documents, their clustering property, their inter-relationships (or citation relationships) but it also can display in detail any particular document. Our system also allows user to retrieve and view the actual document from the local storage or from the original Internet. Detail of the visualization and interaction is further described at section 4.

3 CAVis

The *CAVis* system involves a numbers of related subsystems that run independently. The collection, conversion, text mining and clustering processes are required a large computational power. Therefore, they are operated before the visualization because the speed is important for human interaction. These processes are only operated once and the information is then recorded for the

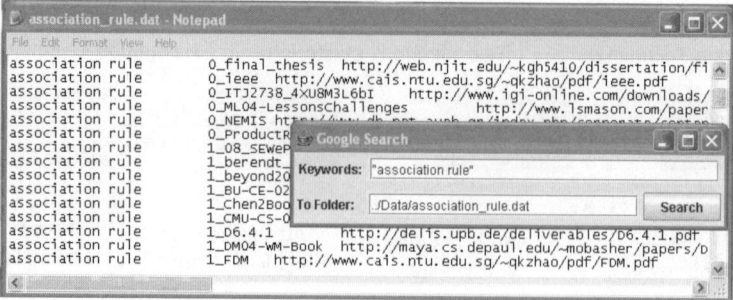

Fig. 2. An example of the search interface

visualization and interaction. This section describes the technical detail of three most important processes including text mining algorithm, clustering algorithm and the visualization.

3.1 Text Mining

We modify a text mining algorithm from Liu et al. [12] to extract keywords from search articles and generate a similarity matrix. We first use a customized stop-list in computer science to filter out the irrelevant words from the collected articles. We then use a statistical method to extract keywords from filtered articles based on the method described by Andrade and Valencia [1]. The extraction generates a list of frequencies of words in each article. The estimation of the significance of words is worked by comparing the frequency of words from the word lists of all articles. The results of the list of frequencies are than normalized to ensure the consistency of size among the documents. Finally, a similarity matrix is generated based on the list of frequencies from all documents. Detail of this algorithm can be found at [12].

3.2 Clustering

Given a similarity matrix, an effective approach to document classification is cluster analysis. Clustering methods can be traditionally classified into four categories including partitioning, hierarchical, density-based, and grid-based methods. Among the four kinds of methods, hierarchical methods can be directly applied to process the similarity matrix without needing the original data. To provide a comparison and demonstrate the compatibility of the layout method, two clustering methods are selected to classify the documents. The first one is the classic hierarchical agglomerate clustering (HAC) method and the second one is *FACADE* [18]. The two clustering methods merge data in different ways. While HAC considers the similarity value between pair of documents, FACADE considers a group of documents and uses group density information to merge data hierarchically.

3.3 Visualization

We use a modified version of the *EncCon* algorithm [16] so that the visualization can be able to handle clustered graph. We use colors to visualize the articles based on their hierarchical structure and clustering property. The use of visualization techniques aims to provide not only a two-dimensional graphical visual interface for viewing the entire collection and a particular document, but also an interface where users can learn, interact and retrieve the corresponding articles (see those from Figure 4 to Figure 6).

In our visualization system, nodes are used to represent the article and edges are representing the relationships among these documents, i.e. the citation and co-citation. Therefore, the edges are used to show not only the hierarchical relations, i.e. connections between articles and a categories, but also a citation between articles. In addition, an article can belong to one or more category.

Technically, the layout algorithm first groups nodes at the same cluster within a sub-graph as a subgroup for partitioning purpose. This partition ensures that nodes in the same cluster are positioned close together and are painted with the same colors for easy identification (see Figure 4 and Figure 5).

The modified *EncCon* layout algorithm is used to present the structure of the articles. This algorithm was chosen and used in our implementation to take its advantages of efficient utilization of geometric space, fast calculation, and aesthetical niceness. Figure 3 shows an example of the visualization of a very large cluster graphs using the modified *EncCon* layout algorithm. This algorithm can lay out the articles hierarchically based on their category and cluster property. We next briefly describe some technical detail of the *EncCon* layout algorithm.

4 Article Visual Analysis

4.1 Layout

Although there have been many proposed innovative techniques both in visualization and navigation to facilitate the design of interactive visualization, most of these techniques, however, do not consider all the aspects involved in interactive visualization design. Currently only a few interactive visualization solutions satisfy the multiple design requirements such as 1) space utilization, 2) fast computation, and 3) minimization of human cognitive process.

The original *EncCon* layout technique [16] used a fast enclosure+connection algorithm to calculate the geometrical layout of large hierarchies. This technique is later extended to make it be able to handle general and clustered graph in a two-dimensional space. It essentially inherits the advantage of space-filling techniques that maximize the utilization of display space by using area division for the partitioning of sub-graph and nodes. Note that the issue of space utilization becomes significantly important when visualizing large graphs with hundreds or thousands of nodes and edges because of the limitation of screen pixels. It is similar to the original *Squarified* Tree-Maps [3] that uses a rectangular division method for recursively positioning the nodes. This property aims to provide

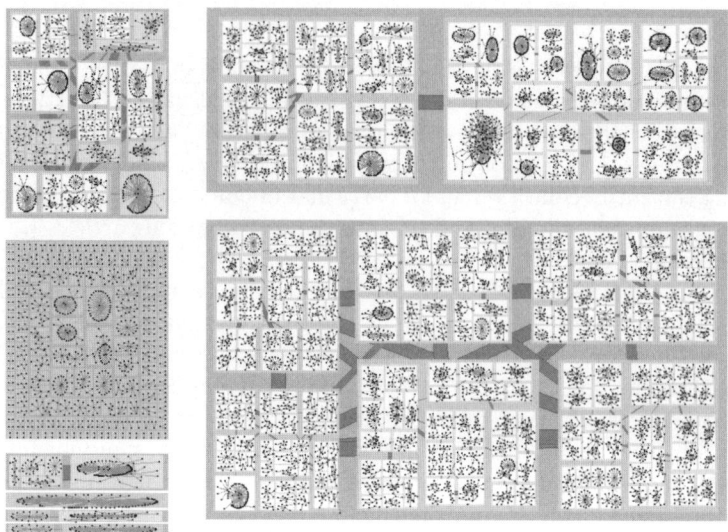

Fig. 3. An example of the visualization of a large dataset using modified EncCon layout algorithm

users with a more straightforward way to perceive the visualization and ensures the efficient use of display space.

In order to address the specific criteria of *EncCon* drawing, we use 'squarified' rectangles for the area division. The EncCon drawing ensures that all hierarchical sub-graphs are inside rectangular geometric local regions. Thus, there is no overlapping between rectangular local regions of nodes and their sub-trees. However, our area division algorithm is different from the *Squarified Tree-Maps* algorithm [3]. In *Squarified Tree-Maps*, the partitioning is accomplished through the horizontal-vertical manner. In *EncCon*, this is achieved in the circular manner, in which all rectangles are placed in the north-east-south-west order around four sides of the parent rectangle. Both of the above partitioning algorithms ensure the efficiency of space utilization. The *EncCon* visualization also uses a node-link diagram to present the hierarchical structure explicitly.

The EncCon technique has been extended from [16] to handle general graphs. This technique first analyzes and finds community structure among a dataset and it then generates a clustered graph. The clustering process works independently from the layout algorithm so that it does not affect the reaction time which the speed is important for the real time interaction. This clustering algorithm is also a modification from Newman [15]. We next visualize the hierarchical-clustered graphs using a combination two algorithms. The clustered subgroups are laid out using an *enclosure+connection* partitioning algorithm to ensure the space-efficiency. And the leaf nodes from a subgroup are laid out using a number standard graph layout algorithms such spring force-directed graph layout algorithm [7] and circular drawing, and any simple layout algorithm. This is because the space-efficiency is not a key issue for such a small number of leaf

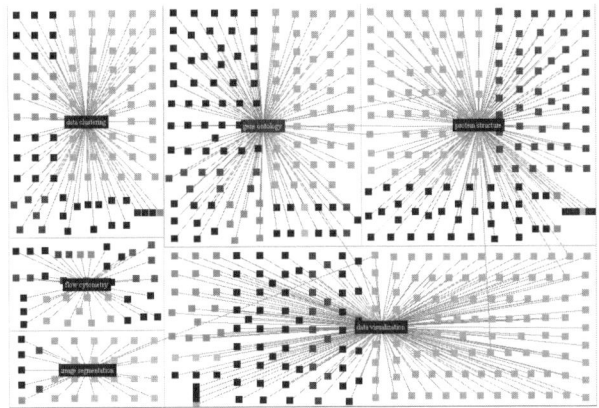

Fig. 4. An example of the visualization of an entire collection of articles using FACADE clustering algorithm

nodes and the aesthetical niceness is more important. Figure 3 shows an example of the visualization of a large graph using the modified *EncCon* layout algorithm.

4.2 Navigation and Interaction

The semantic zooming is used for navigating entire article collection. This technique enlarges the display of a focused sub-category with more details (see Figure 4 and Figure 5). When being selected by a left mouse-click (zoom in), the selected node moves forward to the position of the root, i.e. center of the rectangular display area. The display region of the selected node now expands to the entire display area. In other words, we only visualize the sub-graph of selected

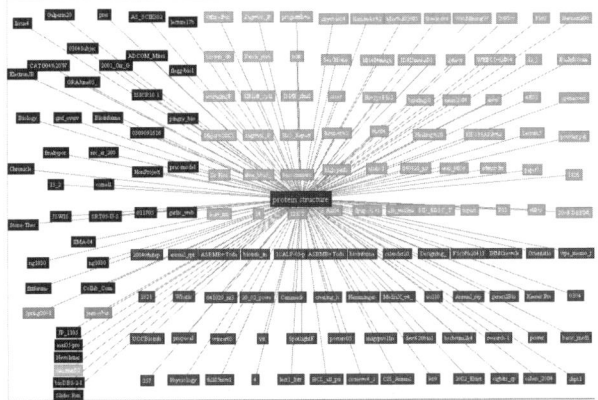

Fig. 5. An example of the visualization using FACADE clustering algorithm and focusing on the topic "protein structure"

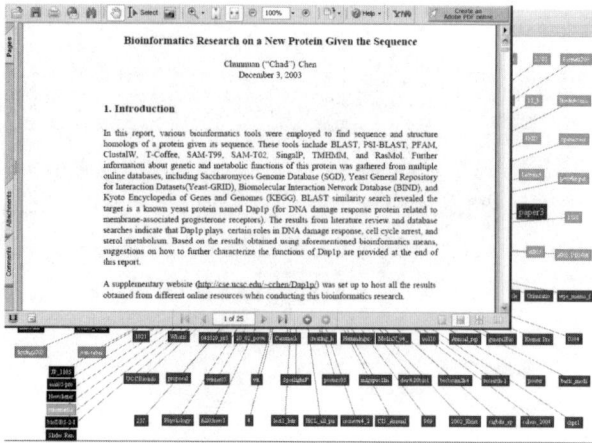

Fig. 6. An article retrieval

node. This viewing technique requires the recalculation of positions of all vertexes in a sub-tree at a time in corresponding to the left-click. The system displays back the father's hierarchy (zoom out) when the user clicks on the right button. Double-clicking on a node will trigger the system to retrieve and display the actual article in their original format (see Figure 6).

5 Citation Visualization

Our technique first use a fast clustering algorithm to discover community structure in the bibliographic networks of scientific papers. The clustering process partitions an entire network into relevant abstract subgroups so that the visualization can provide a clearer and less density of display of global view of the complete graph of citations. We next use the modified *EncCon* algorithm to archive the optimization of graph layout.

Similar to [9], our technique also employs rich graphics to enhance the attributed property of the visualization including publication years and number of citations. In the citation visualization, nodes are used to represent articles and edges between nodes are representing citations among these articles. Each node in our visualization is associated with a color corresponding to its publication year. Our program automatically matches colors for publication years based on an array of default colors. Our visualization also presents the number of citations for articles using bright-background-color circles in which the area of a circle is proportional to the number of papers citing to the article (see Figure 7 and Figure 8).

Figure 7 illustrates a an example of the citation visualization on a group of papers. The visualization shows the scientific papers in an abstract manner as well as the relationship among them. Figure 8 presents the enlarge view when a selected subgraph from Figure 7 is selected to zoom in. This figure indicates that

Visualization of Relational Structure Among Scientific Articles 423

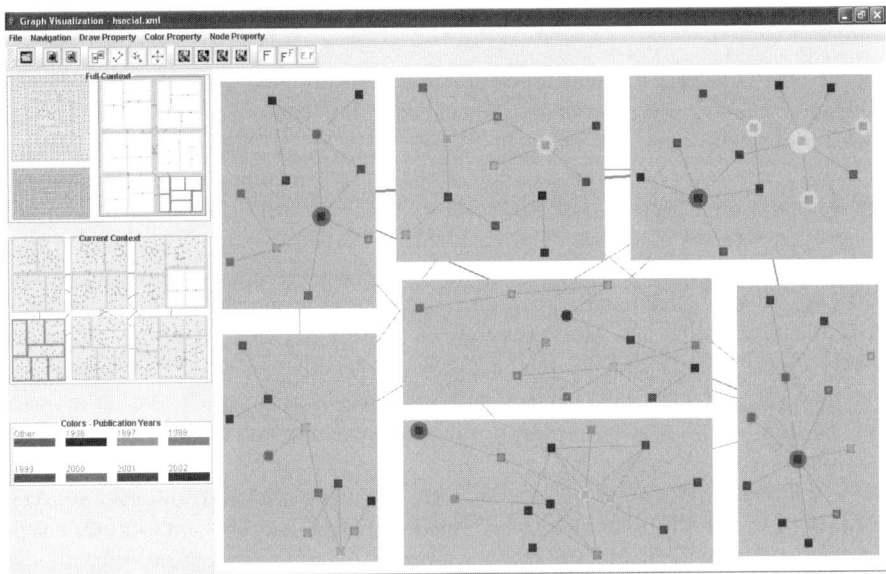

Fig. 7. An example of the visualization of a collection of research papers

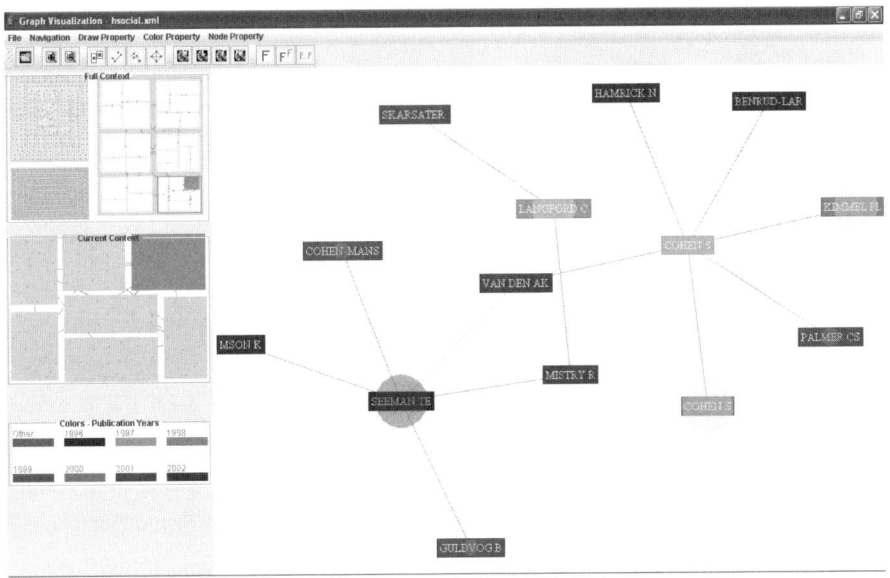

Fig. 8. An example of a navigational view of a selected subgraph from Figure 7

the most cited article is from the author Cohen 1997 as well as some good cited articles from Seemante 1996, Cohen 1998, Langford 1997 and Kimmel 1998.

6 Conclusion and Future Work

We have presented a method for collecting, text-mining, clustering and visualizing scientific articles in information science. Our technique first automatically search and retrieve articles using the Google Web APIs. These articles are stored at local hard-drives and they are converted to text-format for further analysis. The system then analyzes and mines those articles to find the similarity matrix. Clustering algorithms are applied to identify the similar articles. We also provide a visualization and interaction interface where users can view the overall structure as well as navigate through the collection or articles. Our system is also able to visualize the citation relations among articles. Although this work is only at the early state, its promising result has shown a feasibility of a system that can automatically collect, analyze and visualize the scientific articles. This technique can operate without the manual collection of articles as well as citations or co-citations.

We will extend our automatic search and article retrieval from other sources, including from IEEE or ACM digital library, as well as documents in other type of formats. This will ensure a better comprehension of the collection. We will also try to investigate new mining algorithms to extract associated information the articles such as title, authors and cited references and others. This information will be used to automatically visualize the cross citing relevance among the articles.

Acknowledgments. The authors would like to thank Yu Qian and Kang Zhang for providing and advising the text mining program. This project is partly supported by Australia ARC Discovery research grant #DP0665463.

References

1. Andrade, M., Valencia, A.: Automatic Extraction of Keywords from Scientific Text: Application to the Knowledge Domain of Protein Families. Bioinformatics 14, 600–607 (1998)
2. Brandes, U., Willhalm, T.: Visualization of Bibliographic Networks with a Reshaped Landscape Metaphor. In: VisSym 2002, Proceedings of Joint Eurographics - IEEE TCVG Symposium on Visualization, pp. 159–164. ACM Press, New York (2002)
3. Bruls, M., Huizing, K., van Wijk, J.J.: Squarified Treemaps. In: Proceeding of joint Eurographics and IEEE TCVG Symposium on Visualization, Vienna, Austria, pp. 33–42. Springer, Heidelberg (2000)
4. Chen, C.: Searching for intellectual turning points: progressive knowledge domain visualization. In: Proceedings of the Arthur M. Sackler Colloquium of the National Academy of Sciences. The National Academy of Sciences of the USA, Irvine, CA (May 9-11 2003), pp. 1–8 (2003)
5. Chen, C.: CiteSpace II: detecting and visualizing emerging trends and transient patterns in scientific literature. Journal of the American Society for Information Science and Technology (to appear 2006)

6. Chen, C.: Visualising semantic spaces and author co-citation networks in digital libraries. Information Processing and Management 35, 401–420 (1999)
7. Eades, P.: A heuristic for graph drawing. Congressus Numerantium 42, 149–160 (1984)
8. Elmqvist, N., Tsigas, P.: CiteWiz: A Tool for the Visualization of Scientific Citation Networks. Technical Report 2004-05, Chalmers University of Technology and Göteborg University, Sweden (2004)
9. Huang, M.L., Nguyen, Q.V: Visualization of Large Citation Networks with Space-Efficient Multi-Layer Optimization. In: ICITA 2007, Proceedings of 4th Int'l Conf. in IT and Applications, Harbin, China, pp. 489–495 (2007)
10. Jain, A.K., Murty, M.N., Flynn, P.J: Data clustering: a review. ACM Computing Surveys 31(3), 264–323 (1999)
11. King, B.: Step-Wise Clustering Procedures. Journal of Am. Stat. Assoc 69, 86–101 (1967)
12. Liu, Y., Navathe, S.B., Civera, J., Dasigi, V., Ram, A., Ciliax, B.J, Dingledine, R.: Text mining biomedical literature for discovering gene-to-gene relationships: a comparative study of algorithms. IEEE/ACM Transactions on Computational Biology and Bioinformatics 2, 62–76 (2005)
13. Mackinlay, J.D., Rao, R., Card, S.K.: An organic user interface for searching citation links. In: Proceedings of CHI 1995, Denver, Colorado, pp. 67–73. ACM Press, New York (1995)
14. Modjeska, D., Tzerpos, V., Faloutsos, P., Faloutsos, M., BIVTECI,: a bibliographic visualization tool. In: CASCON 1996, Proceedings of the 1996 conference of the Centre for Advanced Studies on Collaborative Research, (Toronto, Ontario, Canada, November 12-14, 1996), pp. 28–37. IBM (1996)
15. Newman, M.E.J.: Fast algorithm for detecting community structure in networks. Journal of Phys. Rev. E 69. 66133 (2004)
16. Nguyen, Q.V., Huang, M.L.: EncCon: an approach to constructing interactive visualization of large hierarchical data. Information Visualization Journal 4(1), 1–21 (2005)
17. Noel, S., Chu, C.H., Raghavan, V.: Visualization of document co-citation counts. In: IV 2002 Proceedings of Sixth International Conference on Information Visualisation, July 10-12, 2002, pp. 691–696. IEEE Computer Society Press, Los Alamitos (2002)
18. Qian, Y., Zhang, G., Zhang, K.: A Fast and Effective Approach to the Discovery of Dense Clusters in Noisy Spatial Data. In: Proceedings of ACM SIGMOD 2004 Conference, Paris, Franc, June 13-18 2004, pp. 921–922. ACM Press, New York (2004)
19. Schneider, J.W.: Naming clusters in visualization studies: parsing and filtering of noun phrases from citation contexts. In: Proceedings of 10th International Conference of the International Society for Scientometrics and Informetrics (ISSI 2005), Sweden, July 24-28, 2005, pp. 406–416. Karolinska University Press (2005)
20. Sneath, P.H.A., Sokal, R.R.: Numerical Taxonomy. Freeman, London, UK (1973)
21. Ward, J.H.JR.: Hierarchical Grouping to Optimize an Objective Function. Journal of Am. Stat. Assoc. 58, 236–244 (1963)
22. White, H.D., McCain, K.W.: Visualizing a discipline: an author co-citation analysis of information science. Journal of the American Society for Information Science 49(4), 327–355 (1998)
23. Google Web APIs (accessed December 12, 2006), http://www.google.com/apis/
24. PDFBox - Java PDF Library (accessed December 12, 2006), http://www.pdfbox.org/

3D Model Retrieval Based on Multi-Shell Extended Gaussian Image

Dingwen Wang[1,2], Jiqi Zhang[2], Hau-San Wong[2], and Yuanxiang Li[1]

[1] Comupter Science School, Wuhan University
430079 Wuhan, China
wdingwen@gmail.com, yxli@whu.edu.cn
[2] Department of Comupter Science, City University of Hong Kong
999077 Hong Kong, China
jzhang@cs.cityu.edu.hk, cshswong@cityu.edu.hk

Abstract. In this paper, we consider a new shape representation for 3D object, called multi-resolution Multi-Shell Extended Gaussian Image (MSEGI) which eliminates the major drawback of EGI for not containing any direct distance information. MSEGI decomposes a 3D mesh model into multi-concentric shells by the normal distance of the outward surfaces to the origin, and captures the surface area distribution of a 3D model with surface orientation in each concentric shell. Then this distribution function is transformed to spherical harmonic coefficients which can provide multi-resolution shape descriptions by adopting different dimensions. Experimental results based on the public Princeton Shape Benchmark (PSB) dataset of 3D models show that the MSEGI significantly improves EGI, and outperforms CEGI and some popular shape descriptors.

Keywords: 3D Model Retrieval, Extended Gaussian Image, Spherical Harmonics.

1 Introduction

With the general availability of 3D digitizers and scanners, we have witnessed an explosion of the number of available 3D models. As a result, it becomes a great challenge to categorize and retrieve 3D models in the multimedia and manufacturing industries so that 3D-model retrieval attracts more and more research interests. A variety of methods for characterizing 3D-models have been proposed in recent years, such as Light Field Descriptor, Spherical Harmonic Descriptor, Radialized Spherical Extent Function, Extended Gaussian Image, Voxel, D2 Shape Distribution and so on [2], [6], [11], [4], [8], [12]. Extended Gaussian Image (EGI) is reported as one of the most well-known and important methods for distinguishing man-made objects from natural objects among 12 shape descriptors in [9] by recording the variation of surface area along surface orientation. However, a major drawback of EGI is that it doesn't contain any direct distance information. Hence, it's difficult to determine if the 3D object is convex or non-convex. Fig. 1 illustrates this issue. To eliminate this drawback, we propose multi-shell Extended Gaussian Image.

The remainder of the paper is organized as follows. The next section contains a summary of related work. Section 3 gives the detailed descriptions of the proposed approach. The experimental results are presented in Section 4. Section 5 addresses the issue of the sampling rate of EGI based methods. Finally, the conclusion and future work are presented in Section 6.

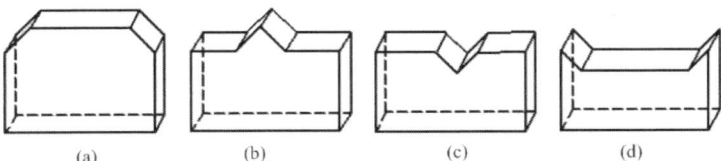

Fig. 1. Convex (a) and non-convex (b), (c), (d) objects represented by the same EGI

2 Related Work

Horn firstly presents the Extended Gaussian Image which computes the probability distribution of the surface normal of a 3D object over a Gaussian Sphere [4]. Then Ikeuchi applies the EGI representation to convex object recognition [3]. Sun and Sherrah apply the EGI to detect the symmetry of 3D object [10]. Horace and Wong apply the EGI to 3D head models retrieval [14]. It is also reported that the EGI is useful for pose determination by Brou [1].

The major advantage of the EGI is translation invariance, because it only records the variation of surface area with surface orientation. However, this major advantage introduces its major drawback: not containing any direct distance information. To eliminate this deficiency, several extensions of EGI have been proposed to deal with non-convex objects, e.g., H. Matsuo proposes More Extended Gaussian Image [7] that can represent non-convex objects, but it isn't suitable to 3D model retrieval; Kang presents the Complex Extended Gaussian Image which is a complex-valued spherical function giving the distribution of normals and associated normal distance of points on the surface [5].

3 Multi-Shell Extended Gaussian Image

3.1 Overview of Method

First, the 3D models are triangulated and normalized in the same canonical frame. Then, we uniformly decompose a 3D model into multi-concentric shells, divide the surface of each shell into cells, and capture the surface area distribution of a 3D model along surface orientation in each concentric shell. At last, the feature vectors are extracted by adopting spherical harmonics transform of this surface area distribution function, and multi-resolution MSEGI descriptors are obtained through choosing the dimensions of feature vectors. We also compare the performance of MSEGI extracted by two different properties of spherical harmonics.

3.2 Normalization

The normalization step transforms the 3D models into a uniform canonical frame. The goal of this procedure is to eliminate the effect of transforming a 3D object by a different scale, position, rotation, or orientation. Pose normalization can improve the performance of the shape descriptors. For a given 3D triangular mesh model, we denote it by a set of triangles

$$T = \{T_1, \cdots, T_m\}, \quad T_i \subset \mathbb{R}^3 \tag{1}$$

Where T_i is consisted of a set of three vertices

$$T_i = \{P_{A_i}, P_{B_i}, P_{C_i}\} \tag{2}$$

For the center and normal of triangle T_i, we denote them by g_i and $\boldsymbol{n_i}$ respectively. For the surface area of triangle T_i, it is represented by S_i, while the total area of the mesh is S:

$$g_i = (P_{A_i} + P_{B_i} + P_{C_i})/3 \tag{3}$$

$$\boldsymbol{n_i} = (P_{C_i} - P_{A_i}) \times (P_{B_i} - P_{C_i}) \tag{4}$$

$$S_i = |(P_{C_i} - P_{A_i}) \times (P_{B_i} - P_{C_i})|/2, \quad S = \sum_{i=1}^{m} S_i \tag{5}$$

Then we perform the following processes to fulfill the object pose normalization.

(i) Translation: in general, models are translated to the center of mass. In this paper, we adopt a different method (Equ. 6) to compute the geometric center.

$$O_I = (O_x, O_y, O_z) = \frac{1}{S} \sum_{i=1}^{m} S_i \boldsymbol{n_i} \tag{6}$$

(ii) Rotation: we apply the principle component analysis method to normalize the 3D models for orientation. The eigenvectors and associated eigenvalues of the covariance matrix C_I are obtained by integrating the quadratic polynomials $P_i \cdot P_j$, over the centers on the surface of all polygons. The three eigenvectors sorted by decreasing relevant eigenvalues are the principal axes and can be used to fix the models after rotation. The ambiguity between positive and negative axes is resolved by choosing the direction of the axes so that the area of model on the positive side of the $x-$, $y-$, and $z-$axes is greater than the area on the negative side.

$$C_I = \frac{1}{12S} \sum_{i=1}^{m} (f(P_{A_i}) + f(P_{B_i}) + f(P_{C_i}) + 9f(P_{g_i}))S_i \qquad (7)$$

$$f(v) = (P_i - O_I) \cdot (P_i - O_I)^T \qquad (8)$$

(iii) Scale: the average distance from all points on the surfaces of all polygons to the center of mass. This value can be used to fix the models for isotropic scales.

3.3 Sampling

After normalizing the object, we uniformly decompose a 3D model into concentric shells with radii $r_c = 1, 2, \cdots, N_s$ ($c = 1, 2, \cdots, m$) and divide the surfaces of each shell into $N \times N$ cells:

$$\theta_i = \frac{2\pi}{N} i, \quad i = 0, 1, 2, \cdots, N \qquad (9)$$

$$\phi_j = \arccos\left(1 - \frac{2i}{N}\right), \quad j = 0, 1, 2, \cdots, N \qquad (10)$$

$$r_i = \lceil \frac{d_k - \min(d_k)}{\max(d_k) - \min(d_k)} \times N_s \rceil, \quad i = 1, 2, \cdots, m \qquad (11)$$

Where d_k is the normal distance of the surface to the origin in the direction of the surface normal. Then we perform two steps of mapping. First, we map the normal distance of surface triangle on the object to a concentric shell by Equ. 11. Second, for each surface triangle, we treat its area as the weight of normal and map the area to the cell (θ_i, ϕ_j) on the shell which is the most approximate one to the normal direction; finally we can get the surface area distribution function $f(r_c, \theta_i, \phi_j)$ along each concentric shell.

3.4 Spherical Harmonic Transform

Spherical harmonics transform is introduced as a useful tool for 3D model retrieval in [13]. The theory of spherical harmonics says that any spherical function $f(\theta, \phi)$ can be composed as the sum of its harmonics:

$$f(\theta, \phi) = \sum_{l=0}^{\infty} \sum_{m=-l}^{m=l} \tilde{f}_{l,m} Y_l^m(\theta, \phi) \qquad (12)$$

Where $\tilde{f}_{l,m}$ denote the Fourier coefficient and $Y_l^m(\theta, \phi)$ is the spherical harmonic base, calculated by certain products of Legendre functions and complex exponentials. Fig. 2 illustrates the spherical harmonic $Y_l^m(\theta, \phi)$ which are complex functions defined on spherical, are up to degree 3.

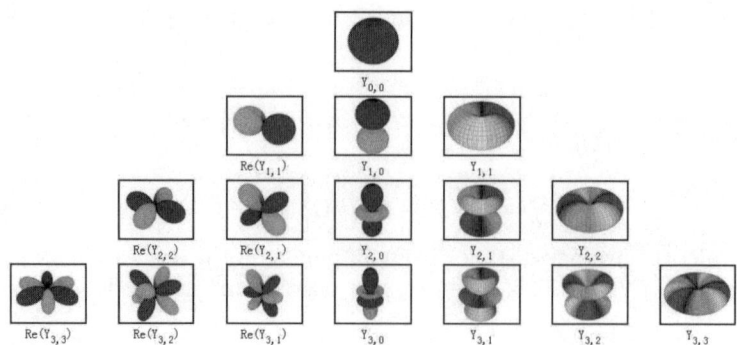

Fig. 2. Spherical Harmonics basis functions $Y_l^m(\theta, \phi)$: a visualization of complex functions defined on spherical, are up to degree 3

The area distribution function $f(r, \theta, \phi)$ is to perform spherical harmonics to extract the feature vectors of MSEGI. Since $f(r, \theta, \phi)$ is band-limited with bandwidth $N/2$, then we can express $f(r, \theta, \phi)$ as:

$$f(r, \theta, \phi) = \sum_{l=0}^{N/2} \sum_{m=-l}^{m=l} \tilde{f_{l,m}} Y_l^m(r, \theta, \phi) \quad (13)$$

Where l is the degree of spherical harmonics. In the practical application, feature vectors can be extracted from the first $l+1(l < N/2)$ rows of coefficients, which can provide the multi-resolution feature descriptors for 3D models by choosing different degree l. Two important properties of spherical harmonics, which are used to extract different 3D shape feature vectors of based on a function on the spherical S^2, are the following:

Property 1. Let $f \in L^2(S^2)$ be a real-valued function, i.e., $f : S^2 \to \mathbb{R}$. Then, the following symmetry between the coefficients exists:

$$\tilde{f_{l,m}} = (-1)^m \overline{\tilde{f_{l,-m}}} \Rightarrow |\tilde{f_{l,m}}| = |\tilde{f_{l,-m}}| \quad (14)$$

Where, $\tilde{f_{l,m}}$ and $(-1)^m \overline{\tilde{f_{l,-m}}}$ are conjugate complex numbers.

Property 2. A subspace of $L^2(S^2)$ of dimension $2l + 1$, which is spanned by the harmonics $Y_l^m(-l \leq m \leq l)$ of degree l, is invariant with respect to rotation of the sphere S^2.

We can use the absolute values of $\tilde{f_{l,m}}$ as components of our feature vectors to obtain the feature vector F_1. According to Property 1, if $f(\theta, \phi)$ is a real-valued function, then the symmetry relationship between the coefficients exists. Thus, the feature vector F_1 is composed by the first $l+1(l < N/2)$ rows of the obtained coefficients.

$$F_1 = (|\tilde{f_{0,0}}|, |\tilde{f_{1,0}}|, |\tilde{f_{1,1}}|, \cdots, |\tilde{f_{l,0}}|, \cdots, |\tilde{f_{0,0}}|) \quad dim(F_1) = l(l+1)/2 \quad (15)$$

In addition, we can also obtain the feature vector F_2 with rotation invariance according to Property 2, without normalizing the orientation of a 3D model.

$$F_2 = (\|f_0\|, \cdots, \|f_{dim-1}\|), \quad \|f_i\| = \sqrt{\sum_{m=-l}^{m=l} |\tilde{f_{l,m}}|^2}, \quad dim \leq N/2 \qquad (16)$$

However, F_2 lose the information on degree l, which decrease the retrieval performance of the feature vector. Moreover, EGI is defined over a tessellation of the unit sphere in which the surface area of each cell is not uniform. As a result, the irregular sampling limits the rotation invariance of the shape descriptors. It will be interesting to investigate how the regular sampling affects the performance. The comparison of F_1 and F_2 is showed in section 4.

4 Experiment

We adopt the public Princeton Shape Benchmark (PSB) dataset of 3D models [7] which contains 1814 models, and compare their performance of MSEGI, CEGI, EGI, EXT, SHELL and SECTOR on the subset of PSB, which contains 1280 objects spanning 108 categories including plane, ship, car, human, animal, plant, furniture and so on. MSEGI-F_1 denotes 3D shape feature vector transformed by Equ. 15 that is not rotation invariant, but can represent more 3D object information. MSEGI-F_2 denotes 3D shape feature vector extracted by Equ. 16 which needn't be applied PCA as a pre-process to save time and is lower dimension feature vector, but lose some information on the degree. The parameters of each shape descriptor are as follows. All EGI-based shape descriptors are sampled on 32×32 spherical grid and the harmonic coefficients of the descriptor are up to

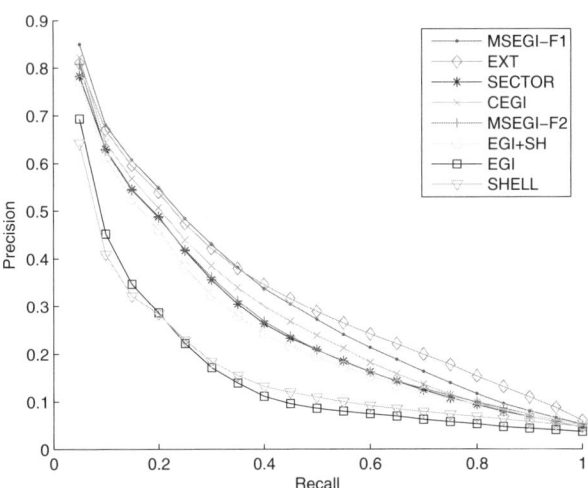

Fig. 3. Precision-recall curves of MSEGI and other six shape descriptors

Table 1. Performance of MSEGI and other six shape descriptors

Descriptors	Length	NN	FT	ST	E-Measure	DCG
MSEGI-F_1	180	56.1%	25.8%	34.9%	20.8%	57.8%
EXT	136	55.0%	24.7%	36.5%	21.7%	58.8%
SECTOR	136	51.5%	21.4%	30.3%	17.6%	54.4%
CEGI	36	50.2%	23.1%	32.2%	18.8%	55.6%
MSEGI-F_2	40	47.7%	21.8%	30.7%	18.1%	54.1%
EGI+SH	36	45.2%	20.4%	29.7%	17.1%	52.5%
EGI	1024	33.8%	9.5%	13.7%	8.0%	41.9%
SHELL	30	23.4%	11.1%	18.7%	9.9%	43.2%

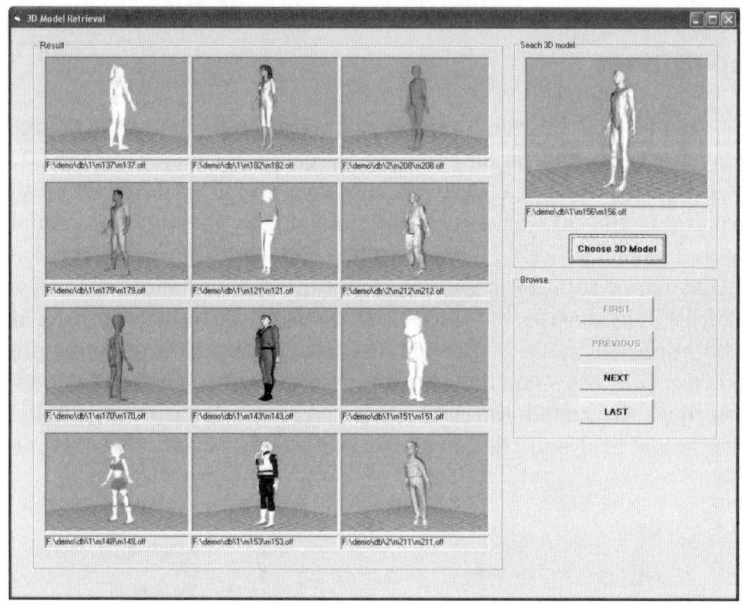

Fig. 4. The retrieval result of a human as the query

degree 8. MSEGI is decomposed into 5 concentric shells. EXT and SECTOR are computed on a 64 × 64 spherical grid, which harmonic coefficients are up to degree 16. The sampling issue of the EGI-based shape descriptors is further discussed in Section 5.

We use precision vs. recall curves, a standard evaluation technique for retrieval systems, to compare the effectiveness of our algorithms (Fig. 3), and also evaluated these shape descriptors using (i) Nearest-Neighbor (NN) measure, which represents the percentage of the closest matches that belong to the same class as the query; (ii) First-tier and Second-tier, which represent the percentage of models in the query's class that appear within the top K matches, where K depends on the size of the query's class, Specifically, for a class with $|C|$ members,

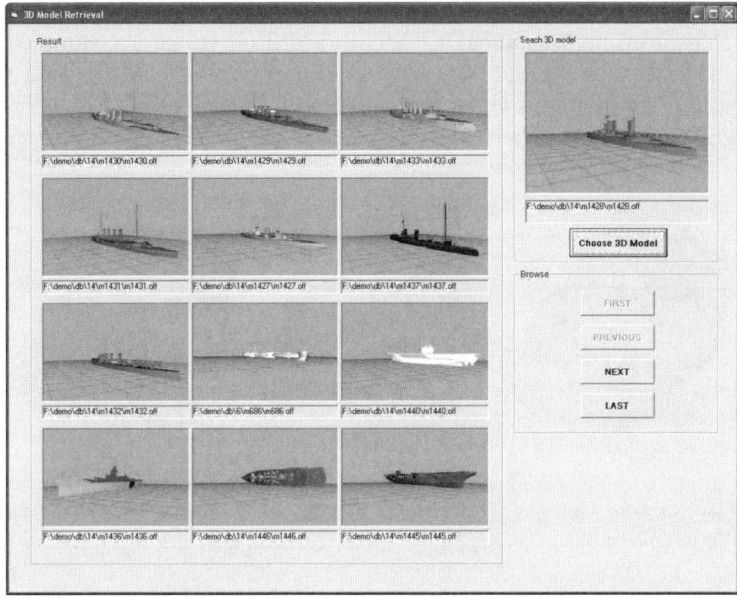

Fig. 5. The retrieval result of a ship as the query

$K = |C| - 1$ for the first tier, and $K = 2(|C| - 1)$ for the second tier; (iii) E-Measure, which represents a composite measure of the precision and recall for a fixed number of retrieved results; (iv) Discounted Cumulative Gain (DCG), which represents a statistic that weights correct results near the front of the list more than correct results later in the ranked list under the assumption that a user is less likely to consider elements near the end of the list. The results are summarized in Table 1.

These evaluation results indicate that MSEGI significantly improve EGI shape descriptor, and outperforms CEGI and some popular methods. Table 1 show MSEGI-F_1 is more effective than MSEGI-F_2. In practical application, we can balance the retrieval performance and time cost to choose MSEGI-F_1 or MSEGI-F_2.

We further implement 3D model retrieval system based on the MSEGI descriptor. Fig. 4 and Fig. 5 illustrate some retrieval results on our 3D model retrieval system when the ship and human model are given as queries.

5 Discussion

The EGI have some characteristic different from the other shape descriptors. The EGI descriptor is only relevant to the surface area and normal of a 3D object, and the number of the faces of the mesh. For most of the shape descriptors such as SECTOR, EXT, SHD, GEDT and so on, denser sampling can represent the models more accurately and improve the performance of retrieval. However, it isn't suitable to the EGI-based shape descriptors. Fig. 7 and Fig. 8 respectively show the retrieval performance of EXT and MSEGI which are sampled on the 16 × 16, 32 × 32, 48 × 48 and 64 × 64 spherical grids. Fig. 7 shows that the performance of EXT increases when the sampling grid increases. However, MSEGI sampled on the 32 × 32 spherical grid is more effective than the other sampling.

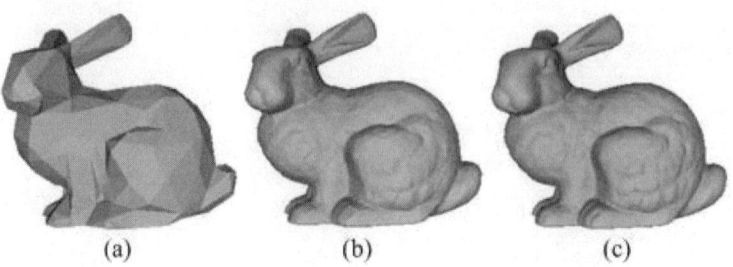

Fig. 6. The Stanford Bunny (a), (b) and (c) respectively contains 575, 4838 and 23581 faces (from left to right)

To further investigate the sampling issue, we extract the EXT and EGI feature vector of the Stanford Bunny (Fig. 6) without spherical harmonics on different spherical grid sampling, and use their cosine distance to measure their difference. Cosine distance (Equ. 17) takes the norm of the feature vectors into account and thus gets rid of the influence of the size of the feature vectors, and its value is between 0 and 1. Table 2 shows that the distances between Stanford Bunny (a) and (b), (c) of EGI are 70.61% and 67.07% on 64 × 64, which are longer than

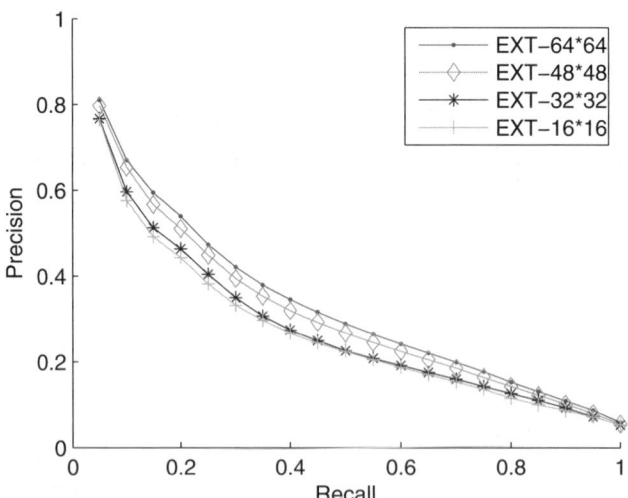

Fig. 7. Precision-recall curves for the shape feature vectors of EXT based on different spherical grids

them on EXT, and are possibly longer than the cosine distance between different 3D models. In other words, the Stanford Bunny (a), (b) and (c) are not at all similar by compare their feature vectors of EGI.

$$Cosine\,distance: \quad 1 - \frac{\sum_{i=1}^{n} x_i y_i}{\sqrt{\sum_{i=1}^{n} x_i^2}\sqrt{\sum_{i=1}^{n} y_i^2}} \qquad (17)$$

From above discussion, we can conclude that EGI is very sensitive to the different levels-of-detail of a 3D object. We calculate the statistics about the number of the faces of all 3D models on PSB, and they range from 16 to 316498, and the ratio of 3D models which number of faces are below 900, between 900 and 4200, and over 4200 on PSB, is respectively 33.4%, 33.4% and 33.2%. So it is unsuitable for the 3D models with small number of faces to be sampled on 32×32 and 64×64 spherical grids. For the PSB dataset of 3D models, we should choose small spherical grids measurably. However, it should be sampled on large spherical grids for spherical harmonics in order to obtain good effect. There

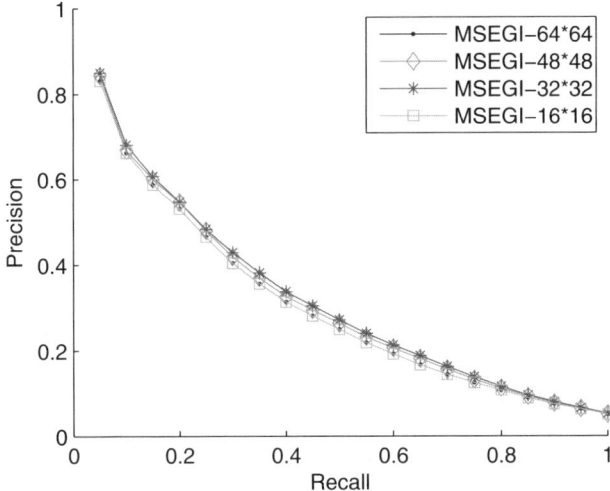

Fig. 8. Precision-recall curves for the shape feature vectors of MSEGI based on different spherical grids

Table 2. Performance of EGI and EXT on different sampling grids

Spherical Grid	EGI			EXT		
	a-b	a-c	b-c	a-b	a-c	b-c
16×16	20.38%	19.65%	2.04%	5.87%	6.39%	0.86%
32×32	47.76%	45.51%	6.91%	6.54%	7.72%	1.40%
48×48	59.88%	58.95%	15.21%	6.40%	7.53%	1.78%
64×64	70.61%	67.07%	23.73%	6.71%	7.52%	1.56%

should be a trade-off to balance these two requirements. In our experiments, it is suitable for MSEGI to sampling on 32×32 spherical grids on PSB. We also conclude that EGI-based shape feature vectors are more suitable to the 3D models with approximately equal number of surface, e.g. EGI is applied to head model retrieval [14].

6 Conclusions and Future Work

In this paper, we propose a new 3D shape descriptor, called multi-resolution Multi-Shell Extended Gaussian Image. The main contribution of MSEGI is that it eliminates the major drawback of the original EGI for not containing any distance information to some extent. MSEGI can provide the multi-resolution feature vectors with different dimensions, and provide different feature vectors by balancing the retrieval performance and time cost. In addition, we discuss the sampling issue that MSEGI and other EGI-based shape descriptors are sensitive to levels-of-detail and different tessellations of 3D models. The experiments on the PSB database show that MSEGI significantly outperforms EGI, and is more effective than CEGI and some popular descriptors.

In the future, we will explore the sampling issue of EGI and more robust normalization approaches to further improve the performance of MSEGI method.

Acknowledgement. The work described in this paper was supported by Hong Kong Research Grants Council (RGC) Competitive Earmarked Research Grant (CERG) Project no. 9041033.

References

1. Brou, P.: Using the Gaussian Image to Find the Orientation of Objects. The International Journal of Robotics Research 3, 89–125 (1984)
2. Chen, D.Y., Tian, X.P., Shen, Y.T., Ming, O.Y.: On visual similarity based 3D model retrieval. Computer Graphics Forum 22, 223–232 (2003)
3. Hebert, M., Ikeuchi, K., Delingette, H.: A spherical representation for recognition of free-form surfaces. IEEE Transactions on Pattern Analysis and Machine Intelligence 17, 681–690 (1995)
4. Horn, B.K.P.: Extended Gaussian images. In: Proceedings of the IEEE 72, 1671–1686 (1984)
5. Kang, S.B., Ikeuchi, K.: Determining 3-D object pose using the complex extended Gaussian image. In: IEEE Computer Society Conference on Computer Vision and Pattern Recognition, pp. 580–585 (1991)
6. Kazhdan, M., Funkhouser, T., Rusinkiewicz, S.: Rotation invariant spherical harmonic representation of 3D shape descriptors. In: Proceedings of the 2003 Eurographics/ACM SIGGRAPH symposium on Geometry processing, Aachen, Germany, vol. 43, pp. 156–164 (2003)
7. Matsuo, H., Iwata, A.: 3-D object recognition using MEGI model from range data. In: Proceedings of the 12th IAPR International Conference on Pattern Recognition, vol. 1, pp. 843–846 (1994)

8. Osada, R., Funkhouser, T., Chazelle, B., Dobkin, D.: Matching 3D models with shape distributions. In: International Conference on Shape Modeling and Applications, pp. 154–166 (2001)
9. Shilane, P., Min, P., Kazhdan, M., Funkhouser, T.: The Princeton Shape Benchmark. In: Shape Modeling International, Genova, Italy, pp. 167–178 (2004)
10. Sun, C.M., Sherrah, J.: 3D symmetry detection using the extended Gaussian image. IEEE Transactions on Pattern Analysis and Machine Intelligence 19, 164–168 (1997)
11. Vranic, D.V.: An improvement of rotation invariant 3D-shape based on functions on concentric spheres. In: International Conference on Image Processing vol. 3, pp. 757–760 (2003)
12. Vranic, D.V.: 3D Model Retrieval, University of Leipzig, Germany (2004)
13. Vranic, D.V., Saupe, D., Richter, J.: Tools for 3D-object retrieval: Karhunen-Loeve transform and spherical harmonics. In: IEEE Fourth Workshop on Multimedia Signal Processing, pp. 293–298 (2001)
14. Wong, H.S., Cheung, K.K.T., Ip, H.H.S.: An evolutionary optimization approach for 3D human head model classification. In: Proceedings of the 5th ACM SIGMM international workshop on Multimedia information retrieval, Berkeley, California, pp. 94–101. ACM Press, New York (2003)

Neurovision with Resilient Neural Networks

Erkan Beşdok

Erciyes University, Engineering Faculty, Photogrammetry Division, Soft-Computing and Computational-Vision Lab., 38039, Kayseri, Turkey
ebesdok@erciyes.edu.tr

Abstract. A *Neurovision System* can be defined as an artificial tool that sees our physical world. The purpose of this paper is to show a novel tool to design a 3D artificial vision system based on Resilient Neural Networks. Camera Calibration (CC) is a fundamental issue for Computational-Vision. Classical CC methods comprise of taking images of objects with known geometry, extracting the features of the objects from the images, and minimizing their 3D backprojection errors. In this paper, a novel implicit-CC model based on Resilient Neural Networks, CR, has been introduced. The CR is particularly useful for 3D reconstruction of the applications that do not require explicitly computation of physical camera parameters in addition to the expert knowledge. The CR supports intelligent-photogrammetry, photogrammetron. In order to evaluate the success of the proposed implicit-CC model, the 3D reconstruction performance of the CR has been compared with two different well-known implementations of the *Direct Linear Transformation* (DLT). The proposed method is also robust sufficiently for dealing with different cameras because it is capable of fusion of the image coordinates sourced from different cameras once the neural network has been trained.

1 Introduction

This paper presents a novel approach for Camera Calibration (CC) using Resilient Neural Networks. CC is employed in computational-vision for camera pose estimation and requires an optimized solution of complex system of imaging-equations [1, 2, 3, 4, 5, 6, 7, 8, 9]. CC is the procedure of explanation of mapping the theoretical camera-model to the actual imaging system and determining the 3D rotations and spatial position of the camera with respect to a predefined geodetic reference system.

CC techniques can be classified into two major groups: linear methods and nonlinear methods. Computational efficiency is the major advantage of linear camera calibration methods but they suffer from accuracy and robustness due to their sensitivity of noise, which affects image coordinates. Nonlinear camera calibration methods supply extremely better solutions but they are computationally-burden and require good initial values for optimization processes. CC is a critical phase to recover the 3D information from 2D images. It addresses the problem of determining intrinsic and extrinsic camera parameters using 2D images and their corresponding 3D geometries.

Due to the noise-influenced image coordinates, most of the existing CC techniques are unsuccessful aspect of robustness and accuracy. By using neural networks, it becomes unnecessary to know the physical parameters [1,2,3] of the imaging systems for 3D reprojection of objects from their 2D images. CC is conventionally realized by imaging some special patterns with known geometric properties [1,2,3,4,5,6,7,8,9].

In this paper, a novel method based on Resilient Neural Networks (CR) [10,11,12], is presented to alleviate the problems with the existing CC techniques [2,3]. With the proposed method, a Neural Network [10,11,12] is employed to explain the mapping between 3D object coordinates and the 2D image coordinates and a planar pattern has been observed at seven different orientations for the establishment of training and test data sets of the artificial neural network. The proposed method is robust sufficiently for dealing with different cameras because it is capable of fusion of the image coordinates sourced from different cameras once the neural network has been trained. Furthermore, the proposed method is able to recover 3D-shape of the observed object without having optimized-explicit parameters [1,2,3,5] of imaging model if there are at least two images of the related object.

The rest of the paper is organized as follows: *Resilient Neural Networks* are explained in Section 2. *Proposed Method* and *Experiments* are given in Section 3 and Section 4, respectively. Finally, *Conclusions* are given in Section 5.

2   Resilient Neural Networks

Artificial Neural Network (ANN) [8,9,10,11,12] is an advanced learning and decision-making technology that mimics the working process of a human brain. Various kinds of ANN structures and learning algorithms have been introduced in the literature [8,9,10,11,12]. In this paper, RNs have been used for the CC.

In contrast to other gradient algorithms, this algorithm does not use the magnitude of the gradient. It is a direct adaptation of the weight step based on local gradient sign. The RN generally provides faster convergence than most other algorithms [10,11,12]. The role of the RN is to avoid the bad influence of the size of the partial derivative on the weight update. The size of the weight change is achieved by each weight's update value, $A_{ji}(k)$, on the error function $E(k)$, which is used to calculate the delta weight as in Equation 1.

$$\Delta w_{ji}(k) = \begin{cases} -A_{ji}(k) & if\ B(k) > 0 \\ +A_{ji}(k) & if\ B(k) < 0 \\ 0 & else \end{cases} \quad (1)$$

where $B(k)$ is $\frac{\partial E}{\partial w_{ji}}(k)$ and

$$A_{ji} = \begin{cases} \eta A_{ji}(k-1), & B(k-1)B(k) > 0 \\ \mu A_{ji}(k-1), & B(k-1)B(k) < 0 \\ A_{ji}(k-1), & else \end{cases} \quad (2)$$

where $B(k-1)$ is $\frac{\partial E}{\partial w_{ji}}(k-1)$, η and μ are the increase and decrease factors, respectively where $0 < \mu < 1 < \eta$.

More details about the algorithm can be found in [10, 11, 12].

A number of calibration methods implementing intelligent agents have been introduced recently [4, 6, 8, 9]. The intelligent agents based camera calibration methods introduced in the literature generally require a set of image points with their 3D world coordinates of the control points and the corresponding 2D image coordinates for the learning stage. The CR can merge multisource camera images, for the reconstruction of a 3D scene, due to its high flexible properties of learning from examples.

3 Proposed Method

In this paper, the algorithm of the CR and the preparation steps of the Learning Data and Validation Data used in the training RN structure of the CR are explained below:

1. Determine the Image Coordinates of the Control Points (u,v), which are over the Calibration Patterns as seen in Fig.1 and obtain 3D world-coordinates, (X,Y,Z), of the related (u,v). Each Calibration Pattern plane involves 11x7=77 points, hence, totally 539 calibration points over the seven Calibration Patterns planes have been used in this study.
2. Select 300 Control Points randomly for training and use the remaining 239 Control Points for validation of training in order to avoid from over-training disaster of RN structures.
3. Train the RN-based neural structure that achieves a mapping from 2D (u,v) image space to 3D world-space (X,Y,Z) (Fig.1). The input data of the neural structure are the image coordinates (u,v) of the Control Points and the output data are the corresponding (X,Y,Z) world-coordinates of the input data (u,v).
4. Apply the image coordinates, (u,v), to the neural structure in order to compute the world coordinates, (X,Y,Z), of the related image points.

Extensive simulations exposed that accuracy of 3D reconstruction of the neural network systems increases with the increasing number of the cameras used in the imaging system. Therefore, the imaging system used in this paper comprises four *virtual* Fire-i IEEE 1394 cameras. Consequently, neural structures have eight inputs $(u_1, v_1, u_2, v_2, u_3, v_3, u_4, v_4)_p$ and three outputs $(X, Y, Z)_p$ for each point p.

The maximum epoch of the training phase was predefined as 250 000 for the neural structure but in order to control the computational burden of the proposed method, the training phase ended when the error goal of the RN was smaller than 0.001. Extensive simulations have been realized by using different initial conditions for neural structure and the best solution of problem of RN has been used in this paper. In the proposed method, the *Purelin Training Function* of the Matlab® [10] has been used for the neural structures.

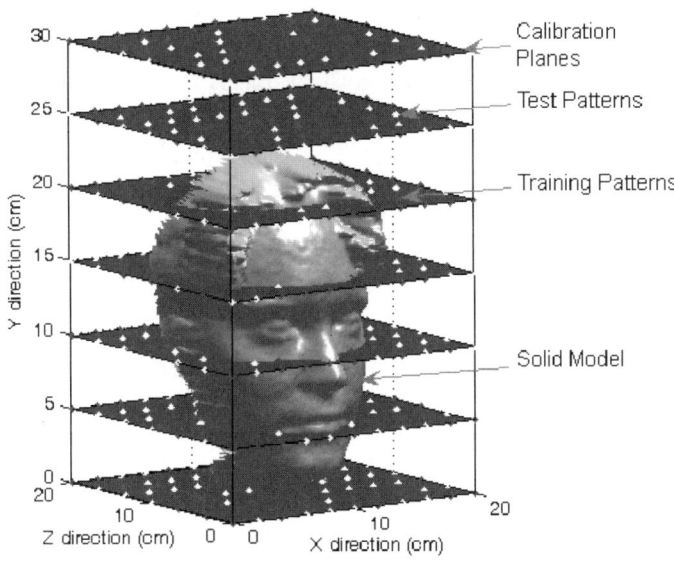

Fig. 1. The seven Calibration Patterns and the Solid model of the Point Cloud which was obtained by using the proposed method for Test Pattern

4 Experiments

In this paper, a set of real images have been employed in the experiments of the CR. The obtained results have then been compared with both DLT [1] and M-DLT [5] methods. Four *virtual* Fire-i IEEE 1394 cameras have been used in the experiments. Distortion models of the virtual cameras have been obtained by using the well known camera calibration toolbox given in [13]. The neural structure of the proposed method has $4_{Camera} * 2_{Image\ Corrdinates} = 8$ inputs. Each of the Calibration Pattern Planes have 11*7=77 Control Points. Image Coordinates (u,v) of the Control Points were acquired with subpixel accuracy for [13]. 17.430 measurements have been acquired over the Test Object. The seven Calibration Patterns and The Point Cloud obtained by using the proposed method for the Test Pattern was illustrated in Fig.1. The Calibration Patterns

Table 1. Mean-Squared-Error (MSE) values of the proposed CR, DLT [1], and M-DLT [5] methods

	MSE of X	MSE of Y	MSE of Z
DLT (Train)	0.0025	0.1372	0.0008
DLT (Validation)	0.0027	0.1537	0.0010
M-DLT (Train)	0.0025	0.1374	0.0008
M-DLT (Validation)	0.0027	0.1543	0.0010
CR (Train)	0.0021	0.1255	0.0008
CR (Validation)	0.0023	0.1482	0.0009

have been used for setting Calibration Data sets of the proposed method, DLT and M-DLT. Planimetric and depth reconstruction accuracies were evaluated in Mean-Squared-Error (MSE) as seen in the Table 1. Extensive simulations show that CR supplies statistically acceptable and more accurate results as seen in Table 1. The method proposed in this paper was implemented by using the Image Processing, Image Acquisition and Neural Network toolboxes of Matlab® v7.3.0.

For the analysis of the CR, the images obtained from the camera have been employed directly. The image coordinates of the control points have been extracted with *subpixel accuracy* and no deformation corrections have been applied to the images.

Before using the DLT methods (DLT and M-DLT), the required distortion corrections have been applied to the coordinates obtained as a result of the image matching. On the other hand, no distortion corrections have been applied to the corresponding coordinates in the CR method.

Solid model of the Point Cloud which was obtained by using the proposed method for the Test Pattern was illustrated in Fig.1. Solid models obtained from 3D measurement systems are often noisy. Therefore, mesh smoothing is required for realistic mesh models. In this paper a variation of the method proposed in [14] has been used as a smoothing operator for illustration of solid model.

4.1 Statistical Analysis

In the Manova [15], mean vectors of a number of multidimensional groups are compared. Therefore, the Manova is employed to find out whether the differences in the results of CR, DLT and M-DLT are statistically significant or not.

The tested null hypothesis is;

H_0: $\mu_1 = \mu_2 = ... = \mu$
H_1: at least two μ's are unequal

where μ_1 is the mean vector of the group # 1, μ_2 is the mean vector of the group # 2 and μ is the population mean vector.

As a result of the implemented Manova, no statistically significant difference has been found between the results of CR, DLT and M-DLT. That means the null hypothesis cannot be rejected. This hypothesis test has been made using Wilk's Λ and χ^2 tests. The details of both of the tests and Manova can be found in [15].

For the Wilk's Λ test, the test and critical values are computed as 0.9953 and 0.7244, respectively, given that α significance level is 0.05 and degrees of freedom for the sum of squares and cross-products are 29997 and 29999, respectively. Due to the condition of $\Lambda_{test} > \Lambda_{critic}$, the null hypothesis cannot be rejected.

For the χ^2 test [15], the test and the critical values are computed as 2.00 and 6.00, respectively, given that α significance level is 0.05 and degrees of freedom is 3. Due to the condition of $\chi^2_{test} < \chi^2_{Critic}$, the null hypothesis cannot be rejected.

That is to say that there is no statistically significant difference between the results of CR, DLT and M-DLT.

This outcome statistically verifies the advantages of the CR method in various perspectives.

5 Conclusions

A Resilient Neural Network based camera calibration method for 3D information recovery from images is proposed in this paper. The obtained results have been compared with the traditional DLT [1] and M-DLT [5].

The main advantages of the CR are as follows: It does not require the knowledge of complex mathematical models of view-geometry and an initial estimation of camera calibration, it can be used with various cameras by producing correct outputs, and it can be used in dynamical systems to recognize the position of the camera after training the ANN structure. Therefore, the CR is more flexible and straightforward than the methods introduced in the literature.

The advantages of the CR may be summarized as follows:

- Offers high accuracy both in planimetric (X,Y) and in depth (Z).
- The CR is flexible and significantly easier to implement than the CC methods presented in the literature.
- Simple to apply and fast after training.
- Suitable for robotic neurovision systems.
- The CR can use same or different type cameras together or separately.
- Does not have any camera and zoom lens type restriction.
- Does not use physical model of vision systems.
- Users do not need to know complex mathematical models of the view-geometry, therefore, proposed CR does not need to have detailed technical knowledge on CC.
- An approximated solution for initial step of CC is not employed.
- Optimization algorithms are not employed during 3D reconstruction in contrary to the same well-known 3D acquisition methods.
- No image distortion model is required.

Acknowledgments

This study has been supported by the Research Foundation of Erciyes University under the project entitled "3D Laser-Scanner Design by Using Soft-Computing", 2007 and the project codes of FBT-05-31, FBA-05-25.

References

1. Hatze, H.: High-Precision Three-Dimensional Photogrammetric Calibration and Object Space Reconstruction Using a Modified DLT Approach. J. Biomechanics 21, 533–538 (1988)
2. Tsai, R.: A Versatile Camera Calibration Technique for High Accuracy 3d Machine Vision Metrology Using Off-the-Shelf TV Cameras and Lenses. IEEE Journal of Robotics and Automation 3(4), 323–344 (1987)
3. Çivicioglu, P., Beşdok, E.: Implicit Camera Calibration By Using Resilient. In: King, I., Wang, J., Chan, L., Wang, D. (eds.) ICONIP 2006. LNCS, vol. 4233, pp. 632–640. Springer, Heidelberg (2006)

4. Jun, J., Kim, C.: Robust camera calibration using neural network. In: Proceedings of the IEEE Region 10 Conference, TenCon, vol. 99(1), pp. 694–697 (1999)
5. Xiong, Y., Hu, H.: 3D reconstruction approach in neurovision system. In: Proc. Optical Measurement and Nondestructive Testing: Techniques and Applications, SPIE, vol. 4221, pp. 331–342 (2000)
6. Qiang, J., Yongmian, Z.: Camera Calibration with Genetic Algorithms. IEEE Transactions on Systems, Man, and Cybernetics- Part A: Systems and Human 31(2), 120–130 (2001)
7. Pan, H.P., Zhang, C.S.: System Calibration of Intelligent Photogrammetron. International Archives of Photogrammetry and Remote Sensing 34(2) (2002)
8. Beşdok, E., Çivicioğ, P.: Adaptive Implicit-Camera Calibration in Photogrammetry Using Anfis. In: Gabrys, B., Howlett, R.J., Jain, L.C. (eds.) KES 2006. LNCS (LNAI), vol. 4251, pp. 606–613. Springer, Heidelberg (2006)
9. Wei, G.Q., Hirzinger, G.: Hirzinger, Multisensory visual servoing by a neural network. IEEE Transactions on Systems, Man and Cybernetics, Part B 29(2), 276–280 (1999)
10. Mathworks Inc., Matlab, Neural Networks Toolbox, Image Poccessing Toolbox, Image Acquisition Toolbox, Mathworks (2005)
11. Reidmiller, M.: Rprop- Description and Implementation Details, Technical Report, Germany. University of Karlsruhe (1994)
12. Reidmiller, M., Braun, H.: A Direct Adaptive Method for Faster Backpropogation Learning: The Rprop Algorithm. In: Proceedings of the IEEE Int. Conf. On Neural Networks, San Francisco, CA, pp. 586–591 (1993)
13. Bouguet, J.Y.: (2004), http://www.vision.caltech.edu/bouguetj/calib_doc/
14. Çivicioglu, P.: Using Uncorrupted Neighborhoods of the Pixels for Impulsive Noise Suppression with ANFIS. IEEE Transactions on Image Processing 16(3), 759–773 (2007)
15. Rencher, A.C.: Methods for Multivariate Analysis. Wiley-Interscience, John Wiley-Sons, Inc (2002)

Visual Information for Firearm Identification by Digital Holography

Dongguang Li

School of Computer and Information Science, Edith Cowan University, 2
Bradford Street, Mount Lawley, WA 6050, Australia
d.li@ecu.edu.au

Abstract. In digital holography a CCD camera records optically generated holograms which is then reconstructed numerically by a calculation of scalar diffraction in the Fresnel approximation. The digital photography facilitates real time transmission of the message via traditional communication methods. In this paper the principle of digital holography and its application to the 3D image encryption-decryption are reviewed. The experimental results of firearm identification recording using digital holography and their numerical reconstruction are presented.

Keywords: holography; firearm identification; image processing.

1 Introduction

Holographic techniques allow preserving and visualizing three-dimensional information of objects. In principle, a hologram records the distribution of the complex amplitude of the wave front scattered by an object so that reconstruction from the hologram produces a complete visual illusion of the object [1].

Conventional holography records holograms on photographic films that chemical development and optically reconstruction of the recorded wave is needed to display the object image. This procedure is time consuming that prevents the real time application. The storage of large amount of films is also an issue.

With the development of solid-state image sensors and computer capacities, it is possible to direct record holograms by charge coupled device (CCD) camera and numerical reconstruction of the 3-D image by computer [2].

Since the idea of computer reconstruction of a hologram was proposed, various configurations of digital holography have been demonstrated in different areas such as surface contouring [3], holographic microscopy [4], vibration analysis [5], image encryption [6] and 3-D object recognition [7].

In our case the security is the most important in image transmission as we are dealing with the firearm identification system. Characteristic markings on the cartridge and projectile of a bullet are produced when a gun is fired. Over thirty different features within these markings can be distinguished, which in combination produce a "fingerprint" for identification of a firearm [8]. Various firearm identification system based on 2-D digital image has been developed. The information recorded by the 2-D image limits the firearm identification. The security of the image transmission over the Internet is also an issue. If one can use the digital holography to record the 3-D information of used firearms combined with optical encryption it is possible to improve the firearm identification system in efficiency and security.

In this paper we first review the principle of digital holography and its application to image encryption and decryption. Then we present the initial results of recording a used bullet using digital holograph.

2 Principle

A basic set-up of off-axis digital holography is presented in Figure 1. The laser beam is split into two beams, reference beam and object beam. The reference beam direct to the CCD and the object beam illuminate the object. The reference beam and the diffusely reflected object beam are interfering at the surface of CCD. The micro-interference pattern or hologram is recorded by CCD electronically. The real image can be reconstructed from the digitally recorded hologram if the diffraction of the reconstructing wave at the microstructure of the hologram is carried out by numerical methods.

The object wave $H_O(x, y)$ at the CCD plane combined with a reference wave having the complex amplitude $H_R(x, y)$ to produce the intensity distribution

$$I_H(x,y) = |H_R(x,y) + H_O(x,y)|^2 \\ = |H_R|^2 + |H_O|^2 \\ + H_R^* H_O + H_R H_O^* \quad (1)$$

where * denotes complex conjugate.

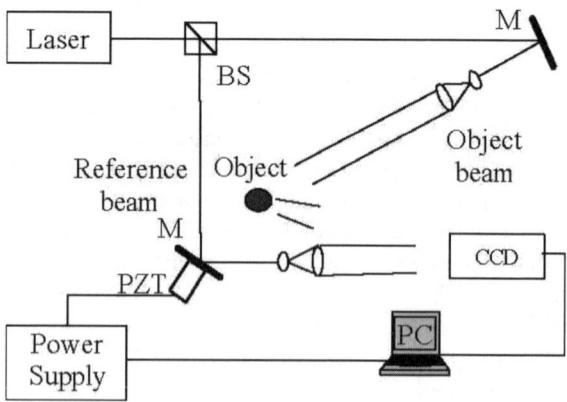

Fig. 1. Basic set-up of off-axis digital holography, BS - Beam Splitter, M - Mirror, CCD - Charge Coupled Camera

This intensity distribution is recorded by a black-and-white CCD camera as digital hologram. The digital hologram I_H (k, l) results from two-dimensional spatial sampling of $I_H(x, y)$ by the CCD:

$$I_H (k, l) = I_H (k\Delta\zeta, l\Delta\eta) \qquad (2)$$

where $\Delta\zeta$ and $\Delta\eta$ are the pixel sizes of the CCD in horizontal and vertical direction. For recording an hologram by a CCD, the incidence angles q_x and q_y between object wave and the reference wave have to fulfill the following condition [1]

$$\theta_x < \frac{\lambda}{2\Delta x}, \quad \theta_y < \frac{\lambda}{2\Delta y}$$

In conventional holography reconstruction is conducted by illuminating the hologram with a replica of the reference wave, $H_R(x, y)$. The transmitted wave from the hologram is proportional to $H_R(x, y) I_H(x, y)$.

$$T(x, y) = H_R(x,y) I_H(x,y)$$
$$= H_R |H_R|^2 + H_R |H_O|^2 + |H_R|^2 H_O + H_R^2 H_O^* \qquad (3)$$

On an observation plane a three-dimensional image of the object is formed from the transmitted wave. The first two terms of Eq. (3) correspond to the zero order diffraction, the third term to the virtual image, and the forth term to the real image.

In the off-axis holography the objective wave and reference wave arrives in the hologram plane with separate directions and the different terms of the reconstructed wave front will propagate along different directions, owing to their different spatial frequencies.

The field in the image plane x-h at a distance d from the hologram plane x-y is obtained by calculating through Fresnel-Kirchoff integral [1]. Assuming the reconstruction reference wave is a plane wave represented at hologram plane by

$$H_R(x,y) = C \exp[i(k_x x + k_y y)] \qquad (4)$$

where C is a constant. The image wave in the image plane is

$$IMG(\xi,\eta) = C \exp[-\frac{i\pi(\xi^2 + \eta^2)}{\lambda d}]$$
$$\iint I_H(x,y) \exp[-\frac{i\pi(x^2+y^2)}{\lambda d}] \exp[i\frac{2\pi(x\xi+y\eta)}{\lambda d}] dx dy \qquad (5)$$

The digital reconstruction simulates the physical reconstruction. We consider the discrete finite form of Eq. (5), by taking into account the two-dimensional sampling of the hologram $I_H(x, y)$ by the CCD. The discretisation is given by the CCD pixel size *(Dx, Dy)*. The continuous variables *x* and *h* in the image plane are replaced by *mDx* and *mDh* and the variables *x* and *y* are replaced by *kDx* and *lDy*, where *m, n, k* and *l* are integers. If we consider M•N pixels, the discrete representation of Eq. (5) can be written in the following form:

$$IMG(m\Delta x, n\Delta y) = C\exp[-i\pi\lambda d(\frac{m^2}{M^2\Delta\xi^2} + \frac{n^2}{N^2\Delta\eta^2})]$$
$$\sum_{k=1}^{M}\sum_{l=1}^{N} I_H(k,l) \exp[-\frac{i\pi(k^2\Delta x^2 + l^2\Delta y^2)}{\lambda d}]$$
$$\exp[i2\pi(\frac{mk}{M} + \frac{nl}{N})],$$
$$m = 1, 2, ..., M; \quad n = 1, 2, ..., N. \tag{6}$$

Eq. (6) shows that the reconstructed wave is determined by the two-dimensional discrete inverse Fourier transform of the multiplication of the hologram $I_H(k, l)$ by the quadratic phase part

$$\exp[-\frac{i\pi(k^2\Delta\xi^2 + l^2\Delta\eta^2)}{\lambda d}]$$

The standard algorithm of fast Fourier transformation can be applied here. The intensity $I(m, n)$ and the phase $f(m, n)$ of the reconstructed image are given, respectively, by

$$I(m,n) = |IMG(m\Delta\xi, n\Delta\eta)|^2 \tag{7b}$$

$$\phi(m,n) = \arctan\left(\frac{\text{Im}[IMG(m\Delta\xi, n\Delta\eta)]}{\text{Re}[IMG(m\Delta\xi, n\Delta\eta)]}\right) \tag{7b}$$

A typical example of recorded hologram is shown in Figure 2a. Figure 2c represents the numerically reconstructed image according to Eq. (6) and Eq. (7a). It is clear that the reconstructed image suffers particularly from the bright zero-order diffraction (dc-term) which affects the quality of the reconstructed real image.

Since the off-axis geometry introduces a spatial modulation of the interference as indicated in Eq. (3), the method of spatial filtering can be used to remove the zero-order term and even the virtual image term as did in section 4.

3 Phase-Shifting Algorithm

The zero-order and virtual image can also be removed by the phase-shifting-hologram method in which four holograms are recorded with the same reference wave but with phase differences $\pi/2$.

We use for reference waves that are in quadrature with one after another, namely $H_{R1}=H_R(x, y)$, $H_{R2}=H_R(x, y)exp(ip/2)=iH_R$, $H_{R3}=H_R(x, y)exp(ip)=-H_R$ and $H_{R4}=H_R(x, y)exp(i3p/2)=-iH_R$. Then we can record the corresponding four holograms with intensity distributions $I_{H1}(x, y)$, $I_{H2}(x, y)$, $I_{H3}(x, y)$ and $I_{H4}(x, y)$. According to Eq. (1) we have

$$I_{Hn}(x,y) = |H_{Rn}|^2 + |H_O|^2 + H_{Rn}^* H_O + H_{Rn} H_O^* \quad (8)$$
$$n=1, 4$$

Illuminating the above four holograms with the corresponding reference waves and adding together the transmission waves, we have

$$H_{R1}I_{H1} + H_{R2}I_{H2} + H_{R3}I_{H3} + H_{R4}I_{H4} = 4|H_R|^2 H_O \quad (9)$$

Comparing Eq. (9) with Eq. (3), we can see that the terms of zero-order and virtual image part are cancelled out, leaving a term of pure term which is identical to the original object wave except for a coefficient. The reconstruction procedure described in the previous section (see Eq. (6)) can be used here to obtain the complex amplitude of the reconstructed field in the image plane.

In the experimental set-up a continuous phase shift can be obtained by pushing the mirror M in the reference beam by applying specific voltage to the PZT mounted at the bottom of the mirror. In this way, the reference beam travels a short optical path that corresponds to a relative phase change.

4 3D Image Encryption

An experimental set-up for 3D image encryption is shown in Figure 2. It is similar to the phase-shifting digital holography in Figure 1 except for a phase mask inserted in the reference beam and a removable mirror inserted in front of the object.

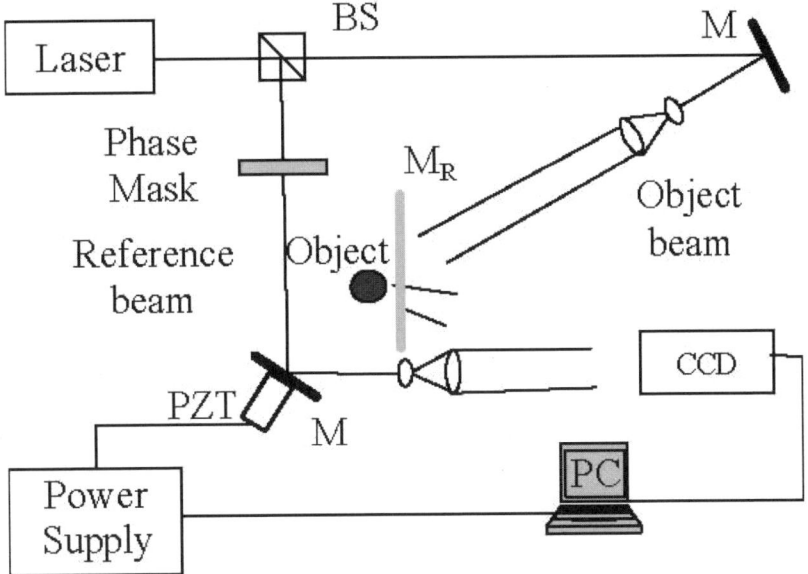

Fig. 2. Basic set-up of 3D image encryption, BS - Beam Splitter, M - Mirror, CCD - Charge Coupled Camera, M_R – Removable mirror

The principle of encryption-decryption is simple. As in the phase-shifting holography we record four holograms with the same reference wave but p/2 phase difference with each other. At this time the phase mask is embedded into the reference beams. If we want to reconstruct the image we have to know the numerical expression of the 4 reference waves, according to the Eq. (9). This numerical expression of the reference wave with phase mask embedded serves as encryption key. Without this key we can not reconstruct the image.

The key information can be recorded using the same set-up by inserting a mirror before the object. This mirror stops the light illuminating the object but reflects the light towards the CCD. Using the Eq. (9) we obtain the transmission wave from the phase mask which will serves as a reference wave in the decryption. The constant in the right side of Eq. (9) only affect the brightness of the reconstructed image which can be adjusted in the numerical reconstruction procedure.

5 Initial Experiments

In the initial experiment, we used a CCD camera with pixel sizes of 7x7 μm^2 and 480x640 elements. The object is placed 500 mm away from the CCD camera. Figure 3(a) is a digital hologram of a damaged bullet recorded with the set-up presented in Figure 1. Interference fringes are clearly observable in this image. The computed two-dimensional Fourier spectrum is presented in Figure 3(b) where we can see that the difference interference terms produce well separated spectrum contributions due to the off-axis configuration. The central bright spots represent the spectrum of reference wave and reflected object wave intensity. The interference terms $H_R^*H_O$ and $H_R H_O^*$ are located symmetrically with respect to the centre of the image. It is that these two terms contribute to the reconstruction of real and virtual images.

The reconstructed amplitude image from the hologram of Figure 3(a) is shown in Figure 3(c). The image consists of a real image, a virtual image and zero order diffraction that is the bright spot in the centre. The second order diffractions of real and virtual images can also be seen in the image. The reconstructed image is a bit messy due to the very bright zero-order diffraction and virtual image. These can be removed by using phase-shifting algorithm as described above. This is the future experimental we are planning. For the time being we remove these components by using filtering. The mask used for the spatial filtering is a transparent window which keeps only components that contribute to the real image in the Fourier spectrum. To avoid suppression of the high-frequency components of the interference terms, this window must be chosen as large as possible.

Figure 3(e) presents the Fourier spectrum after conducting spatial filtering to the Fourier spectrum in the Figure 3(b). The inverse Fourier transform of the filtered spectrum results in the filtered hologram in Figure 3(d). It can be seen that the bright background in Figure 3(a) disappears in Figure 3(d) and the visibility of interference fringes is enhanced. The inverse Fourier transform is used to numerically reconstruct the image. The reconstructed image is shown in Figure 3(f), which is much cleaner than the first reconstructed image in Figure 3(c).

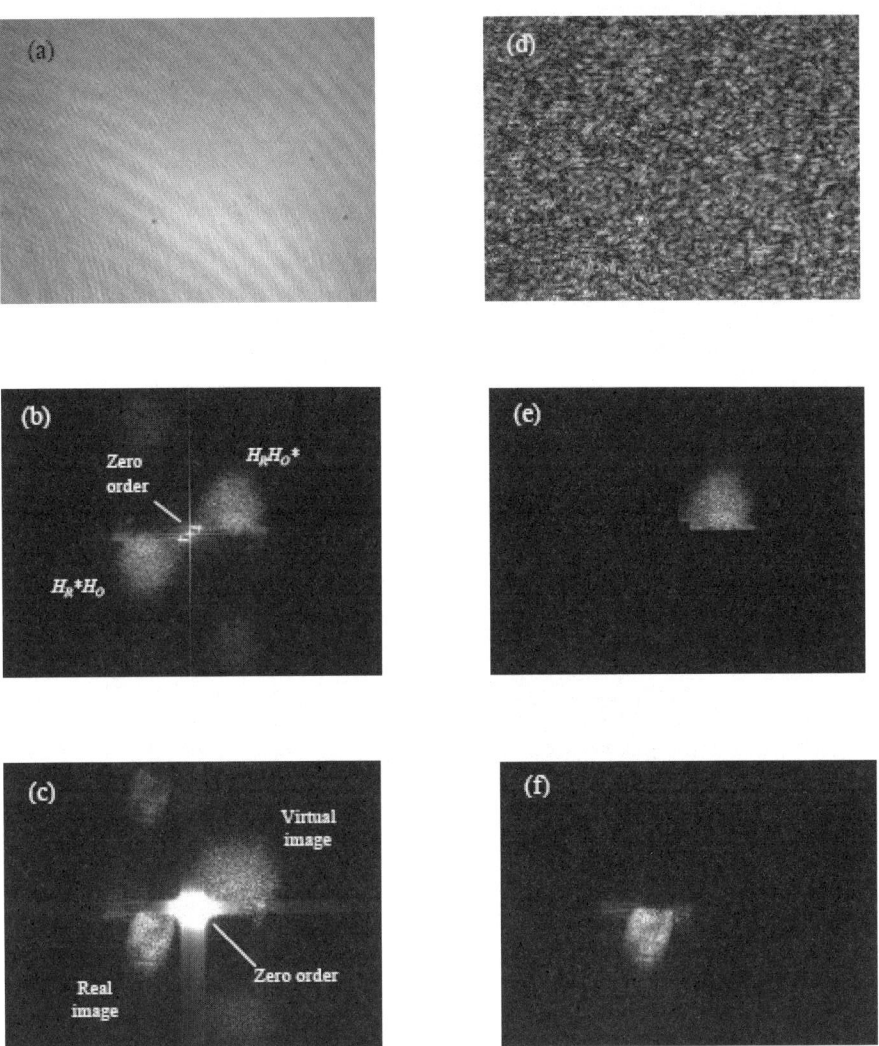

Fig. 3. Digital filtering the zero order diffraction. (a) Original hologram, (b) two dimensional Fourier spectrum of the original hologram, (c) amplitude image reconstructed from the original hologram, (d) the hologram after filtering, (e) two dimensional Fourier spectrum of the filtered hologram, (f) amplitude image reconstructed from the filtered hologram.

6 Conclusions

The digital holography has been demonstrated to be possible to record the 3-D information of a used bullet. The digital filtering techniques are adopted to clean the digital reconstruction image. Phase-shifting digital holography and 3D information encryption-decryption experiments are in progressing.

References

1. Yaroslavskii, L.P., Merzlyakov, N.S.: Methods of Digital Holography. Consultants Bureau, New York (1980)
2. Schnars, U., Juptner, W.: Direct recording of holograms by a CCD target and numerical reconstruction. Appl. Opt. 33, 179–184 (1994)
3. Yamaguchi, I., Ohta, S., Kato, J.: Surface contouring by phase-shifting digital holography. Opt. and Laser Eng. 36, 417–428 (2001)
4. Yamaguchi, I., Kato, J., Ohta, S., Mizuno, J.: Image Formation in Phase-Shifting Digital Holography and Applications to Microscopy. Appl. Opt. 40, 6177–6186 (2001)
5. Pedrini, G., Zou, Y.L., Tiziani, H.J.: Digital Double Pulse-Holographic Interferometry for Vibration Analysis. J. Mod. Opt. 42, 367–374 (1995)
6. Lai, S., Neifeld, M.A.: Digital wavefront reconstruction and its application to image encryption. Opt. Comm. 178, 283–289 (2000)
7. Frauel, Y., Javidi, B.: Neural network for three-dimensional object recognition based on digital holography. Opt. Lett. 26, 1478–1480 (2001)
8. Smith, C.I., Cross, J.M.: Optical imaging techniques for ballistics specimens to identify firearms. In: Proceedings of the IEEE International Carnahan conference on security technology, pp. 275–280. IEEE Press, New York (1995)
9. Cuche, E., Marquet, P., Depeursinge, C.: Spatial Filtering for Zero-Order and Twin-Image Elimination in Digital Off-Axis Holography. Appl. Opt. 39, 4070–4075 (2000)

GIS-Based Lunar Exploration Information System in China

Sheng-Bo Chen[1,2] and Shu-Xin Bao[1]

[1] College of Geoexploration Science and Technology, Jilin University,
Changchun, 130026, China
chensb@jlu.edu.cn
[2] State Key Laboratory of Remote Sensing Science, Institute of Remote Sensing
Applications, CAS / Beijing Normal University, Beijing 100101, China

Abstract. The Moon is Earth's only natural satellite. The Moon is becoming a prime destination for efforts by many nations seeking to expand into space. The project of China's lunar exploration, Chang'E Project, was approved to set up, opening the curtain of China's deep space exploration. In this study, a GIS-based lunar exploration information system in China is first built before the launch of Chang'E -1. By the system, the lunar exploration information in the past and near future can be managed and analyzed. And the client can browse some lunar exploration results in the internet, such as Apollo landing sites, and lunar craters. Thus, it is available to integrate and compare the lunar exploration results of China to the past lunar exploration under the system.

Keywords: WebGIS; Lunar; Exploration Information System; Apollo.

1 Introduction

The Moon is Earth's only natural satellite. The visible nearside of the Moon has been mapped in detail with its bright highlands and dark maria since Galileo turned the first telescope to the skies in 1609. It was not until the Space Age began that we got our first glimpse of the lunar farside from the Soviet Luna 3 spacecraft in 1959. The farside has virtually none of the dark maria that give the nearside its appearance, and it contains the biggest and deepest impact basin in the entire solar system, the Aitken Basin. Up to now, the Moon remains the most popular target for any spacefaring nation's first adventurous steps into planetary exploration. It was surely the target of the space race between the United States and Soviet Union in the 1960s and 1970s. The most recent mission to the moon was the Unite States' Prospector and Europe's SMART-1. The Moon is becoming a prime destination for efforts by many nations seeking to expand into space, such as China (Chang'E), Japan (SELENE), and India (Chandrayaan-1)[1][2]. NASA begins now its new lunar exploration activities that enable sustained human and robotic exploration that perform orbital reconnaissance and demonstrate capabilities for sustainable solar system exploration [3].

In January 2004, the project of China's lunar exploration was approved to set up, opening the curtain of China's deep space exploration. China's lunar exploration program, Chang'E Project, comprised three development phases: orbiting about, landing on and returning from the Moon. The first phase of the project is to launch a lunar orbiter round the Moon in 2007 to carry out lunar exploration. The second phase is to launch a lunar rover to realize soft landing on the lunar surface from 2007 to 2010. The third phase is to send a lunar rover onto the lunar surface and then return to the Earth after exploring and collecting lunar samples. The first step is to develop and launch the lunar orbiter "Chang'E -1" and set up an initial system of the lunar exploration engineering at the same time. The scientific objectives of Chang'E -1 are as follows: (1) to obtain three-dimensional images of the lunar surface; (2) to analyze the distribution of 14 useful elements and materials below the lunar surface; (3) to probe the features of lunar soil; (4) to explore the space environment between the Moon the Earth, and above the lunar surface[4][5].

Based on the past results of the lunar exploration, it is necessary to integrate and compare them to the lunar exploration of China. In this study, a GIS-based lunar exploration information system in China was first built before the launch of Chang'E -1. The structure, data flow and interaction of the system are first designed. And the functional modules are implemented based on a Web-based GIS software, ESRI ArcIMS. Finally, some application examples are also presented.

2 System Design

The objectives of the lunar exploration information system are the management and analysis of the lunar exploration information. Specially, the comparison between the past exploration results and Chinese exploration information and their integration will be conducted by the system.

2.1 Structure

To achieve the objectives, the prototype structure of lunar exploration information system follows as Figure 1, where the system is made up by the Server and Browser. The lunar exploration information is saved in the Server. The processing and analysis of all the exploration information are also conducted in the server. All the results can be Web-based browsed by the distributed users. A Web-based platform software for geographic information system (GIS), ESRI ArcIMS, will be employed for the development for the system (Fig.1).

2.2 Data Flow

The source data in the lunar exploration information system include the past exploration results and the Chinese future exploration information. The past exploration results are mainly involved in topographic elevation contour, geological information, and free-air gravity data and their attributes from the United States Geological Survey (USGS)[6][7]. All the results are geo-referenced to the same reference system, GCS_Moon_2000 of USGS. Especially, the landing sites and lunar craters from lunar exploration results, including their location (latitude/longitude) and some pictures, have been collected. Thus, the landing sites will be managed to be point information.

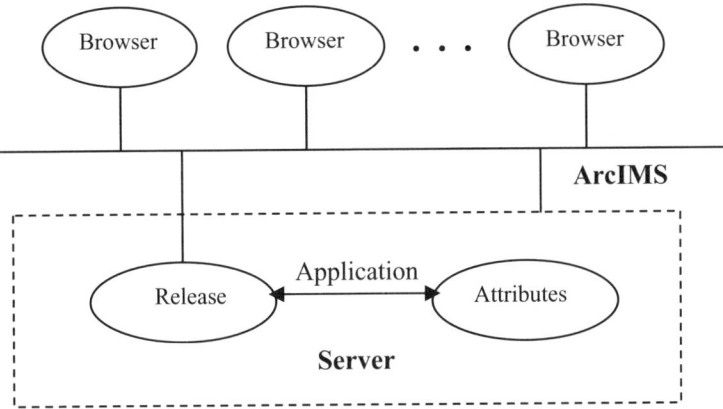

Fig. 1. Prototype structure of lunar exploration information system. Based on ArcIMS, the users can browse the information in the server, and the analysis or processing results can be released to the browsers by the server.

Furthermore, the chemical components and thickness of regolith (lunar soil) are geo-referencing to the same reference system. With the advance of Chinese lunar exploration program, more and more data will be also referenced to this system (Fig.2).

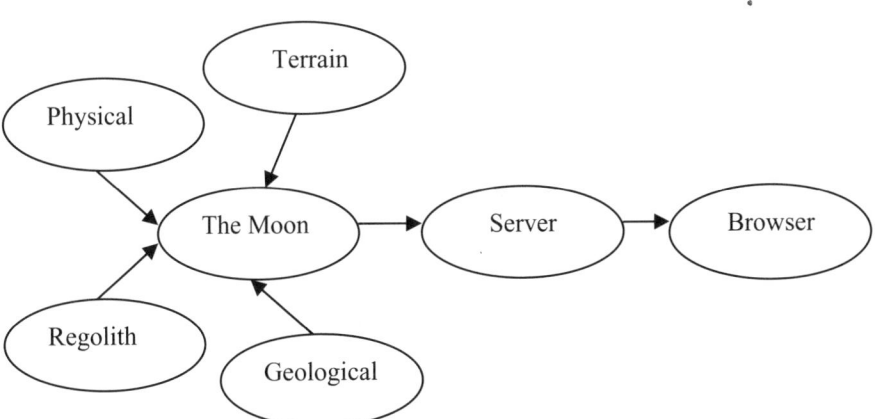

Fig. 2. The data flow in the lunar exploration information system. The information is involved to physical and geological exploration information, the results on lunar terrain and regolith (components and thickness), and so on.

2.3 Interaction

The interactions between the browsers and server can be controlled by a mouse or keyboard. The requests of the users are accessible to the browsers, and the messages are sent to the server under the Internet by ArcIMS. The messages are processed and analyzed by the server, and the responds are returned to the browsers. For example, the users can check the altitudes of any location on the moon, when the mouse clicks on the topographic map in the system (Fig.3).

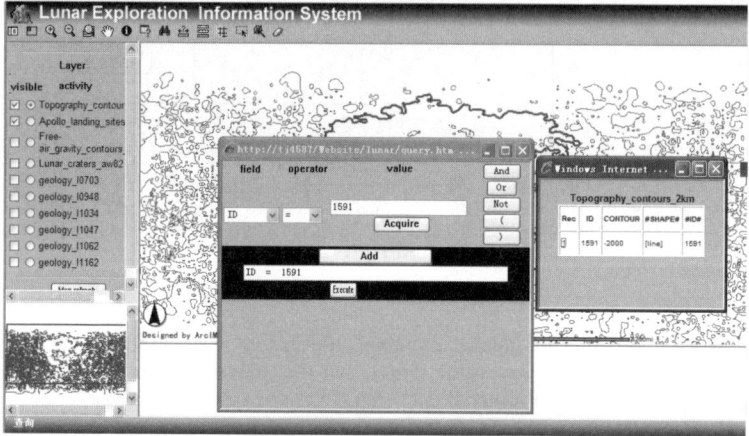

Fig. 3. The contour line is chosen by the mouse. And its attributes, such as altitude, color, and thickness, are highlighted in the topographic map from the server.

3 System Implementation

Lunar exploration information system is implemented based on ESRI GIS software, ArcIMS (Fig.4). The functional modules are built.

3.1 ArcIMS

ArcIMS is the solution for delivering dynamic maps and GIS data and services via the Web. It provides a highly scalable framework for GIS Web publishing that meets the needs of corporate Intranets and demands of worldwide internet access. ArcIMS services have been used by a wide range of clients, including custom Web application, ArcGIS desktop, and mobile and wireless devices [8].

Data are saved in spatial server, including query server, feature server, image server, geocode server. The clients are customized by HTML (Hyper Text Markup Language). The request is encapsulated by ArcXML, one of XMLs (Extensible Markup Language), where the geographic information, such as layer information, service information, is marked. The ArcXML request is transmitted to Application server by Servlet connector. The returned ArcXML from the server are parsed by JavaScript in the client. And the pictures can be downloaded and highlighted from the server (Fig.5).

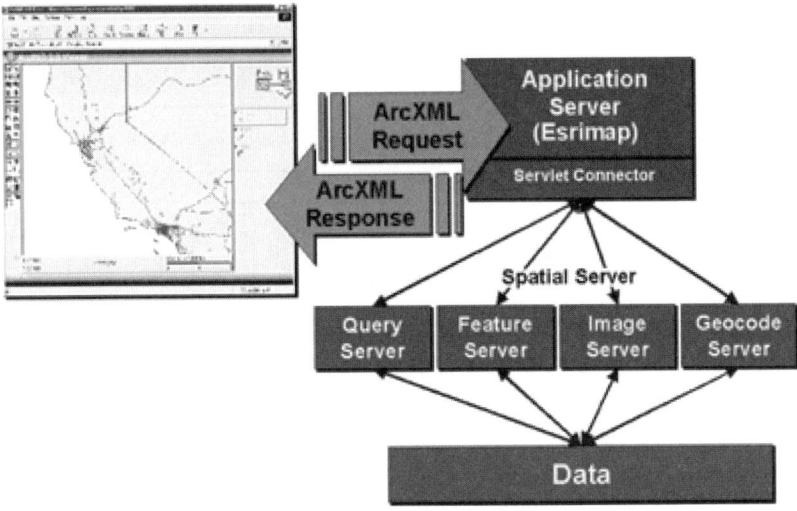

Fig. 4. The procedure to implement the lunar exploration information system. It is based on ArcIMS, one of the products of ESRI GIS.

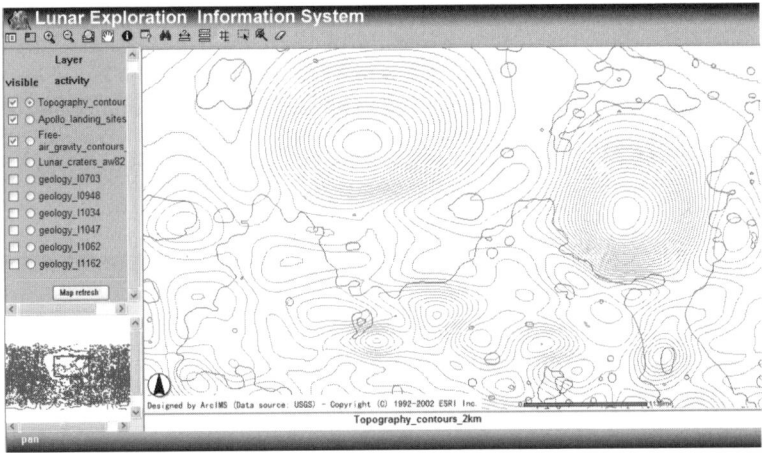

Fig. 5. The Interface of Lunar Exploration Information System. There are three main windows: map visual window, map full extent window, layer list window.

3.2 Functional Modules

The functional modules of the lunar exploration information system comprise of map operation and layer operation modules. The map operation includes map pan, zoom in and out, full extent, and legend functions. And the layer operation includes layer list, find, query, measure, select, and so on (Fig.6).

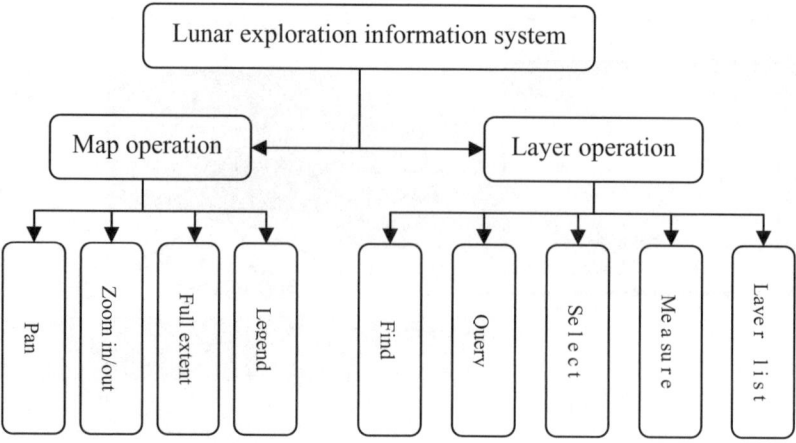

Fig. 6. Functional modules of the lunar exploration information system. Map operation and layer operation are two main modules of the system.

4 Applications

In China, the lunar orbiter "Chang'E -1" has not now been launched. The lunar exploration information system is based on the past lunar exploration results. The lunar exploration information and its reference system are from USGS, which will be taken as the framework for the Chinese lunar exploration information. Thus, different resources of lunar exploration information can be overlaid to the same reference system, such as the topographic contour, geological maps, free-air gravity contour, and so on (Fig.4).

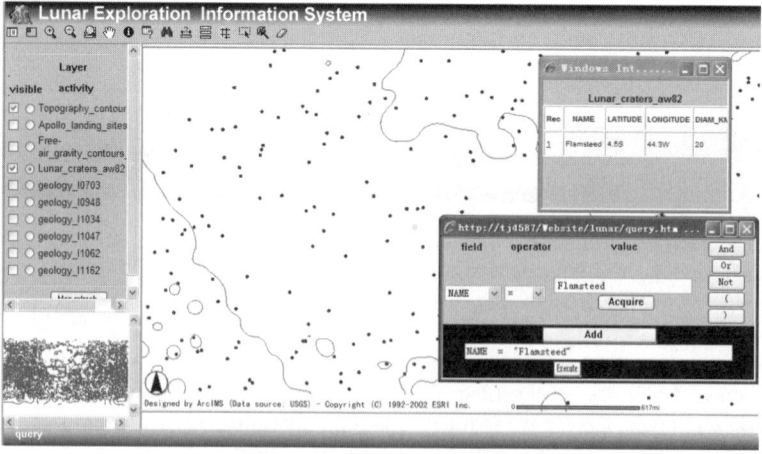

Fig. 7. The lunar craters can be queried by the system. And their related information, like their sizes and location, are presented in the system.

Some point information, like Apollo landing sites and lunar craters, have been saved in the system. On the Fig. 7, the lunar craters can be found by their location, sizes, and other parameters. On the Fig. 8, the Apollo landing sites can be found by their location. And the related pictures of the Apollo landing sites are presented by the system.

Furthermore, the lunar exploration information by Chinese Chang'E project will be managed and analyzed by the system in the near future with the launch of Chang'E -1 orbiter.

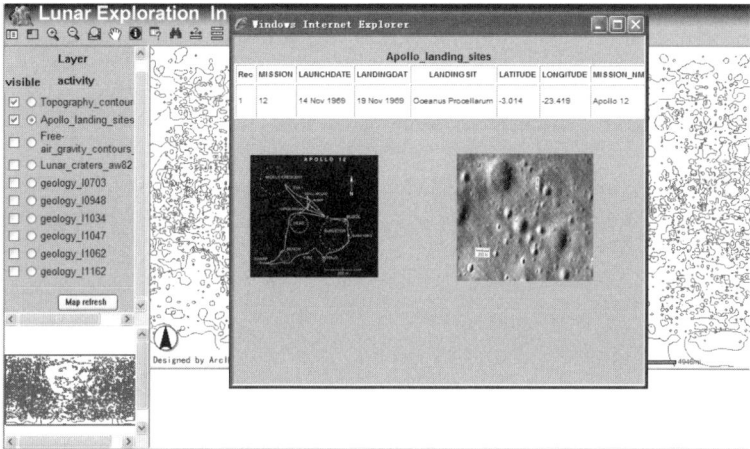

Fig. 8. The Apollo landing sites can be queried by the system. And their related information, like the picture and location, are presented in the system.

5 Conclusions

With the launch of Chinese Chang'E-1 satellite in 2007, it is necessary to have a lunar exploration information system to manage and analyze those exploration results. Specially, there have been large amount of exploration information in the past lunar exploration. The comparison between these exploration results and calibration of the Chinese lunar exploration information are important definitely to the Chinese lunar exploration. Based on ArcIMS, a lunar exploration information system are designed and implemented. By the system, a large amount of lunar exploration information in the past and near future will be managed and analyzed. The client can browse these exploration results in the internet, like the Apollo landing sites and lunar craters information.

Acknowledgments. The research was sponsored by the National Natural Sciences Foundation of China (NSFC) (No.40471086) and Open Fund of Remote Sensing Science (No.SK050006), State Key Laboratory of Remote Sensing Science, Jointly Sponsored by the Institute of Remote Sensing Applications of Chinese Academy of Sciences, and Beijing Normal University.

References

1. The Planetary Society-Making You a Part of the Next Age of Exploration, http://planetary.org/ explore/topics/the_moon
2. Binder, A.B.: Lunar Prospector: Overview. New Series 281, 1475–1476 (1998)
3. Utilization of Lunar Resouces to Enhance Space Missions, http://www.mars-lunar.net/Reality.or.Fantasy
4. China's Lunar Probe Chang'e, http://en.cast.cn/ShowSpecial.asp?SpecialID=10
5. Yongliao, Z., Ziyuan, O., Chunlai, L.: Progress of Exploration and Study of the Moon. Chinese Journal of Space Science 20, 93–103 (2000)
6. Moon General Image Viewer, http://webgis.wr.usgs.gov/website/moon%5Fhtml/viewer.htm
7. Lunar and Planetary Institute, http://www.lpi.usra.edu
8. ArcIMS-GIS and Mapping Software, http://www.esri.com/software/arcgis/arcims/index.html

Semantic 3D CAD and Its Applications in Construction Industry – An Outlook of Construction Data Visualization

Zhigang Shen[1], Raja R.A. Issa[2], and Linxia Gu[3]

[1] W145 NH, University of Nebraska-Lincoln, Lincoln, NE 68588-0500
zshen2@unl.edu
[2] RNK 304 / Box 115703, Universities of Florida, Gainesville, FL 32611-5703
raymond-issa@ufl.edu
[3] CM 170, South Dakota School of Mines and Technology, Rapid City, SD 57701
linxia.gu@sdsmt.edu

Abstract. In response to the need of using electronic design data directly in construction management applications, many CAD developers have started implementing semantic data models in their CAD products using industry foundation classes (IFCs). While helpful, these semantic CAD applications have limitations when used in actual construction practices. The case studies in the thesis indicated that: 1) the semantics of the current data model (IFC as an example) is not rich enough to cover the richer details of the real trade practices; 2) the current implementation of the semantic data model lacks the mechanism to provide multiple trades views at various detailed levels. This paper also provided suggestions for the future development of semantic data model of construction industry.

Keywords: 3D CAD, IFC, Semantic, Data Visualization, Construction.

1 Introduction

As a major source of construction project data, CAD applications have traditionally been used as a drafting tool in the AEC industry, instead of as an integrated design information database for supporting production activities. As the result, information retrieval from design documents remains a manual process even though most design documents are in electronic format. This situation is one of the important factors that discourage contractors from using CAD in their operations (Figure 1).

A construction manager usually needs to coordinate many subcontractors, architects and engineers. In addition, frequent design changes are typical in the construction project. Thus, manually searching for design data and updating becomes a heavy burden for the project manager. In recent years the development of BIM (Building Information Model) started to address the existing barriers, which prevent 3D CAD application in construction industry.

The intention of BIM is to integrate the visualization capacity of the 3D CAD with the construction information data to allow construction managers to digitally retrieve and analyze design data directly from the CAD application. Some parametric CAD products [2] [3] incorporated IFC [4] product models in their internal data structure, which allowed these CAD applications to export semantically rich IFC data for reuse by other third-party application. These parametric AEC CAD applications not only allow designers to do quick 3D conceptual designs but they are also able to do quantity survey automatically.

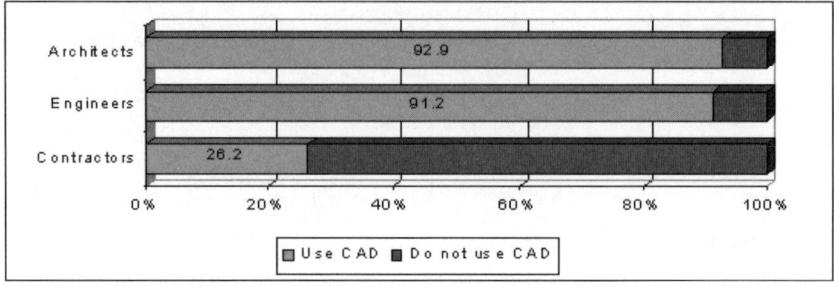

Fig. 1. CAD usage in AEC industry [7]

Figure 2 illustrates the differences between the traditional geometric 3D CAD model and the parametric semantic 3D model. Figure 3 shows the differences of the internal data structure between the regular graphic-based CAD application and the IFC based CAD application. On top of the raw geometry data, the IFC model adds many construction semantic parameters, which intend to provide construction applications with a platform of semantically consistency.

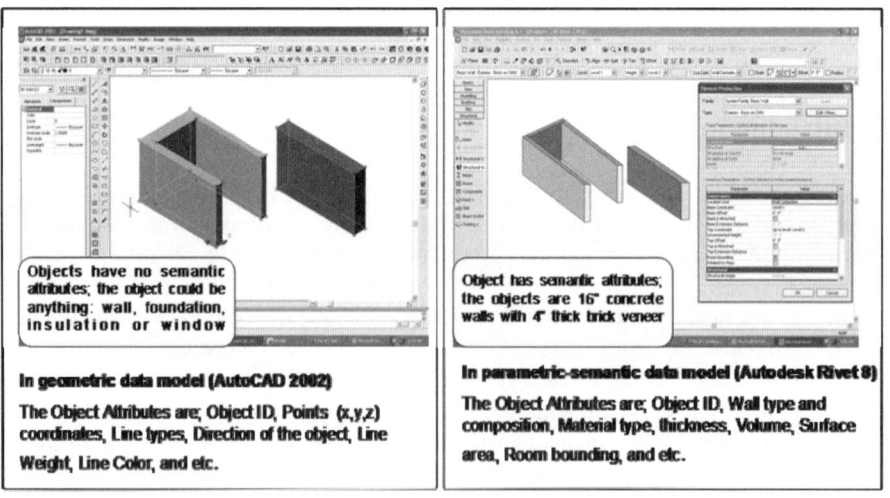

Fig. 2. Attributes of geometric data model vs. parametric-semantic data model

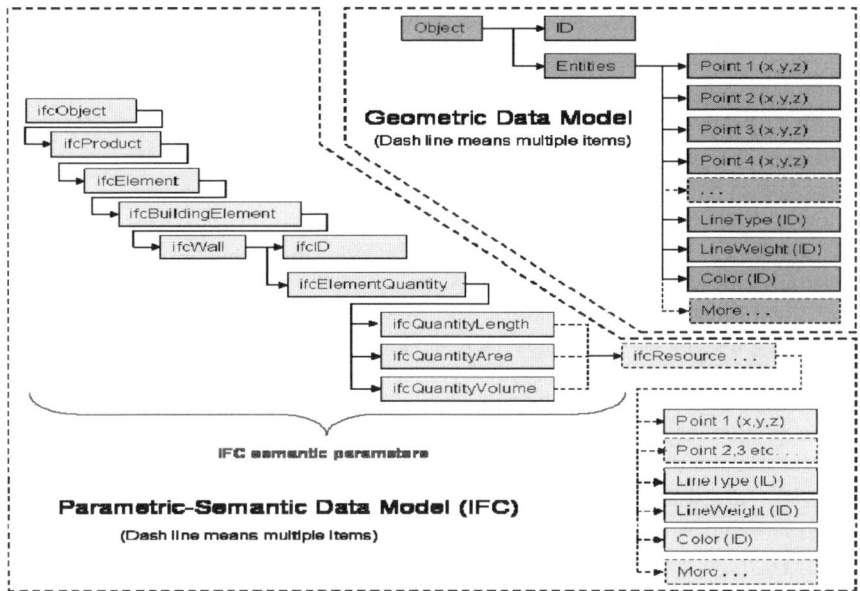

Fig. 3. The data structure of geometric data model vs. parametric-semantic data model

2 Limitations of Using IFC 3D CAD Application in Construction Quantity Survey

Cost estimates are crucial for contractors to find out expected project cost, to create their budget and to evaluate the impact of the design changes. Because cost estimating is a quite time-consuming task many tools were developed to help the estimator to get the job done quicker. Those tools range from manually using color markers to using digitizing boards, to on-screen digitizing and to the latest CAD-based automatic estimating software.

In recent years, the IFC product-process model [4] has become more mature and was implemented by a few CAD vendors and construction software vendors. A few software applications implementing the IFC model have shown the capability to provide material quantities based on the CAD design documents. While helpful, the CAD based estimating applications are limited by the IFC semantic data model when used in calculating detailed quantities. The major limitation is that the current IFC model is not semantically rich enough to cover all the construction process data and job conditions.

The detailed construction estimate includes quantities that are much more complex than just simple material quantities. Many quantitative values depend on the construction process/construction methods and the very specific job conditions. In the current IFC model the process data structure is quite limited.

There are numerous construction methods and products and each construction method will have its own domain-specific views to the data model. When further considering the broad scope, uniqueness and the complexity of the construction project, it is very unlikely, as many researchers agree, that the single IFC model can cover all the construction aspects needed for supporting detailed estimates [5] [6].

3 Concepts of Product/Procurement Quantity (PPQ) and Process Quantity (PCQ) in Construction Quantity Survey

In order to further discuss the existing problems in the current CAD based estimating applications, it is necessary to first clarify the concept of quantities in the detailed estimating practice. The job quantities and cost are always the focal points of construction management throughout the project. The construction manager obtains physical job quantities and design features from the design documents. The construction manager comes up with the unit cost for this particular quantity amount based on the specific design feature and the applied construction methods. The specific cost of a specific portion of the project is calculated by multiplying the specific physical quantities and the unit cost. The overall project cost is the summation of the costs of each piece of the different specific features.

Project quantity is an over-generalized term when we consider detailed estimates. Project quantities can be further divided into Product/Procurement Quantities (PPQ) and Process.

Quantities (PCQ) and their corresponding costs are Product/Procurement Cost (PPC) and Process Cost (PCC).

The PPQ, as implied in its name, refers to physical material quantities, which will become part of the final building product when the project is completed. The PPQ, which includes the bill of materials, is often used for procurement. The PPQ are contained in the design documents and are relatively stable over the overall project duration. Since the material costs are relatively stable and will not vary much from job to job, both PPQ and PPC can be considered as static, as long as there is no significant design change.

On the other hand, PCQ refers to quantities that are closely-related to the construction methods, sequence and job conditions. PCQ differs significantly from one company to another, from construction method to construction method, even for the exact same project.

A particular PCQ depends on the particular construction method, the particular labor skills, the particular construction sequence and many other job conditions. For example, the formwork quantity of a concrete wall, which by definition is included in the PCQ, varies significantly between the cast-in-place method and the pre-cast method. Not only does the PCQ depend on the particular job condition, but the unit price of PCQ also depends on the particular job conditions. For example, the unit price of painting a dark wall color will cost more than the unit price of painting a light wall color.

3.1 The PCQ Case Example

To further explain the concepts of PCQ and PPQ, a simple case example of brick veneer is presented in this section to illustrate how construction methods and trade knowledge affect the detailed quantity estimating. Figure 1-7 shows a tilt-up concrete wall with brick veneer that is to be built as a retail building envelope. Figure 7(a) shows how this tilt-up panel is cast on the casting bed. Figure 7(d) shows how the panel is cast using the brick veneer as an embedded casting back. The task is to estimate the cost of the brick veneer, i.e. material and installation.

As stated in section 1.4, the PPQ, which is the brick material quantity in this case, looks easy to find by just adding areas of surface A and surface B and multiplying by the waste factor.

This gives us the square footage of the brick veneer of this panel. However, if the estimator knows how the contractor casts the tilt-up panel then the estimator would know this quantity is wrong.

The actual procedure of casting a brick veneer tilt panel is as follows:

1. The brick veneer is laid face-down on the casting bed.
2. Then the panel rebar is put on top of the brick veneer.
3. The concrete is poured on the back of brick veneer.
4. The panel is tilted up and put it in its place.

In this process the brick veneer is embedded by the gravity of the concrete. However, the brick veneer of the B side of this panel as shown in Figure 4(c) cannot be installed the same way because the form board of the panel on the B side cannot hold the brick veneer in its place during the pouring of concrete. So the brick veneer of surface B must be installed later on after the panel is tilt up and put in place.

Because of this detailed tilt panel construction method, the installation cost and waste factors of surface B are significantly higher than those of surface A. So in order to get the correct cost of the veneer the estimator has to calculate the A area and the B area separately and apply their corresponding waste factors and unit installation cost separately. In this case both the A area and the B area are PCQ, which is determined by this particular tilt panel construction method and job condition.

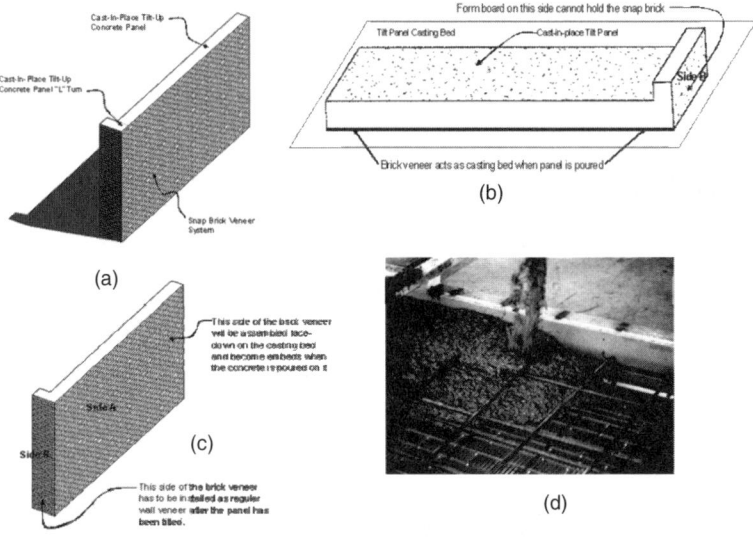

Fig. 4. Estimate snap break veneer on a tilt-up concrete wall

This example illustrates how the quantities are broken down into small pieces because of the impact of a particular construction method and a different unit price is applied to each piece based on their different installation methods. Without knowing the trade specific knowledge, it is unrealistic to expect one to estimate the right quantities and the right cost especially for the PCQ and PCC in this example.

3.2 Concept of Multi-dimensional Construction Views/Level of Details

Buildings are very complex and unique: no two buildings are the same in the world. One project could include underground utilities, concrete products, a parking garage, elevators, boilers, chillers, a helicopter platform, sculptures, swimming pools, roller coaster etc. Very few if any products in the world have so broad a scope of works.

The diversity of the building project participants reflects the degree of the building's complexity and uniqueness. Since there are so many participants from different backgrounds and with different interests in every building project, it is very natural that each participant has their own view of the project. Figure 5 shows a multi-view example of an actual building component.

Construction management is about quantity and process management. When it comes to construction quantities, every project participant has specific quantities they are interested in, which reflect their specific construction information access needs. These quantities may not have the same level of details.

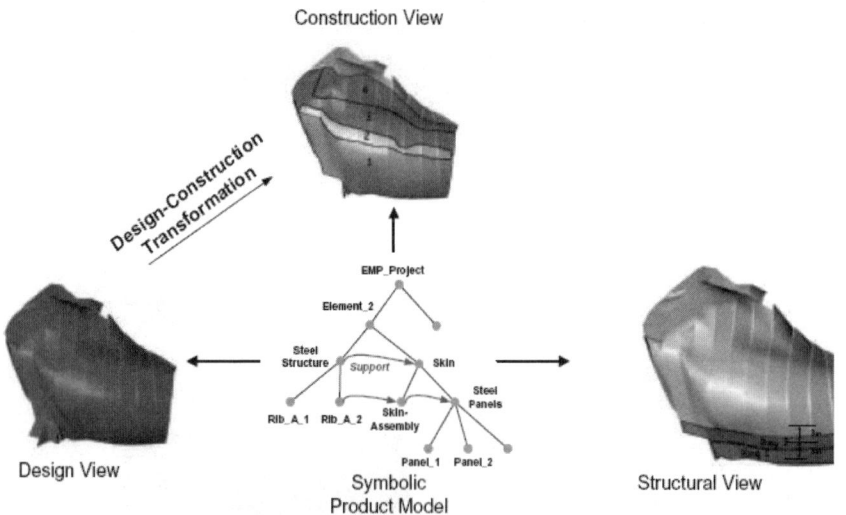

Fig. 5. Multi-dimensional example of a building component [1]

4 Conclusions

In summary, although the current BIM models and its implementations have improved their capacity of accommodating the construction management requirements, the limitations still exist.

The fundamental remaining question is that how to create a construction data retrieval application using the current 3D visualization applications to reflect the contractors' perspectives.

It's the author's view that the solution will rely on the more mature interpretation tools of the current 3D semantic data model. These interpretation tools shall address the particular trades' perspectives.

References

1. Akbas, R., Fischer, M.: Examples of Product Model Transformations in Construction. In: Lacasse, M.A., Vanier, D.J. (eds.) Durability of Building Materials and Components 8. Institute for Research in Construction, Ottawa ON, K1A 0R6, Canada, pp. 2737–2746 (1999)
2. ArchiCAD (2007), http://www.graphisoft.com/products/archicad
3. AutoDesk Revit (2007), http://revit.downloads.autodesk.com
4. IFC (2006), http://www.iai-international.org/index.html
5. Issa, R., Flood, I., O'Brien, W. (eds.): 4D CAD and Visualization in Construction: Developments and Applications, A.A. Balkema, Lisse, The Netherlands (2003)
6. O'Brien, W., Issa, R., Shen, Z., Xie, H.: Configurable environments: a vision for future project information technologies. In: Kazi, S. (ed.) Knowledge Management in the Construction Industry: A Socio-technical Perspective, pp. 339–358. Idea Group Publishing, Hershey, PA 17033, USA (2004)
7. Rivard, H.: A Survey on the Impact of Information Technology on the Canadian Architecture, Engineering and Construction Industry. ITcon 5, 37–56 (2000)

A Fast Algorithm for License Plate Detection

Vahid Abolghasemi and Alireza Ahmadyfard

Shahrood University of Technology, Shahrood, Iran
vahidabolghasemi@yahoo.com, ahmadyfard@shahroodut.ac.ir

Abstract. In this paper we propose a method for detection of the car license plates in 2D gray images. In this method we first estimate the density of vertical edges in the image. The regions with high density vertical edges are good candidates for license plates. In order to filter out clutter regions possessing similar feature in the edge density image, we design a match filter which models the license plate pattern. By applying the proposed filter on the edge density image followed by a thresholding procedure, the locations of license plate candidates are detected. We finally extract the boundary of license plate(s) using the morphological operations. The result of experiments on car images (taken under different imaging conditions especially complex scenes) confirms the ability of the method for license plate detection. As the complexity of the proposed algorithm is low, it is considerably fast.

Keywords: License Plate Recognition, Image processing, Edge detection, Mathematical morphology.

1 Introduction

During recent years number of moving vehicles in roads and highways has been considerably increased. Hence, today traffic management and Intelligent Transportation Systems (ITSs) design has taken into account [1]. One of the effective solutions to control the traffic of vehicles is employing a License Plate Recognition (LPR) system. A LPR system has numerous applications such as traffic law enforcement, parking management, ticketing, control of restricted areas, etc [1, 2].

In general the LPR system consists of three major parts [3]: License plate detection, character segmentation and character recognition. A desired LPR system has to work under different imaging conditions such as low contrast, blurring and viewpoint changes. It is also supposed to perform properly in complex scenes and bad weather conditions. The car plate detection is the most important and crucial task in a plate recognition system [3]. In addition response time is another restriction in real time applications such as license plate tracking.

During last years different approaches have been proposed including edge analysis [4], neural networks [6], color and fuzzy maps [7, 8], vector quantization, texture analysis [9], etc. Applying some restrictions on input images, a method can achieve an acceptable result. However the most algorithms of existing methods fail in real scenario. Hence always designing a versatile algorithm has been concerned.

The techniques based on edge statistics are of the first methods in this area. Due to sufficient information from edges in the plate region, these techniques yield promising results. The advantage of edge-based methods is their simplicity and speed [1]. But using edge information alone the rate of success is low especially in complex (appearance of unwanted edges) and low contrast (missing plate edges) images. In spite of this, combining edge information with morphological operations, improves the performance, although it decreases the speed factor [4].

In some edge-based Methods the aim is to detect boundaries of a car plate Using Half Transform (HT) [10]. This approach has difficulty in extracting distorted or dirty plates. It also has known as a complicated approach [1].

Texture analysis is another useful approach for detecting license plates. This approach takes the advantage of existing homogenous and frequent texture-like edges in the plate region. Gabor filters have been one of the major tools for texture analysis [11]. Using these filters, the process is independent of rotation and scaling. It has the ability of studying images in an unlimited number of directions. But it is a time consuming and complex method specially when applied to large images.

During the last many years developments dealing with simple images have been achieved acceptable results. But recent researches have been directed to processing complex images with unconstrained conditions [12]. Therefore the current paper focuses dealing with such images.

In this paper an edge-based method on License Plate Detection is presented. The proposed method has considerably reduced the problem of unwanted edges appeared in complex images. We can accomplish this by applying a special filter which describes the plate model. After this step only a few connected regions (at most 4) has been left. It should be noted that this process is extremely fast. From now on the processes would be done on these small regions instead of whole image. We then apply some simple morphological operations and employ plate features including aspect ratio, shape and edge density for offering the probable region as a plate candidate.

The rest of the paper is organized as follows. Section 2 explains the proposed method on license plate detection. Then experimental results are presented in section 3. At last the paper is concluded in section 4.

2 Methodology

By observing license plates in images, two main features are noticed. First, horizontal edges around a car plate are relatively strong and dominant. Second density of vertical edges across a car plate are significant, while background edges are usually either long curves or very short. These two important features and also low complexity for edge-based analysis motivate us to use edge information for the car plate detection.

In the proposed method we first estimate the locations with significant density of vertical edges. These locations are good candidates for plate regions. To filter out false candidates in the obtained result, we design a match filter which gives a strong response to the instance of the license plate in the image. Now we explain in detail the proposed algorithm for license plate detection.

First, using Sobel operator we measure the gradient of the input image. Then an appropriate threshold is applied on the result to extract the image edges as a binary image. We aim to detect locations with significant local density of edges which are good candidates for car license plates. In order to estimate the local edge density we blur the binary edge image using a Gaussian kernel. Then we apply the plate model filter to this blurred image. A threshold selected close to maximum gray level value of this image is performed. This yields a binary image including a few Regions. These regions have been analyzed using some simple morphological operations to locate the boundary box of the probable candidates as license plates.

2.1 Estimation of Vertical Edge Density

As a fast and simple operator, we used Sobel to obtain the gradient of the input image in vertical direction (Eq. 1).

$$h = \begin{bmatrix} -1 & 0 & 1 \\ -2 & 0 & 2 \\ -1 & 0 & 1 \end{bmatrix} \quad (1)$$

The selection of a proper threshold to extract strong edges from the gradient image and prevent to miss important edge information is relatively difficult. By a high threshold level we may miss plate edges, whereas a low level for the threshold results lots of weak edges in the clutter part of the scene. An example of extracted edges is shown in Fig. 1. Fig. 1b, c show binarized images with low and high thresholds, respectively. Although local thresholding may alleviate the sensitivity of edge detection to the threshold value, it is relatively complex.

Since we are not aware of the strength of vertical edges at the license plate, a relatively low threshold is used for the edge detection (Fig. 1b). Thus we can be sure about preserving plate edges even in bad illumination condition. From Fig. 1 one can be seen that the upper and lower outer parts of the plate region in both images (Fig. 1b, c) consist of weak vertical edges. So we apply the global thresholding and set the threshold value to the one third of the strongest edges in the image.

It is worth to note that the plate region features significant density of vertical edges. In order to detect candidate regions for the car license plate we estimate edge

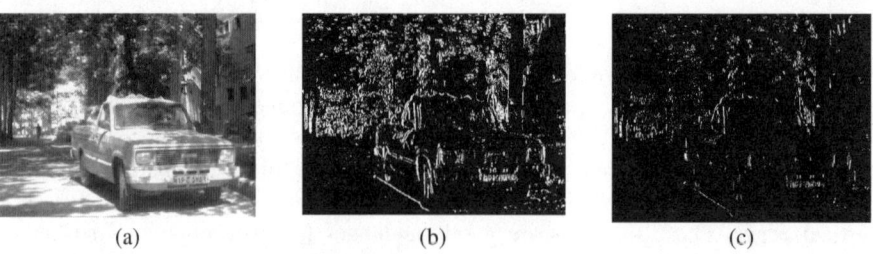

Fig. 1. (a) A car image in a complex scene. The vertical edges extracted from the gradient image using (b) low and (c) high threshold values.

density across the edge image by applying a Gaussian kernel on it. Fig. 2 shows the result of applying Gaussian kernel on Fig. 1b. As can be seen from this figure, regions with high density of vertical edges, such as the plate region, appear as bright regions.

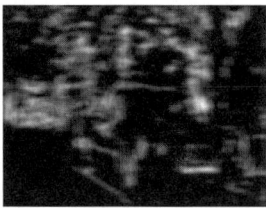

Fig. 2. The result of applying Gaussian kernel on the edge image in Fig. 1 (b)

2.2 Designing a Match Filter

In this stage we aim to detect candidates as car plate from the provided edge density image. We construct a model for car plates by observing license plate examples in edge density images.

We simply modeled intensity of image (Fig. 2) at license plate region using a mixture of Gaussian functions. A 3D plot of this model is shown in Fig. 3a. Using this model we emphasis that intensity remains constant in the plate region along horizontal direction. While the intensity along vertical direction at the middle of region is maxima and it reduces near the borders (Fig. 2). This function models the dark strips above and below of the car plate. We construct this model using three Gaussian functions. The mathematical equation of this model is:

$$h(x,y) = \begin{cases} A.\exp\left(-x^2 / 0.2\sigma_x^2\right) & \text{for } 0 \le x < \frac{m}{3} \text{ or } \frac{2m}{3} < x \le m, 0 \le y < n \\ B.\exp\left(-x^2 / 2\sigma_x^2\right) & \text{for } \frac{m}{3} \le x \le \frac{2m}{3}, 0 \le y < n \end{cases} \quad (2)$$

The function demonstrates a rectangular m*n mask. As can be seen from Fig. 3a, the filter varies only in vertical (i.e., x) direction. Thus the symbol σ_x is the variance of the main lob toward x direction. It should be noted that variance of the side lobs is 0.1 of main lob variance. It is for this reason that the plate region itself occupies a wider space compared to above and below part of it. And also two side lobs are designed to feature two horizontal edges of plate boundary. A, B (A < 0, B > 0) factors are designed so that they cause mean value of the filter to be zero.

We use the designed model as a match filter and apply it to the edge density image to find instance of license plates. We expect this filtering to provide strong response at plate-like regions and weak response at uniform or non-plate regions (such as asphalt, trees, etc). It is worth noting that the mean value of the model is zero, so it responses zero when it is applied to a uniform region. The result of experiments in section 4 confirms this concept.

The result of above filtering is compared against a predefined threshold to find a few candidates as license plates. Note that because of having strong response in plate-like regions, this process is not sensitive to the threshold value. We set the threshold value equal 80 percent of maximum intensity value in the filtering result image i.e., Fig. 2. The detected regions are processed during the next stage in order to find the boundary of plates.

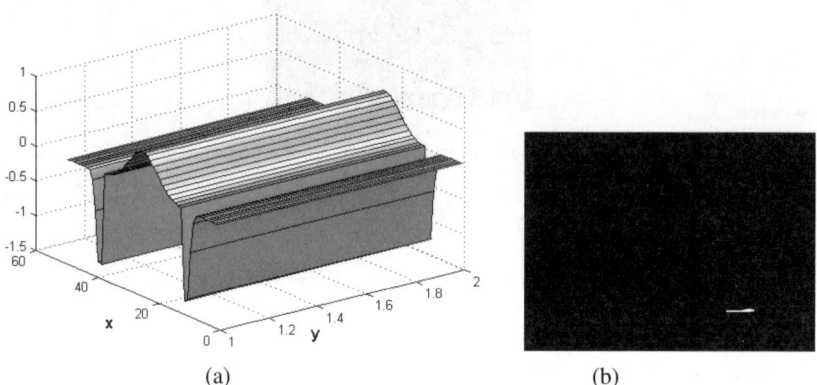

Fig. 3. (a) A 3-D plot of the plate model filter. (b) The result of filtering and thresholding on Fig. 2.

2.3 Region Extension Procedure

The candidate regions may have an irregular shape with different size. Assume a candidate region indicates the true license plate; we are not sure either it encompasses the entire plate or a portion of it. The thing we are sure about is the position (coordinate) of this hypothesis plate regard to whole image. To be sure the entire plate is surrounded by the candidate region; we extend the extracted candidates twice as the biggest expected license plate in the image. Fig. 4a, b, c depict the related region before and after this procedure, respectively.

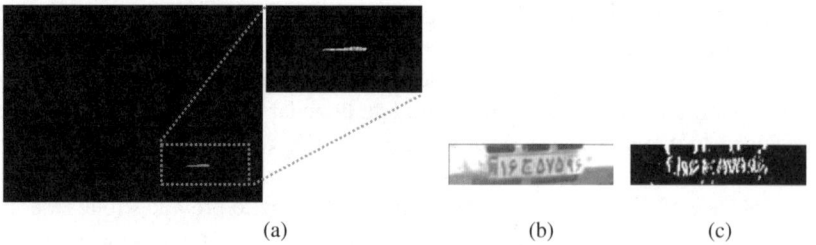

Fig. 4. (a) Image shown in Fig. 3(b). (b) Describing portion of the license plate. (b) Result of extending procedure and (c) its corresponding edges.

2.4 License Plate Extraction Using Morphological Processing

Morphology is a technique of image processing based on shapes. The value of each pixel in the output image is based on a comparison of the corresponding pixel in the input image with its neighbors. By choosing the size and shape of the neighborhood, we can construct a morphological operation that is sensitive to specific shapes in the input image [3, 13].

We use morphology to find the license plate(s) from candidates regions in the image. Although morphological operations are generally slow if being applied on whole image, processing of a few candidate regions is performed relatively fast. Since within a plate region several characters are placed, the density of vertical edges is high in compare with background. Using this property we apply morphological closing to connect characters in the license plate. As seen from Fig. 5a applying the closing operation on Fig. 4c yields connected regions. We use a rectangle with height 3 and width 10 pixels as structuring element. In order to remove thin regions, we apply the morphology opening next. The structuring element for this operation is a rectangle smaller than the character size on the plate image. It can be found from Fig. 5b that the latter process removes clutter regions.

Fig. 5. (a) Closed edge image shown in Fig.4 (c). (b) Result of opening on Fig. 5(a).

Fig. 6. (a) Final defined region and (b) its corresponding g\ray scale image

Some features such as shape, aspect ratio and size of constructed region(s) are used to detect car plate(s) in the image. The result is given in Fig. 6. It is worth noting that our experiments shows using these three criteria are enough to find the car plate region from the candidates.

3 Experimental Results

The experiments have been conducted using more than 500 images of cars with Iranian license plates (Fig. 7). The images have been taken from natural scenes mainly with complex background and under different illumination and weather conditions. The license plates in the images have been viewed from different angles, and distances. The images size is 480*640. The proposed algorithm has been applied on intensity images. Some input images and the result of plate detection has been shown in Fig. 8.

In this experiment, 452 license plates out of 500 have been correctly extracted. The promising detection rate 90.4 % has been achieved with low processing complexity so the algorithm is performed fast.

We can make a tradeoff between accuracy and speed by altering the introduced threshold at filtering stage. The experimental results with different thresholds are given in Table 1. A comparison between the proposed method and some well-reported methods in the literature is given in Table 2.

The result of experiment confirms the ability of the proposed algorithm for detecting the location of plate in spite of significant variation of size and viewing angle. Meanwhile the performance of the algorithm is promising when it is applied on low contrast images. It is wroth noting that even in complex scenes only a few region(s) (less than three regions) are detected as plate candidates.

Fig. 7. Samples of license plates

From Table 2 it can be seen that the proposed method outperforms the methods reported in [5, 8, 14] from the detection rate points of view. Although the Cano et al [9] method yields better detection rate, it is considerably slower than the proposed method.

The experimental results also shows the Cano et al [9] method fails to detect license plate when its image is taken from extreme viewing angles.

Table 1. Different threshold results

Normalized threshold level (out of 1)	Number of images with correct detected plates (out of 500)	Speed (sec)
0.9	442 (88.4%)	~1
0.8	463 (92.6%)	~2
0.7	487 (97.4%)	~3
0.6	498 (99.6%)	~4

Table 2. A comparison of some methods

Ref.	Processor	Detection Rate (%)	Time (sec)	License plate Detection Method
The Proposed	MATLAB 6.0, P IV 3.0GHz	90.4	~1.4	Edge statistic and Morphology
[8]	P III, 1 GHz	89.2	~0.3	Combination of Color and Shape information of plate
[5]	MATLAB 6.0, P IV 3.0GHz	82.7	~3	Edge statistic and Morphology
[14]	AMD K6, 350 MHz	80.4	~4	Mathematical Morphology
[9]	AMD Athlon 1.2 GHz	98.5	~34	A supervised classifier trained on the texture features

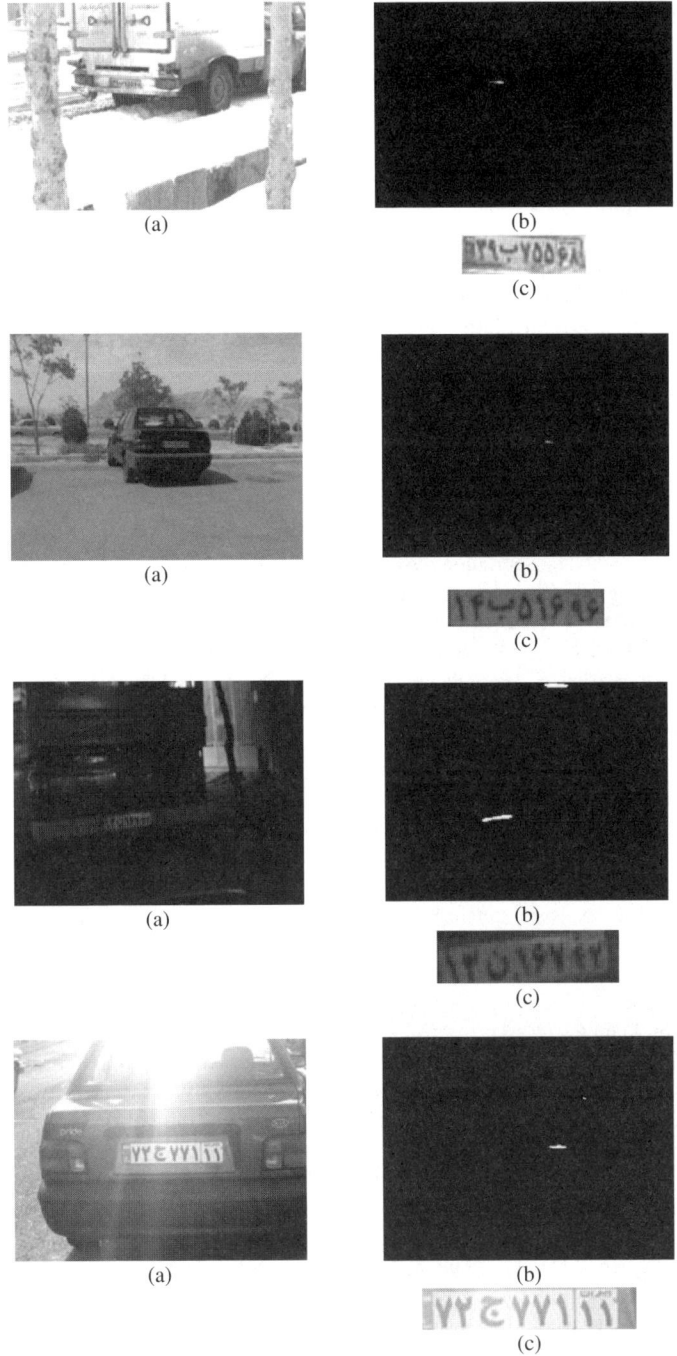

Fig. 8. Some test images and their results (a) Input images (b) Result after filtering (c) Final detected boundary

4 Conclusion

An algorithm for detecting car license plates was proposed. The method works based on presence of vertical edges in plate region. In order to lower the risk of missing plate region a relatively low threshold is set for edge detection. Then we designed a match filter to model the edge density in the plate zone. This filtering yields a few regions as candidates for license plate. The key role of using this filter is, reducing the size of search region for the late. Thus the computational complexity of the algorithm is considerably low.

Finally by applying some morphological operations, on the candidate regions and using simple geometrical plate features such the most probable region(s) is declared as license plate(s).

References

1. Anagnostopoulos, C.N.E., Anagnostopoulos, I.E., Loumos, V., Kayafas, E.: A License Plate-Recognition Algorithm for Intelligent Transportation System Applications. IEEE Trans. Intelligent Transportation Systems 7(3) (2006)
2. Sirithinaphong, T., Chamnongthai, K.: The recognition of car license plate for automatic parking system. In: Proc. of 5th Int. Symp. Signal Processing and its Applications, pp. 455–457 (1998)
3. Mahini, H., Kasaei, S., Faezeh Dorri, Fatemeh Dorri.: An Efficient Features–Based Licence Plate Localization Method. In: Proc. of IEEE Int. Conf on Pattern Recognition ICPR (2006)
4. Parisi, Claudio, E.D.D., Lucarelli, G., Orlandi, G.: Car plate recognition by neural networks and image processing. In: Proc. IEEE Int. Symp. Circuits and Systems, vol. 3, pp. 195–198. IEEE Computer Society Press, Los Alamitos (1998)
5. Zheng, D., Zhao, Y., Wang, J.: An efficient method of license plate location. Pattern Recognition Letters 26, 2431–2438 (2005)
6. Draghici, R.S.: A neural network based artificial vision system for license plate recognition. Int. J. Neural Systems 8, 113–126 (1997)
7. Chang, S.-L., Chen, L.-S., Chung, Y.-C., Chen, S.-W.: Automatic License Plate Recognition. In: Proc. IEEE Int. Conf Intelligent Transportation Systems, IEEE Computer Society Press, Los Alamitos (2004)
8. Shi, X., Zhao, W., Shen, Y.: Automatic license plate recognition system based on color image processing. In: Gervasi, O., Gavrilova, M., Kumar, V., Laganà, A., Lee, H.P., Mun, Y., Taniar, D., Tan, C.J.K. (eds.) ICCSA 2005. LNCS, vol. 3483, pp. 1159–1168. Springer, Heidelberg (2005)
9. Cano, J., Perez-Cortes, J.C.: Vehicle license plate segmentation in natural images. In: Perales, F.J., Campilho, A., Pérez, N., Sanfeliu, A. (eds.) IbPRIA 2003. LNCS, vol. 2652, pp. 142–149. Springer, Heidelberg (2003)
10. Duan, T.D., Hong Du, T.L., Phuoc, T.V., Hoang, N.V.: Building an automatic vehicle license plate recognition system. In: Proc. Int. Conf. Comput. Sci. RIVF, pp. 59–63 (2005)
11. Kahraman, F., Kurt, B., Gökmen, M.: License plate character segmentation based on the gabor transform and vector quantization. In: Yazıcı, A., Şener, C. (eds.) ISCIS 2003. LNCS, vol. 2869, pp. 381–388. Springer, Heidelberg (2003)

12. Matas, J., Zimmermann, K.: Unconstrained License Plate and Text Localization and Recognition. In: Proc. 8th Int. IEEE Conf. Intelligent Transportation Systems, IEEE Computer Society Press, Los Alamitos (2005)
13. González, R.C., Wood, R.E.: s: Digital Image Processing, Ch. 9. Prentice Hall, Englewood Cliffs (2002)
14. Martín, F., García, M., Alba, L.: New methods for automatic reading of VLP's (Vehicle License Plates). In: Proc. IASTED Int. Conf. SPPRA (2002)

Applying Local Cooccurring Patterns for Object Detection from Aerial Images

Wenjing Jia[1], David Tien[2], Xiangjian He[1], Brian A. Hope[3], and Qiang Wu[1]

[1] Faculty of Information Technology, University of Technology, Sydney
PO Box 123, Broadway, NSW 2007, Australia
{wejia,sean,wuq}@it.uts.edu.au
[2] School of Information Technology, Charles Sturt University,
Panorama Avenue Bathurst, NSW, 2795, Australia
dtien@csu.edu.au
[3] Department of Lands, NSW, Australia
Tony.Hope@lands.nsw.gov.au

Abstract. Developing a spatial searching tool to enhance the search capabilities of large spatial repositories for Geographical Information System (GIS) update has attracted more and more attention. Typically, objects to be detected are represented by many local features or local parts. Testing images are processed by extracting local features which are then matched with the object's model image. Most existing work that uses local features assumes that each of the local features is independent to each other. However, in many cases, this is not true. In this paper, a method of applying the local cooccurring patterns to disclose the cooccurring relationships between local features for object detection is presented. Features including colour features and edge-based shape features of the interested object are collected. To reveal the cooccurring patterns among multiple local features, a colour cooccurrence histogram is constructed and used to search objects of interest from target images. The method is demonstrated in detecting swimming pools from aerial images. Our experimental results show the feasibility of using this method for effectively reducing the labour work in finding man-made objects of interest from aerial images.

Keywords: Local cooccurring patterns, colour cooccurrence histogram, swimming pool detection.

1 Introduction

Developing a spatial searching engine to enhance the search capabilities of large spatial repositories for Geographical Information System (GIS) update, which is able to automate the process of identifying objects from high-resolution aerial images and in turn update the vector data, has now attracted more and more attention of researchers from both the area of data mining and the areas of image analysis and pattern recognition. One of the typical goals of such searching engine is to automatically search the required image for an object and return a list of possible matches for that item. Content-based image retrieval (CBIR) techniques which combine both spatial and textual features of images have been widely used. In a typical spatial searching engine system, users specify one of the objects they are looking for as a basis of the searching; the system searches the required image for the object and returns a list of possible matches for that item. The positions of the matched objects can then be used to update the vector data of GIS system [1,2]. The recent interest is mainly in identifying objects from high-resolution aerial images, thanks to the technical development on high-resolution remote sensor. The major interest for such purpose is in identifying small structures, such as individual road segments and buildings. Objects that need to be identified include various man-made buildings, housing, properties, swimming pools, lakes, dams, roads, and vegetation.

Recognition of visual objects is a fundamental task in vision. Typically, objects to be recognized are represented by many local features or local parts. Testing images are processed by extracting local features which are then matched with the object's model image. Most existing work that uses local features assumes that each of the local features is independent to each other. However, in many cases, this is not true. For instance, the idea presented in [3], [4], and [5], uses the colour-based local features only. Hence, it is found to be sensitive to the selection of the training set and may lead to detection failures when there are other areas which have similar colour distribution or when the objects have larger colour variance with the training set. Many existing methods for building extraction use the shape information only. Hence, it is hard to differentiate them from other similar objects which apparently have different colours and edge distributions from the target. For our system, we propose applying the cooccurring patterns of multiple local features, such as colour distribution and edge-based shape features, for object detection [6].

In this paper, a way of disclosing the cooccurring relationships between local features such as colour and edge is demonstrated in swimming pool detection. The local cooccurring pattern (LCP), which consists of characteristic local features and the statistical cooccurrence relationship between them, is then applied to recognise and locate swimming pools from aerial images. Note that the objects of interest in this research, i.e., the swimming pools, may have various shapes and large variance in their colours values. To utilise the local cooccurring patterns robustly, we introduce a histogram-based object detection method together with a robust histogram matching algorithm. A histogram representation, called colour cooccurrence histogram (CCH), is introduced and utilised as the key to search through the input image and retrieve the regions of interest (ROIs), i.e., regions that have similar CCHs with the model histogram. During detecting process, a scanning window moves through the input image. At each position, the CCH histogram of the region covered by the scanning window will be calculated and compared with the model CCH histogram. Regions that have a higher matching score with the model CCH will be kept as the regions of interest (ROIs). Regions with low matching scores will be marked as background. To further refine the detection result, regions segmented by an edge-preserving image smoothing technique that matches with the retrieved ROIs are returned as the final detection regions.

The remaining parts of this paper are organised as follows. Section 2 first introduces the edge-preserving image smoothing procedure that are used to segment the input image. Then, colour cooccurrence histogram (CCH), which reveals the coccurring patterns of colour and shape features in a simple way, are constructed and compared in Section 3 in order to find regions of interest. Experimental result of detecting swimming pool from aerial images is presented in Section 4. This paper is concluded in Section 5. Our future plan to continue this research is also mentioned in this section.

2 Image Segmentation Based on Edge-Preserving Smoothing

Before extracting colour-based features, it is common that images are smoothed to obtain a smaller colour map and finer edge map. This can be achieved via two major ways: reducing the number of colours and filtering the input images.

The operation for reducing colour number is also called *colour quantisation* in references. In [7], colour quantisation was simply performed in RGB colour space using a k-means nearest neighbour algorithm. However, the RGB colour space is known to be sensitive to colour variation. In [8], an optimal colour quantisation algorithm based on variance minimisation in CIE Luv space [9] is employed by introducing only slight colour distortion into the quantised images. However, unlike the case mentioned in [8] where the colours of interested objects, vehicle number plates, only take a small number of colours, the colours of objects in an aerial image can be very large. Hence, using fixed number of colours to quantise an aerial image will unavoidly introduce large perceptible difference in quantised images for those small objects, which usually is the interest for many GIS applications. In [10], a perceptual colour naming method was employed in order to handle colour variation. The standard ISCC-NBS Colour Names Dictionary [11] was employed to quantise and name the colour space. Each pixel is assigned to the standard name of the closest ISCC-NBS centroid colour, according to the Euclidean distance in the CIE Lab colour space [12]. This method uses a single colour set for all images. Since it is needed to find a best match in the standard colour dictionary for each colour represented in the aerial image, it usually takes extremely long time to complete the task.

On the other hand, filtering, or smoothing, image is perhaps the most fundamental operation of image processing and computer vision. In the broadest sense of the term "filtering", the value of the filtered image at a given location is a function of the values of the input image in a small neighborhood of the same location [13]. This idea indiscriminately blurs the image, removing not only noise but also salient information. As edge-based shape features are equally important as colour features in our project, we need to preserve edge information as much as possible. Discontinuity-preserving smoothing techniques adaptively reduce the amount of smoothing near abrupt changes in the local structure, i.e., edges. In order to smooth the input image while keeping important edge information as much as possible to enable the extraction of both the color and edge based local features, we study the two most popular edge preserving smoothing techniques: bilateral filtering and mean shift filtering.

The bilateral filtering was introduced by Tomasi and Manduchi in [13] in 1998. Recently, with the extensive usage of the bilateral filtering in other areas than image denoising, such as demosaicking, image abstraction, image retinex, optical flow estimation, etc, a fast algorithm which uses signal processing approach to approximate the standard bilateral filter has been proposed by Paris and Durand in [14].

The basic idea underlying bilateral filtering is to do in the *range* domain of an image what traditional filters do in its *spatial* domain. Two pixels can be *close* to each other, that is, occupy nearby spatial location, or they can be *similar* to one another, that is, have similar values, possibly in a perceptually meaningful fashion [13].

Bilateral filtering can be achieved via a convolution of the image brightness function f with a spatial filter k_s and a range filter k_r in a small area (called *convolution window*) surrounding each reference pixel of which the radius is determined by the spatial bandwidth σ_s. It will take a long time to carry on the convolution processing if the convolution window is large.

Another edge-preserving smoothing technique which is also based on the same principle, i.e., the simultaneous processing of both spatial and range domains, is mean shift filtering, proposed by Comaniciu and Meer in [15]. It has been noticed that the bilateral filtering uses a static window in both domains. The mean shift window is *dynamic*, moving in the direction of the maximum increase in the density gradient. Therefore, the mean shift filtering has a more powerful adaptation to the local structure of the data. In our algorithm, an image segmentation method based on mean shift filtering technique is used to obtain a finer detection result. The basic procedure of this method is illustrated as follows. For more details, please refer to [15].

Let $\{x_i\}_{i=1,2,...,n}$ be the original image points, $\{z_i\}_{i=1,2,...,n}$ be the points of convergence, and $\{L_i\}_{i=1,2,...}$ be a set of labels indicating different segmented regions.

1. For each image point $\{x_i\}_{i=1,2,...,n}$, run the mean shift filtering procedure until convergence and store the convergence point in $z_i = y_{i,c}$, as shown below:
   (a) For each image point $\{x_i\}_{i=1,2,...,n}$, initialise $j = 1$ and $y_{i,1} = x_i$. The first subscript i of $y_{i,j}$ denotes the ith image point, and the second subscript j denotes the jth iteration.
   (b) Compute $y_{i,j+1}$ according to Equation 1 until convergence of $y_{i,c}$.

$$y_{i,j+1} = \frac{\sum_{i=1}^{n} x_i g(\|\frac{x-x_i}{h}\|^2)}{\sum_{i=1}^{n} g(\|\frac{x-x_i}{h}\|^2)} \quad (1)$$

(c) Assign $z_i = (x_i^s, y_{i,c}^r)$, which specifies the filtered data z_i at the spatial location of x_i^s to have the range components of the point of convergence $y_{i,c}^r$.

2. Delineate the clusters (suppose there are m clusters), denoted by $\{C_p\}_{p=1,2,\ldots,m}$, in the joint domain by grouping together all z_i which are closer than h_s in the spatial domain and h_r in the range domain under a Euclidean metric, i.e., concatenate the basins of attraction of the corresponding convergence points.

3. For each $i = 1, 2, \ldots, n$, assign $L_i = \{p | z_i \in C_p\}$.

Thus, using the mean shift procedure in the joint spatial–range domain, the input colour aerial images are segmented into many regions, each of them are represented with different colours. Parameters h_r and h_s are set experimentally.

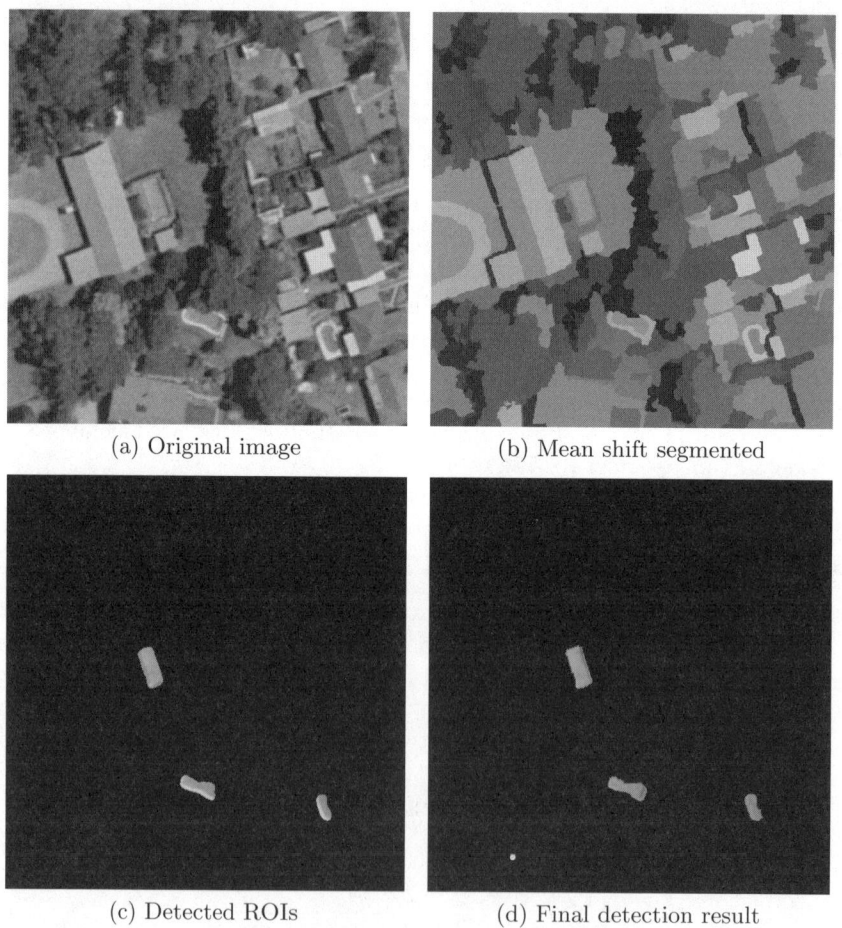

(a) Original image (b) Mean shift segmented

(c) Detected ROIs (d) Final detection result

Fig. 1. The procedure of the algorithm

Figure 1(a) shows an example of aerial image. Figures 1(b) shows the segmented image using a mean shift filter ($(\sigma_s, \sigma_r) = (7, 5)$). As it can be seen from Figure 1(b), applying the edge-preserving smoothing techniques, image noise appearing in the original image has been smoothed out, and the colours of objects in the image becomes much more uniform. This may lead to a better edge detection result, which in turn leads to a better shape representation, and simpler colour-based feature representation. On the other hand, if we look at the edges, such as the road boundary or the buildings' boundaries, the edges are very well kept and become sharper.

3 Detecting ROIs Using Colour Cooccurrence Histogram

3.1 Constructing Colour Cooccurrence Histogram (CCH)

Each element of the CCH histogram of a 2-D colour image $f(x, y)$ records the frequencies of occurrence of pairs of pixels that are at a distance away and take specified colour values.

Let us denote the colour value at point p_1 as c_1, and the colour value at point p_2 as c_2. The corresponding bin of the CCH histogram, denoted by $h(c_1, c_2, l)$, is counted as:

$$h(c_1, c_2, l) = size\left(\left\{(p_1, p_2) \middle| \begin{array}{l} p_1, p_2 \in f, \text{ and} \\ \|p_1 - p_2\| = l \end{array}\right\}\right), c_1, c_2 \in \mathbf{C}, \quad (2)$$

where $\mathbf{C}$ is the colour set of the image, and the function $size$ counts the number of elements in a set. In our experiments, the l is set as 4 and $\|p_1 - p_2\|$ uses Euclidean distance.

All image points in the whole image plane are scanned. The frequency of occurrence of the same colour pairs in the image at a distance away is recorded and summed in the corresponding bin of the CCH histogram.

Finally, the CCH histogram is normalised using the total number of points in the image.

3.2 Comparing CCH Histograms

After the CCH histogram is constructed, the next problem is how to compare two CCH histograms.

Normalised L_1-norm distance and the histogram intersection have been proposed in the past for comparing histograms. However, both metrics demand an identical colour matching in an exact bin-to-bin way between two histograms and thus are very sensitive to colour variance. In practice, the colours of real world images of same kind of objects can be varied both in the scene itself and in the image capturing process. This can be seen from the swimming pool images used in this research. Some examples are shown in Figure 2. Hence, images with same visual information but with different colour intensities may degrade the similarity level significantly when the conventional HI is used.

Fig. 2. Some examples of swimming pool images in our experiments

In order to overcome this problem, Wong et al. [16] proposed a *merged-palette histogram matching* (MPHM) method. The essence of the method is to extend the intersection from bins of *identical* colours to bins of *similar* colours. In their algorithm, as long as the distance between two colours is less than a fixed threshold, say Th, the intersection between the bins of these two colours will be calculated. This algorithm has produced more robust image retrieval results for images captured under various illumination conditions. However, it assumes an *identical weight* of the contribution between colours which have different similarities with the given colour. When applied to swimming pool images, the matching scores exhibited to be very sensitive to colour variations.

In [17], a Gaussian weighted histogram intersection (GWHI) method has been proposed which overcomes the above problems and has demonstrated to be much less sensitive to colour variations compared with the conventional HI method and Wong et al.'s MPHM method [17]. In this method, instead of applying identical weights to different colours, a weight function of colour distance has been employed to differentiate the matching between colours which have different distances to the given colour.

The GWHI method has shown to be less sensitive to colour variations. However, this method only addresses the intersecting of histogram bins between two colours. For the purpose of comparing two CCH histograms, where there are two pairs of colours involved in the intersection, the GWHI method is further extended to facilitate intersection of two pairs of colours.

In the following two subsections, the principles of the GWHI method is first introduced. Then, how to use the extended GWHI method to compare CCH histograms is described.

3.3 Gaussian Weighted Histogram Intersection (GWHI)

Assume all colours of pixels in the model image are taken from a colour set, denoted by $\mathbf{C}_M$, and all colours of pixels in the target image are taken from another colour set, denoted by $\mathbf{C}_T$ (the subscripts "M" and "T" denote for "model" and "target" respectively). Note that, the colour sets of two images can be either identical or different.

In [17], Jia et al. proposed applying a Gaussian weight function of colour distance to the conventional HI method in order to describe the relationship between the colour distance and the weight as:

$$\eta = \sum_{c_i \in C_M} \sum_{c_j \in C_T} \min\left(h_M(c_i), h_T(c_j)\right) \cdot \exp\left(-\frac{\parallel c_i - c_j \parallel^2}{2\sigma^2}\right), \quad \parallel c_i - c_j \parallel \leq BW \tag{3}$$

where $h_M(\cdot)$ and $h_T(\cdot)$ are each element in the model histogram H_M and the target histogram H_T. The parameter σ and the bandwidth BW in this equation can be selected empirically. In order to be consistent with Wong et al.'s MPHM algorithm, the contributions of identical colours in two methods are set as the same, and the overall weights in two methods are approximately the same. 99.9% energy of Gaussian function is used to approximate its total energy in infinite space. It has been derived that $\sigma \doteq 0.8Th$ and $BW \doteq 3.3\sigma \doteq 2.6Th$ [17] [18], where Th is the threshold used in [16], as shown in Figure 3. Note that, from the below Equation 4, it can be seen that $\parallel c_1 - c_2 \parallel$ never takes negative values. However, for an intuitionistic understanding, in this figure, weight functions when $\parallel c_1 - c_2 \parallel < 0$ are also plotted.

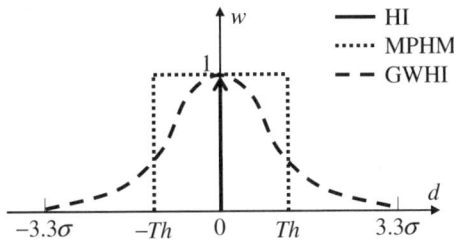

Fig. 3. Weight functions in three histogram intersection methods, where $d = \parallel c_1 - c_2 \parallel$

The colour distance between two colours is defined as the perceptual difference in CIE Luv colour space as:

$$\parallel c_1 - c_2 \parallel = \sqrt{(l_1 - l_2)^2 + (u_1 - u_2)^2 + (v_1 - v_2)^2} \tag{4}$$

where $c_1 = (l_1, u_1, v_1)$ and $c_2 = (l_2, u_2, v_2)$ are two colours represented in CIE Luv colour space. l takes a value in the range of $[0, 100]$, u takes a value in the range of $[-83, 175]$, and v takes a value from $[-134, 107]$. The bandwidth in our experiments is set as $Th = 15$, which is same as in [16]. Thus, we have $\sigma \doteq 16$ and $BW \doteq 52.8$.

3.4 Comparing CCH Histograms Using Extended GWHI

In order to compare two CCH histograms in perceptually similar colour bins, the following operations are performed.

Initialise the matching ratio η as 0. For each element $h_M(c_i, c_j, l)_{|c_i, c_j \in C_M}$ in the CCH histogram of model image, scan each entry $h_T(c_m, c_n, l)_{|c_m, c_n \in C_T}$ in the CCH histogram of target image and perform the following operations (1) through (3) until all entries in the model CCH histogram has been scanned:

1. Compute the colour distances $d_1 = \|c_i - c_m\|$ between the first pair of colours c_i and c_m, and $d_2 = \|c_j - c_n\|$ between the second pair of colours c_j and c_n, as defined in Equation 4. Take the larger distance as the distance between two pairs of colours, denoted by d, i.e., $d = \max(d_1, d_2)$.
2. If $d \leq BW$, where BW is the bandwidth of Gaussian weight function (see Equation 3), decide the intersected section between two CCH histograms and apply the Gaussian weight function of distance as:

$$\gamma = \min(h_M(c_i, c_j, l), h_T(c_m, c_n, l)) \times \exp(-\frac{d^2}{2\sigma^2}), \qquad d \leq BW,$$

where $c_i, c_j \in C_M$, and $c_m, c_n \in C_T$. The selection of σ and the bandwidth BW in the Gaussian weight function are discussed in previous subsection.
3. Find the sum of all the weighted intersections as:

$$\eta \leftarrow \eta + \gamma$$

The final η is exported as the matching rate between two CCH histograms. Since both histograms have been normalised to $[0, 1]$, the resultant η takes value in the range of $[0, 1]$. The larger the matching rate is, the higher similarity the two images have, when the CCH histograms are concerned.

Note that in above procedure, the larger distance of d_1 and d_2 is taken to decide the weight. The reason for this is that for any two pairs of colours in two images respectively, as long as there is one pair of colours being viewed as "dissimilar", the CCH between these two pairs of colours should not be matched. This design further guarantees that the intersection between CCH histogram bins only happens on "similar" colour pairs with the condition that both colour pairs must be viewed as "similar" before an intersection is computed.

4 Experiments and Discussion

4.1 Selection of Model and Parameters

As mentioned earlier in this paper, a model CCH histogram is needed to detect regions of interest. The selection of the model CCH can take either of the following two ways. According to our experimental results, there is not significant difference in the detection results.

The first way is a straightforward method. From the training data, i.e., the images of sample swimming pools, manually pick up one image which colour intensities are perceptually not too bright nor too dark compared with other training sample images. For instance, we have tried using the sixth sample in the first row of Figure 2 as the model image. The CCH histogram of the model image

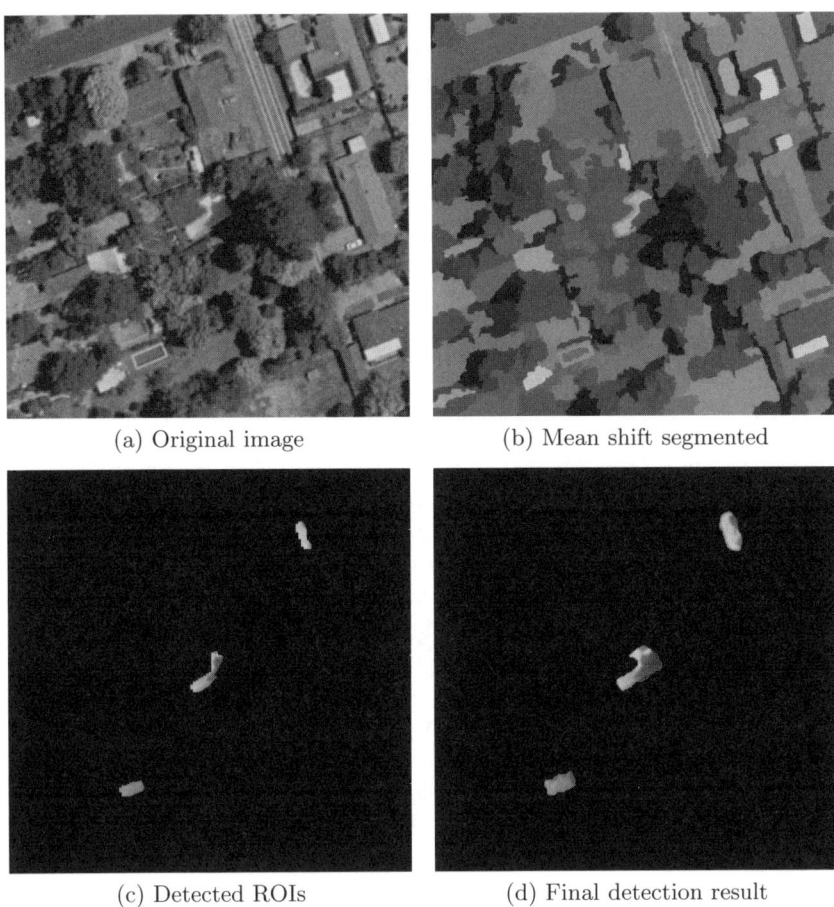

Fig. 4. Applying local cooccurring patterns for swimming pool detection

is then computed as the model CCH, and this histogram is used to compare with CCH histograms computed at each position while the scanning window is moving around the image. Regions that give a higher matching score are kept as ROIs.

The second method is based on statistical analysis on the training data. For each training sample, the CCH histogram is firstly computed. Then, a single statistical average CCH histogram is obtained based on all of the CCH histograms of training samples. This average histogram is used as the model CCH histogram. Same searching procedure as in the first method is performed to find the ROIs.

The size of the scanning window in this algorithm is selected considering the minimum size of possible swimming pools in images that the algorithm is able to detect. In our experiments, the scanning window is a 8×8-pixel square window. The moving step is normally set as half of the width and half of the height of the scanning window, i.e., 4-pixel. However, a finer ROI result can be obtained with smaller step value at the cost of more iterations and hence longer searching time.

Another very important parameter is used to decide how similar a region's CCH is to the model CCH so that the region can be taken as a region of candidate. Again, this can be decided via statistical analysis on the training samples. First, the matching score of each training sample to the model in terms of CCH histogram is computed. Then, the minimum value of the matching scores can be set as the threshold of deciding whether a region can be taken as a candidate. In our experiments, regions that have a matching score with the model CCH histogram larger than 0.5 are reserved as ROIs.

4.2 Experimental Results

In this algorithm, we first use the CCH histogram to search ROIs. Then, the segmented regions based on the colour and edge information which are covered by ROIs are returned as the detection result. One of our experimental results is shown in Figure 1.

Figure 4(a) shows another example of a partial aerial image. Figure 4(b) shows the image segmentation result using a mean shift filter, where different regions are painted in different colours. Figure 4(c) is the detected ROIs via comparing the CCH histograms of regions covered by scanning window and the model CCH, where the background are marked as black points. The final detected swimming pools are shown in Figure 4(d), where finer detection results are retrieved from Figure 4(b) which are covered by ROIs in Figure 4(c).

5 Conclusions and Future Work

In this paper, we present the idea of using local cooccurring patterns for object detection from aerial images. For this application, we propose to use the cooccurrence patterns of local features, including colour features and edge-based shape features, for object detection purpose. A semi-automatic method to detect any object from aerial images is presented. In this method, the users can use the mouse to click any objects of interest, the systems then automatically computes and picks up an area and uses this as a training set.

Our future work in the next step is to study more complex shape features and combine the cooccurring patterns associated with the colour-based features to train a classifier for detecting randomly selected objects.

Acknowledgments. The work has been supported by the Cooperative Research Centre for Spatial Information, whose activities are funded by the Australian Commonwealth's Cooperative Research Centres Programme.

References

1. Mena, J.B.: State of the art on automatic road extraction for gis update: a novel classification. Pattern Recognition Letters 24(16), 3037–3058 (2003)
2. Wang, H., Miller, P.: Disvoering the local co-occurring pattern in visual categorization. In: Proceedings of the IEEE International Converence on Advanced Video and Signal based Surveillance Ssystem, IEEE Computer Society Press, Los Alamitos (2006)

3. Mena, J.B., Malpica, J.A.: An automatic method for road extraction in rural and semi-urban areas starting from high resolution satellite imagery. Pattern Recognition Letters 26(9), 1201–1220 (2005)
4. Mena, J.B., Malpica, J.A.: Color image segmentation using the dempster-shafter theory of evidence for the fusion of texture. In: Proceedings of the ISPRS Workshop XXXIV-3/W8, pp. 139–144 (2003)
5. Mena, J.B., Malpica, J.A.: Color image segmentation based on three levels of texture statistical evaluation. Applied Mathematics and Computation 161(1), 1–17 (2005)
6. Jia, W., Tien, D.: Discovering local cooccurring patterns from aerial images. In: Proceedings of the International Conference on Information Technology and Applications, pp. 300–305 (2007)
7. Chang, P., Krumm, J.: Object recognition with color cooccurrence histograms. In: Proceedings of the 1999 IEEE Computer Society Conference on Computer Vision and Pattern Recognition, vol. 2, pp. 498–504. IEEE Computer Society Press, Los Alamitos (1999)
8. Jia, W., Zhang, H., He, X., Wu, Q.: Image matching using colour edge cooccurrence histograms. In: SMC 2006. Proceedings of the 2006 IEEE International Conference on Systems, Man, and Cybernetics, pp. 2413–249 (2006)
9. Wu, X.: Efficient statistical computations for optimal color quantization. Graphics Gems 2, 126–133 (1991)
10. Crandall, D., Luo, J.: Robust color object detection using spatial-color joint probability functions. In: Proceedings of the 2004 IEEE Computer Society Conference on Computer Vision and Pattern Recognition, vol. 1, pp. 379–385. IEEE Computer Society Press, Los Alamitos (2004)
11. Kelly, K.L., Judd, D.B.: Color universal language and dictionary of names. National Bureau of Standards special publication 440. Washington, DC: U.S. Department of Commerce, National Bureau of Standards (1976)
12. Giorgianni, E.J., Madden, T.E.: Digital Color Management: Encoding Solutions. Addison-Wesley, Reading, MA (1997)
13. Tomasi, C., Manduchi, R.: Bilateral filtering for gray and color images. In: Proceedings of the Sixth International Conference on Computer Vision, pp. 839–846 (1998)
14. Paris, S., Durand, F.: A fast approximation of the bilateral filter using a signal processing approach. In: Proceedings of the European Conference on Computer Vision (2006)
15. Comaniciu, D., Meer, P.: Mean shift analysis and applications. In: Proceedings of the Seventh IEEE International Conference on Computer Vision, vol. 2, pp. 1197–1203. IEEE Computer Society Press, Los Alamitos (1999)
16. Wong, K., Cheung, C., Po, L.: Merged-color histogram for color image retrieval. In: Proceedings of the International Conference on Image Processing, vol. 3, pp. 949–952 (2002)
17. Jia, W., Zhang, H., He, X., Wu, Q.: Gaussian weighted histogram intersection for license plates classification. In: Proceedings of the 18th International Conference on Pattern Recognition, pp. 574–577 (2006)
18. Jia, W., Zhang, H., He, X., Wu, Q.: Refined gaussian weighted histogram intersection and its application in number plate categorization. In: Proceedings of the 3rd International Conference on Computer Graphics, Imaging and Visualization, pp. 249–254 (2006)

Enticing Sociability in an Intelligent Coffee Corner

Khairun Fachry[1,2], Ingrid Mulder[3], Henk Eertink[3], and Maddy Janse[4]

[1] User System Interaction, Eindhoven University of Technology, The Netherlands
[2] Archives and Information Studies, University of Amsterdam, The Netherlands
k.n.fachry@uva.nl
[3] Telematica Instituut, Brouwerijstraat 1, Enschede, The Netherlands
{ingrid.mulder, henk.eertink}@telin.nl
[4] Philips Research, High Tech Campus 34, Eindhoven, The Netherlands
maddy.janse@philips.com

Abstract. In our project on the design of intelligent applications for an office environment we focus on the coffee corner as it is the place in an office where professional and social interactions intermingle. In this article the user-centered design of tangible interfaces, i.e. i-Candies and i-Bowl, which have been used as input devices to access company information in a coffee corner, is described. The i-Candies and the i-Bowl illustrate how everyday objects can be part of a context-aware system. By picking up one of the i-Candies and placing it on the i-Bowl, the i-Candy controls which information is presented on the wall display in the coffee corner. In our evaluation, we found that the office workers not only were enticed to interact with the i-Candies, but also that the i-Candies provided mediums that stimulate sociability among office workers.

Keywords: User-centered Design, Evaluation, Context-aware System, Tangible User Interface.

1 Introduction: Defining for Sociability

Context-aware systems and intelligent devices such as wireless LAN, (indoor and outdoor) location sensing technology, and short range network functions are increasingly being used to support people's social life. A system is context-aware if it uses the context to provide relevant information and/or services to the users, where relevancy depends on the user's task [6]. Here, we search for a system that supports and elicits social interaction in a shared office area. When exploiting technologies for context awareness in an office environment, we are challenged with the question of how to combine professional and social interactions in one system. On the one hand, the system should maintain the professional atmosphere in the office by taking into account situated interaction and privacy. On the other hand, the system should support sociability among colleagues. The company in focus is a research company that employs 110 employees situated in two connected buildings, which have 4 floors each. The employees who work in different projects are spread throughout the office environment where on every floor there is a coffee corner. The coffee corner in focus is equipped with intelligent devices such as device-discovery, sensors, Bluetooth, RFID reader, camera, speaker and advanced displays (Figure 1). Using these devices, the current coffee corner system allows the office workers to locate their colleagues within the office. The detection of location is implemented using a Context Management Framework (CMF) [16]. The CMF is a distributed system that aggregates context-information about identities (users, devices and places).

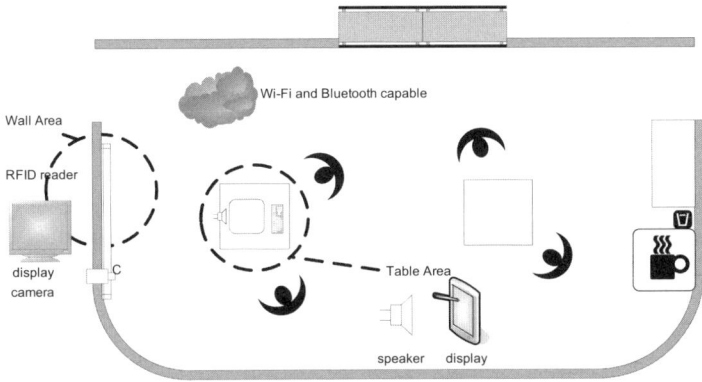

Fig. 1. The Observed Coffee Corner

Elaborating upon current research on context-aware systems, and inspired by the coffee corner available "in the house", we are particularly interested in the following research questions. How is the coffee corner being used by office workers? And how can the sociability among office workers be enhanced? To investigate these questions, we applied the usability engineering design lifecycle [13].

Sociability is defined as the design of sociable places through physical conditions that enable them to bring people together and permit them to socially interact with each other [19]. Furthermore, the concept of social affordance has been applied in this work. Social affordance is the relationship between the properties of an object and the social characteristics of a group that enable particular kinds of interaction among members of the group [2]. Although many researches have different perspectives on the objectives of social affordance, the social affordance objectives defined by [11] is considered the most relevant for this work.

Table 1. Objective of Social Affordances [11]

Social affordances	Awareness of others in their activities
Action afforded	Stimulate informal communication
Interaction aimed at	Communication

2 User Study

User requirements for a system that supports sociability were collected by means of among others direct observation, collecting ideas from users, coffee break brainstorming and an expert design session.

2.1 Direct Observation

As an exploratory study, we investigated how the office workers behave in the intelligent coffee corner with open-ended and unstructured direct observations [15].

Results

Rich group dynamics were observed in the coffee corner. During two weeks of observation, every day a group of 6-10 people gather for a coffee break twice a day. They usually synchronize their breaks by having one person approached the others in their offices or via MSN. In informal meetings like coffee corner breaks, we found out that in most cases people who are having breaks together are because they are on the same floor, they work together or they are doing work closely related.

It was observed that a shared physical environment promotes informal social communication [18]. Although the shared area is located in a professional environment, the office workers tended to perform informal and semi-formal communication while they were in the coffee corner. The conversation went around project topics for example identifying new projects or opportunities for collaboration, or knowledge and technology sharing. Aside of the projects topics, topics such as upcoming events, weekends, sports, and weather were several out of many topics that the office workers discussed about during their breaks.

The usage of technology in the coffee corner was centered on looking for their colleagues. As many as 22 people opted in to the system so that they can be located while they are in the office area. The other people can see where their colleagues are and their current activities. Another use of technology was also for internet browsing.

Furthermore, we noticed that the office workers' behaviors in the coffee corner vary from person to person. Therefore, it is important to know the type and characteristics of the office workers. We could categorize three groups of users[1]:

Frequent users: They have coffee regularly in the observed coffee corner and perform socially with each other.

"Looking familiar" users: They have coffee occasionally in the observed coffee corner. Their visits are incidental. They could be project partners from other companies or universities.

Guest users: They are visitors of the employees. In most cases they are always accompanied by the visited employee due to their unfamiliarity with the area.

It was observed that in order to share information, the office workers brought along printed documents, pictures or parts of newspaper to the coffee corner. Office workers brought along their personal electronic devices such as mobile phones or PDA's to show their information, for example pictures.

---

[1] Both office workers and users are used interchangeably.

As for the interaction with the system, a keyboard and a mouse were used as input devices. This way, the office workers were forced to take turn in interacting with the system by passing around the input devices. For the most part, one frequent user was in charge of the input access and the other office workers watched or gave instructions.

2.2 Collecting Ideas from Users

An idea 'submitter' was created and displayed on the coffee corner display to capture inspiring ideas for an application that can enhance sociability born at the coffee table. The idea submitter is an electronic version of the classic idea box.

Results
Most of the received ideas are examples of applications that the office workers would like to have and technologies that could be used in developing those applications. In total 23 different ideas of applications were received. All ideas were analyzed and clustered based on similarities that resulted in the following potential applications:

Company Awareness. The office workers would like to have an application that makes them aware of the events that are happening within the company.
Personalized Information. The office workers would like to share their selected personal information with their colleagues.
Games. The office workers would like to have short and interactive games during breaks to play with their colleagues.
Idea Publisher. The office workers would like to have a system that supports exchanging ideas among the office workers.

2.3 Coffee Break Brainstorming

Since the office workers are in the coffee corner during their breaks and the intelligent devices are installed in the coffee corner, we chose the coffee corner as the place to brainstorm with the office workers. An invitation email was sent to all office workers to join brainstorming sessions. A preset script of questions was used as a framework to keep the discussion running. This gave us the opportunity to get a deeper insight in potential applications, to unravel concerns and to discuss possible solutions to overcome those concerns.

All four possible applications were visualized in short scenarios. We chose to visualize the scenario with animation because animations can capture the office workers attention, and therefore, trigger discussions among the office workers. The visualization was uploaded on the coffee corner display. A moderator was present in the coffee corner and structured the discussions. The comments were noted on post-it notes by two note keepers. Since we had four ideas for applications, we presented one idea each morning. Furthermore, when there happened to be a crowd of office workers in the coffee corner at other times, the moderator used this opportunity to acquire more input. Post-it notes, pens and a flip board were always placed next to the display so the office workers could see others comments.

Results

Every brainstorming session took 30-45 minutes and was attended by 10-14 office workers resulting in a considerable amount of qualitative feedback. While the idea submitter merely answered the kind of interactions that might happen in a coffee corner, the brainstorming sessions gave more insight in the submitted ideas.

To illustrate our approach, an example of the brainstorming quotes is presented:

"No typing for keyboard, but for example scanning your badge for information"

The office workers were not in favor of having keyboard or mouse as their interaction input devices. They claimed that keyboard and mouse only allow one person access. Moreover, keyboard and mouse give them the feeling of being behind their computer again during their break time (requirement 4 and 5).

We adopted affinity diagrams [8] to organize and analyze the data from the brainstorming sessions resulting in a list of user requirements for the design of applications for the intelligent office coffee corner. The requirements that are generic for all scenarios are listed below:

1. The content should be semi-formal information e.g. upcoming presentation, new employees, etc.
2. The system should support personalization by providing a possibility to present the office workers' selected personal information.
3. The system should entice curiosity of the user by providing them daily updated information and interesting interaction input devices.
4. The system should support access from multiple users so that the user does not need to take turn in interacting with the system.
5. The interaction should be easy and fun to use and should not remind the office workers of their work place.

2.4 Experts Design Session

An experts design session was organized to collaboratively select and design scenarios that will be implemented for the intelligent coffee corner. Six experts with various expertise backgrounds were invited. These experts work in the office in focus and are therefore also potential users of the system. The session started by reflecting the results from the brainstorming sessions. The experts were asked to discuss the best scenarios for implementation based on the result of the brainstorming sessions.

Results

The session resulted in a decision to implement the company awareness scenario for several reasons. First of all, this scenario appeared to be the most favorite by the participants of the brainstorm sessions; indicated by the number of positive feedbacks (written on the post-its notes). Secondly, it supports social interaction among the office workers. It is a 'giving and receiving' system, which enables exchange of semi-formal information among the office workers. Finally, it gives the possibility to entice the office workers to interact with the system by giving them 'unexpected' feedback. This is an important aspect as Agamanolis [1] found that half the battle in designing an interactive situated or public display is designing how the display will invite that interaction. In keeping with Churchill et al. [4] users need constant encouragement and demonstration to interact with the interactive public display. An ideal company awareness system should show daily information based on the received information from the office workers. Thus, by showing updated information, we aim to entice curiosity of the office workers so that they will interact with the system.

3 Designing for Sociability

The basic concept is that the system is used to broadcast general information (requirement 1). General information supporting company awareness can be published and accessed by every office worker. In this way, the system acts as an interactive broadcasting system.

3.1 Technology Alternatives

As mentioned in the observation result, the current input devices forced users to take turns for interaction with the system. To make the interaction access in the coffee corner more "social" by allowing multiple people to access the system (requirement 4), several technologies were considered, i.e. speech, RFID and Bluetooth technologies. These technology alternatives were sketched on paper and their benefits and limitations were being discussed with two experts who work in the office in focus and know the constraints of the coffee corner area. RFID tags which are covered in interesting forms are chosen (named as interaction tags for the rest of the paper). Such interaction tags have several advantages. They can be spread on the table so that users can easily choose and pick them. This enables multiple users to decide what they want to display and it gives them the same access rights to the system. Users can choose the interaction and they can interrupt each other. This process mimics the conversation process that occurs in social interaction.

3.2 Interaction Tag Design

My momma always said: "Life is like a box of chocolates. You never know what you're gonna get." (Forrest Gump, 1994).

By introducing a new way of interaction, we attempt to entice curiosity of the users to interact with the system (requirement 3). For that purpose the interaction tags should be designed in such a way that they are easy and fun to use and do not remind the office workers of their personal PCs (requirement 5). Taken all these requirements into account candies on a sweet dish inspired us for the design of the input device. The input device, called i-Candies, has been designed in the form of candies to make the interaction playful and interesting. The metaphors of candies and bowl match the environment of the coffee corner, as there are cups and tea bags on the table. A collection of available candies is placed on the table and the reader is embedded on the bowl. This form factor requires the user to pick up a candy out of the candy box and place it into the bowl.

3.3 Information Design

Pictures of the i-Candies are presented on the display as navigational cues. The user chooses the category that they want to present by placing the candy on the plate. The display then presents the requested information, updates its screen and presents the next possible presentation (Figure 2).

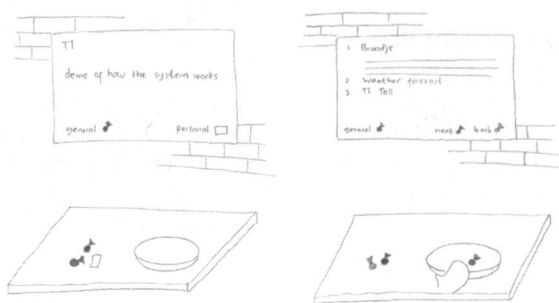

Fig. 2. Form Factor Candies and Navigational Cues on the Display

4 The prototype: A Sweet Dish for the Coffee Corner

The prototype has been implemented as a web-based application running on the coffee corner display. The prototype consists of three main parts: 1) get context, 2) retrieve the latest tag and 3) retrieve content (Figure 3(a)). The context information needed by the coffee corner system is provided by the input device identification i.e. the RFID identification. The RFID identification is facilitated by the CMF. In the coffee corner system, the CMF provides data regarding the interaction between the user, the RFID identification numbers and the RFID reader on the coffee table. It provides the attributes location, time, and RFID tag number and stores them in a database of the application server.

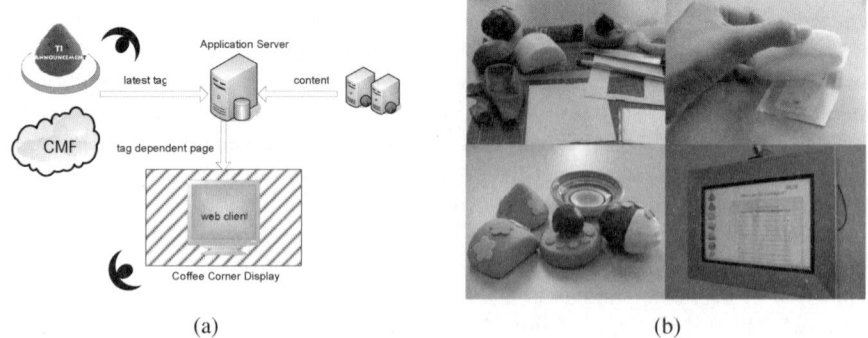

Fig. 3. (a) System Architecture and (b) the Making of the I-Candies

New content is provided if a new RFID tag is detected, using a polling mechanism. It checks every second if there is a new RFID tag. It detects the tag number and retrieves the page associated with the tag number. The candies are designed in different shapes to allow differentiation of functions and to attract the users. For navigation purposes icons illustrating the i-Candy are placed on the sidebar (Figure 3(b)). The i-Candies icons are position at the same column provide the same access level. A selected i-Candy is shown by transparently overlapping a picture of the i-Bowl on the selected i-Candy icon.

All content is gathered from existing information uploaded by the office workers in current offices services and external information that could be interesting for the office workers. Because the content is updated automatically by the system, the system does not need any maintenance. Following information is presented:

6. *Company Announcements* are taken from the internal knowledge management.
7. *Latest Publications* presents the most recent publication that the office workers have published.
8. *Where are Our Colleagues* presents the location of their colleagues.
9. *What is Happening in the World* presents the international news.
10. *How is the weather?* presents the weather forecast.
11. *Anything Funny* presents the comic of the day.

In order to entice the users to interact with the system, a distraction mode has been implemented to trigger attention. The idea is to have a screen saver-like mechanism that presents the content of the system. The distraction mode is designed in such way that when nobody is interacting with the system, the system shows its content page by page every 5 seconds.

5 Evaluation Study

The goal of the evaluation study was to investigate whether the i-Candies system enhanced the sociability among the users. Other than sociability aspects of the system, we also evaluate whether the system fulfills the user requirements. Data have been collected by means of questionnaires, observation and contextual interviews.

Questionnaires. A questionnaire has been designed to evaluate whether the system fulfills the research objectives and the user requirement. The questionnaire consists of 9 items (see Table 2). The first part (Q1-Q4) assesses the user requirements. The second part (Q5-Q9) assesses the sociability of the system, adopted from the sociability scale [12]. The sociability scale is a self-reporting questionnaire for measuring the perceived sociability [12]. For all questions, the judgments were made on 5–point Likert scales (1 = strongly disagree; 5 = strongly agree). The questionnaire, was administrated electronically (using SF Survey World). It consists of a pre i-Candy evaluation and a post i-Candy evaluation questionnaires. In this way, a comparison before and after the i-Candies system deployed can be measured, thus can lead us to analyze if the system enhances the sociability.

Table 2. The Evaluation Questionnaire

No	Item
1	I think the system was easy to use
2	I think the interaction with the system is fun
3	I use the system frequently
4	The system evokes my curiosity
5	The system allows spontaneous informal communication
6	This environment allows non task-related conversation
7	This environment enables me to take a good work relationships with my colleagues
8	I feel comfortable in this environment
9	This environment enables me to make close relationship with my colleagues

The pre evaluation questionnaire has been used to define the baseline of the current coffee corner. A baseline survey was sent by e-mail to the office workers before the system was deployed, thus before the i-Candies prototype was introduced into the organization. Thus, the questionnaire evaluates the system that was already available in the coffee corner, i.e. the location detection system for colleagues. A post i-Candy evaluation questionnaire was sent to the office workers with the intention to know the effect after the i-Candy system is deployed. The same set of question as in the pre-evaluation questionnaire has been used.

Observation and Interview. The purpose of the observation was to capture important remarks regarding the system that will be used for the contextual interview. Other than that, the observation approach was used to know if the content of the conversation has been influenced by the new application. In-depth interviews are used to solicit their experiences and remarks.

Results

From the completed questionnaires of the pre evaluation (N=33) and post evaluation (N=28), 9 participants answered both the evaluation questionnaires (paired participants). This might happen due to the holiday period which has caused many participants who involved in the pre evaluation were not in the office during the post evaluation or vice versa. A score of 3 or higher indicates that the respondents agreed with the statements in the questionnaire. In other words, participants found the i-Candies were easy to use (M=4.44), that the interaction with the i-candies was fun (M=4.33), they did not use the system frequently (M=2.67), the i-candies did evoke their curiosity (M=4), the system allowed for spontaneous informal (M=3.89) and non task-related conversations (M=4.33) and the users felt comfortable in the coffee corner (M=4.11). Respondents were neutral about the system enabling good work relationships with their colleagues (M=3) and negative towards the system enabling close relationships with colleagues (M=2.67). The result shows that in keeping with our expectations the mean values of post evaluation have increased for most of the questions except for the mean values of Q4 (curiosity) which remains constant. To test whether these increases are large enough, a student t-test was performed. The ease of use of the i-Candies has been significantly increased (t=-3.09, $p<0.05$). The fun factor while interaction with the system has also been increased significantly (t=-3.59 $p<0.05$). Additionally, the system allowed for more spontaneous informal conversations (t=-2.68, p<0.05), and also the environment allows for more non task-related conversations (t=-3.27, p<0.05) compared to the initial systems available in the coffee corner. Although the mean scores for the i-Candies regarding to system use (Q3), curiosity (Q4), enabling relationships (Q7, Q9), and for feeling comfortable (Q8) are increased, these values do not differ significantly (p>0.05) with those scores found in the baseline study.

Fig. 4. Sociability in the Coffee Corner

To summarize, from the observation as well as from the interviews, it was found that people were chatting or were discussing with each other while interacting with the system. They started up a new discussion based on the information that was presented on the display. Furthermore, multiple users were curious about the system. Put differently, the system enticed people to interact with it.

6 Discussion and Conclusion

In keeping with Mark Weiser's vision of ubiquitous computing where a world in which computational services can be naturally and "invisibly" integrated into our physical environment [17], the i-Candies and the i-Bowl illustrate how everyday objects related to drinking coffee can be part of a context-aware coffee environment. Computers in the coffee corner have very well-defined input interfaces i.e. mouse and keyboard. In the current project, we therefore, focused on making this input interface "blended" in the environment and more playful for the users. The interaction tags of candies is a part of the evolving line of tangible interface work [3,10], which determined to investigate interactive techniques for input devices. Candies are something that is consumed during informal meetings. The nature of a candy as something that is sweet is being replicated to human feelings. Candy-like system could make people feel sweet and rejuvenated during the coffee break. In our system, the displayed information are interesting news, comics, company information that make people think and smile. Also the candies are being shared by people as the nature of the information that is presented in the coffee corner.

The information that the candies control has been studied before. The Notification Collage [7] portrays items posted by users within a workgroup by creating a collage, rather than a single item at a time. Placing one i-Candy at a time presents a single item which provides content that is more easily digestible for the users. The Awareness Module [9] utilizes a large display in a public area or an office; however it requires the users to actively create content of the display. In our project, we re-used user-generated content, because we think that users in the public area should receive the information without needing to submit anything. In this way, maintenance is reduced and might be not needed at all.

Furthermore, work on designing technology-rich spaces to attract user interaction and enhance human-to-human sociability has been a promising research. GROUPCAST [14] and Intellibadge system [5] are applications that present information in a large display in a shared area. In the current project we elaborated upon these ideas by combining information with 'sweet' input devices and providing ways to make the passersby to be curious with the system.

The main goal of the project was to support sociability among office workers. From our evaluation study it was shown that this goal has been met. Interestingly, by showing the recently published results on the public screen, colleagues were encouraged to archive their work digitally in the company's system instead of doing so because they felt compelled. Another not trivial result of the i-Candies was that it influenced the interaction among the research engineers in a positive way. Mainly because of the fact that our prototype derived upon existing ideas, and relies upon existing work done by colleagues on context-aware systems, our implementation in the available coffee corner was established in many related projects and therefore also affected many colleagues' work. The positive effect of playing with the system was that they get more insight in each other contributions and their perspectives about how to reuse the existing system. And consequently, results in constructive collaborative design.

Another interesting result is that even after the evaluation period, the system is still in use in the office. The users gave positive feedback and proposed many ideas of applications that the company awareness system could be tapped into. As for the generic deployment of the system, we recommend the system should combine both general and company-specific content. In our study for example, we combined general information which could be interesting for any company such as world news, company announcement, with content such as scientific publications which could be interesting for research institutes or universities.

Acknowledgments. The project was performed in the context of AMIGO Project (IST 004182) and was conducted in Telematica Instituut. The system uses a context management framework, developed in the Freeband Awareness project (http://awareness.freeband.nl).

References

1. Agamanolis, S.: Designing Displays for Human Connectedness. In: O'Hara, K., Perry, M., Churchill, E., Russell, D. (eds.) Public and situated displays: Social and interactional aspects of shared display technologies, Kluwer, Dordrecht (2003)
2. Bradner, E., Kellogg, W., Erickson, T.: The adoption and use of 'Babble': A field study of chat in the workplace. In: Bodker, S., Kyng, M., Schmidt, K. (eds.) Proceedings of the 2001 international ACM SIGGROUP conference on supporting group work, pp. 154–161. ACM Press, New York (1999)

3. Camarata, K., Do, E.Y.-L., Johnson, B.D., Gross, M.D.: Navigational blocks: Navigating information space with tangible media. In: Proceedings of the International Conference on Intelligent User Interfaces, San Francisco, USA, pp. 31–38 (January 2002)
4. Churchill, E.F., Nelson, L., Denoue, L., Helfman, J., Murphy, P.: Interactive systems in public places: Sharing multimedia content with interactive public displays: a case study. In: DIS2004. Proceedings of the 2004 Conference on Designing Interactive Systems, Cambridge, MA, pp. 7–16 (August 2004)
5. Cox, D., Kindratenko, V., Pointer, D.: IntelliBadge™: Towards Providing Location-Aware Value-Added Services at Academic Conferences. In: Dey, A.K., Schmidt, A., McCarthy, J.F. (eds.) UbiComp 2003. LNCS, vol. 2864, pp. 264–280. Springer, Heidelberg (2003)
6. Dey, A.K.: Understanding and using context. Personal and Ubiquitous Computing 5, 20–20 (2001)
7. Greenberg, S., Rounding, M.: The Notification Collage. In: CHI 2001 Proceedings, pp. 514–521. ACM Press, New York (2001)
8. Hackos, J.T., Redish, J.C.: User and Task Analysis for Interface Design. John Wiley & Sons, USA (1998)
9. Huang, E.M., Tullio, J., Costa, T.J., McCarthy, J.F.: Promoting Awareness of Work Activities through Peripheral Displays. In: CHI 2002. 2002 Conf. in Human Factors in Computer Systems Extended Abstracts, pp. 648–649 (2002)
10. Ishii, H., Mazalek, A., Lee, J.: Bottles as a Minimal Interface to Access Digital Information. In: Extended Abstracts of CHI 2001, pp. 187–188 (2001)
11. Kreijns, K., Kirschner, P.A, Van Buuren, H., Jochems, W.: Determining sociability, social space and social presence in (a)synchronous collaborative groups. Cyberpsychology and Behaviour 7(2), 155–172 (2004)
12. Kreijns, K., Kirschner, P.A., Jochems, W.: The sociability of computer-supported collaborative learning environments. Journal of Education Technology & Society 5(1), 8–22 (2002)
13. Mayhew, D.J.: The Usability Engineering Lifecycle. Morgan Kaufmann, San Francisco, CA (1999)
14. McCarthy, J.F., Costa, T.J., Liongosari, E.S.: UNICAST, OUTCAST & GROUPCAST: Three Steps toward Ubiquitous Peripheral Displays. In: Abowd, G.D., Brumitt, B., Shafer, S. (eds.) Ubicomp 2001: Ubiquitous Computing. LNCS, vol. 2201, pp. 332–345. Springer, Heidelberg (2001)
15. Taylor-Powell, E., Steele, S.: Collecting evaluation data: Direct observation (G-3658-5). Madison: University of Wisconsin-Extension, Cooperative Extension (1996) (accessed June 20, 2006), http://cecommerce.uwex.edu/pdfs/G3658_5.PDF
16. van Sinderen, M.J., van Halteren, A.T., Wegdam, M., Meeuwissen, H.B., Eertink, E.H.: Supporting context-aware mobile applications: an infrastructure approach. IEEE Communications Magazine 44(9), 96–104
17. Weiser, M.: The Computer for the 21st Century. Scientific American 265(3), 94–104 (1991)
18. Whittaker, S., Frohlich, D., Daly-Jones, O.: Informal communication: What is it like and how might we support it? In: Proceedings of CHI 1994 (1994)
19. Whyte, W.H.: The Social Life of Small Urban Spaces. Conservation Foundation, Washington, DC (1980)

Geometric and Haptic Modelling of Textile Artefacts

Fazel Naghdy, Diana Wood Conroy, and Hugh Armitage

School of Electrical, Computer and Telecommunications Engineering,
University of Wollongong, Australia
fazel@uow.edu.au

Abstract. Geometric modelling and haptic rendering of textiles is an area of research in which interest has significantly increased over the last decade. A haptic representation is created by adding the physical properties of an object to its geometric configuration. While research has been conducted into geometric modelling of fabrics, current systems require textile data to be manually entered into the computer simulation by a technician. This study explores the possibility of automatic generation of geometric and haptic models of real world textile samples. The development of a scalable and generic methodology for geometric and haptic modelling of plain weave textiles made from wool yarn is reported. This system has been successfully implemented using a step-wise procedure. Initially, an image of the textile artefact is captured. Then the critical features of the image are extracted from the image and deployed in a finite element model. The geometric model is augmented by adding physical properties of the textile and developing the haptic model. Two different haptic rendering procedures are implemented based on Reachin Application Programming Interface 3.2 (API). The developed methodologies are described and results obtained are provided.

Keywords: Geometric model, haptic rendering, finite element model, textile artefact, weave style.

1 Introduction

Geometric modelling and haptic rendering of textiles is an area of research in which interest has significantly increased over the last decade. A haptic representation is created by adding the physical properties of an object to its geometric configuration. While research has been conducted into geometric modelling of fabrics, current systems require textile data to be manually entered into the computer simulation by a technician. This study explores the possibility of automatic generation of geometric and haptic models of real world textile samples.

This study addresses an existing gap in IT and multi-modal interaction with computers by providing tactile rendered tools to touch textile artefacts on the computer. Textiles create knowledge of texture and touch as the primary emotive forces of an artefact and textile practitioners are experts in haptic knowing. The ancient world drove its civilisation through innovative craft, tracked in archaeology through material culture analysis. A better technique, a more subtle skill in spinning threads differently could raise a city to richness, and magical craft objects were the subjects of innumerable myths of transformation and growth [1], [2]. Through 'techne', skill, the inventive ability of the craftsman Dedalus, who conceived the prototypical computer network - the labyrinth – enabled Icarus to fly, a seemingly impossible aspiration.

Such an interface will also enhance the effectiveness of computer-based perception and selection of fundamental weaves, especially plain weave and basket weave in varieties of texture and sett (fineness and coarseness). The hunger to understand and work with the rich variety of fibres now available – from polypropylene and smooth plastics to every kind of plant fibre and recycled 'found' materials – drives craftspeople from all cultural and ethnic backgrounds to create and experiment.

This system will also have particular potential for applications in museum practice and the textile auction houses. Museums/dealers with rare textile exhibits will be able to display (interactive) virtual images of their holdings with tactile experience of them built in.

In this paper, the development of a scalable and generic methodology for geometric and haptic modelling of plain weave textiles made from wool yarn is reported. This system has been successfully implemented using a step-wise procedure. Initially, an image of the textile artefact is captured. Then the critical features of the image are extracted from the image and deployed in a Finite Element Model developed for this purpose in ANSYS 9.0 (ANSYS Inc.).

The geometric model is augmented by adding physical properties of the textile and developing the haptic model. Two different haptic rendering procedures are implemented based on Reachin Application Programming Interface 3.2 (API) (Reachin Technologies AB). The first process uses a surface-based approach, where a polygonal haptic representation is attributed to the geometric model. The second design maps the textile image directly to a simple geometric representation of the surface.

The paper is organised as follows. Initially, a review of the previous work related to this project is carried out. The methodology developed for geometric modelling will be then described. This will be followed by the procedure deployed for haptic rendering of the geometric model. A typical model produced by this approach will be presented and some conclusions will be drawn.

2 Background

The literature does not reveal any previous work on the application of haptic rendering to textile artefacts in a virtual environment. The closest are the studies conducted in haptic texture rendering which has recently received increasing attention. In this approach, micro-geometry-scale features are added to the surface of an object. The work has mostly had its focus on modelling and rendering techniques [3]. [4], [5].

In the latest works, Crossan et al have applied granular synthesis – an established technique for synthesising audio, to haptic rendering of texture [6]. The results show that the same area of texture manifests different feeling at different times due to the stochastic nature of the approach. In the work conducted by Weisenberger and

Polling, three sensory modalities including visual, auditory and haptic have been used for virtual texture discrimination tasks. The results have not been consistent across all stimuli and indicate that the visual, auditory and haptic sensory information have a complex interaction and their effectiveness depends on the stimulus condition [7]. Domingues-Ramfrez and Parra-Vega have applied a constrained Lagrangian-based oriented framework to dynamically calculate the contact forces generated in haptic interaction with a deformable virtual object.

Overall, the main challenge faced in haptic rendered texture seems to be the unrealistic nature of perceived feeling in many instances [8]. This has been due to the complexity of the haptic rendering pipeline and the somatosensory system of the human. Choi and Tan have studied this unrealistic sensation and referred to it as "perceived instability" [9]. Through psychophysical experiments, they have observed three types of instability referred to as buzzing, aliveness and ridge. They suggest that such instabilities are significantly caused by the two stages of collision detection and force computation [10].

Study of the literature indicates that work of texture haptic rendering has not been extended to include fabric and textile. The closest identified work is the development of an on line intelligent database to retried fabrics that match the manufacturers' preferences in colour, drape and system, representing a high quality and best value [11]. A fuzzy clustering method is applied to objectively estimate the fabric property and select it from the database [12].

In the preliminary work conducted by this research group, an effective framework to generate a geometric model of a textile from plain weave samples was developed. Our findings indicate that the geometric haptic model should investigate an algorithm which is able to produce accurate yarn diameter results for multi toned textile samples and to differentiate different weave styles [13]. Traditionally, woven structures were represented and communicated through the grid diagrams of pattern drafting, which relate to digital models used to develop the haptic interface.

3 Approach and Methodology

Communicating the experience of touch through haptic interfaces that will render the textures of textiles displayed on a computer terminal has been the primary target of the project. Communication through texture and touch is fundamental to craftspeople and to textile designers making prototypes for industry.

Touching the texture of a fabric by fingers or body creates a complex multi-sensory emotional and cognitive experience [14]. The emotional aspects of this primary sensation are obvious in the linguistic complexity of touch. This is well understood and defined by textile professionals in terms of qualitative parameters. Developing a quantitative evaluation of these parameters and an objective understanding of the associated processes and relationships, particularly in the context of generating them through interaction with a virtual reality system, are the challenges pursued in this project. The work towards the aim of the project has proceeded in two stages:

(a) Geometric modelling which provides visualisation of the fabric in the virtual environment.
(b) Haptic modelling associating physical and mechanical properties with the geometric model.

3.1 The Basic Concept

The design of the geometric model has been kept as close as possible to the structure of the woven fabric to produce a generic and scalable model. The structure of weaves represented in grid diagrams forms three main groups, plain weave, twill and satin weaves which each have their own textural attributes in relation to the direction of warp (vertical threads) and weft (horizontal threads) [15]. The myriad patterns of woven textiles such as Basket and Leno are all derived from these three main structures (Table 1). Such plain weave weft-face structures are found in knotted rug and tapestry weaving. Basket weaving is a warp-faced structure. The grid structure of the warp and weft in woven textiles mimics the pixilation of the digital image. Textile construction is essentially a pixillated grid, with its multiplicity of interlocking threads.

Table 1. Weave Styles [16]

Plain	Twill	Satin
Basket	Leno	

Based on this concept, a textile artefact can be treated as a discrete mesh of points [16]. A typical representation for plain weave is illustrated in Figure 1. Similar correspondence can be established between other types of weaves and mesh structure. The characteristics of the discrete mesh depend on the properties of the thread used in warp and weft. In addition, each textile has a 'count' of warp threads and weft threads to the centimetre, ranging from 5 warp threads per cm in a rug or tapestry, to 150 per cm in a fine silk fabric. This is reflected in the mesh structure.

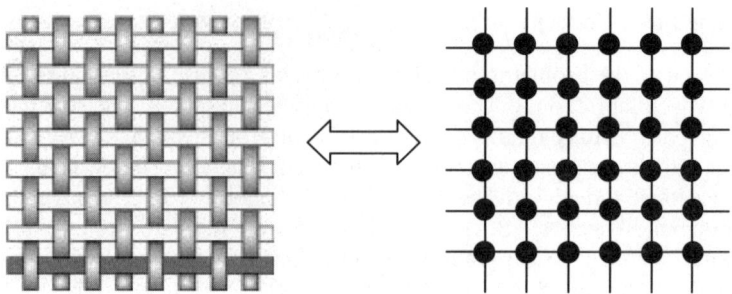

Fig. 1. Correspondence between plain weave and a discrete mesh

The modelling process assumes that the thread is inextensible and when the fabric is deformed there is no slippage at the crossings between warp and weft. The haptic modelling augments the geometric model with physical characteristics as explained later.

3.2 Generating Geometric Model

In this stage of the study, building the geometric model was based on plain weave textiles (Figure 2). The process started by capturing the image of textile using a still digital camera with a resolution of 4 Mega pixels. The pictures were taken at a consistent distance of 200mm from the textile sample. All images were taken with the flash to provide a high degree of clarity.

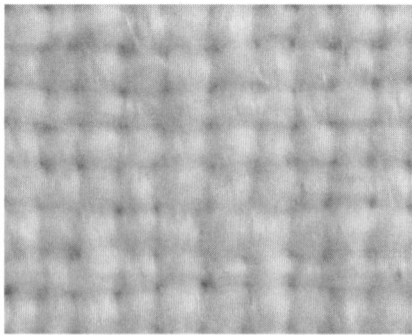

Fig. 2. Plain weave textile sample

Using digital image processing, the widths of the warp and weft yarns were derived. It was assumed that yarns were circular in shape and the spaces between parallel yarns were small enough in comparison to the yarns to be ignored.

The geometric modelling is completed by generating a deformable surface from the mesh structure. There are different approaches available to generate a deformable surface from the mesh structures produced from woven textile. Ng and Grimsdate [17] have provided a comprehensive survey of methods developed for geometrical modelling of cloth. In a more recent work, Liayan et al [18], have developed efficient algorithms for segmentation and parameterisation of arbitrary polygon meshes.

In this stage of the work, the concept is implemented using the finite element software ANSYS (by Leap Australia). The digital image of the fabric has been segmented to derive the diameters of the warp and weft. The measured parameters are fed into the generic model of the textile implemented in ANSYS. The model consists of three layers: key-points, splines and areas. In ANSYS, key-points are defined as dimensionless entities and are only a single point within a three-dimensional coordinate space.

Splines interconnect the key-points and are curved lines with only one dimension; length. Areas have two dimensions: width and length, and are comprised of key-points and splines, creating a modelling hierarchy of required structures to form the polygonal surface. The model is formed by reiteratively defining a sub-section of the fabric, which is created by plotting a spline over five key points. This process defines the basic shape of the yarn. The finite element model produced for the plain wave is illustrated in Figure 3.

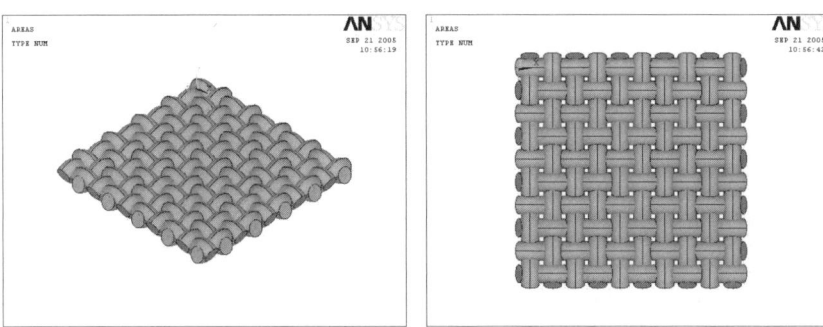

Fig. 3. Generation of Finite Element model of the textile from the digital image of the textile

The finite element model generated can adapt itself to the size of yarns, as well as the spacing between yarns. This is shown for uneven yarns with large spacing in Figure 4.

Further work was carried out to improve the realism of the model. A generic approach was developed to model a multi-ply yarn formed by multiple smaller yarns. Another enhancement was to apply twist to a single yarn. Examples of the produced models are illustrated in Figure 5.

4 Haptic Rendering

In haptic modelling, the mechanical characteristics of the fabric are added to the geometric model. In such models the dynamic properties of the fabric and its draping behaviour should be accurately included to provide accurate interaction with the user. The modelling is carried out based on a software package called Reachin API, which enables integration of visual and haptic rendering of a scene. Programming is carried out in VRML, Python and C++. Specialised nodes provided by this software will realise specific functionality and speed up simulator implementation.

The haptic rendering of the textile artefact has been carried out based on Reachin API. This application enables fast haptic model development, utilising VRML surface representations with Python script and C++ custom nodes to facilitate dynamic model realisation and manipulation.

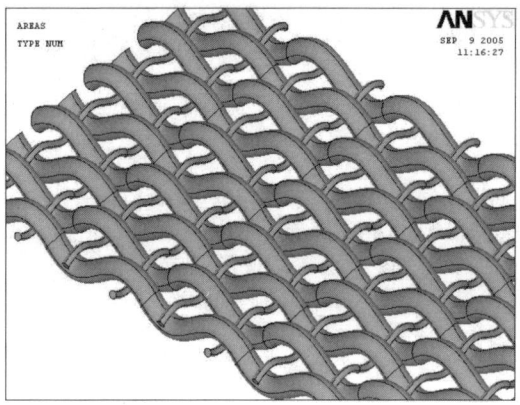

Fig. 4. Model with Uneven Yarns and Large Spacing

Two different approaches have been used to create the haptic model in the Reachin API. In the first method, a 2D image texture can be imported into Reachin where it is represented both visually and haptically by specifying a URL field. The latter uses a Reachin specific *BumpMapSurface* node to transform the image texture into a surface whose depth varies according to pixel intensity. This provides a tangible surface produced from the texture by mapping the pixels at different heights according to their colour index; where white (255, 255, 255) is the highest and black (0, 0, 0) is the lowest value.

In the second approach, a geometric model represented by Virtual Reality Modelling Language (VRML) script is rendered both visually and haptically in the Reachin API. VRML is a higher level language than other virtual reality languages, allowing for rapid program development. However, it is has more advanced functionality than other modelling languages. VRML script enables the production of simple, cross-platform model representation. Haptic properties can be added to any structure, specifically, *Shape* nodes, within the VRML file in the Reachin API.

ANSYS is able to export models in a VRML format (version 1.0), which can be easily redefined in the required format (version 2.0).

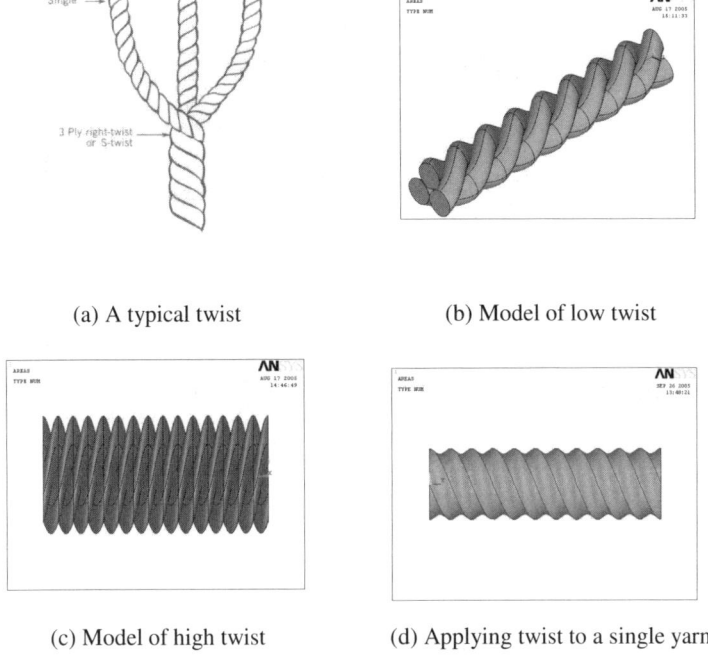

(a) A typical twist (b) Model of low twist

(c) Model of high twist (d) Applying twist to a single yarn

Fig. 5. Geometric Model Enhancement

In this work, the operator interacts with the haptic rendered model of the textile artefact and develops a subjective feeling of its properties. The type of interaction will depend on the interface that can be created between the human operator and the model. Based on the current and developing haptic interfaces, it is feasible to envisage that interaction will be limited to touch and multiple-user strokes. Future developments in haptic devices might make other types of interaction a possibility.

The VRML files produced by ANSYS are usually quite large. For example, the model for plain weave model developed in the work has a size of 96.8 MB in VRML 1.0 format. The file size is expanded to 101MB when it is converted to VRML 2.0 format. It is possible to reduce the file size by applying an optimisation process. An optimisation algorithm called VIZUP {Vizup Technology) was applied to plain weave model. The results of the optimisation process at four different levels are illustrated in Figure 6.

It can be seen that there is little difference in appearance between the 85% and 90% reduced models. However once the reduction in the number of triangles reaches 95%, holes in the surface topology become apparent. Such holes would alter the feel of the model. For this reason the 90% model was selected for haptic rendering as it provides a good compromise between visual (aesthetic) and haptic fidelity. The significant decrease in program size increases the program's code traversal and therefore screen update rates. Increasing the update rate of the process prevents the system becoming unstable and terminating the application.

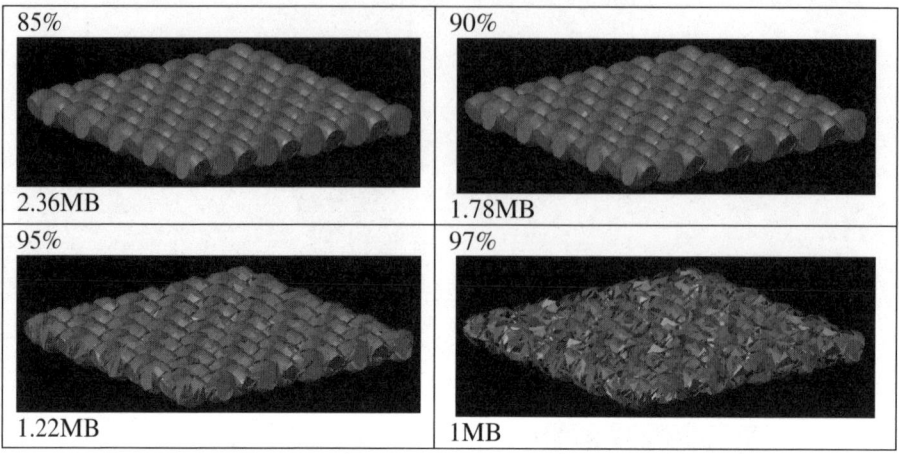

Fig. 6. Haptic rendering models after optimisation

5 Conclusions

The development of a generic framework for geometric and haptic modelling of textile artefacts was reported in this paper. The basic features of the geometric model are derived from the warp and weft yarn diameters obtained through digital image processing of the textile samples. These features are then used in a scaleable finite element model to produce a high quality geometric model of the textile. The result is a realistic visual representation of the original textile, adaptable to any given warp or weft diameters.

In order to add kinaesthetic sensation to the visual representation of the fabric geometric model, two haptic rendering methods were deployed. Optimisation methods were applied to reduce the size of the haptic rendered model for real-time interaction with the model. The overall model is a virtual object which the operator can manipulate and feel the roughness of the surface using a haptic device.

In the second method of haptic rendering, a surface node is applied directly to the geometric model. This surface whilst visually and haptically convincing, does not truly represent the real texture of the textile. Overall, the first haptic rendering method provides a more accurate representation of warp and weft of textile samples.

The work so far has been focussed on plain weave which represents the simplest style of textile. The future work should pursue two major endeavours. The first is developing effective algorithms to generate scalable geometric models for both mono and multi- toned textile samples and different weave styles. In the second focus, more realistic, stable and effective haptic rendering algorithms should be developed to accurately model the physical behaviour of the textile as it is manipulated by the user. The latter should include many complex physical characteristics exhibited by textile.

References

1. Wood-Conroy, D.: A silver bowl for Artemis. In: Zimmer, J., Victoria, C. (eds.) Contemporary Crafts Review, Melbourne 25(95) (September 19, 1995)
2. Wood Conroy, D.: Oblivion and metamorphosis: Australian weavers in relation to ancient artefacts from Cyprus. In: Rowley, S. (ed.) Re-inventing Textiles Tradition and innovation, vol. 1, pp. 111–131. Telos Art Publishing, Winchester (2000)
3. Massie, T.H.: Initial haptic explorations with the Phantom Virtual Touch through point interaction. In: Master's Thesis, Dept. Manufacturing, MIT Press, Cambridge (1996)
4. Ho, C., Basdogan, C., Srinivassan, M.A.: Efficient Point Based Rendering Techniques for Haptic Display of Virtual Object. Presence 8(5), 477–491 (1999)
5. Crossan, A., Williamson, J., Murray-Smith, R.: Haptic granular synthesis: targeting, visualisation and texturing. In: Proc. 8th Int. Conf. on Information Visualisation, pp. 527–532 (2004)
6. Weisenberger, J.M., Poling, G.L.: Multisensory roughness perception of virtual surfaces: Effects of correlated cues. In: Proc. 12th Int. Symp. on Haptic Interfaces for virtual Environment and Systems, pp. 161–168 (2004)
7. Choi, S., Tan, H.Z.: Toward Realistic Haptic Rendering of Surface Textures, pp. 40–47. IEEE Computer Society Press, Los Alamitos (2004)
8. Choi, S., Tan, H.Z.: An Experimental study of perceived instability during haptic texture rendering effect of collision detection algorithm. In: Proc. 11the Int'l Symp. Haptic Interfaces for Virtual Environment and Teleoperator Systems, pp. 197–204 (2003)
9. Choi, S., Tan, H.Z.: Aliveness: Perceived instability from passive haptic texture rendering system. In: Proc. IWWW/RS Int'l Conf. Intelligent Robots and Systems, pp. 2678–2683. IEEE Computer Society Press, Los Alamitos (2003)
10. Gider, A.: An Online Fabirc Database to Link Fabric Drape and End-user Properties. Master of Science Thesis, p.3 (2004)
11. Chen, J., Chen, Y., Zhang, B., Giders, A.: Fuzzy linear clustering for fabric selection from online databases. In: Proc. Fuzzy Information Processing Society, NAFIPS (2002)
12. Armitage, H.: Creating a Geometric Model of Textile Artefacts in a Virtual Environment. Honours Thesis, School of Electrical, Computer and Telecommunications Engineering, University of Wollongong (2005)
13. Moody, W., Dillon, P., Baber, C., Wing, A.: Factors underlying fabric perception. In: Proc. Eurohaptics (2001) cited at, http://www.eurohaptics.vision.ee.ethz.ch/2001.shtml
14. Tovey, J.: Weaves and Pattern Drafting. BT Batsford Ltd, London and New York, pp. 10–11 (1969)
15. Netcomposites, http://www.netcomposites.com/education.asp?sequence=42
16. Aono, P., Denti, D.E., Breen, M.J.: Fitting a woven cloth model to a curved surface: Dart insertion. In: IEEE Journal of Computer Graphics and Applications, pp. 60–70 (1996)
17. Ng, H.N., Grimsdate, R.I.: Computer graphics techniques for modelling cloth. IEEE Journal of Computer Graphics and Applications, 28–41 (September 1996)
18. Zhang, L., Liu, S., Wu, X., Zhou, L.: Segmentation and parametrisation of arbitrary polygon meshes. In: Proc. Geometric modelling and processing (2004)

A Toolkit to Support Dynamic Social Network Visualization

Yiwei Cao, Ralf Klamma, Marc Spaniol, and Yan Leng

Informatik 5 (Information Systems), RWTH Aachen University,
Ahornstr. 55, 52056 Aachen, Germany
{cao, klamma, spaniol, leng}@i5.informatik.rwth-aachen.de

Abstract. In this paper we introduce the design, implementation and evaluation of the Dynamic Visualization Toolkit (DyVT) to support complex dynamic social network visualization. Dynamic aspects of social networks such as spatiotemporal as well as personalized information can be visualized in a common toolkit. To that end, an XML-based target language DyVTML is an extension of existing schemata enabling expression, storage and interchange of rich animated social network data. With the language and the available tool support, even less-experienced users can visualize temporal data in animations and spatial data in maps and personalize it with icons and colors. The prototype is evaluated by the visualization of large mailing list data sets.

Keywords: Information systems, Information visualization, Social network analysis, XML.

1 Introduction

Social networks present a determinable structure that shows how people know each other either directly or indirectly. They are a special kind of networks in which nodes are entities that have values in a social context and the related social relationships. Although one of the key characteristics of social relationships is changes, most social network visualization is static due to the high complexity of social network data. For instance, the temporal attributes are often neglected which indicate when social relations take place or are discarded. However, understanding networks from a dynamic perspective is essential, because it facilitates reasoning real objects such as complex dynamic systems that evolve over time in the real world.

Besides just creating some nice graphics, social network visualization can generate learning situations [14]. It also provides investigators with new insights into network structures and helps them communicate with others [5]. So far, the research community has developed a number of tools for building, analyzing and visualizing social networks. However, these existing tools have several problems. First of all, they have "structural bias" which implicitly denies much of the dynamic nature of social relations [9]. The static data lacks other network data attributes such as spatial data [10]. Researchers are more and more interested in how networks develop and change timely and spatially. Secondly, the visualization results are not intuitive enough. Users are not allowed to choose the appearance of social network visualization. Finally, there is no specified language that covers a wide range of dynamic aspects for social network visualization in terms of interoperability, yet.

Our research intends to surmount these limitations. An XML-based target language is specified to be used in a metadata repository where temporal, spatial metadata are integrated into a uniform XML file. Furthermore, this target language also supports visualization. Based on this XML language, we have developed a Dynamic Visualization Toolkit (DyVT) to give end users a better insight into their social networks. It uses animations to visualize temporal data that shows how relationships emerge over time. Moreover, graphs with map backgrounds are used to represent spatial data. Users' personalized information is also represented dynamically as well.

In sum, DyVT aims to visualize temporal, geospatial and personalized information by extracting, integrating, and processing data from diverse data sources. The results of the system can be also transformed into various multimedia formats. In the prototype of DyVT, we use data from the mailing lists of the EU Network of Excellence project PROLEARN (www.prolearn-academy.org) focusing on technology enhanced learning and professional training within Europe.

The rest of this paper is organized as follows. Section 2 gives an overview of the state of the art in the field of dynamic social network visualization. Section 3 is devoted to the architectural design of DyVT. Section 4 describes the corresponding implementations and presents the prototype. Finally, we summarize the research work and give the perspective of future work in Section 5.

2 Related Work

Social network concepts are the background knowledge for social network visualization. We also observe the diverse media in the Internet which builds up a digital social environment [12]. Visualization methods are discussed with regard to temporal, geospatial and personalized data. Furthermore, several XML based interchange formats are introduced, owing to the requirement of defining a target language to express rich social network data.

2.1 Basic Social Network Concepts

Social networks are based on an assumption of the importance of relationships among interacting units. The social network perspective encompasses theories, models, and applications that are expressed in terms of relational concepts or processes. The important concepts of social networks are as follows [4]. *Actors* and their actions are viewed as interdependent rather than independent and autonomous units. Relational ties or linkages between actors are channels for transfer or flow of either material or nonmaterial resources. *Network models* focus on individuals and consider the network structural environment as opportunities for or constraints on individual actions. Network models conceptualize social, economic, and political structures etc. as lasting patterns of relations among actors.

In practices, the Internet is increasingly serving as a mediator of social activities. It facilitates the social processes of communities in many categories of environments and mechanisms. The *digital social environment* can be defined as social environments that provide diverse forms of support to the social processes. It includes a variety of systems ranging from very explicit and centralized community systems that directly support people's interactions, to some decentralized community systems that support peer-to-peer mode of interaction and that are directly controlled by their users [12]. The communication links are key objects to be visualized in social network visualization.

Centralized media support one-to-many mode including forums, Wikis, newsletters, and centralized mailing lists etc. *Forums* are public online open spaces for discussion in which the communication link exits between every two members participating in the same online forum. *Virtual community systems* are a community of people sharing common interests, ideas, and feelings over the Internet or other collaborative networks. The communication link exists between two members if two persons share resources. In *newsletters* the communication link exists between the administrator and every other member who is subscribed to the newsletter. A *centralized mailing list* is a list of people who subscribe to a periodic mailing distribution on a particular topic. When an email message is sent to the mailing list, it is automatically forwarded to all addresses in the list. So senders have communication links to every member in the list. *Wikis* allow users freely creating and editing Web page content in Web browsers. The communication link between two persons could be two members who edit the same Web page.

Decentralized media support peer-to-peer mode including Weblogs, emails, and mailing lists etc. For *emails*, a communication link is defined between two persons if they send emails to each other. Compared to the centralized mailing list, in *decentralized mailing lists* a communication link is similar to email. *Weblogs* are basically a journal published on the Web. They have two kinds of communication links. One is that one member makes comments to the Weblog entry of another member. The other is that two members make comments to the same Weblog entry.

In conclusion, all these communication links on the Internet is valuable social network data to research on forming and development of communities and on social processes within as well as cross communities.

2.2 Visualization of Social Networks

Social network analysts use two kinds of tools from mathematics to represent information about patterns of ties among social actors: node-link graphs and matrixes.

The more common visualization is node-link graphs which consist of nodes (actors) connected by edges (ties). The main goal is optimization of graphs' layout for comprehensibility and aesthetics. For example, certain mechanism is used to place nodes in adjacent positions according to topological and structural criteria [1]. Considering various types of social network data, temporal, geospatial, and personalized visualization are discussed respectively in the following.

Temporal visualization. Much recent research shows that the most static network images do a poor job of understanding how networks develop and change. In order to represent the structure of social network efficiently, at least two dimensions are needed. Thus, there is no dimension to visualize time. Based on the literature survey, temporal visualization approaches based on node-link graphs are represented by *animation layout*. It is considered as the most suitable way to represent changes of underlying network structures. It is complicated to determine the effectiveness of animation, which is influenced by many factors. Among them, two important factors are stated in [3]. *Readability* of visualization depends on aesthetic criteria to get visualization comprehensive. And *mental map preservation* means those nodes that exist in the series of networks remain in the same positions. An intuitive dynamic visualization should balance both criteria.

Compared with the static layout, animation layout considers much on dynamic social network relations which usually have two dimensions [10]. One is the relational pace concerning the rate of change in relations. The pace of change information can be described with regard to levels, changes or stabilities. The other is sequence which focuses on the order of relations. A better understanding of network changes depends on identifying such order of relations. With the both aforementioned criteria in mind, multiple images of networks are produced according to the relational pace and then these series of images are placed in sequences. In addition, sliders or other controls are often used to navigate the animation loop directly.

Geospatial visualization. With the development of network and information technology, the spatial data collection, sharing and analyzing is becoming more and more important. Spatial objects have spatial relationships such as overlap and containing [11], which adds complexity to social network data. Compared to traditional visualization methods, background maps are a common way to visualize and understand geospatial data of social networks. Moreover, Web map services are effective approaches to geospatial data visualization.

Personalized visualization. Another interesting category of visualization is to represent users' preference of data visualization. We call it personalized visualization here. It could visualize icons, sizes, and colors of nodes, colors, weights, line types of links, even the layout of social network according to various layout algorithms.

2.3 Languages for Dynamic Social Network Data

Languages for social network data are used to describe the structures and contents of the sets of observations [6]. One of the most general characteristics of social network data is that they have values in a social context. Referred to languages for dynamic social network data, the focus is how to represent attributes that change over time. Besides, the principle types of dynamic social network data are relational data and attribute data.

Relational data are connections defined by the different rules in digital social environments. Relational data focuses on the investigation of the social network structure. *Attribute data* is about attributes, opinions and behaviors of nodes. Both relational and attribute data can be gathered as a whole from various data sources, such as questionnaires, direct observation, written records, and experiments etc.

In order to represent relational data and attribute data, three important XML based markup languages for graphics have been surveyed.

GraphML stands for Graph Markup Language and is a graph exchange format that aims to represent either relational data or attribute data. GraphML is also the only published format that supports manipulation of dynamic graph data. In addition, GraphML is supported in large number of graph analysis and visualization software, such as JUNG, yFile, etc. Its major features are listed as below [2]:

- *Simplicity.* It is easy to parse and interpret for both humans and machines.
- *Generality.* There is no limitation with respect to the graph models such as hyper graphs and hierarchical graphs etc.
- *Extensibility.* It is possible to extend the format in a well-defined way to represent additional data required by arbitrary applications or more sophisticated use, e.g. sending a layout algorithm together with the graph.
- *Robustness.* Systems that are not capable of handling the full range of graph models or added information can be easily recognized. Models can be extracted to the subsets that can be handled in the systems.

The graph model used in GraphML is: $G = (V, E, D)$, Where V is the set of nodes, E is the set of edges and D represents data labels. A valid GraphML data file has two parts: the header is used to define some basic features of the GraphML, such as XML standard, XML Schema and a root element. The graph topology is a central part of the GraphML including definition of both nodes and edges.

The GraphML elements can contain any number of graphs. Edges and nodes may be ordered arbitrarily. For instance, it is unnecessary to list all nodes before all edges. The complexity of GraphML is low, since the space requirement for saving a graph with n nodes and m edges in GraphML is only $O(n+m)$.

KML is a file format used to display geographic data in a special browser for maps, such as Google Earth, Google Maps, and Google Maps for mobile. A KML file is processed in the same way as HTML (and XML) files are processed by Web browsers. Like HTML, KML has a tag-based structure with names and attributes used for specific display purposes. Thus, Google Earth and Google Maps are the common browsers for KML files [8].

DyNetML is an XML-based social network language to address the needs of data interchange, developed at Carnegie Mellon University [13]. DyNetML represents dynamic network data as sets of time slices. Each of the time slices is a descriptive snapshot of the organization at a given time. A dynamic network element is defined as a sequence of *MetaMatrix* elements representing a snapshot of the organization for one time period. Each of the *MetaMatrix* elements consists of the following objects. *TimePeriod* allows clear identification of each time slice. *Properties and measures* represent data about all of the time slices. *Node* contains one or more node sets. *Networks* contain all networks in one time slice. And *Anthrop* facilitates the link of network data to anthropological data.

Summarily, DyNetML can be used to describe temporal data, while KML format can be used to represent geospatial metadata. Since GraphML is designed to be extended easily, the implementation of logical integration of these three concepts can be performed by extending GraphML.

3 System Design of DyVT

With reference to the state-of-the-art technologies, suitable methods are chosen to develop the DyVT. In this section the design issues of DyVT is discussed. It begins with the requirement analysis, followed by the system concepts and the data model.

3.1 Requirements Analysis

Both functional and non-functional requirements are analyzed. The main motivations are both the limitations of existing social network visualization systems and user-specified requirements.

The limitations of the existing social network tools are critical. First, these tools lack rich social network data representation. The interoperability oriented target languages are rarely expressive enough to fully represent the rich social network data. Second, current dynamic social network visualization forms are quite restricted. Line graphs simulate the social network changes as lines in the graphs. However, it is impossible to represent the global change of network over time, because such summary statistics provide information on a single dimension of a network structure. The other approach is to examine separate images over time, whereas this approach lacks readability. The reason is that it is impossible to identify the sequence linking node position in one frame to the position in the next. Third, the function of user customization is often missing. Customization refers to user preferred data about social network visualization. Fourth, the dynamic nature of social relations is ignored. The dynamic nature refers to contextualized information such as changing spatial, temporal and personalized information from users. A survey of the existing dynamic social network visualization tools shows that normally only one of these dynamic natures is visualized.

Based on the state of the art and potential users, functional requirements can be divided into four categories.

1. Integration and interoperability of metadata. A target language records all kinds of data including raw data, temporal data and spatial data as well. Thus, it allows expressing and exchanging rich social network data.
2. Visualization of temporal data. The system should provide appropriate animation visualization for end users to know how data changes along time.
3. Visualization of spatial data. The user should be able to obtain the graph on maps where nodes and edges are located according to their real geographical location, if the social network data is location relevant.
4. Visualization of personalized data. The system should enable the user to have flexible choices to define the graphs in their own ways by selecting colors, sizes or icons etc.

3.2 System Concepts and Data Modeling

The main concept of DyVT is depicted in Figure 1. Social network data together with temporal, spatial, and appearance or personalized data are the input to DyVT. In DyVT temporal, spatial and personalized visualization are supported via an XML-based target language. An XML-based target language, so-called DyVTML, is defined to integrate metadata from different sources. It enables the system to visualize temporal, spatial and personalized data in a highly collaborative unified interface. XML is chosen because of its simplicity, extensibility, interoperability, and openness.

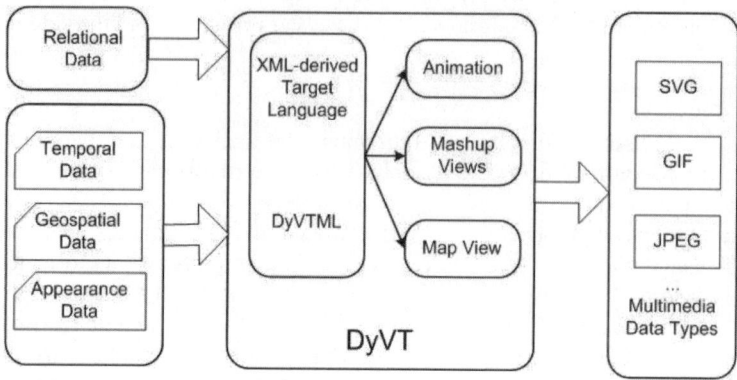

Fig. 1. System concepts of a Dynamic Visualization Tool (DyVT)

Users can make good use of temporal data by an animation that shows how relationships emerge over time. At the same time, users get spatial insight into their data with the layout of these nodes on the map according to the geographic information. Users have also flexible options to specify the graph styles. It is possible for users to choose their favorite colors, icons to represent graph. In addition, the visualization results are changed dynamically by the users' different choices. Furthermore, with the help of the DyVTML, the visualization results are compatible to various multimedia formats such SVG, GIF and JPEG etc. Consequently, they can be well exported into the diverse formats.

After presenting the system concepts, we introduce DyVT on its data level. The data used in DyVT mainly comes from three sources: mailing list databases, spatial data, and personalized data which refers to user chosen appearance data. The mailing list data is stored in relational database, spatial data is in some spatial databases and user chosen appearance data is defined by users through a graphical user interface. The DyVTML is used to interoperate data extracted from different data sources and to transform them into data elements with the corresponding predefined tags.

However, the major drawback of XML-based format stems from the size and complexity of XML files. The growth in size is dictated by needs for rigorous markup, as every data element requires a number of delimiter tags to describe its function to the parser. In order to solve this problem, we extract some redundant data elements from the DyVTML and define a new data format called Appearance Data Markup Language (ADML).

DyVTML is also used to store social network data for visualization component. The major characteristic of these data is independence of visualization tier which includes relational social network data, temporal data and spatial data. Relational social network data is extracted from the mailing list data and contributes to build the structure of the social network. Temporal data is also from the mailing lists and provides temporal attributes to define both nodes and edges. Spatial data is extracted from some existing geospatial databases and used to determine the location of each node. Summarily, in the context of the mailing list the data stored in DyVTML describes email communication events with attributes about when and where they occur.

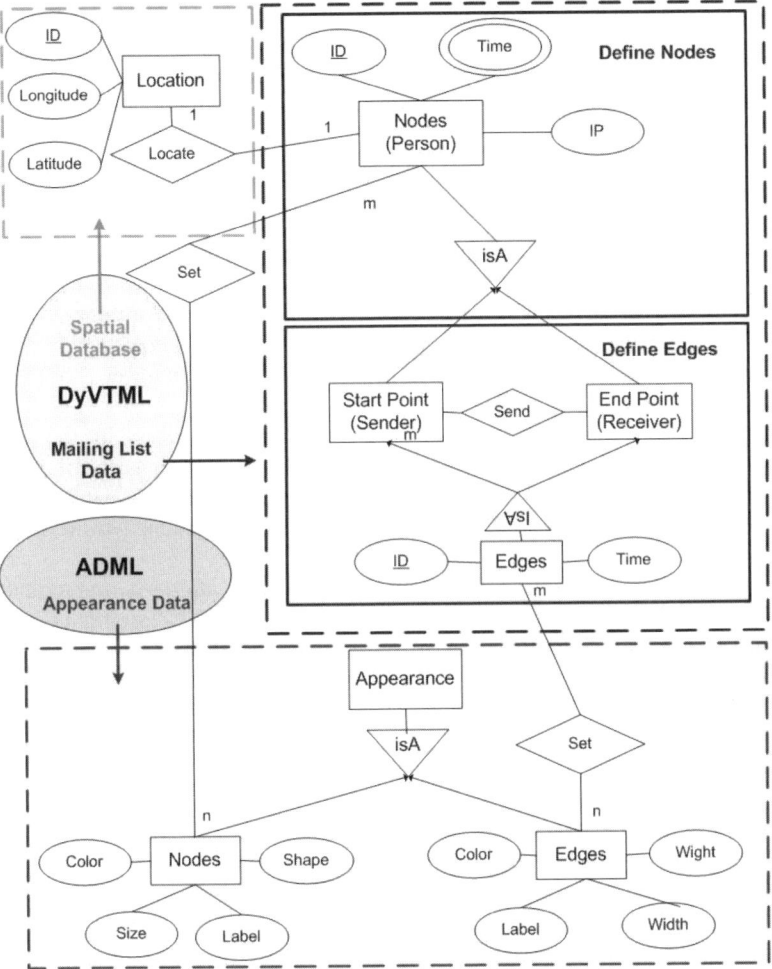

Fig. 2. Data modeling of DyVT in the entity relationship diagram

ADML is used to store user chosen appearance data. It comprises node appearance data such as colors, sizes, shapes, and label colors, etc. as well as edge appearance data such as colors, widths, weights, and label colors, etc. correspondingly.

There are two advantages to use these two data formats separately. On the one hand, redundancy is reduced. As the number of nodes increases, it results in great complexity to define appearance for each node and edge in the system. It leads to redundant expression for describing both nodes and edges in DyVTML. On the other hand, processing time is saved. In order to get a better visualization result, DyVT enables users to change settings of graphical appearance for both nodes and edges. An updating process will be called several times based on the same social network data, when the user changes the settings. Usually, the appearance data is much smaller than the underlying social network data. Using this method, the processing time will be reduced by only parsing the updated appearance data rather than the whole social network data.

The relationship of ADML and DyVTML as well as the data entities are illustrated in Figure 2. This data model is the guideline for the system implementation.

4 Implementation of DyVT

The implementation of DyVT is based on three tiers: the database tier, the metadata enrichment tier and the visualization tier.

The *database tier* contains raw data sources. In DyVT, we have three data sources: the mailing list database of the EU project PROLEARN, the GeoLiteCity database to get the mapping information between IP addresses to cities, and user chosen network appearance data.

Since DyVT reads data from different data sources, then in the *metadata enrichment tier* we need to refine the raw data. Raw data consists of two types: relational data and attribute data. The former can be extracted from database by some constraints such as content or time etc. The latter can be attained from the graphic user interfaces.

To integrate these two types of data efficiently, the *metadata refining* module deals with this piece of integration work through defining an XML-based data format for metadata storage. On the other hand, in the *metadata modeling* module, a built-in parser makes it possible to extract key fields from the XML-based DyVTML data file created in the previous modules for the further use. In the *visualization tier,* a resulting metadata model is responsible for containing all the information needed to draw a visual representation of the data. Two handlers are used, the *temporal view handler* and the *map view handler*. They are used to create visual features according to their algorithms. Contents of the visual abstraction are put into three types of views including animation view, static view and map view by users' choices. These visualization results can be exported into several formats such as SVG, GIF, and JPEG etc.

Moreover, like the basic visualization model, we provide user interaction in this tier as well. DyVT provides two levels of user interactions. One is for details of each time window in animation views. User can extract any static graph generated in any time window. The other interaction mainly serves the needs of experts who focus on studying the detail of layout algorithms. From the user interfaces, users can see and edit more details on configuration parameters for each algorithm. Two layout algorithms are currently implemented. The cycle layout arranges nodes within the predefined radius. The Kamada-Kawai layout (KK layout) is based on the algorithm by Kamada and Kawai [7] which optimizes a graph with the goal of the lowest energy sum. The energy is defined as the topological distances between each pair of nodes.

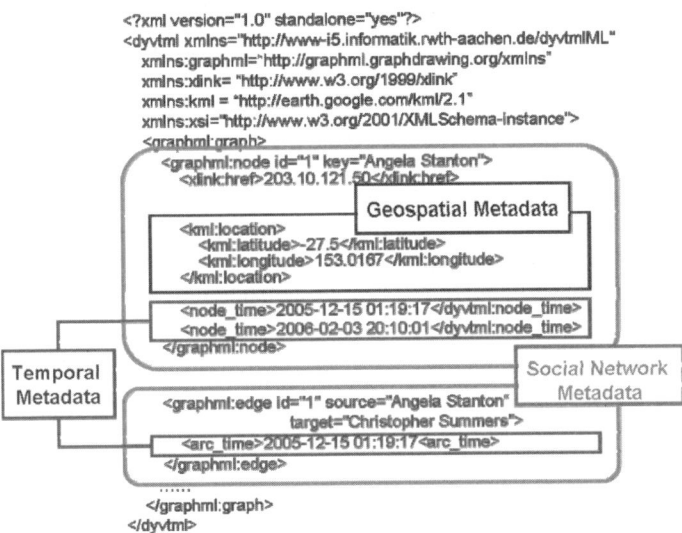

Fig. 3. An instance of DyVTML

In short, after we have captured the all needed kinds of metadata in the database tier and have refined and remodeled this metadata in the metadata enrichment tier, the visualization component is also executed in this tier. The actual rendering of the metadata onto views is done, too. These views can provide varying perspective onto data. For example, the use of maps shows the visualization with geographical attachment, and animations show the dynamic temporal data visualization.

The DyVTML (cf. Figure 3) plays an important role to interchange the data among the three tiers to visualize the dynamic social network. The DyVT screenshot in Figure 4 illustrates the relationships among senders and receivers of the PROLEARN mailing list within a certain time period. The location of each sender or receiver is also visualized via Google Map API. A user-friendly interface is available by means of a set of wizards.

The evaluation is performed by a task-oriented usability testing within the PROLEARN user communities. We defined tasks and asked users to finish the tasks and to record the time. Tasks are selected to get better feedback from users' point of view of functional and non-functional requirements of the system. Evaluation results show that DyVT helps users build and specify the social network data visualization effectively via the dynamic views.

5 Conclusions and Future Work

As a prototype for dynamic social network visualization, DyVT provides the possibilities to realize temporal visualization with animation in multiple views, spatial visualization, and personalized visualization in one uniform platform. So DyVT is innovative at unifying the three types of visualization in one solo toolkit. The XML-based language DyVTML enhances data interoperability among various multimedia data formats.

Besides, according to users' feedback some improvement work can still be done in future. Currently, only email exchange between two persons are implemented, which is a single relationship among actors. However, social relations among actors are usually more complicated, in which actors are connected in multiple ways simultaneously. We call this kind of network multi-relational network representing a heterogeneous set of nodes and edges. Therefore, there exist not only multiple entities of varying types, but also different ways or semantics by which these entities are connected.

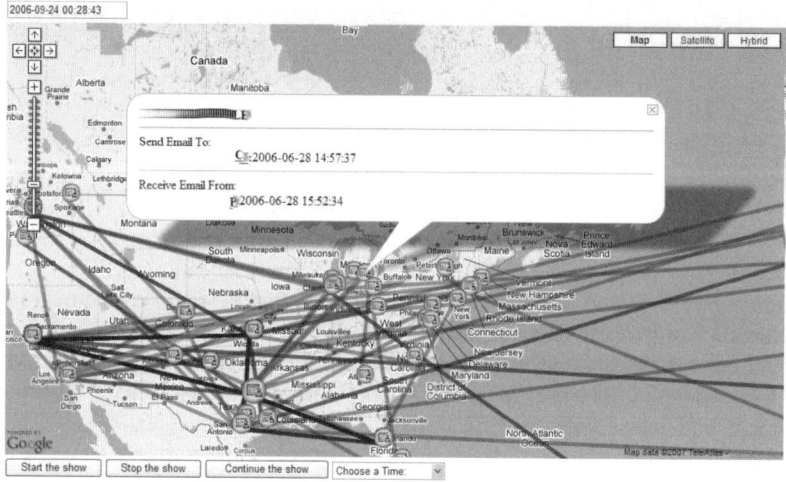

Fig. 4. The screenshot DyVT for geospatial visualization in continuous time slices and with user chosen icons

In addition, only node-link graphs are used to visualize both temporal and geospatial network data. With regard to the usability testing results, the readability of node-link graph deteriorates significantly, when the size of the graph increases. Hence, a node-link diagram is only suitable for small graphs. To visualize large graphs, other representation should be employed such as matrix-based representation. Compared with node-link graphs, matrix has no links overlapping problems and is less affected by the increase of the number of nodes.

With regard to the layout algorithms, there existed several well-developed layout algorithms for traditional static layout. It also provides many possibilities to either modify these algorithms to be suitable for dynamic visualizations or develop new layout algorithms for dynamic social networks.

Moreover, the evaluation of the results in DyVT depends on the users themselves. It lacks the objective view of result. Statistical analysis tools should be integrated to give objective evaluation results.

Now DyVT is used to visualize the mailing list data. However, mailing lists are just one pattern in the digital social environments. A series of other patterns choosing and executing components should be added in the database tier.

Not limited in social network, visualization is an intuitive way to get a better insight into underlying data. With the aforementioned improvement possibilities, DyVT is a useful visualization toolkit to be integrated into other applications easily and quickly in future.

References

[1] Bertini, E.: Social network visualization: A brief survey. The Blog of Enrico Bertini (October 2005), http://www.dis.uniroma1.it/ bertini/blog/bertini-socialnetvis-2.pdf
[2] Brandes, U., Eiglsperger, M., Herman, I., Himsolt, M., Marshall, M.S.: GraphML progress report: Structural layer proposal. In: Mutzel, P., Jünger, M., Leipert, S. (eds.) GD 2001. LNCS, vol. 2265, pp. 501–512. Springer, Heidelberg (2002)
[3] Erten, C., Kobourov, S.G., Le, V., Navabi, A.: Simultaneous graph drawing: layout algorithms and visualization schemes. Journal of Graph Algorithm and Applications 9(1), 165–182 (2005)
[4] Wasserman, S., Faust, K.: Social network analysis: methods and applications. Cambridge University Press, Cambridge (1994)
[5] Freeman, L.C.: Visualizing social network. Journal of Social Structure 1(1) (2000), http://www.cmu.edu/joss/content/articles/volume1/Freeman.html
[6] Hanneman, R.A., Riddle, M.: Introduction to social network methods Riverside, CA: University of California, Riverside. (2005) http://faculty.ucr.edu/ hanneman/
[7] Kamada, T., Kawai, S.: An algorithm for drawing general undirected graphs. Information Processing Letters 31, 7–15 (1989)
[8] KML home page (March 14, 2007) http://earth.google.com/kml/index.html
[9] Milgram, S.: The small world problem. Psychology Today 1, 61–67 (1967)
[10] Moody, J., McFarland, D., Bender-deMoll, S.: Dynamic network visualization. American Journal of Sociology 110(1), 1206–1241 (2005)
[11] Morris, A.J., Abdelmoty, A.I., El-Geresy, B.A., Jones, C.B.: A Data-Flow Approach to Visual Querying in Large Spatial Databases. In: Chang, S.-K., Chen, Z., Lee, S.-Y. (eds.) VISUAL 2002. LNCS, vol. 2314, pp. 175–186. Springer, Heidelberg (2002)
[12] Nabeth, T.: Unders tanding the identity concept in the context of digital social environments. Project report of INSEAD CALT - FIDIS (January 2005) http://www.calt.insead.edu/Project/Fidis/documents/ 2005-fidis-Understanding_the_Identity_Concept_in_the_Context_of_Digital_Social_Environments.pdf
[13] Tsvetovat, M., Reminga, J. and Carley, K.M.: DyNetML: Interchange Format for Rich Social Network Data, Technical report, Institute for Software Research International School of Computer Science, Carnegie Mellon University, CMU-ISRI-04-105 (January 2004)
[14] Viegas, F.B., Donath, J.: Social network visualization: can we go beyond the graph. In: Workshop on Social Networks for Design and Analysis: Using Network Information in CSCW (2004)

The Predicate Tree – A Metaphor for Visually Describing Complex Boolean Queries

Luca Paolino, Monica Sebillo, Genoveffa Tortora, and Giuliana Vitiello

Dipartimento di Matematica e Informatica, Università di Salerno
via Ponte don Melillo, 84084, Fisciano(SA) -Italy
Tel.: +39 089 963324
{lpaolino, msebillo, tortora, gvitiello}@unisa.it

Abstract. In this paper, we describe a visual language, based on the so-called *Predicate Tree Metaphor*, which allows users to visually build complex sentences for querying commonly used search engines. By using this visual language, no parentheses have to be applied, and no precedence rules have to be known. Promising results about the usability of the proposed interface are reported, on the basis on an experimental between-group study, performed on a Yahoo-based prototype of the proposed graphical environment.

Keywords: Visual metaphors, sum of product expression, search engines, usability evaluation.

1 Introduction

Looking for information, whether is it for finding an article, discover a path to a restaurant, book a trip by airplane, or any other intentions, is something that many of us do in our daily lives. In order to find out the information we need, we have to make the query appropriately. For example, in searching for a paper, parameters that we might involve in the query are title, author, keywords, references to other papers and the file type. The search is then conducted by checking the results, adding more filters, e.g., for search refinement, until we discover the target information.

However, in many cases, the process of data discovering is so difficult and so long that users are not able to succeed. Often, this problem is due to the fact that users are not adequately able to compose queries according to the language which underlies the search engine. In other words, they are not able to properly combine logical conditions, also taking into account precedence rules.

Additionally, textual languages are not so proper for inexpert users because some differences exist between natural language and Boolean logic, with respect to the meaning associated with AND/OR combined conditions. For example, in English a user may pose the query:

"Find all the documents that have pdf format and those that have doc as format"

If this query was translated directly from English into a query language, the constraint clause would take the form:

format ="pdf" and format ="doc"

The results of this query would always be an empty set because a document cannot be in pdf and doc format at the same time - the constraints of the query should have been ORed together. To overcome these limits and allow users to compose complex queries in a simple and intuitive way, visual languages and interfaces seem to be interesting solutions.

In this paper, we propose a visual language, based on the so-called *Predicate Tree Metaphor*, which allows users to visually build complex sentences for querying commonly used search engines. By using this visual language, no parentheses have to be applied, and no precedence rules have to be known. A prototype of the proposed graphical environment has been implemented based on the Yahoo search engine, in order to overcome present limits of the Yahoo's access method. As a matter of fact, we show that the adopted Predicate Tree Metaphor provides users with the ability to exploit the engine in a deeper way.

The paper is organized as follows: in Section 2, the *Predicate Tree* language is presented together with its environment. Here, the main concepts concerning the construction of the visual representations are given. The algorithm for translating a tree instance into a Yahoo query form is also described. In Section 3, we report on an experimental between-group study, performed on a Yahoo-based prototype of the proposed graphical environment, involving two groups of ten subjects each. Section 4 makes a short discussion on related work. Finally, Section 5 concludes the paper with some final remarks and discussion on future work.

2 The Predicate Tree Language

In the search engine field, the management of sentences composed by predicates connected by *AND*, *OR* and *NOT* operators is very common. However, in most cases, their employment is underutilized and often incorrect because it requires a deep knowledge of Boolean logic and a specific knowledge of the used search engine. As an example, Yahoo™ and Google™ differently interpret the query *P1 AND P2 OR P3 AND P4*. Yahoo selects documents which satisfy the *(P1 AND P2)* and *(P3 AND P4)* expressions, while Google™ looks for documents respecting *(P1 AND P2 AND P4)* or *(P1 AND P3 AND P4)*. In practice, the correct interpretation depends on the precedence rules the search engine applies. Moreover, not all the search engines rely on the complete Boolean algebra, limiting the resulting expressive power. This is the case for Google, which only uses simple Boolean expressions with no parentheses. Figure 1 shows some known search engines and the corresponding set of Boolean operators [8].

In this intricate world of search engines and Boolean operators, the inexpert user who needs to perform a medium/hard query may get lost and needs to pass through several trials and errors before achieving the required information. In this context, visual between the user and the machine by alleviating problems coming from the inability to manage logical connectors, parentheses and other specific query language structures.

Search Engines	Boolean	Default
Google Review	-, OR	and
Yahoo! Review	AND, OR, NOT, ( ), -	and
Ask Review	-, OR	and
Live Search Review	AND, OR, NOT, ( ), -	and
Gigablast Review	AND, OR, AND NOT, ( ), +, -	and
Exalead Review	AND, OR, NOT, ( ),-	and
WiseNut Review	- only	and

Fig. 1. The principal Boolean operators working on some search engines

In this section, we propose a novel interface, based on the *Predicate Tree (PT)* visual language, able to hide the complexity of Yahoo™ query composition by means of a simple and intuitive structure.

It provides users with the ability to define the complex queries without knowing how the Yahoo's logical operators work thanks to a tree structure where:

1. the root represents a *True* predicate,
2. nodes represent simple conditions which can be visually created,
3. edges connect predicates which should be valid at the same time in the resulting documents (logical AND)
4. nodes on different paths are logically OR-connected conditions.

The *Predicate Tree* is used to select all the documents which satisfy at least one complete path on the tree instance. For example, the tree depicted in Figure 2 requires to select the documents which make true at least one among *(P1 AND P2), (P1 AND P3), P4*.

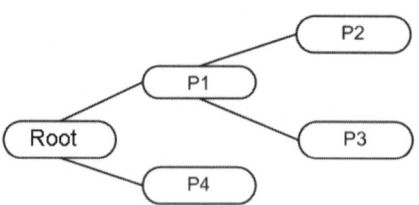

Fig. 2. An example of query according to the Predicate Tree language

In order to prove the effectiveness of our interface, we show one typical example of complex query involving the most important Yahoo's fields.

Before starting the example, we need to specify which predicates we implement into the structure. For the sake of simplicity and brevity, we decided to manage just a subset of those possible. Anyway, the model can be easily extended in order to completely simulate the language.

Table 1 shows a summary of the symbols presently managed.

Table 1. Advanced query symbols of the Yahoo Search Engine

Symbol	Meaning
+	If a common word is essential to get the results you want, you can include it by putting a "+" sign in front of it.
-	You can exclude a word from your search by putting a minus sign ("-") immediately in front of the term you want to exclude from the search results.
OR	Yahoo search supports the Boolean "OR" operator. To retrieve pages that include either word A or word B, you can use an uppercase OR between terms.
AND	Google search supports the Boolean "AND" operator. To retrieve pages that include both word A and word B, you can use an uppercase AND between terms.
Date	The *"date:"* query prefix will filter the results to include only documents that were inserted in the Google database within the specified Julian date range.
File Extension	The *"file type:"* query prefix will filter the results to only include documents with the extension specified..

Let us suppose *we want to find out documents describing either* JavaScript *syntax or* HTML *syntax. The first kind of document must be exclusively in* MSWORD *format while the second one must be either in* PDF *format if it was updated during 2006 or in any format if it was updated in 2007.*

According to the rules specified in Section 2, we need to build up a *Predicate Tree* where each complete path selects a part of the total result we want. First of all, the sentence requires that we have to select two kinds of documents, either those containing the JavaScript word or those containing the HTML word.

We want to find out documents describing either JavaScript *syntax or* HTML *syntax.*

This corresponds to the Predicate Tree as shown in Figure 3.

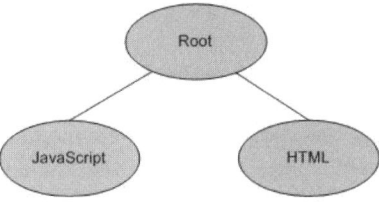

Fig. 3. The first query composition step

The second part of the sentence asks for filtering JavaScript hits by indicating those in DOC format, namely:
The first kind of document must be exclusively in MSWORD *format* ....
The resulting tree is shown in Figure 4.

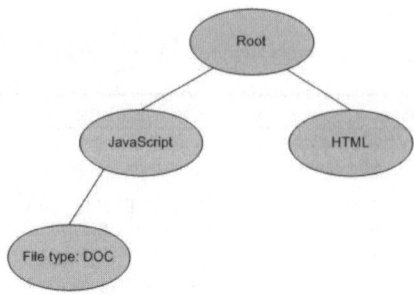

Fig. 4. The second query composition step

Finally, the rest of the sentence, ... *while the second one must be either in* PDF *format if it was updated during 2006 or in any format if it was updated in 2007,* may be translated as shown in Figure 5.

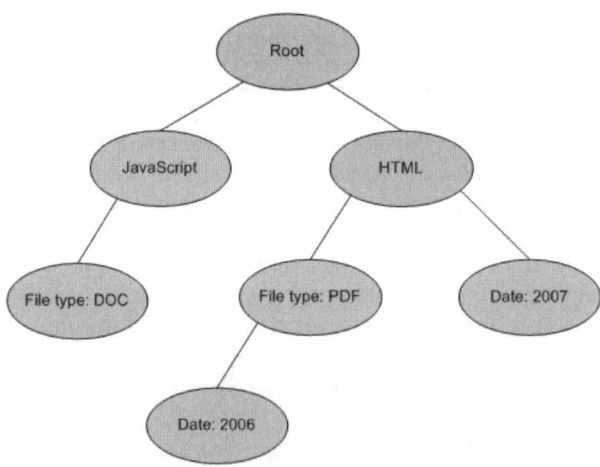

Fig. 5. The third query composition step

Figure 6 shows the *Tree Metaphor* as implemented by means of the HTML and JAVASCRIPT languages into an Internet page.

The root node represents the search engine where the request will be submitted. In the present paper, we choose Yahoo™ but other engines may be selected through the menu located on the left-upper corner. The nodes located below the root indicate either the filters applied on the current request or filters which are not yet specified and are not currently parsed for the final request. Such a difference is highlighted by assigning different colors to each set, namely green for the former and blue for the latter set. As a matter of fact, each time the user makes a choice through the node's menu, the node is automatically green colored and, at the same time, two empty blue nodes are created in order to suggest users the positions where new filters can be set. When starting, just the root and a blue node connected to the root appear on the interface. When the user applies a new filter to this node, two new nodes are automatically created, one connected to the root and the other to the filtered node which becomes the father node. This is to say, nodes are added where new filters may be applied. Each time a new filter is set on a blue node, that node becomes green and two new empty nodes are created as its sibling and child, respectively.

In order to set nodes with the required filters, each node is provided with a two level menu. In the first one, the user may select the filter category s/he needs. Currently, the Yahoo search engine provides selection on the 'last update date', the 'extension', the 'web domain', the 'language' and specific strings. Once selected the category, the set of sub-choices appears into a second level menu. As an example, if we select the Extension menu, the list of accepted file formats would appear on the screen in the bottom-right corner, as shown in Figure 6. In case the user needs to specify a textual pattern, the interface provides a specialized input text field. In order to resemble the user choice, the final selection appears on the node as either an icon resembling such a choice or a string. A question mark is visualized next to every empty node, to highlight the current path selection. As a matter of fact, each question mark visualizes a natural language description of the path from the leaf next to the question mark to the tree root. This functionality allows users to improve the query comprehension and to make specific improvements or correction whenever required.

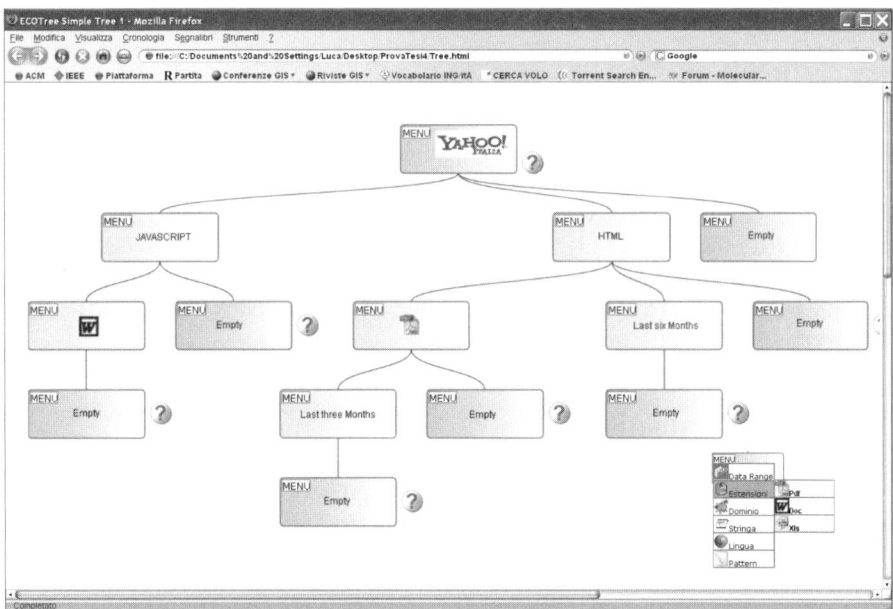

Fig. 6. The visual composition of the example query into the Predicate Tree

Once the tree is built, the second step is to determine the algorithm for translating the tree instances into the language understood by the search engine, in this case Yahoo. As described in [8], Yahoo' queries are in Sum of Product (SOP) form, namely the precedence rules follow the sequence NOT, AND, OR. In case we need to force this precedence, we might use parentheses.
The algorithm is shown in Figure 7.

```
1 String query
2  for each path in tree
3     for each node in path
4        if node is not root then
5           query concat node
6           if node is not leaf then query concat "AND"
7  if path is not the last then query concat "OR"
```

Fig. 7. The algorithm performing the translation from the Predicate Tree language to the Yahoo query language

In the first row a query string is defined in order to contain the query textual representation. Successively, a *for* cycle starting at 2 and ending at 7 is defined in order to return a reference to each path. According to the previous example, it returns the (Root, JavaScript, filetype:doc), (Root, HTML, filetype:pdf, date:2006) and (Root, HTML, date:2007) paths, sequentially. Within this cycle, a new *for* starting at Row 3 and ending at Row 6 returns a reference for each node belonging to the considered path. Once obtained the node reference, the system checks if it is different from the root, and in that case, the algorithm inserts in the query string its textual representation at Row 5. Moreover, if the system verifies that the node reference is not a leaf then the AND string will be concatenated to *query* at row 6. At the end of the first cycle of Row 3, *query* will contain the (JavaScript AND filetype:doc) string. Finally, the last statement of Row 2 cycle verifies if the referenced path is not the last one, in this case, the OR string will be concatenated and the Row 3 *for* performs another cycle, otherwise the algorithm stops the execution. According to the example, at the end of the execution, the final query string will be (JavaScript AND filetype:doc) OR (HTML AND filetype:pdf AND date:2006) OR (HTML AND date:2007).

3 Usability Evaluation

In the present section, we report on the evaluation process, which has been carried out on the visual search language based on the Predicate Tree Metaphor, as described in the previous section.

The evaluation strategy we have adopted has been to perform an experimental study, meant to compare the Yahoo Query Language (YQL) and the Predicate Tree language.

The study was settled by carrying out the following preliminary activities:

- Task analysis: meant to identify the relevant tasks which the users involved in the experiment should perform
- Definition of the parameters we should be able to measure and how to measure them. We had to choose the properties which we considered significantly influential for the major usability factors, namely efficiency, efficacy and user's satisfaction.

The *learnability* of the considered language was soon identified as one of the most important criteria against which the two languages should be compared. This usability factor refers to the ease with which new users can learn how to effectively use the search language, achieving a high performance degree [1]. Thus, the typology of tasks we considered for our experiment was 'query composition'.

Given such a typology of tasks, we decided that learnability should be measured by collecting a score on the subjects' performance, based on:

- the error rate of each subject during query composition
- the kind of errors made by the subjects.

Another crucial criterion, *user's satisfaction*, was measured by combining a 'think-aloud' evaluation approach with questionnaires. The think-aloud technique consisted in encouraging subjects' comments during their query composition tasks, and making appropriate annotations. Upon task completion, the subjects were invited to complete questionnaires, meant to estimate users' satisfaction with respect to each of the considered languages.

The subjects were divided into two groups of 10 people each. Some of the subjects had programming experience while others were non-programmers who were familiar with common web search tasks. They were equally distributed between the two groups, in order to improve the dependability of the achieved results.

A training session on each of the two languages preceded the experiment. During such session the subjects were instructed on how to compose a complex query with either languages. Each group was assigned a set of 5 search tasks expressed in natural language. The two sets were pair wise comparable, involving similar attributes and logical expressions. Proposing similar search tasks to both groups, we were able to perform a between-groups evaluation. The subjects in one group were asked to express the searches in the form of YQL queries, while the others were asked to use the Predicate Tree language, so that a comparison between the languages was possible. Biases due to knowledge transfer within each group were reduced by changing the order in which each subject performed his/her tasks. The list of submitted search tasks with the corresponding language syntax is reported in the Appendix.

The subjects query composition tasks were scored by exploiting the error classification initially proposed by Reisner in [7], namely, (C) for completely correct, (D) for minor data error,(M) for minor language error,(S) for error of substance, (F) for error of form, (N) for query not attempted.

The results are shown in Table 2. They show that:

- The group performing Predicate Tree queries correctly answered more than 20% than the group performing YQL queries. However, the percentage of correct queries are quite similar, namely 31 for the YQL and 39 for the PT. A significant growth has been instead reported for the number of minor language error, e.g. misspellings and punctuation, as well as for the number of queries with minor data error, namely queries where data is not supplied completely as required.
- People were encouraged to try the composition of the query. As a matter of fact, all the subjects belonging to the PT group tried to compose the queries whereas 9 percent of queries have not been answered by the YQL group.
- As we expected, the percentage of errors of form are notably decreased, resulting 48 from the YQL group and 27 from the PT group.

In order to monitor the overall user satisfaction, subjects were asked to answer some questions after having performed the tasks. Questions mainly concerned with three arguments, namely, general reactions to the language used, specific comments on the performed search tasks and on the difficulties encountered and support achieved during query composition.

Table 2. Percentage of query responses in each category

Response Category	Group YQL	Group Predicate Tree
C (Correct)	31	39
D (Minor data error)	4	8
M (Minor language error)	7	17
Total essentially correct	*42*	*64*
S (Error of substance)	2	9
F (Error of form)	48	27
N (Not attempted)	9	0
Total incorrect	*58*	*36*

As for the first argument, answers may be divided into four parts according to the external subdivision (YQL and PTL) and internal subdivision (programmers and non-programmers) of the groups. Programmers of both groups found no difficulty in composing the queries but those belonging to the PTL group observed that a notable support came in task expenditure thanks to the use of the tree metaphor. As a matter of fact, although textual languages are more concise with respect to visual languages people generally prefer to compose queries in a visual way rather than in a textual way. According to the particular answers we received, programmer-subjects particularly appreciated the fact that they do not have to address their efforts to correctly write tags or to use parentheses.

On the other side, non-programmers considered YQL very hard to use and remember, and reported that they felt uncomfortable in performing the most complex search tasks. Differently, those belonging to the PTL group observed that the given visual environment encouraged them to carry out the assigned complex tasks, thanks to an adequate feedback and to the ability to recover from wrong actions.

4 Related Work

Information can be extracted from data sources either by browsing, i.e. going from a high level view through progressive refinement until finding the desired data, or by querying, i.e. specifying some criteria that describe properties of the data desired. The first method is better when the user does not know what exactly s/he is looking for, or does not know the data source schema, so s/he navigates or explores the data until s/he sees some relevant information. In addition, browsing is the only means to extract data when there is no query language or schema associated to the data, as is the case of the web. The second method addresses the needs of users who are looking for specific information, which means they have to use expressions to describe data properties.

In our work, we tried to apply the querying methodology to unstructured data, trying to visually describe expressions specified according to search engine languages. For this reason, the work more closely related to our proposal, is concerned with visual search systems for databases, such as Filter/flow [3], Kaleidoquery [4] and FindFlow [9].

In Filter/flow, users adopt the pipe metaphor to describe Boolean logic. Each condition is like a filter for the water flow, if two conditions should be satisfied at the same time (AND) then they are located as a sequence of cocks, while if at least just one should be satisfied then the flow is divided in two minor flows which may be interrupted by cocks, namely the conditions. Downstream the cocks, flows are newly connected.

Kaleidoquery specifies AND and OR by using a representation similar to that used to represent the operation of AND and OR gates in electronics. The inputs of the gate are represented as switches, if the constraint holds true then the instance will pass through the switch.

More recently, the FindFlow interface has been presented in [9]. FindFlow creates a tree representation for searching data where each node contains the partial result of the query specified using filters on arcs necessary to reach the node from the root.

Our approach differs from the above under three aspects, namely the goal it was conceived for, the results presentation, and the user's support for the definition of new conditions.

As a matter of fact, our work was conceived for web search activities. Users may compose queries involving the most important search options such as looking for constant sentences, the last update time, the document format, the document's language and so on. Moreover, the Predicate Tree language uses nodes to define filters and edges to represent logical operators. According to our definition, acyclic graphs are not allowed in order to avoid misinterpretation by users. In case, a filter is common to two or more paths, it must be duplicated or located along the paths so that no cycles are composed.

Finally, in the Predicate Tree environment, when the user defines new conditions (filters) for search refinement, s/he is provided with a mechanism which suggests where new filters might be added. This feature makes the visual composition even easier to use, supporting predictability of the interface.

Another tree-based approach to query composition, was presented in [2]. Here, queries are represented by trees within an environment named Geographical Visual

Query Composer (GVQC). In this environment, logical operators are represented by some specific nodes labelled AND or OR. They are used to indicate how to combine nodes representing conditions. However, the use of logical operators into the visual representation may cause problem with people which interpret the AND operator in the natural language way, this is to say as an OR.

A completely different approach for querying the Internet is by browsing. Two popular examples are Quintura [6] and PageBull [5]. The Quintura search engine offers a visual search using a map of tags or hints contextually related to a search query. Every time we select a tag, the map is updated visualizing the most related values. The selection is recursively repeated until the user reaches his/her goal. Another commercial visual query tool is PageBull. It allows users to specify generic patterns into an input text form and then visualize the most important hits by presenting their thumbnails.

5 Conclusion and Future Work

In this paper, we presented the Predicate Tree Metaphor, namely a visual language able to support users for the construction of complex Boolean expressions.

In the future, we plan to extend the prototype in order to make the search parametric with respect to the most common web search engines. Then, further controlled experiments will be carried out in order to verify the goodness of the proposed approach, as well as to compare different search engines against search specific properties.

References

1. Dix, A., Finlay, J., Abowd, G., Beale, R.: Human-Computer Interaction, 3rd edn. Prentice Hall, Englewood Cliffs (2004)
2. Guo, D.: A Geographic Visual Query Composer (GVQC) for Accessing Federal Databases. In: National Conference for Digital Government Research, Boston, MA, pp. 1–4 (2003)
3. Young, D.: A graphical filter/flow representation of Boolean queries: a prototype implementation and evaluation source. Journal of the American Society for Information Science archive 44(6), 327–339 (1993)
4. Murray, N., Paton, N.W., Goble, C.A.: Kaleidoquery: A Visual Query Language for Object Databases. In: ACM Working Conference on Advanced Visual Interfaces, L'Aquila, Italy, pp. 247–257. ACM Press, New York
5. PageBull, available at http://www.pagebull.com
6. Quintura, available at http://www.quintura.com/
7. Reisner, P., Boyce, R.F., Chamberlain, D.D.: Human Factors Evaluation of Two Database Query Languages–Square and Sequel. In: AFIPS Proceedings, pp. 447–452. AFIPS Press, NJ (1975)
8. Search Engine, available at http://www.searchengineshowdown.com/features
9. Shizuki, B., Hansaki, T., Misue, K., Tanaka, J.: FindFlow: A visual interface for information search based on intermediate results. In: Asia-Pacific Symposium on Information Visualization, Tokyo, Japan, vol. 60 (2006)

Appendix

In this appendix we show the queries we submitted to the subjects during the experiment phase. We propose ten queries, five for the Predicate Tree writing and five for the YQL writing, labelled with Query 1, 2, etc having an incremental difficult level. For the sake of simplicity and brevity, we decided to manage just a subset of those possible. Anyway, the model can be easily extended in order to completely simulate the language.

Table 3. Query writing. Users are given a question stated in natural language and required to write a query in the Predicate Tree query language.

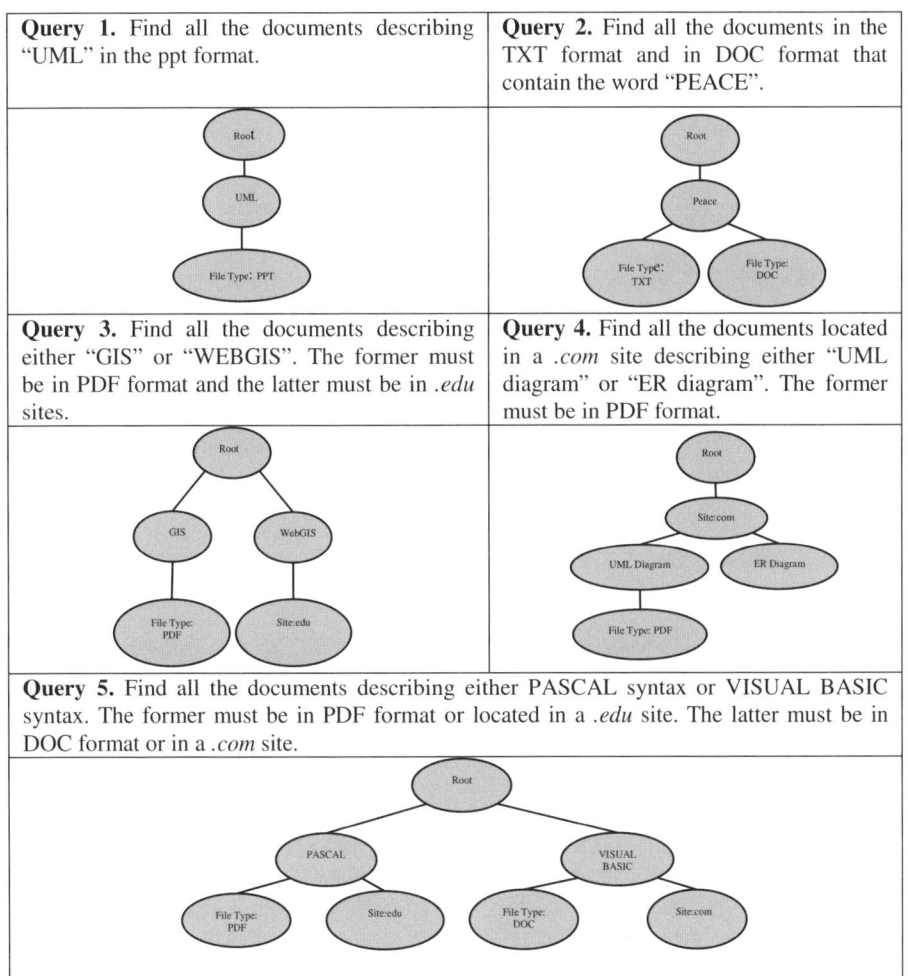

Table 4. Query writing. Users are given a question stated in natural language and required to write a query in the Yahoo query language.

Query 1. Find all the documents describing the Pascal language in the TXT format located on a *.edu* site	**Query 2.** Find all the documents in the PDF format and in PPT format that contain the word "Environment".
"Pascal" AND originurlextension:txt AND site:edu	"Environment" AND originurlextension:pdf OR originurlextension:ppt
Query 3. Find all the documents describing either "JAVA" or "JSP". The former must be in PDF.	**Query 4.** Find all the documents located in *.com* site describing either "GIS" or "WEBGIS".
JSP OR JAVA AND originurlextension:pdf	(gis OR webgis) AND site:com
Query 5. Find all the documents describing either tangent or cotangent function. The former must be in PDF format. The latter must be in XLS format and located in *.edu* site.	
tangent AND originurlextension:pdf OR cotangent AND originurlextension:xls AND site:edu	

Potentialities of Chorems as Visual Summaries of Geographic Databases Contents

Vincenzo Del Fatto[1,2], Robert Laurini[1], Karla Lopez[1], Rosalva Loreto[3], Françoise Milleret-Raffort[1], Monica Sebillo[2], David Sol-Martinez[4], and Giuliana Vitiello[2]

[1] LIRIS, INSA-Lyon, France
[2] DMA, Università di Salerno, Italy
[3] Benemérita Universidad Autónoma de Puebla, México
[4] Tecnológico de Monterrey, Puebla, México
Vincenzo.Del-fatto@insa-lyon.fr, Robert.Laurini@insa-lyon.fr,
Karla.Lopez@insa-lyon.fr, rloreto13@yahoo.com.mx,
Francoise.Raffort@insa-lyon.fr, msebillo@unisa.it,
davidr.sol@gmail.com, gvitiello@unisa.it

Abstract. Chorems are schematized representations of territories, and so they can represent a good visual summary of spatial databases. Indeed for spatial decision-makers, it is more important to identify and map problems than facts. Until now, chorems were made manually by geographers based on the own knowledge of the territory. So, an international project was launched in order to automatically discover spatial patterns and layout chorems starting from spatial databases. After examining some manually-made chorems some guidelines were identified. Then the architecture of a prototype system is presented based on a canonical database structure, a subsystem for spatial patterns discovery based on spatial data mining, a subsystem for chorem layout, and a specialized language to represent chorems.

Keywords: Geographic Information Systems, chorem, geographic generalization, semantic generalization, database summarizing.

1 Introduction

For many decisions, visual tools are necessary, and especially for spatial decision making for which cartography is an essential tool. When it is the cartography of facts, usually decision-makers are satisfied, but when it deals with visualization of problems, conventional cartography is rather delusive. Indeed it seems more interesting to locate problems and perhaps to help discover new problems or hidden problems. In other words, relevant database summaries can be a good approach [Saint-Paul et al., 2005 also for geographic databases.

A research program was launched between several research institutions in order to test whether cartographic solutions based on chorems can be more satisfying. Invented by [Brunet 1986, 1993], chorems can be defined as a schematized representation of territories. By schematized, one means that the more important is a sort of short global vision emphasizing salient aspects. This definition can be a good starting point to construct maps for spatial decision making. In other words, the goal of this research project is starting from existing databases, to analyze them so that to extract chorems by spatial data Mining [Pech Palacio et al. 2002] and visualize them.

This paper will be organized as follows. First chorems will be studied essentially as a new tool for visualizing and summarizing geographic information. Then the description of the architecture of a prototype system will be given.

2 What Are Chorems?

2.1 From Conventional Cartography to Chorem Maps

As previously said, according to Brunet, chorems are a schematized representation of a territory. In the past, chorems were made manually by geographers, essentially because they had all the knowledge of the territory in their mind. This knowledge was essentially coming from their familiarity of the territory, its history, the climatic constraints and the main sociological and economic problems.

Fig. 1. Conventional administrative map of France

Figure 1 shows a conventional map of France emphasizing administrative divisions. Figure 2 gives an example of a chorem map of France, in which the following aspects are stressed:
- The geometric shape is simplified,
- Only big cities are mentioned (Paris, Lyon, Marseilles and Lille),

- Only important mountains are shown, Alps as a frontier towards Italy, Pyrenees towards Spain, and the Massif Central forcing traffic to follow the Rhone axis,
- Major traffic axes and seas are depicted,
- And the French territory is divided in two parts, Eastern part the more developed, and Western part the less developed.

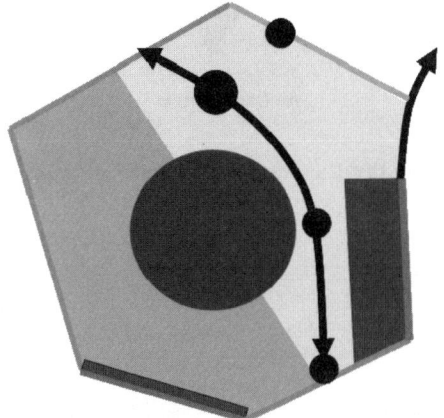

Fig. 2. A chorem map of France

We claim that such a map is much more informative about the difficulties of France than a flat administrative map.

Another example is issued from the water problem in Brazil. Indeed, a conventional map only showing main rivers as illustrated in Figure 3, does not lead to the solution of various problems such as:

- locations of places lacking water
- locations of the places with too much water
- locations of aquatic resources
- locations of humid zones
- locations of the water resources
- locations of the deserts,
- etc.

Lastly, let us examine the chorem given in Figure 4 schematizing the US population in which the country is represented by a rectangle. The exterior arrows show emigration from foreign countries and inner arrows the more important flows coming from North East. The country is split into two zones, the Eastern more populated, and the Western part less populated.

Bearing in mind all those examples, we claim that those chorem maps are much more informative and helpful to decision-makers. Those chorem maps can be seen bothr as the layout of geographic knowledge, and a kind of summary for geographic databases characterized by:

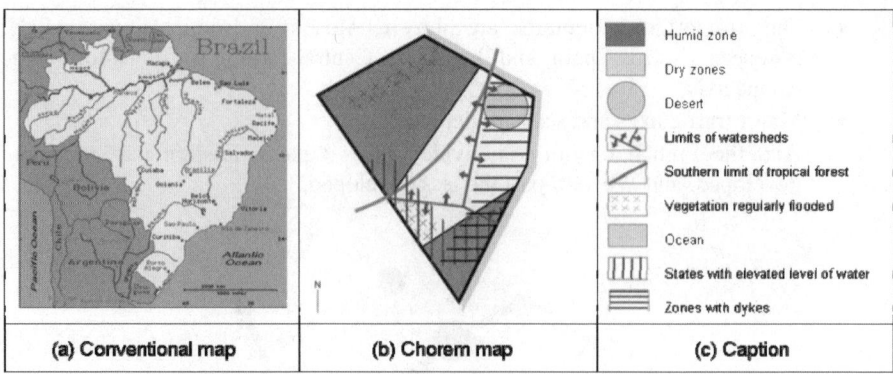

Fig. 3. The water problem in Brazil using a conventional river map (a) and a chorem map (b) issued from [Lafon et al. 2005]

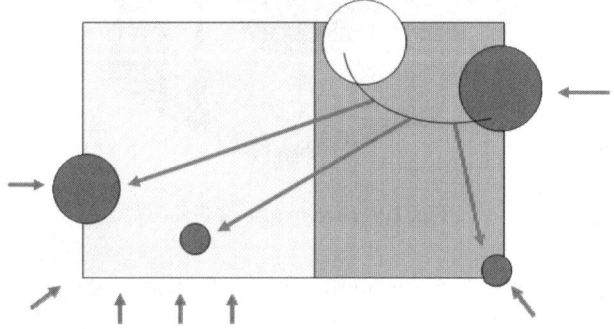

Fig. 4. Chorem for the US population [USA]

- a geographic generalization in order to simplify the shape of the territory under study,
- and a semantic generalization in order to select the more salient aspects of the non-spatial attributes of the geographic database.

Now the question is how to represent visually those information chunks? Two possibilities exist:

- either to define a complete vocabulary (by means of icons) which can be used in any situation (this was the Brunet's attitude when defining his chorems by means of a table),
- or to let the user define himself his own vocabulary by providing a caption.

2.2 Results of a Study of Existing Manually-Made Chorem Maps

A study was conducted about the chorems as they were used in several maps. Approximately 50 manually-made chorem maps were studied. The results are:

- even if the chorem concept is used by a lot of geographers, the Brunet's vocabulary is not very used;
- generally the users define their own chorem vocabulary,
- usually less than 10 chorems are used in a map,
- the more used patterns can be regrouped into main categories such as (1) main cities (which can be retrieved by SQL SELECTs), (2) main regions which can be retrieved by clustering and (3) main flows which can be retrieved by both clustering and SELECTs.

2.3 Towards Summarizing Geographic Databases

To conclude this paragraph, it appears that chorems in addition to the initial definition (*schematized representation of territories*) can be potentially used for other goals such as:

- visually summarizing spatial database contents,
- global vision of a spatial database [Shneiderman, 1997, Laurini et al. 2006]
- and representing visual geographic knowledge.

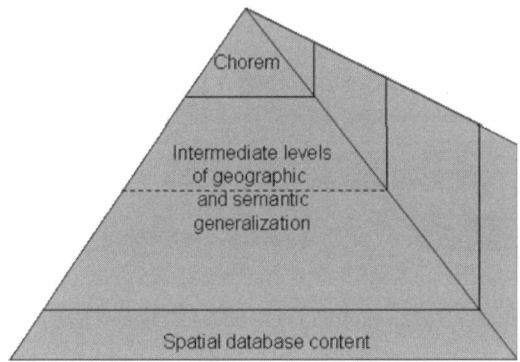

Fig. 5. A pyramid of contents

As a chorem can be seen as a visual summary, some other layers of visual schematization can be defined from the database contents. So a sort of pyramid can be defined in which the apex is the chorem map, and the basement the database contents. At intermediate levels, several levels of geographic and semantic generalization can be defined. See Figure 5 for such a pyramid.

To explore those new possibilities, some prototypes must be designed, implemented and tested. Let us examine a proposed architecture.

3 Architecture of the System

An explorative system has been designed according to the main following specifications (Figure 6):

1 – chorem discovery based on spatial data mining, the result being a set of geographic patterns or geographic knowledge (upper part of Figure 6),
2 – chorem layout including geometric generalization, selection, algorithms for visualization (lower part of Figure 6).

To facilitate spatial data mining and extract relevant semantics, a canonical database structure has to be defined. As an intermediary between chorem discovery and chorem layout, a language has be defined, named ChorML.

Let us examine all of those issues.

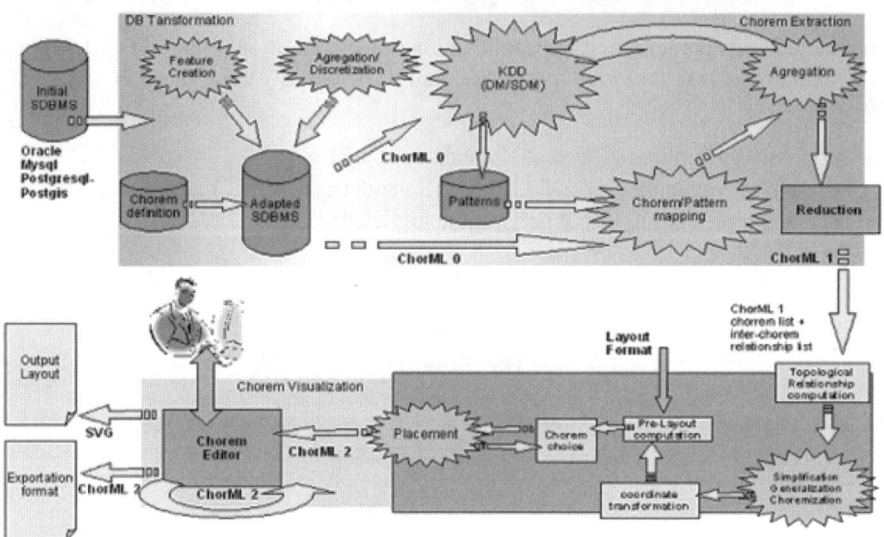

Fig. 6. Architecture of an explorative system for chorem discovery and layout

3.1 Canonical Database

The system begins by a database to be mined in order to extract spatial patterns. However, the data mining algorithms are not flexible enough to deal with any kind of spatial databases. In order to solve this problem, or in other words to avoid the problem of interoperability between our system and any kind of geographic databases, a structure has been designed, named canonical database. A canonical database is defined as a fixed structure of a geographic database so that any data mining algorithm must be applied without modification. Thus, the users must transform their initial database into this structure, either by a list of views, or by creating new tables with this structure.

Another problem is the vicinity of the territory. Indeed, in several encountered manual chorem maps, external information can be added, such as the names of sea, of adjacent countries and so one. In order to provide this information, which is currently not in the initial database, a special table of the canonical database was defined. For instance, a canonical database (spatial and non-spatial) at country level will include:

- basic information such as cities, regions, main hydrology, main roads, mountains, etc.
- more elaborated information such as networks, flows, barriers
- external information such as boundary types, names of seas and of adjacent countries
- etc.

3.2 Spatial Pattern Discovery

As previously said, spatial patterns will be extracted from spatial data techniques. See [Ester et al, 1997] or [Holder-Cook, 2005] for details. However, in data mining it is well known that a lot of patterns can be retrieved. Two problems exist:

- setting of list of techniques to be used taking our context into account,
- selecting chorems from patterns.

So, among the relevant techniques, we have chosen to use first clustering and aggregation procedures together with SELECTs.

The next phase is how to identify chorems from spatial patterns, taking into consideration that a maximum of 10 chorems must be chosen. Those ten chorems must correspond to the more important spatial patterns. At this point, we have no clear-cut solution to reduce the number of patterns. In our first prototype we have decided not to implement an automatic solution: for that a visual interface will help the user choose the more important patterns (chorems) for the layout phase.

3.3 Chorem Layout

Once the list of chorems and the set of constraints among them are obtained from the Chorem Extraction Subsystem, they are sent to the Visualization Subsystem in order to derive a visual representation of chorems and chorem maps, both in terms of layout and semantic content.

As shown in Figure 6, five different tasks are performed by this subsystem, namely chorem drawing, coordinate translation, best-placement of chosen chorems, pre-layout computation and chorem editing. As for the chorem drawing, it is performed through three, not necessary interconnected, steps, named *simplification, choremization* and *generalization*, where some procedures and spatial operators are invoked. In Figures 7-9, such transformations are illustrated. In particular, the *simplification* step determines a simplified version (see fig. 7(b)) of the data geometry, by reducing the number of vertices of the original shape (see fig. 7(a)).

As for the *generalization* step, which is a well known set of techniques in cartography [Buttenfield-McMaster, 1991], it may be invoked to group features that share some common properties, both geometric and descriptive, and generate a unique geometric representation of the involved elements. Figures 11(a) and 11(b) depict such a transformation.

The *choremization* phase associates a regular shape (see fig. 10(b)) with the possible simplified geometry of data (see fig. 10(a)).

One of the problems which may arise when simplifying and generalizing chorems, is related to the possible loss of crucial spatial constraints among elements of the original map. Thus, when the boundary is simplified, cities such as harbours which are located along the boundary must move with the boundary; otherwise, harbours would be positioned in the middle of the sea, or in the middle of the land. In order to preserve the spatial consistency among geographic elements, topological constraints are checked and, if a violation occurs, the Visualization Subsystem modifies the city location, accordingly.

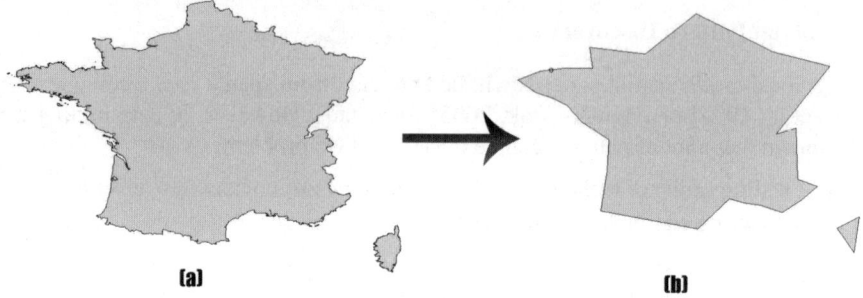

Fig. 7. An example the simplification process

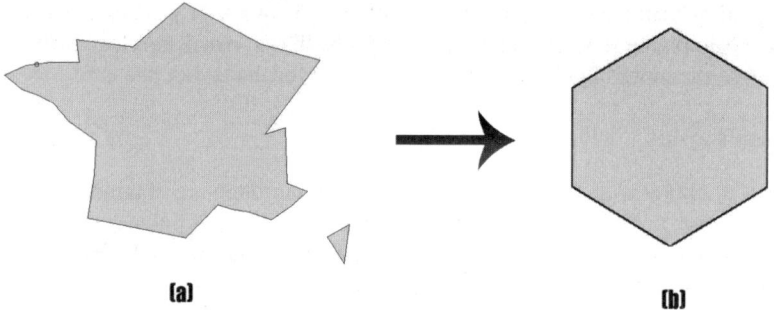

Fig. 8. An example the choremization process

Fig. 9. An example the generalization process

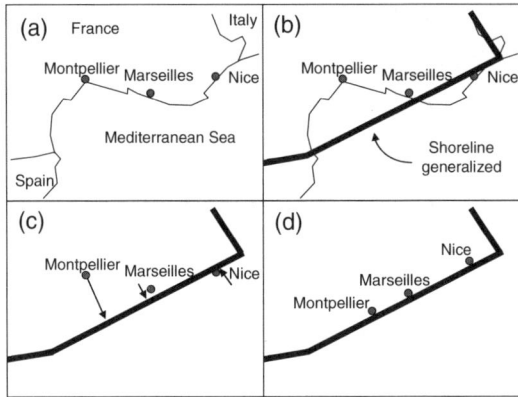

Fig. 10. Projecting harbors onto generalized shoreline. (a) situation before generalization. (b) generalized shoreline. (c) harbors must be moved. (d) final layout.

See Figure 12 for an example along the French Mediterranean shoreline.

It is worth to notice that in order to both preserve topological constraints and properly apply spatial operators, an underlying geographic reference system is maintained during the chorem drawing phase.

Once the drawing of the expected chorem is obtained, users are asked to specify details about the output map, such as the number of colours and the final layout format (for instance A4). The latter affects the number of chorems that can be introduced onto a map, since it is necessary to guarantee the readability requirement.

Based on the information provided by users, the next phase translates the chorem coordinates, acquired with respect to the original geographic reference system, into new coordinates defined with respect to a reference system local to the chosen visualization format.

At this stage, chorems extracted by the Chorem Extraction Subsystem are associated with a locally georeferenced visual representation. The goal of next step consists of aggregating chorems onto the output map. This is accomplished by a multi-agent system whose aim is to spatially arrange chorems onto the chosen visualization format and determine their best placement [Jones, 1989], preserving structural and topological constraints among them. It is worth to point out that in order to guarantee the best placement requirement, independent sets of interrelated chorems may be aggregated onto different maps, in order to provide users with more intuitive and readable chorem maps.

Anyway, some difficulties can occur regarding chorem placement and layout, as well as further refinements affecting semantic and graphic properties may be required by users. To this aim, users are provided with a tool for chorem editing which allows them to refine the expected output map. In particular, the Chorem Editor may perform the following tasks:

- import of a list of chorems positioned onto a chorem map;
- chorem display starting from the information derived from the previous steps;
- modification of both visual representation and semantic structure of chorems, without loss of consistency between them; in order to solve problems regarding chorem placement and layout the Chorem Editor can change chorem positions, colours and shape;
- generation of a graphical representation based on SVG (Scalable Vector Graphics) [SVG];
- export of both a graphical representation (SVG) and a proper ChorML-based representation of chorems.

Figure 11 shows the visual interface of the Chorem Editor, which has been built as an extension of the Magelan Graphics Editor, an open source 2D vector graphics editor, based on Java programming language. The Chorem Editor consists of two working areas, namely a property window and a visualization window, and a toolbar containing both a set of buttons and a tabbed list by which functionality may be invoked. In particular, the property window allows users to interact with and modify chorem properties, also affecting the visual representation. Analogously, the visualization window, which is meant at displaying the chorem map under construction, allows users to manipulate its graphic components, also affecting properties displayed into the property window.

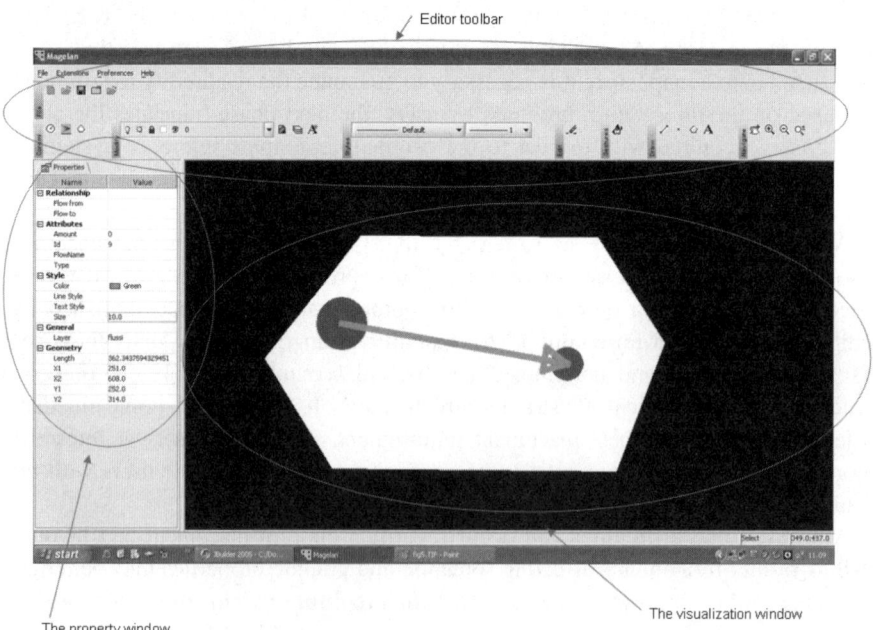

Fig. 11. The visual interface of the Chorem Editor

3.4 ChorML

A language was designed to store chorems. Based on XML, ChorML presents several levels:

- level 0 corresponds to the initial database in GML (Geographic Markup Language) [GML],
- level 1 corresponds to the list of extracted patterns
- level 2 corresponds to positioned chorems
- level 3 is a subset of SVG.

For instance, in level 1, the feature coordinates can be longitude/latitude and feature attributes, whereas in level 2, we deal with pixel coordinates, line styles, colors and textures.

4 Final Remarks

The objective of this paper was to give some elements for the visual summarizing of spatial databases based on automatic discovery and layout of chorems. After a rapid analysis of existing manually-made chorems, some guidelines were exhibited, so that a prototype architecture can be proposed.

Regarding architecture, some modules have already been written and tested (for instance the chorem editor) whereas the specifications of the ChorML language and of the canonical database structure must be finalized.

Present experimentations are processed based on ORACLE 10g using data from Italy. Next study will extract chorems from an historical database of the Mexican city of Puebla during the XVIIth and the XVIIIth centuries.

References

Brunet, R.: La carte-modèle et les chorèmes. Mappemonde 86(4), 4–6 (1986)
Brunet R.: Les fondements scientifiques de la chorématique. In: La démarche chorématique, Centre d'Études Géographiques de l'Université de Picardie Jules Verne (1993)
Buttenfield, B., McMaster, R.: Map Generalization: Making Rules for Knowledge Representation. Longman, London (1991)
Ester, M., Kriegel, H.-P., Sander, J.: Spatial Data Mining: A Database Approach. In: Scholl, M.O., Voisard, A. (eds.) SSD 1997. LNCS, vol. 1262, pp. 47–66. Springer, Heidelberg (1997)
GML, http://www.opengis.net/gml/
Holder, L.B., Cook, D.J.: Graph-based Data Mining. In: Wang, J. (ed.) Encyclopaedia of Data Warehousing and Mining, Idea Group Publishing (2005)
Jones, C.B.: Cartographic Name Placement with Prolog. IEEE Computer Graphics and Applications 9(5), 36–47 (1989)
Lafon, B., Codemard, C., Lafon, F.: Essai de chorème sur la thématique de l'eau au Brésil (2005), http://histoire-geographie.ac-bordeaux.fr/espaceeleve/bresil/eau/eau.htm

Laurini, R., Milleret-Raffort, F., Lopez, K.: A Primer of Geographic Databases Based on Chorems. In: Meersman, R., Tari, Z., Herrero, P. (eds.) On the Move to Meaningful Internet Systems 2006: OTM 2006 Workshops. LNCS, vol. 4278, pp. 1693–1702. Springer, Heidelberg (2006)

Pech Palacio, M., Sol Martinez, D., González, J.: Adaptation and Use of Spatial and Non-Spatial Data Mining. In: Proceeding of International Workshop Semantic Processing of Spatial Data (GEOPRO 2002), Centre for Computing Research, Instituto Politécnico Nacional, México (December 2002)

Shneiderman, B.: Designing the User Interface, 3rd edn. Addison-Wesley, Reading (1997)

SVG, http://www.w3.org/Graphics/SVG/

Saint-Paul, R., Raschia, G., Mouaddib, N.: General Purpose Database Summarization. In: Int. Conf. on Very Large Databases (VLDB 2005), Trondheim, Norway, pp. 733–744. Morgan Kaufmann Publishers, San Francisco (2005)

USA, http://lettres-histoire.info/lhg/geo/geo_usa/cartes_usa_geo.htm

Compound Geospatial Object Detection in an Aerial Image

Yi Xiao, Brian A. Hope, and David Tien

Charles Stuart University, Australia
Department of Lands, NSW, Australia
yix@ee.usyd.edu.au

Abstract. This paper introduces a knowledge based approach that can be used for the identification of jetty/bridge locations in aerial imagery. With the proposed method, the semantic network formalism to represent declarative knowledge embodied in a jetty/bridge image and the appropriate procedural knowledge, the control procedure was established. A knowledge based system was then introduced through image analysis and interpretation, aiming at accurately locating the desired objects from primary vague identification. With the advanced image processing techniques proposed here, the complexity of using knowledge based system for image analysis is reduced and the proposed method can effectively locate the compound geospatial objects of jetties and bridges.

Keywords: compound geospatial object detection, knowledge based system.

1 Introduction

Compound geospatial objects, such as jetties, harbours, airports and factories, are characterised by several sub-objects and their spatial layout. For example, harbours contains boats and sea water while a parking lot contains cars, ground and trees, both with a distinct spatial arrangement. The conventional recognition methods such as statistical techniques and object model analysis such as template matching, shape and texture analysis [1, 2] focus on internal characteristics of an object. They are limited for objects with simple, unique geometric or shape models, e.g. buildings and roads [3-6]. Bhagavathy and Manjunath [7] proposed a model combining space and frequency analysis techniques developed in the framework of texture analysis or the detection of compound geospatial objects. The limitation is the difficulty of detecting compound objects with various spatial and frequent relations. To recognise efficiently compound geospatial objects that have complex, various spatial structures, knowledge about their surrounding environments (context) is required. A number of knowledge based methods have been proposed for describing and identifying the constituents and layout of compound geospatial objects. These methods usually divide an image into primitives (closed regions, lines, etc.) through image segmentation or edge detection/linking. Spatial relations between the primitives are analysed using different models constructed on the knowledge of identifying the target object. For example, in ACRONYM [8], models of objects are symbolically represented by "frames" for detecting complex 3-D objects like an airplane. SIGMA [9] recognises simple artificial objects such as a regular residential area. It consists of three experts: the low-level vision expert for knowledge-based image segmentation, the model selection expert for appearance model selection, and the geometric reasoning expert for spatial reasoning. Nicolin and Gabler [1] describe a knowledge-based system for the automatic interpretation of panchromatic aerial images of suburban scenes, emphasising on knowledge representation and control strategy. MOSES [10] reports the recognition of buildings and parking lots in aerial images. Its models are automatically refined by using knowledge gained from topographical maps or GIS data.

A knowledge-based object recognition system is object-oriented, not only class-specific knowledge but also site-specific knowledge and knowledge of corresponding image processing techniques are required. In this paper, we propose a knowledge-based object recognition system for the location of geospatial objects over water such as jetties and bridges. The internal structure and context aspect of these objects are analysed while going through the recognition processing. The internal structure of an object is presented and identified in primitive extraction stage with advanced image processing techniques and its context aspect is handled in the semantic network. The fusion of multiple features is carried out for the extracted image segments. The semantic network is built up with the knowledge that leads to a specialised scene of a jetty/bridge description. Besides of knowledge representation, the valuation processes as well will be modelled by a mechanism for mixed bottom-up and top-down interpretation of the image.

2 Object Modelling and Detecting

2.1 Object Definition

Based on the general knowledge, we know that a jetty/bridge in an aerial image is a bright and narrow structure, adjacent/surrounded by water. With the specific knowledge of current image processing capabilities, it is described in a concept as a bright, thin structure embedding in the thin, gulfy/isthmus area(s) of water.

2.2 Initial Segmentation for Primitive Extraction

Segmentation of bright and thin structures
The extraction of bright and thin structures is from the combination of intensity information, I_n and edge information, E_d, with a global [11] thresholding segmentation for I_n and canny edge detection, run-length smearing [12] for E_d. Global thresholding segmentation separates the regions with bright pixels in the image from others. The Canny edge detector locates sharp intensity changes. Two thresholds are used to control the strength of ridge of a point to be extracted. Runlength smearing method links the narrow foreground connected components, with a threshold for the size of small background runs to smear.

The segmentation maps $I_{\_n}$ and $E_{\_d}$ are fused based on the intersection, shown in Eq. (1).

$$R_{\_b} = I_{\_n} \cap E_{\_d} \mid s_i > s_{\_th1} \ (i=1,2,...n) \tag{1}$$

Where s_i is the area of the ith region in $I_{\_n} \cap E_{\_d}$. After this operation, only the bright regions with sharp intensity variation remain.

The extraction of thin structures is based on a morphological 'open' operation [11] on the regions in $R_{\_b}$. Denote a region in $R_{\_b}$ as $r_{\_b}$, the thin structures $r_{\_s}$ in $r_{\_b}$ are obtained by the operation shown in Eq. (2).

$$r_{\_s} = r_{\_b} - (r_{\_b} \ominus B_{\_1}) \oplus B_{\_1} \tag{2}$$

Where $\oplus$ denotes the dilation operation and $\ominus$ erosion operation. $B_{\_1}$ is the structure element in the morphological operation, its size depends on the width of the thin structure to be extracted.

Segmentation of water areas

A combination of colour information, $C_{\_o}$ and textural information, $T_{\_e}$, is used for water area segmentation, with a nearest neighbour rule classification [13] for $C_{\_o}$ and garbo filtering for $T_{\_e}$.

Giving the segmentation maps $C_{\_o}$ and $T_{\_e}$ for an image, water regions are then chosen from $C_{\_o}$ based on the intersection, as is shown in Eq. (3).

$$R_{\_w} = C_{\_o} \cap (1 - T_{\_e}) \mid s_i > s_{\_th2} \ (i=1,2,...n) \tag{3}$$

Where s_i is the area of the ith region in $C_{\_o} \cap T_{\_e}$. With the intersection, regions of trees that share colours with water regions will be discarded.

Gulfy/isthmus area extraction

The extraction of gulfy/isthmus areas are based on a morphological 'close' operation [11]. Denote a region in $R_{\_w}$ as $r_{\_w}$, the gulfy/isthmus areas $r_{\_g}$ in $r_{\_w}$ are obtained by the operation shown in Eq. (4).

$$r_{\_g} = (r_{\_w} \oplus B_{\_1}) \ominus B_{\_1} - r_{\_w} \tag{4}$$

2.3 Context Aspect Representation by a Semantic Network

We use a semantic network as implemented in ERNEST [14] to offer a compact formalism for context aspect representation and tools for valuation of features within the identification process. The semantic network is in the form of directed acyclic graph consisting of nodes and links. The nodes represent the concepts expected to appear in the scene with respect of the object while the links of the semantic network form the relations between these concepts (nodes). In which part links represent the relations between a concept and its components, concrete links connect concepts in different conceptional systems. Data structures called attributes are used to describe the nodes for specifying and storing properties of concepts.

The general structure of the semantic networks of a jetty/ bridge is shown in Fig. 1. It consists of nodes and relation slots. Relations are characterized by attributes.

Objects which occur for identifying a jetty/bridge are described by specifying their concretizations and relations with other objects. For example the concepts water and bright and thin structure are specified as parts of a jetty. Attributes of the water are, for example, water's size, texture and colour. By describing an object with the help of its parts, a hierarchical, structural model is constructed. We specify topological relations such as *overlap* and *cover* for attributions about the neighbouring of objects.

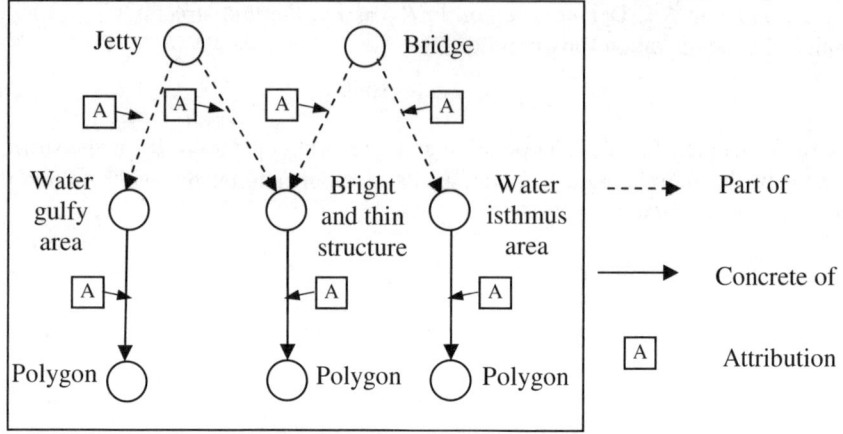

Fig. 1. Semantic network for jetty/bridge detection

The textual descriptions of the concept 'jetty' and 'bridge' are shown in Fig. 2 (a and b). The gulfy areas and the isthmus area are differentiated by *overlap* and morphological *dilation* operation.

```
JETTY
PART: bright and thin structure
   GOAL_NODE:      BRIGHT, THIN STRUCTURE
PART: water gulfy area
   GOAL_NODE:      WATER GULFY AREA
CONCRETE: geometry
GOAL_NODE:       polygon
ATTRIBUT: size of an area
   TYPE_OF_VAL:    REAL
   NUMB_OF_VAL:    1
   RESTRICTION:    comparable to a jetty
   COMP_OF_VAL:    compute_size
   ARGUMENT:       water_gulfy.area, bright and thin
structure.area
TOPOLOGICAL_RELATION:   bright and thin
area_overlap_gulfy area
   VALUE:          1.0
JUDGEMENT:         0.8
```

Fig. 2 (a). Concept of 'jetty'

```
BRIDGE
PART: bright and thin structure
   GOAL_NODE:        BRIGHT AND THIN STRUCTURE
PART: water isthmus area
   GOAL_NODE:        WATER ISTHMUS AREA
CONCRETE: geometry
GOAL_NODE:          polygon
ATTRIBUT:  size of an area
   TYPE_OF_VAL:     REAL
   NUMB_OF_VAL:     1
   RESTRICTION:     comparable to a bridge
   COMP_OF_VAL:     compute_size
   ARGUMENT:        water_isthmus.area, bright and thin
structure.area
TOPOLOGICAL_RELATION:  bright and thin
area_overlap_isthmus area
   VALUE:           1.0
   JUDGEMENT:       0.85
```

Fig. 2 (b). Concept of 'bridge'

2.4 Analysis Processing

In the realisation of the knowledge structure, the control strategy in [14] is applied. It is a mixed bottom-up and top-down search algorithm within the semantic network. The strategy of analysis process is general and problem independent, provided by the semantic network. The analysis starts by creating a goal concept. Following the hierarchy in the semantic network in a top-down way, the concepts on lower hierarchical levels are expanded stepwise until a concept on the lowest level is reached. Since these concepts do not depend on other concepts, its correspondence with an extracted segment from the image can be established and its attributes can be calculated. This is called *instantiation*. Analysis then moves bottom-up to the concept at the next higher hierarchical level. If instances have been found for all parts of this concept, the concept itself can be instantiated. Otherwise the analysis continues with the next concept not yet instantiated on a lower level. After an instantiation, the acquired knowledge is propagated bottom-up and top-down to impose constraints and restrict the search space. Thus, in the analysis process, top-down and bottom-up processing alternates.

The entire analysis process of jetty/bridge search is illustrated in Fig. 3. The initial segmentation of water and bright and thin structure is illustrated in Fig. 3 (b-e). Starting from creating a goal concept of "jetty", the concept depends on two concepts of "water gulfy area" and "bright and thin structure". The analysis moves down onto the level with the concept of "water gulfy area", which is the lowest level in the network. Its correspondence with an extracted segment of water from the image is established and its attributes are calculated. The attributions of water gulfy area are examined in a bottom-up way. If the condition of the rule is satisfied, that is, the water gulfy/isthmus area is a water gulfy area and the water_gulfy.area is within the required range, then the concept of "water gulfy area" is instantiated. Once the concept of water gulfy area is instantiation, the second step is a bottom-up instantiation of expanded concept of bright and thin structure, in a top-down way same to water gulfy area instantiation, where the condition of being a bright and thin structure is checked. If instances have been found for all parts of "jetty", and the topological relation between the parts is matched, the concept itself is instantiated.

Fig. 3. (a)

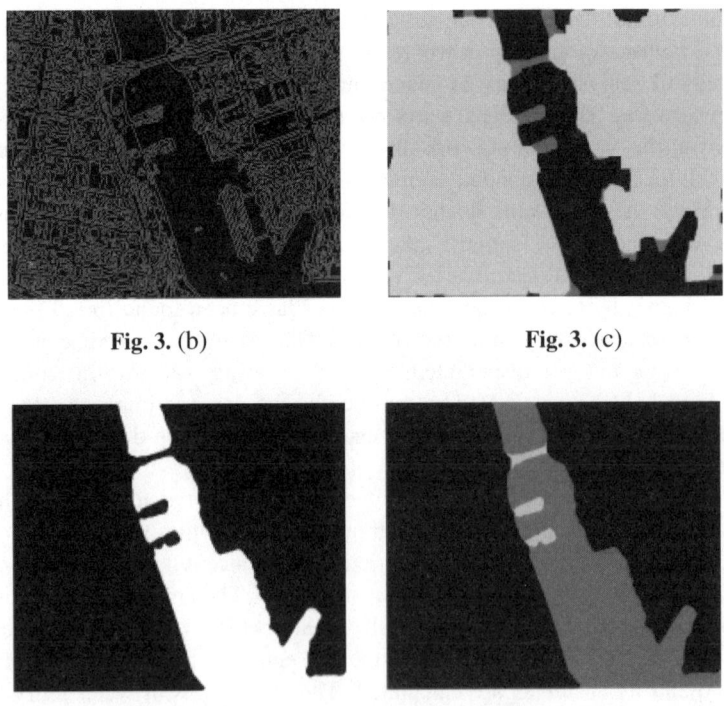

Fig. 3. (b)   **Fig. 3.** (c)

Fig. 3. (d)   **Fig. 3.** (e)

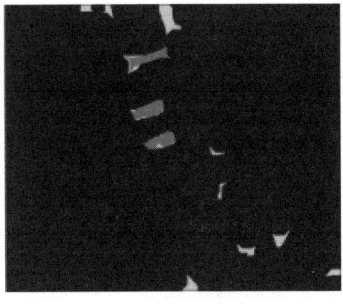

Fig. 3. (f) **Fig. 3.** (g)

Fig. 3. Jetty/bridge search procedure. (a) Testing image; (b) Edge information; (c) Bright and thin structure; (d) Water area; (e) Water gulfy/isthmus area; (f) topological relation computing; (g) Jetty/bridge location.

3 Data Evaluation

True colour aerial images with a capture resolution of *1* m were used for evaluation and included water areas around Raymond Terrace, Tweed Heads, Ulladulla, Wollongong, Newcastle and Sydney in New South Wales, Australia. The parameter values used in the experiment are summarised in Table 1. B_{th} is the thresholding value in global thresholding segmentation. s_{th1} and s_{th2} is set to a value of the minimum size of a jetty and a water region respectively.

Table 1. Parameter setting

Parameter	B_1	s_{th1}	s_{th2}	B_{th}*
Our method	20×20	300	10000	Image dependant

* B_{th} is obtained from the intensity histogram of the image, see the detail in [11].

A variety of jetty and bridge shapes over water were located out of 132 jetties and 7 bridges. Some examples are shown in Fig. 4. Missed jetties mainly occurred where the jetties are too dark and errors occurred in waves close to the shoreline. However, the procedure achieved over 80% detection rate for jetties and bridge identification from the aerial images mentioned above.

There are two main factors affecting the accuracy of the approach: knowledge representation and image processing techniques adopted.

Varieties of image processing techniques have been developed as tools to analyse an image. However, it has been shown that none of the image processing operators is perfect: some important features are not extracted and/or erroneous features are detected. E.g., errors exist in water area segmentation, as in some shallow water areas where colour of water is similar to that of a jetty. For thin structure extraction, when the size of structure element B_1 in the morphological operation increases, some corners of shores might be included; when B_1 decreases, some broad bridges or jetties with boats attached might be missed.

Fig. 4. The locations of jetties extracted are bounded with white lines

The performance of the method also depends on the quality and amount of knowledge it uses. As the approach used didn't cover all the cases, such as waves close to shores and roads extending to water, only some of these objects were extracted in the experiment.

To evaluate the results numerically, the criteria of *completeness* and *correctness* [15] are used to assess the accuracy of the object extraction. The extraction results are classified as true positive (TP), true negative (TN) or false positive (FP). TP is the number of jetties detected manually and by the proposed method, FP is the number of jetties detected by the proposed method but not manually, and TN is the number of jetties detected manually but not by the proposed method. *Completeness* and *correctness* are defined as follows:

$$completeness = \frac{TP}{TP+TN} \qquad (5)$$

$$correctness = \frac{TP}{TP+FP} \qquad (6)$$

completeness measures the ability to find the object regions and *correctness* measures the accuracy of object detection.

completeness and *correctness* of the proposed method for the extraction of jetties were 83% and 92% respectively. Comparison was made to a method using multivariate analysis of spectral, textural, and shape based features [6] (see the results in Table 2).

Table 2. Jetty location accuracy with two different methods

Method	*completeness*	*correctness*
Multivariate analysis [6]	91%	50%
Our method	87%	92%

Both methods showed good abilities in to capturing the objects. Our method excluded a bit more of the dark and small jetties than multivariate analysis did, which is caused by the different colour based segmentation method adopted. However, multivariate analysis method yielded a greater amount of false alarms than our method did. As when using multivariate analysis, it is hard to distinguish between many roads, buildings and jetties because they have similar internal features. Thus, our method is more accurate than multivariate analysis for compound object detection.

4 Conclusion

The combinations of advanced image processing techniques with artificial intelligent techniques provide an efficient way to identify compound geospatial objects in aerial imagery. In the proposed method, task-specific knowledge to represent models of jetties and bridges is given by a semantic network. In the extraction of primitives, more concrete concepts are represented with the advanced image processing techniques, making the knowledge representation based on semantic network and its associated decision process more simple than previous methods.

The performance of the method relies on the quality and amount of knowledge it uses and image processing techniques adopted.

Acknowledgement

The work has been supported by the Cooperative Research Centre for Spatial Information, whose activities are funded by the Australian Commonwealth's Cooperative Research Centres Programme.

References

1. Nicolin, B., Gabler, R.: A knowledge-based system for the analysis of aerial images. IEEE Trans. Geosci. Remote Sens. GRS-25(3), 317–329 (1987)
2. Eakins, J.P.: Towards intelligent image retrieval. Pattern Recognition 35, 3–14 (2002)
3. Fradkin, M., Maitre, H., Roux, M.: Building Detection from Multiple Aerial Images in Dense Urban Areas. Computer Vision and Image Understanding 82, 181–207 (2001)
4. Tupin, F., Roux, M.: Detection of building outlines based on the fusion of SAR and optical features. ISPRS Journal of Photogrammetry & Remote Sensing 58, 71–82 (2003)
5. Peteri, R., Celle, J., Ranchin, T.: Detection and extraction of road networks from high resolution satellite images. International Conference on Image Processing 1, 301–304 (2003)
6. Knudsen, T., Nielsen, A.A.: Detection of buildings through multivariate analysis of spectral, textural, and shape based features. In: Proceedings of IEEE International Geoscience and Remote Sensing Symposium, 2004, vol. 5, pp. 2830–2833 (2004)
7. Bhagavathy, S., Manjunath, B.S.: Modeling and Detection of Geospatial Objects Using Texture Motifs. IEEE Trans. Geosci. Remote Sens. 44(12), 3706–3715 (2006)
8. Brooks, R.A.: Symbolic reasoning among 3-D models and 2-D images. Artificial Intell. 17, 285–348 (1981)
9. Matsuyama, T.: Knowledge-Based Aerial Image Understanding Systems and Expert Systems for Image Processing. IEEE Trans. Geosci. Remote Sens GRS-25(3), 305–316 (1987)

10. Quint, F.: Recognition of structured objects in monocular aerial images using context, in Mapping Buildings, Roads and Other Man-Made Structures From Images. In: Leberl München, F. (ed.) Institut für Photogrammetrie und Fernerkundung. Universität Karlsruhe, Karlsruhe, Germany, pp. 213–228 (1997)
11. Gonzalez, R.C., Woods, R.E.: Digital Image Processing. Addison-Wesley Longman Publishing Co., Inc., USA (1992)
12. Wahl, F., Wong, K., Casey, R.: Block segmentation and text extraction in mixed text/image documents. CGIP 20, 375–390 (1982)
13. Cover, T.M., Hart, P.E.: Nearest Neighbor Pattern Classification. IEEE Transactions on Information Theory IT-13(1), 21–27 (1967)
14. Niemann, H., Sagerer, G.F., Schroder, S., Kummert, F.: ERNEST: A Semantic Network System for Pattern Understanding. IEEE Trans. Pattern Anal. Mach. Intell. 12(9), 883–905 (1990)
15. Heipke, C., Mayer, H., Wiedemann, C., Jamet, O.: Evaluation of Automatic Road Extraction. International Archives of Photogrammetry and Remote Sensing 32(3), 47–56 (1997)

Texture Representation and Retrieval Using the Causal Autoregressive Model

Noureddine Abbadeni

Abu Dhabi University
College of Engineering and Computer Science
PO Box 59911, Abu Dhabi, UAE
noureddine.abbadeni@usherbrooke.ca

Abstract. In this paper we propose to revisit the well-known autoregressive model (AR) as a texture representation model. We consider the AR model with causal neighborhoods. First, we will define the AR model and discuss briefly the parameters estimation process. Then, we will present the synthesis algorithm and we will show some experimental results. The causal autoregressive model is applied in content-based image retrieval. Benchmarking conducted on the well-known Brodatz database shows interesting results. Both retrieval effectiveness (relevance) and retrieval efficiency are discussed and compared to the well-known multiresolution simultaneous autoregressive model (MRSAR).

Keywords: Texture, Content Representation, Image Retrieval, Autoregressive Model.

1 Introduction

The autoregressive (AR) model has been used in different works related to texture classification and segmentation [9], [12], texture synthesis [16] and texture retrieval [10]. Sondge [16] studied different models, among them the AR model, in order to synthesize textures. He considered the simultaneous model (SAR) as well as the separable model, which is a particular case of the SAR model. Kashyap et al. [9] proposed a circular autoregressive model (CAR) in order to take into account rotation invariance. Mao and Jain [12] proposed a multiresolution AR model (MRSAR) which consists in the use of multiple resolutions in order to capture different scales of the texture. They also proposed a rotation-invariant simultaneous autoregressive model (RISAR) to take into account rotation invariance. In the framework of image retrieval, Liu et al. [10] used the MRSAR proposed by Mao and Jain [12] for texture retrieval. They benchmarked the MRSAR model against the Wold model and their results show that the MRSAR gives interesting results even if the Wold model is shown to give slightly better results.

Given their nature as random fields, such models, generally, work better for random textures than regular textures. A common problem with the use of these models is the choice of a neighborhood in which pixels are considered as related to each other. Generally speaking, it is better to choose a small neighborhood for a fine texture and a large neighborhood for a coarse texture. However, in practice there is another problem which arises with large neighborhood. In fact, when the neighborhood is large, and due to the averaging effect phenomenon [12], the discrimination power of the estimated parameters tends to decrease. This phenomenon is related to the curse of dimensionality problem as reported by Mao and Jain [12]. So, such models work better for fine textures. This has also some link with the regularity / randomness of textures. In fact, coarse textures are generally perceived as regular while fine textures are perceived as random, and as it has been already mentioned, the autoregressive model works better, by definition, with random textures.

Another problem associated with such models is the fact that their efficiency is altered with large neighborhoods. In fact, the size of the set of estimated parameters becomes much larger as the neighborhood becomes larger. Also, the more the size of the set of estimated parameters is large, the more the computational cost to extract those parameters is high.

In this paper, we propose to reconsider the SAR model in its causal version without considering multiple resolutions. Causality is not a natural constraint in a bidimensional (2D) space. However, this causality allows to simplify both the model and the synthesis process. Using causal neighborhoods allows to reduce the number of parameters and, thus, the efficiency of parameters extraction. We consider both non-symmetric half-plane (NSHP) and quarter-plane (QP) neighborhoods. Also, we consider small neighborhoods which allows to reduce both the number of parameters and the computational cost necessary to extract parameters. We will show that, when the causal SAR model with small neighborhood is applied in content-based image retrieval, the loss in terms of search effectiveness (relevance) is acceptable since there is an important gain in terms of search efficiency.

The rest of the paper is organized as follows: In section 2, we will define briefly the model; In section 3, we will depict the texture synthesis process; In section 4, some experimental results are presnted and the efficiency of the model is discussed; In section 5, the causal model, with both QP and NSHP neighborhoods, is applied to content-based image retrieval and compared to the MRSAR model used by Liu et al. [10] in terms of search relevance as well as search efficiency; And finally, in section 6, a conclusion is given as well as a brief mention of future investigations related to this work.

2 SAR Definition

The simultaneous (2D) autoregressive model (SAR) is defined as follows:

$$(X_s - \mu) = a_s W_s + \sum_{r \in \Omega^+} a_r (X_{s+r} - \mu) \qquad (1)$$

where s corresponds to position (i, j) on rows and columns, X_s is the grey-level at position s, Ω^+ is the neighborhood on rows and columns of X_s (excluding

X_s itself), $\Omega = \Omega^+ \cup \{s\}$, μ is the local grey-level average in the neighborhood Ω and $[a_s, a_r, r \in \Omega^+]$ are the parameters of the model to be estimated. W_s is a Gaussian white noise, a stationary signal made of non-correlated random variables.

Neighborhood Ω can be defined in different ways. We use causal neighborhoods. The causality constraint means that pixels are ordered in a sequential way (from top to bottom and from left to right). There are two causal neighborhoods: quarter-plan (QP) neighborhood and non-symmetric half-plan (NSHP) neighborhood (pixel (0,0) is called pixel of interest in the rest of this paper). To estimate the set of parameters $[a_s, a_r, r \in \Omega^+]$, we used the well-known least squares error (LSE) method [3].

Finally, let us mention that, in [1], we proposed a perceptual interpretation of the parameters of the AR model. In fact, we propsed to use the mean (or the sum) of these parameters as a measure of the global degree of randomness/regularity of a texture: the more this mean (or sum) of parameters is high, the more the texture is regular; and the more this mean (or sum) is low, the more the texture is random.

3 Texture Synthesis

An interesting propriety of the autoregressive model is the possibility to synthesize the texture using the set of estimated parameters on the same texture. The visual similarity between the original texture and the synthesized one is relatively good (experimental results are shown in section 4). The main steps of the synthesis algorithm can be described as follows (Fig. 1):

- The synthesis algorithm takes as input the set of estimated parameters $[a_r, r \in \Omega^+]$ and a_s as well as the grey-level mean μ of the original image.

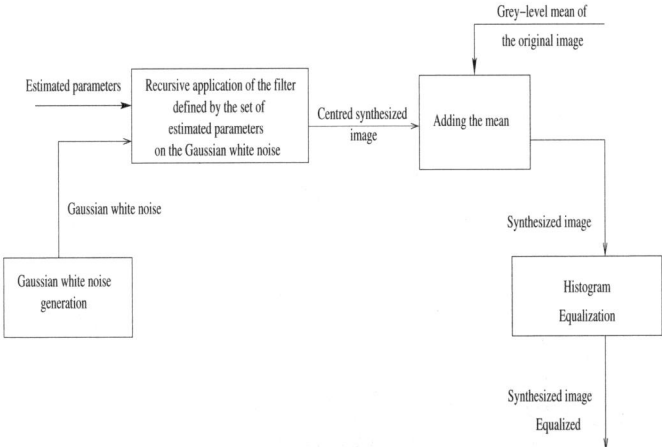

Fig. 1. Main steps in the synthesis algorithm using the set of estimated parameters

- A Gaussian white noise of mean 0 and variance 1 is generated (we can use a different value for the variance).
- The white noise is used in input and is filtered using the set of estimated parameters. The values of pixels are computed in a recursive way according to equation (1).
- At the end of the recursive computations, the grey-level mean of the original image is added to the synthesized image.
- A histogram equalization operation can be applied on the synthesized image for better visualization.

Note that the synthesis process is Gaussian for any neighborhood since the white noise used is Gaussian.

4 Experimental Results

4.1 Estimated Parameters and Synthesized Textures

Tables 1 and 2 give the estimated parameters for a set of 12 original images (from Brodatz database) using QP(1,1) and NSHP(1,1) neignborhoods. Fig. 2 and Fig. 3 show examples of synthesized images according to the synthesis process described above. From these experimental results, we can point out that, when considering larger neighborhoods ((2,2) and (3,3)), for both QP and NSHP versions, values of parameters that are not in the immediate neighborhood of pixel of interest (0,0) are often small compared to those in the immediate neighborhood of pixel (0,0). Of course, pixels that are not in the immediate neighborhood of pixel (0,0) have some importance, but this importance is always less than the importance of pixels in the immediate neighborhood of pixel (0,0). This means that if we consider only pixels that are in the immediate neighborhood of pixel (0,0), we can have the main parameters that capture a large part of the textural information.

Table 1. Estimated parameters for a neighborhood QP(1,1)

Images	$a(0,-1)$	$a(-1,0)$	$a(-1,-1)$
D68	0.4204	0.9621	-0.4036
D36	0.5544	0.7354	-0.3417
D82	0.7713	0.7313	-0.6216
D77	0.6372	0.6071	-0.5619
D95	0.8965	0.7496	-0.6610
D93	0.2799	0.7119	-0.2938
D49	0.9959	0.7038	-0.6987
D110	0.4962	0.6164	-0.2412
D9	0.5265	0.6361	-0.2889
D111	0.6954	0.6236	-0.3765
D24	0.4571	0.7063	-0.2470
D92	0.6560	0.5668	-0.3025

Table 2. Estimated parameters for a neighborhood NSHP(1,1)

Images	$a(0,-1)$	$a(-1,1)$	$a(-1,0)$	$a(-1,-1)$
D68	0.4229	0.0212	0.9449	-0.3995
D36	0.5518	0.1188	0.6110	-0.2979
D82	0.7725	0.0403	0.6986	-0.6179
D77	0.6306	0.1972	0.4195	-0.4631
D95	0.8979	0.1137	0.6133	-0.6326
D93	0.2952	0.4279	0.4145	-0.2391
D49	0.9958	-0.0150	0.7207	-0.7006
D110	0.4922	0.0853	0.5526	-0.2257
D9	0.5225	0.0848	0.5710	-0.2717
D111	0.6853	0.1517	0.4594	-0.3273
D24	0.4577	0.0553	0.6618	-0.2306
D92	0.6512	0.1319	0.4367	-0.2685

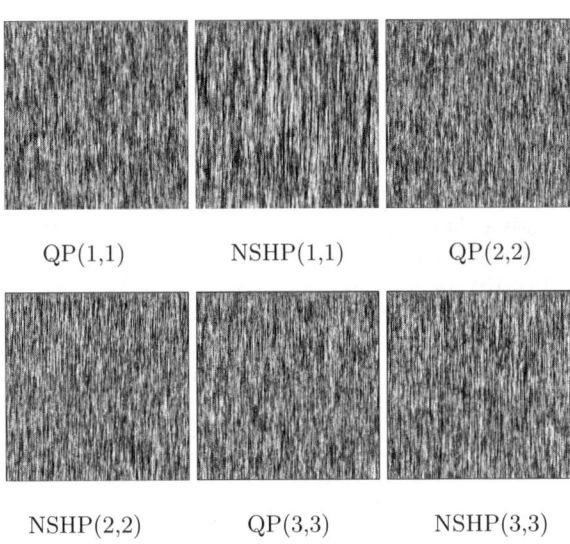

QP(1,1)   NSHP(1,1)   QP(2,2)

NSHP(2,2)   QP(3,3)   NSHP(3,3)

Fig. 2. Synthesized images corresponding to D68 for different neighborhoods

4.2 Efficiency

For a 2D AR model with a neighborhood QP(p,q), the number of parameters is $(p+1)(q+1) - 1$. For a 2D AR model with a neighborhood NSHP(p,q), the number of parameters is $(p+1)(q+1) - 1 + pq$. Table 3 shows the number of parameters to be estimated with respect to the considered neighborhood. For a neighborhood QP(1,1), the number of parameters is 3 while for a NSHP(1,1), the number of parameters is 4. For a neighborhood QP(5,5), the number of parameters is 35 while for a neighborhood NSHP(5,5) the number of parameters

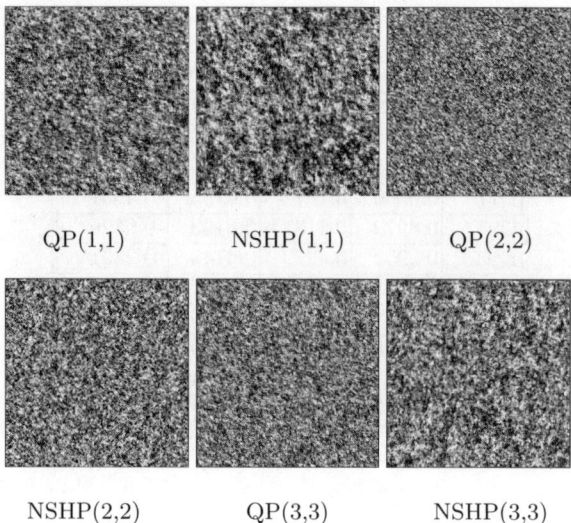

QP(1,1)　　　　　　NSHP(1,1)　　　　　　QP(2,2)

NSHP(2,2)　　　　　　QP(3,3)　　　　　　NSHP(3,3)

Fig. 3. Synthesized images corresponding to D111 for different neighborhoods

is 60. The number of parameters increases in an exponential way with the size of the considered neighborhood.

The time of parameter extraction increases also in an exponential manner with the size of the neighborhood. For example, on a 750MHZ processor, parameter extraction from a 640X640 image with a QP (1,1) model takes less than 1 second while with a QP(10,10) model it takes several minutes.

Table 3. Number of parameters according to different neighborhoods

Neighborhood	QP	NSHP
(1, 1)	3	4
(2, 2)	8	12
(5, 5)	35	60
(10, 10)	120	220
(20, 20)	440	840

When increasing the order of the model, the size of the set of parameters becomes very important and feature extraction takes more time. Increasing the order of the model may achieve some improvement in visual quality of the synthesized images but the price in terms of feature extraction complexity is high. From an image retrieval perspective, it is preferable to consider a small neighborhood for a better efficiency (in terms of number of parameters and parameter extraction complexity). Thus, we decided to use only order (1,1). This means that we have a set of four (4) estimated parameters in the case of NSHP neighborhood and three (3) in the case of QP neighborhood (Table 3). Retrieval experimental results presented later in this paper show that this version of the autoregressive model performs well.

5 Application to Content-Based Image Retrieval

To perform an image retrieval experience, we must dispose of a similarity function in order to compare images. We have used a similarity measure based on Gower's similarity model we have developed in our earlier work [2]. We used it in two versions: a non-weighted version and a weighted version. In the case of the weighted version, the inverse of variance of each estimated parameter was used as a weight for the corresponding feature. In the rest of this paper we use the following notations:

- **QP, QP-V**: QP denotes the non-weighted autoregressive model with a quarter-plan neighborhood and with order (1,1) while QP-V denotes the same model except that each parameter was weighted by the inverse of its variance.

- **NSHP, NSHP-V, DPNS, DPNS-V**: NSHP (or DPNS) denotes the non-weighted autoregressive model with a non-symmetric half-plane neighborhood and with order (1,1) while NSHP-V (DPNS-V) denotes the same model except that each parameter was weighted by the inverse of its variance.

5.1 Retrieval Results

Fig. 4 shows search results obtained for query image D68-1 using the QP-V model. Results are presented in a decreasing order based on the score of similarity with the query image. These results show that the considered models give very good results. The returned images, when they are not in the same class as the query image, generally present a good visual similarity with the query image with respect to some aspects, in particular with respect to the degree of randomness/regularity of images as explained earlier in this paper. For example, results shown in Fig. 4 are all similar in terms of regularity and orientation even if they do not belong to the same classes.

5.2 Recall Measure

Fig. 5 shows the recall (or retrieval) graph for different versions of the AR model (QP, QP-V, NSHP and NSHP-V). The recall plotted in this figure are an average computed over 83 classes among 112 classes in Brodatz database. We have rejected the 29 highly non homogeneous classes, since images within these classes are not visually similar, in order to avoid misleading conclusions. One can point out from that figure that the non-weighted NSHP (or DPNS) version performs better than the other versions. Weighting, using variance, in both cases QP and NSHP, did not allow to improve performance.

5.3 Comparison with the MRSAR Model

Table 4 gives the average recall rate computed over all of the 112 classes and over only 83 classes (excluding 29 highly non homogeneous classes). It gives

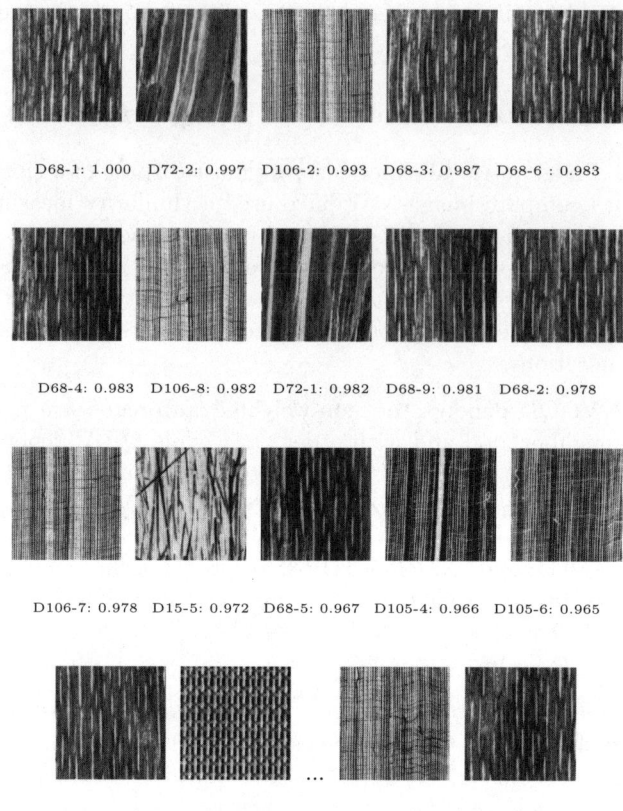

D68-1: 1.000   D72-2: 0.997   D106-2: 0.993   D68-3: 0.987   D68-6 : 0.983

D68-4: 0.983   D106-8: 0.982   D72-1: 0.982   D68-9: 0.981   D68-2: 0.978

D106-7: 0.978   D15-5: 0.972   D68-5: 0.967   D105-4: 0.966   D105-6: 0.965

D68-8: 0.965   D53-3: 0.962   D106-6: 0.953   D68-7: 0.953

Fig. 4. Search results returned for query D68-1 using the AR QP-V model. Images are ordered according to their similarity score with the query image.

also the average recall rate for the main related work, the MRSAR model as benchmarked in [10] . Table 5 gives the time of parameter extraction and the size of the parameter vector for our model and the MRSAR model as benchmarked in [10]. From these two tables, we can draw the following conclusions: 1. In terms of search relevance, the MRSAR model outperforms the causal AR model. The performance of our model is still acceptable, and when we consider positions 50 and 100, for example, the difference between our model and the MRSAR model is not very significant; 2. In terms of efficiency, our model outperforms the MRSAR model in both parameter extraction time and parameter vector size.

In the perspective of application of the AR model in a "real-time" image retrieval application, search efficiency is very important. In this way, it is difficult to apply the MRSAR model in real-time applications since it needs 34 seconds to extract the necessary parameters while our model is applicable can be applied since it takes less than 1 second to extract parameters. Also, the size of the parameter vector has a big importance in retrieving similar images from

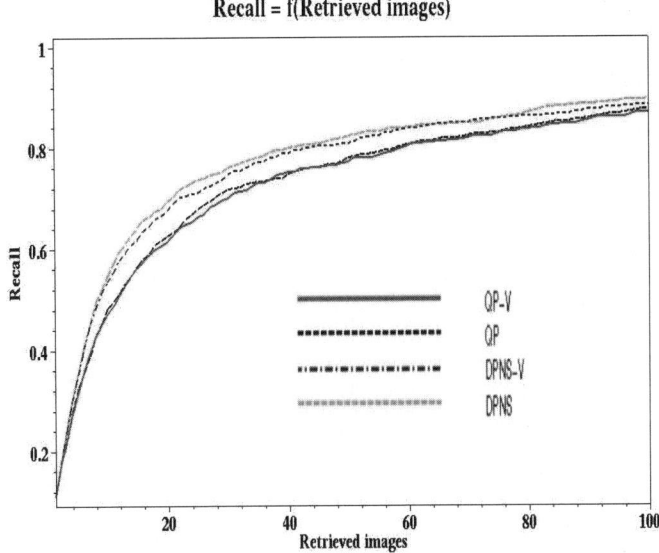

Fig. 5. Recall graph for different versions of the AR model (Note that DPNS corresponds to NSHP)

the database. In fact, some works have already shown that search efficiency are inversely proportional to the size of the parameter vector [18]. The size of parameter vector of our model is 4 against 15 for the MRSAR model. We can say that our model is better than the MRSAR in terms of efficiency and is acceptable in terms of search relevance while the MRSAR model is better in terms of search relevance but is difficult to use in "real time" applications due to its poor efficiency.

Table 4. Average retrieval rate at positions 9, 18, 27, 36, 50 and 100 for the AR NSHP model and the MRSAR model

Model	P9	P18	P27	P36	P50	P100
NSHP (83 classes)	.523	.676	.747	.787	.825	.898
NSHP (112 classes)	.435	.562	.627	.669	.708	.799
MRSAR (112 classes)	.74	.84	.87	.89	.92	.95

Table 5. Parameter extraction time and parameter vector size

Model	Extraction time	Number of parameters
Our causal SAR	< 1 s	4
MRSAR	34 s	15

6 Conclusion

In this paper, we have considered the causal AR model with QP and NSHP neighborhoods. We have briefly defined the model, the parameter estimation scheme and the synthesis process before giving experimental results to show the ability of the model to capture texture content.

We have applied the AR model to CBIR using both the QP and the NSHP versions. Experimental results conducted on the well-known Brodatz database show acceptable effectiveness and very good efficiency by the different versions of the AR model, in particular the non-weighted NSHP version is shown to give the best performance. A comparison is shown with the MRSAR model that shows that the NSHP version is better than the MRSAR model regarding search efficiency while the MRSAR is better in terms of search effectiveness.

Since a perceptual meaning was given to the estimated parameters (measure of randomness/regularity of a texture), the next step we are considering is the development of a user interface that allows flexible formulations of queries by users in order to overcome the severe limitation of almost always using a QBE-like queries in CBIR approaches as well as consider semantic features that can be derived from the estimated parameters since these parameters have perceptual meaning.

References

1. Abbadeni, N.: Perceptual interpretation of the estimated parameters of the autoregressive model. Proceedings of the IEEE ICIP(3), 1164–1167 (2005)
2. Abbadeni, N.: A new similarity matching measure: application to texture-based image retrieval. In: Proceedings of the 3rd International Workshop on Texture Analysis and Synthesis, pp. 1–6 (2003)
3. Abbadeni N.: Recherche d'images basée sur leur contenu. Représentation de la texture par le modèle autorégressif. Research report (in French) No. 216, University of Sherbrooke (1998)
4. Del Bimbo, A.: Visual information retrieval. Morgan Kaufmann Publishers, San Francisco (1999)
5. Brodatz, P.: Textures: A Photographic Album for Artists and Designers, Dover, New York (1966)
6. Dunlop, M.D.: Time, relevance and interaction modeling for information retrieval. In: Proceedings of the International ACM SIGIR Conference, pp. 206–213. ACM Press, New York (1997)
7. Frankot, R.T., Chellappa, R.: Lognormal random-field models and their applications to radar image synthesis. IEEE Transactions on Geoscience and Remote Sensing 25(2) (1987)
8. Gower, J.C.: A general coefficient of similarity and some of its properties. Biometrics Journal 27, 857–874 (1971)
9. Kashyap, R.L., Chellappa, R.: Estimation and choice of neighbors in spatial interaction models of images. IEEE Transactions on Information Theory 29(1), 60–72 (1983)

10. Liu, F., Picard, R.W.: Periodicity, directionality and randomness: Wold features for image modeling and retrieval. IEEE Transactions on Pattern Analysis and Machine Intelligence 18(7), 722–733 (1996)
11. Manjunath, B.S., Ma, W.Y.: Texture features for browsing and retrieval of image data. IEEE Transactions on Pattern Analysis and Machine Intelligence 18(8), 837–842 (1996)
12. Mao, J., Jain, A.K.: Texture Classification and Segmentation Using Multiresolution Simultaneous Autoregressive Models. Pattern Recognition 25(2), 173–188 (1992)
13. Press, W.H., Teukolsky, S.A., Vitterling, W.T., Flannery, B.P.: Numerical recipes in C. Cambridge University Press, Cambridge (1992)
14. Randen T., Husoy J.H.: Least squares image texture analysis and synthesis. Working papers from Hogskolen i Stavanger University, Norway (1994)
15. Solberg A.H.S., Jain A.K.: Texture analysis of SAR images: a comparative study. Research Report, Norwegian Computing Center and Michigan State University (1997)
16. Sondge M.: Synthèse de Textures. Ph.D Thesis, Université Pierre et Marie Curie (Paris VI), Paris (1983)
17. Tuceryan, M., Jain, A.K.: Texture analysis. In: Chen, C.H, Pau, L.F, Wang, P.S.P (eds.) Handbook of Pattern Recognition and Computer Vision, World Scientific (1993)
18. Weber, R., Schek, H.-J., Blott, S.: A Quantitative Analysis and Performance Study for Similarity Search Methods. In: High Dimensional Spaces. Proceedings of the 24th Very Large Database Conference, pp. 194–205 (1998)

An Approach Based on Multiple Representations and Multiple Queries for Invariant Image Retrieval

Noureddine Abbadeni

Abu Dhabi University
College of Engineering and Computer Science
PO Box 59911, Abu Dhabi, UAE
noureddine.abbadeni@usherbrooke.ca

Abstract. In this paper, we present a multiple representations and multiple queries approach to tackle the problem of invariance in the framework of content-based image retrieval (CBIR), especially in the case of texture. This approach, rather than considering invariance at the representation level, considers it at the query level. We use two models to represent texture visual content, namely the autoregressive model and a perceptual model based on a set of perceptual features. The perceptual model is used with two viewpoints: the original images viewpoint and the autocovariance function viewpoint. After a brief presentation and discussion of these multiple representation models / viewpoints, which are not invariant with respect to geometric and photometric transformations, we present the invariant texture retrieval algorithm, which is based on multiple models / viewpoints and multiple queries approach and consists in two levels of results fusion (merging): 1. The first level consists in merging results returned by the different models / viewpoints (representations) for the same query in one results list using a linear results fusion model; 2. The second level consists in merging each fused list of different queries into a unique fused list using a round robin fusion scheme. Experimentations show promising results.

Keywords: Image Retrieval, Invariance, Multiple Representations, Multiple Queries, Fusion.

1 Introduction

Invariance, a difficult problem in computer vision and image analysis, concerns, in the framework of image retrieval, the ability to retrieve all relevant images to a query even if some of them have been transformed according to different geometric and photometric transformations such as rotation, scaling, illumination viewpoint change and contrast change as well as non-rigid transformations [16]. Some works in literature [13] have shown that, when a texture is transformed with some geometric or photometric transformation, human subjects still perceiving it as the same texture. Two images with similar textures must still be similar, for some applications, if we rotate one texture or if we change the contrast of another texture for example. Thus, invariance may be an important problem depending on applications and user's needs.

Several works published in literature deals with the problem of invariance, especially in the case of textures. A good review of such works can be found in [16] and a comparison of some rotation-invariant models can be found in [10]. Invariant texture models can be divided into two main classes [13]: 1. The first class of invariant models consider invariance at the representation level, that is the model is invariant with respect to some transformations; 2. The second class of invariant models consider the invariance many representations or prototypes, for the same image, with different orientations, scales, contrasts, etc.

The work presented in this paper fits within the second approach. We use multiple models / viewpoints to represent textural content of images. At the query level, we use multiple queries to represent a user's need. We use appropriate results fusion models to merge results returned by each model/viewpoint and by each query. Results fusion is articulated into two successive levels: The first level consists in merging results returned by different representations for the same query; 2. The second level consists in merging results returned by each of the multiple queries considered.

The rest of the paper is organized as follows: In section 2, the content representation models / viewpoints are briefly presented and discussed; In section 3, the multiple queries approach is presented and discussed; In section 4, results fusion strategies are presented and discussed; In section 5, The resulting new invariant texture retrieval algorithm is applied to a large texture database and benchmarking in terms of retrieval effectiveness based on the recall (retrieval) measure is presented; And, finally, in section 6, a conclusion and further investigations related to this work are briefly discussed.

2 Multiple Representations

To represent texture content, we use two different models, the autoregressive model and a perceptual model based on a set of perceptual features [5], [2]. The autoregressive used is a causal simultaneous autoregressive model with a non-symmetric half-plan (NSHP) neighborhood with four neighbors. The perceptual model is considered with two viewpoints: the original images viewpoint and the autocovariance function (associated with original images) viewpoint. Each of the viewpoints of the perceptual model used is based on four perceptual features, namely coarseness, directionality, contrast and busyness. So, we have a total of three models / viewpoints, each results in a vector of parameters of size four for a total of twelve parameters.

The features of the perceptual model have a perceptual meaning by construction. The features derived from the autoregressive model have no perceptual meaning by construction; however, we have proposed in [2] a perceptual interpretation of the set of features derived from the autoregressive model. This perceptual interpretation consists in considering those features as a measure of the randomness / regularity of the texture.

These models and viewpoints were not built to be invariant to geometric and photometric transformations. The autoregressive model and the the perceptual model based on the autocovariance function viewpoint have poor performance in the case of invariance due to the fact that both the autocovariance function and the autoregressive model, which is also based on covariance, are very sensitive to variance and transformations that may occur in a texture. The perceptual model based on the original images viewpoint performs well in the case of invariance even if it was not constructed to deal with the problem of invariance. In fact, the measure of directionality proposed is not very sensitive to rotation since we compute a degree of directionality and not a specific orientation. Directionality is, however, sensitive to scaling since it is computed as a number of oriented pixels: the more a texture is coarse, the less is the number of oriented pixels and the less the texture is directional. The measure of coarseness proposed is robust with respect to rotation but it is sensitive to scaling. These same remarks given on coarseness are applicable on busyness. The measure of contrast proposed, obviously, is sensitive to illumination conditions. For more details on the perceptual model, refer to [5], [2] and for more details on the autoregressive model, refer to [1], [2].

3 Multiple Queries

In the framework of content-based image retrieval, we propose a multiple queries approach to tackle the problem of texture invariance. In fact, rather than to try to develop invariant content representation models, which may be very complex and of limited scope, consider the problem of invariance at the query level. Each user's need is represented by a set of query images that are different from each other in terms of orientation, scale, contrast, etc., depending on the user's needs. Search results returned for each query are fused to form a unique results list using appropriate results fusion models. For example, if a user wants to retrieve all textures that are visually similar to a given texture, we can represent this user's need by a set of query images with different orientation(s) and fuse the results returned for each query image to form a unique results list.

Results fusion returned for different queries allows to consider the fact that relevant images to a given query are not necessarily near the query in the feature space, especially in the case of invariance. In fact, since content representation models are not, in general, invariant with respect to different transformations, features computed on the same image which has been rotated, for example, are not the same as the original image and can be very different. So, such features are not in the neighborhood of the query image in the feature space and can be even located in disjoint regions that can be more or less far from each other in the feature space [9]. The use of only one query image will not retrieve, certainly, all the relevant images, according to the degree of variance in the database, since, in this case, these relevant images are not necessarily near the considered query image in the feature space. With this approach, there is computation overhead since we use several queries for the same user's need. However, the size of the vector of parameters does not change for images in the database. Only a user's need is represented with several queries, and thus with several vector of parameters. Thus, efficiency is not altered in an important way. Comparatively to the other approach of tackling invariance at the representation level, and besides its difficulty and its limited scope to handle a wide range of transformations, the number of parameters (features) is generally increased in an important way and this results generally in a notable loss of efficiency.

4 Results Fusion

Results fusion concerns two levels: 1. Results fusion from multiple representations (models / viewpoints) for the same query; 2. Results fusion from different queries to produce one final fused list of results.

4.1 Multiple Representations Fusion (for the Same Query)

Each query is represented by three different sets of parameters corresponding to the three different models / viewpoints. Results fusion returned by each of these three models / viewpoints, for the same query, can be done using appropriate fusion models. We have experimented many results fusion models. The model which gave the best results is the model, denoted Fus_{CL}, and defined as follows:

$$Fus_{CL_{ij}} = \frac{\sum_{k=1}^{K} GS_{M_{ij}^k}}{K} \quad (1)$$

Equation (1) is based on the similarity value returned by the similarity measure $GS_{M_{ij}^k}$ and expresses the merging of results returned by different models / viewpoints M^k as an average of the scores obtained by an image in its different rankings corresponding to different models / viewpoints. K is the number of models / viewpoints considered, i is the query image and j corresponds to images returned for query i according to model/viewpoint M^k. Note that the similarity measure used is based on the Gower coefficient of similarity we have developed in our earlier work [1], [3], [4].

The Fus_{CL} model exploits two main effects: 1. The first effect known as-the chorus effect in the information retrieval community [15]: when an image is returned as relevant to a query by several models / viewpoints, this is a more stronger evidence of relevance than if it is returned by only one model / viewpoint; 2. The second effect called the dark horse effect [15]: when a model/viewpoint ranks exceptionally an image, which is not relevant to a query, in top positions, this can be attenuated by the fused model if the other models / viewpoints do not rank it in top positions (it is very rare for a non-relevant image to be ranked at top positions by several models / viewpoints).

4.2 Multiple Queries Fusion

We can use the same fusion models as in the case of multiple models / viewpoints fusion. However, those models did not give good results in experimental results since the chorus and dark horse effects are not very significant in the case of invariance. In fact, when using multiple queries, each with a different orientation, scale, or contrast for example, the results returned for each query contain, actually, a small number of common images since the content representation models are not invariant. Models that can be used, in this case, are those who are able to take, from each results list (for each query), the best results. This effect is called skimming effect in the information retrieval community. Among such models, we define the MAX model, denoted Fus_{MAX}, as follows:

$$Fus_{MAX_{ij}} = MAX(GS_{M_{ij}^k}) \qquad (2)$$

The MAX model consists in choosing, for the same image ranked in different lists, the one having the highest similarity value. Another fusion model, exploiting also the skimming effect, that can be used is a model called round robin (RR) in the literature [7], [8]. The RR technique makes use of the rank of an image in the returned results list rather than the value of the similarity function. The RR technique consists simply in taking images that are ranked in the first position in each list (corresponding to each query) and give them all the same first position in the fused list, then taking images that are ranked in the second position in each list, corresponding to each query, and give them all the same second position in the fused list and so on. We can stop this process after a threshold of the similarity value and/or the rank or after having retrieved a certain number of images. Obviously, if we find the same image in different lists, which may occur occasionally, only one image is considered.

Note that the RR technique is different from the MAX model. The RR technique exploits the skimming effect in a more effective way than the MAX model does. The RR technique, as mentioned, does not make use of similarity values. If necessary, we can use the rank given to images in the fused list and compute an artificial similarity value for each image in the fused list. To do so, we can use the rank-based similarity measure as defined in [12].

5 Application to Image Retrieval

5.1 Invariant Texture Retrieval Algorithm

A general scheme of the invariant texture retrieval algorithm proposed is given in Fig. 1. The algorithm is based on two levels of results fusion as we have already mentioned. The first level consists in fusing results returned by different models / viewpoints for the same query. The second level consists in fusing results returned for different queries. Multiple models / viewpoints fusion, for the same query, is based on fusion models exploiting the chorus and dark horse effects such as the Fus_{CL} (or simply CL) model while multiple queries fusion is based on fusion models exploiting the skimming effect such as Fus_{MAX} (or simply MAX) and Fus_{RR} (or simply RR) models.

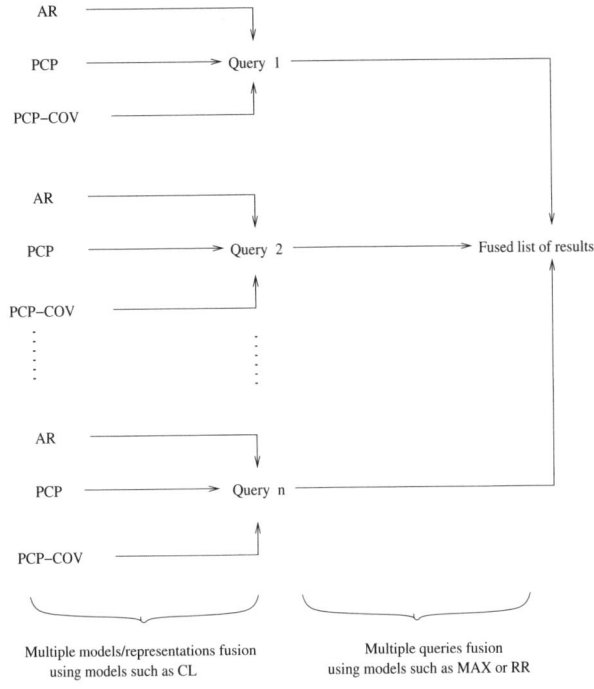

Fig. 1. General scheme of our invariant texture retrieval algorithm

5.2 Benchmarking Database

For experimental results and benchmarking, we have used an image database coming from Ponce's group at UIUC.[1] We considered 22 classes and 40 640x480 images per class for a total of 880 images. Images within the same class have been taken with different photometric and geometric conditions. In each class there is a high degree of variance between images in orientation, scale, contrast as well as non-rigid deformations. In experimental results, the AR, with an NSHP neighborhood, weighted with the inverse of each feature's variance gave the best results among the different versions of the AR models; Both the PCP and the PCP-COV models in their weighted version using Spearman coefficient of rank-correlation gave the best results among the different versions of the perceptual model.

In the rest of this paper, we consider these three best models corresponding to different models / viewpoints. We will use the following notations:

- **AR** or **AR NSHP-V:** The autoregressive model weighted with the inverse of variance.
- **PCP** or **PCP-S:** The perceptual model based on the original images viewpoint and weighted with Spearman coefficients of rank-correlation.

---

[1] http://www-cvr.ai.uiuc.edu/ponce_grp/data/texture_database

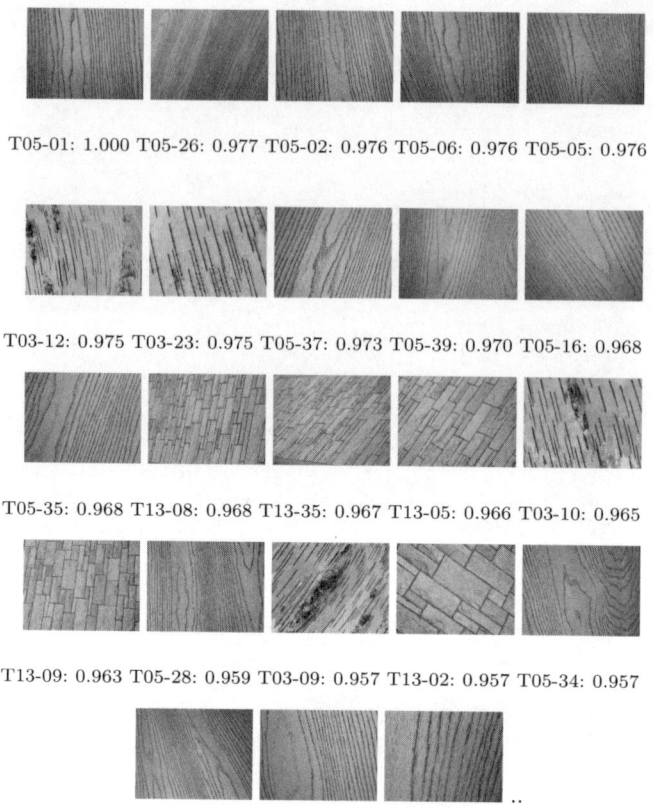

Fig. 2. Search results returned for query T05-01 using the PCP model weighted with Spearman coefficients

- **PCP-COV** or **PCP-COV-S:** The perceptual model based on the autocovariance function viewpoint and weighted with Spearman coefficients of rank-correlation.

5.3 Experimental Results

Fig. (2) shows an example of results obtained with the PCP model weighted with Spearman coefficient of rank-correlation. When examining these results, we can point out that the PCP model, with only four parameters, gave relatively good results even if it was not constructed originally to be invariant with respect to some transformations. The other models, the autoregressive model, and the PCP-COV (perceptual model based on the autocovariance function viewpoint) have less good performance.

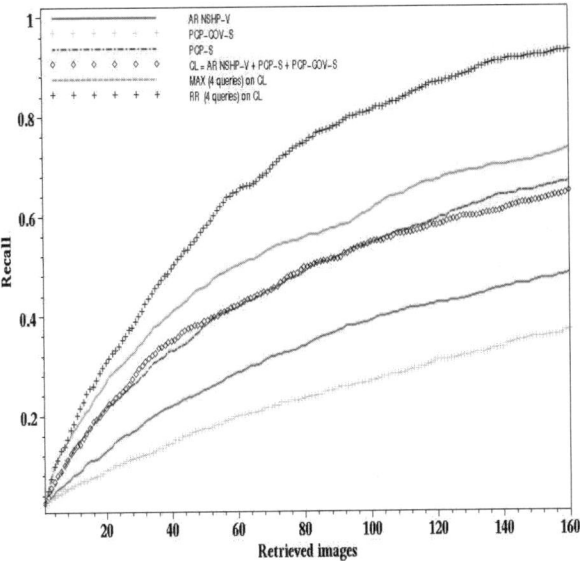

Fig. 3. Average recall graph for different separated models (1 query) and fused model (4 queries)

5.4 Recall Measure

Benchmarking we have done, based on the recall measure, concerns both the use of one query as well as the use of multiple queries:

- When considering one query, we have considered the best version of each separated model/viewpoint. Then we have fused these multiple models / viewpoints, using the CL results fusion model, for the same query.
- When considering multiple queries, based on the fused models / viewpoints for each query, we have fused the results of 4 queries using the MAX model and the RR model as described earlier in this paper. Query images were selected randomly and correspond to images 1, 11, 21 and 31 from each class.

Table 1. Average recall rate at positions 40, 80, 120, and 160 according to different separated and fused models

Model	P40	P80	P120	P160
AR NSHP-V	.219	.34	.426	.485
PCP-S	.331	.489	.593	.67
pcp-COV-S	.145	.234	.31	.37
MAX (4 Queries)	.408	.557	.674	.735
RR (4 Queries)	.495	.747	.867	.932

Fig. 3 shows the average recall graph for different models, both separated and fused, by considering one query and multiple queries. This figure shows that the PCP model weighted with Spearman coefficients performs better than the other separated models. Fusing different models / viewpoints for the same query does not achieve an important improvement in performance since two of the three models / viewpoints consider, namely the AR NSHP-V and the PCP-COV model, have rather poor performance in the case of invariance as explained earlier in this paper. Fusing multiple queries using both the MAX and RR models allow improvement in performance. While the MAX model allows an average improvement in performance, the RR model allows a significant improvement in performance measured in terms of recall. Remember that both the MAX and the RR models exploit the skimming effect while the CL model exploits both the chorus and dark horse effects. Thus, in the case of invariance, the skimming effect, in a multiple queries approach, is more important than both the chorus and dark horse effects. These conclusions can be also drawn by examining table 1, which gives the average retrieval rate at positions 40, 80, 120 and 160 across all the database according to a selection of separated and fused models.

6 Conclusion

This paper presented a multiple models / viewpoints and multiple queries approach to tackle the problem of invariant texture retrieval. Three content representation models / viewpoints were used: the autoregressive model, a perceptual model, based on a set of perceptual features such as coarseness and directionality, used with two viewpoints (the original images viewpoint and the autocovariance function viewpoint). Each of these models / viewpoints results in a parameters vector of size four. These models / viewpoints were not built to be invariant with respect to geometric and photometric transformations. Experimental results and benchmarking conducted on a large database, using the recall measure, show that, when fusing the results returned with each model/viewpoint for the same query, using the CL fusion model, retrieval effectiveness are not significantly improved and are quite similar to the case when we use only the perceptual model based on the original images viewpoint. When fusing results returned by multiple queries, using both the MAX and RR fusion models, retrieval effectiveness are significantly improved especially with the RR fusion model.

The choice of appropriate queries is an open question. In this paper, we have chosen multiple queries in a random way. We believe that, if this choice can be done using some procedure taking into account user's needs, search effectiveness can be improved further.

Aknowledgments. We are thankful to Ponce's group at the University of Illinois at Urbana-Champaign for making available the texture database we used in experimentations presented in this paper.

References

1. Abbadeni, N.: Multiple representations, similarity matching, and results fusion for CBIR. ACM/Springer-Verlag Multimedia Syst. 10(5), 444–456 (2005)
2. Abbadeni, N.: Perceptual interpretation of the estimated parameters of the autoregressive model. Proceedings of the IEEE ICIP(3), 1164–1167 (2005)
3. Abbadeni, N.: Content representation and similarity matching for texture-based image retrieval. In: Proceedings of the 5th ACM SIGMM International Workshop on Multimedia Information Retrieval, pp. 63–70. ACM Press, New York (2003)
4. Abbadeni, N.: A new similarity matching measure: application to texture-based image retrieval. In: Proceedings of the 3rd International Workshop on Texture Analysis and Synthesis, pp. 1–6 (2003)
5. Abbadeni, N., Ziou, D., Wang, S.: Computational measures corresponding to perceptual textural features. In: Proceedings of the 7th IEEE International Conference on Image Processing, IEEE Computer Society Press, Los Alamitos (2000)
6. Abbadeni, N., Ziou, D., Wang, S.: Autocovariance-based perceptual textural features corresponding to human visual perception. In: Proceedings of the 15th IAPR/IEEE International Conference on Pattern Recognition, pp. 3913–3916. IEEE Computer Society Press, Los Alamitos (2000)
7. Belkin, N.J., Cool, C., Croft, W.B., Callan, J.P.: The effect of multiple query representation on information retrieval performance. In: Proceedings of the 16th International ACM SIGIR Conference, pp. 339–346. ACM Press, New York (1993)
8. Berretti, S., Del Bimbo, A., Pala, P.: Merging results for distributed content-based image retrieval. Multimedia Tools and Applications 24, 215–232 (2004)
9. French, J.C., Chapin, A.C., Martin, W.N.: An application of multiple viewpoints to content-based image retrieval. In: Proceeding of the ACM/IEEE Joint Conference on Digital Libraries, pp. 128–130. IEEE Computer Society Press, Los Alamitos (2003)
10. Fountain, S.R., Tan, T.N., Baker, K.D.: A comparative study of rotation invariant classification and retrieval of texture images. In: Proceeding of the British Machine Vision Conference (1998)
11. Jin, X., French, J.C.: Improving Image Retrieval Effectiveness via Multiple Queries. In: ACM Multimedia Databases Workshop, pp. 86–93. ACM Press, New York (2003)
12. Lee, J.H.: Analysis of multiple evidence combination. In: Proceedings of the ACM SIGIR Conference, pp. 267–276 (1997)
13. Lazebnik, S., Schmid, C., Ponce, J.: A Sparse Texture Representation Using Local Affine Regions. Beckman CVR Technical Report, No. 2004-01, University of Illinois at Urbana Champaign (2004)
14. Tuceryan, M., Jain, A.K.: Texture analysis. In: Chen, C.H, Pau, L.F, Wang, P.S.P (eds.) Handbook of Pattern Recognition and Computer Vision, World Scientific (1993)
15. Vogt, C.C., Cottrell, G.W.: Fusion via a linear combination of scores. Information Retrieval Journal 1, 151–173 (1999)
16. Zhang, J., Tan, T.: Brief Review of Invariant Texture Analysis Methods. Pattern Recognition 35, 735–747 (2002)

Author Index

Abbadeni, Noureddine 559, 570
Abolghasemi, Vahid 468
Ahmadyfard, Alireza 468
Armitage, Hugh 502

Badii, Atta 379
Bai, Li 165
Ballan, Lamberto 105
Bao, Shu-Xin 453
Bertini, Marco 105
Beşdok, Erkan 438
Bhattacharya, Prabir 154, 216

Cao, Yiwei 512
Chen, Quqing 205, 322
Chen, Sheng-Bo 453
Chen, Zhibo 205, 322
Chuang, Shun C. 146
Costello, Fintan 391
Crow, Brandon 185

Dai, Zhijun 73
De Witte, Valerie 26
Defée, Irek 49
Del Bimbo, Alberto 105
Del Fatto, Vincenzo 537
Dong, Le 17

Ebright, Patricia 275
Eertink, Henk 490

Fachry, Khairun 490
Fan, Jianping 254
Fang, Shiaofen 275
Fu, Hsin C. 146
Fu, Xianping 223

Gerace, Ivan 242
Gu, Linxia 461
Gu, Xiaodong 205, 322
Guo, Yandong 205, 322
Guo, Yue-Fei 84, 117

Han, Ki-Joon 333
Hao, Jie 344

He, Xiangjian 478
Hong, Dong-Suk 333
Hope, Brian A. 478, 549
Huang, Mao Lin 344, 415

Issa, Raja R.A. 461
Izquierdo, Ebroul 17

Janse, Maddy 490
Ji, Zhen 165
Jia, Wenjing 478
Jo, Hyungeun 358

Kamata, Sei-ichiro 61
Kang, Hong-Koo 333
Kerre, Etienne E. 26
Kim, Joung-Joon 333
Klamma, Ralf 512
Kong, Xiaodong 136

Laaksonen, Jorma 93
Laurini, Robert 537
Leng, Yan 512
Leung, Clement 298
Leung, Maylor K.H. 175
Lew, Michael 1
Lewis, Paul H. 126
Li, Dongguang 445
Li, Fang 175
Li, Hao 117
Li, Qin 154
Li, Shijin 117
Li, Yuanxiang 426
Li, Yun 193
Liang, Dequn 223
Lim, Chang-young 358
Lin, Yung-Chuan 230
Liu, Jiming 298
Liu, Yuee 310
Lopez, Karla 537
Loreto, Rosalva 537
Lu, Bao-Liang 193
Lu, Hong 38, 84, 403
Luo, Hangzai 254
Luo, Qingshan 136

Ma, Jie 84
Mastroleo, Marcello 242
Meade, John 391
Mélange, Tom 26
Milani, Alfredo 242
Milleret-Raffort, Françoise 537
Montoya, Higinio Ariel 185
Moraglia, Simona 242
Mulder, Ingrid 490

Nachtegael, Mike 26
Naghdy, Fazel 502
Naghdy, Golshah 369
Nguyen, Quang Vinh 415
Nunziati, Walter 105

Orgun, Mehmet A. 285

Pao, Hsiao T. 146
Paolino, Luca 524
Phung, Son Lam 369

Rodhetbhai, Wasara 126
Ryu, Jung-hee 358

Satoh, Shin'ichi 254
Schulte, Stefan 26
Sebillo, Monica 524, 537
Shao, Wenbin 369
Shen, Linlin 165
Shen, Yanran 38
Shen, Zhigang 461
Shu, LihChyun 230
Simoff, Simeon 415
Sluzek, Andrzej 5
Sol-Martinez, David 537
Spaniol, Marc 512
Su, Bolan 403
Sun, Huey-Min 230

Tan, Shirley Z.W. 175
Tian, Li 61
Tien, David 478, 549
Tjondrongoro, Dian 310
Tortora, Genoveffa 524

Viitaniemi, Ville 93
Vitiello, Giuliana 524, 537

Wang, Charles 205, 322
Wang, Dingwen 426
Wang, Dongsheng 223
Wang, Hai 267
Wang, Lei 310
Wang, Shouhong 267
Wong, Hau-San 426
Wood Conroy, Diana 502
Wu, Qiang 478
Wu, Yihong 73

Xiao, Yi 549
Xu, Yeong Y. 146
Xue, Xiangyang 38, 254, 403

Yektaii, Mahdi 216
You, Jane 154
You, Qian 275
Yu, Hui 403

Zeng, Guihua 136
Zhang, Jiqi 426
Zhang, Kang 285, 344
Zhang, Ke-Bing 285
Zhang, Xiaozheng Jane 185
Zhang, Yi-Bo 154
Zhong, Daidi 49
Zhu, Meng 379

Printing: Mercedes-Druck, Berlin
Binding: Stein+Lehmann, Berlin

Lecture Notes in Computer Science

Sublibrary 6: Image Processing, Computer Vision, Pattern Recognition, and Graphics

Vol. 4844: Y. Yagi, S.B. Kang, I.S. Kweon, H. Zha (Eds.), Computer Vision – ACCV 2007, Part II. XXVIII, 915 pages. 2007.

Vol. 4843: Y. Yagi, S.B. Kang, I.S. Kweon, H. Zha (Eds.), Computer Vision – ACCV 2007, Part I. XXVIII, 969 pages. 2007.

Vol. 4842: G. Bebis, R. Boyle, B. Parvin, D. Koracin, N. Paragios, S.-M. Tanveer, T. Ju, Z. Liu, S. Coquillart, C. Cruz-Neira, T. Müller, T. Malzbender (Eds.), Advances in Visual Computing, Part II. XXXIII, 827 pages. 2007.

Vol. 4841: G. Bebis, R. Boyle, B. Parvin, D. Koracin, N. Paragios, S.-M. Tanveer, T. Ju, Z. Liu, S. Coquillart, C. Cruz-Neira, T. Müller, T. Malzbender (Eds.), Advances in Visual Computing, Part I. XXXIII, 831 pages. 2007.

Vol. 4815: A. Ghosh, R.K. De, S.K. Pal (Eds.), Pattern Recognition and Machine Intelligence. XIX, 677 pages. 2007.

Vol. 4814: A. Elgammal, B. Rosenhahn, R. Klette (Eds.), Human Motion – Understanding, Modeling, Capture and Animation. X, 329 pages. 2007.

Vol. 4792: N. Ayache, S. Ourselin, A. Maeder (Eds.), Medical Image Computing and Computer-Assisted Intervention – MICCAI 2007, Part II. XLVI, 988 pages. 2007.

Vol. 4791: N. Ayache, S. Ourselin, A. Maeder (Eds.), Medical Image Computing and Computer-Assisted Intervention – MICCAI 2007, Part I. XLVI, 1012 pages. 2007.

Vol. 4781: G. Qiu, C. Leung, X. Xue, R. Laurini (Eds.), Advances in Visual Information Systems. XIII, 582 pages. 2007.

Vol. 4778: S.K. Zhou, W. Zhao, X. Tang, S. Gong (Eds.), Analysis and Modeling of Faces and Gestures. X, 305 pages. 2007.

Vol. 4756: L. Rueda, D. Mery, J. Kittler (Eds.), Progress in Pattern Recognition, Image Analysis and Applications. XXI, 989 pages. 2007.

Vol. 4738: A. Paiva, R. Prada, R.W. Picard (Eds.), Affective Computing and Intelligent Interaction. XVIII, 781 pages. 2007.

Vol. 4729: F. Mele, G. Ramella, S. Santillo, F. Ventriglia (Eds.), Advances in Brain, Vision, and Artificial Intelligence. XVI, 618 pages. 2007.

Vol. 4713: F.A. Hamprecht, C. Schnörr, B. Jähne (Eds.), Pattern Recognition. XIII, 560 pages. 2007.

Vol. 4679: A.L. Yuille, S.-C. Zhu, D. Cremers, Y. Wang (Eds.), Energy Minimization Methods in Computer Vision and Pattern Recognition. XII, 494 pages. 2007.

Vol. 4678: J. Blanc-Talon, W. Philips, D. Popescu, P. Scheunders (Eds.), Advanced Concepts for Intelligent Vision Systems. XXIII, 1100 pages. 2007.

Vol. 4673: W.G. Kropatsch, M. Kampel, A. Hanbury (Eds.), Computer Analysis of Images and Patterns. XX, 1006 pages. 2007.

Vol. 4642: S.-W. Lee, S.Z. Li (Eds.), Advances in Biometrics. XX, 1216 pages. 2007.

Vol. 4633: M. Kamel, A. Campilho (Eds.), Image Analysis and Recognition. XII, 1312 pages. 2007.

Vol. 4584: N. Karssemeijer, B. Lelieveldt (Eds.), Information Processing in Medical Imaging. XX, 777 pages. 2007.

Vol. 4569: A. Butz, B. Fisher, A. Krüger, P. Olivier, S. Owada (Eds.), Smart Graphics. IX, 237 pages. 2007.

Vol. 4538: F. Escolano, M. Vento (Eds.), Graph-Based Representations in Pattern Recognition. XII, 416 pages. 2007.

Vol. 4522: B.K. Ersbøll, K.S. Pedersen (Eds.), Image Analysis. XVIII, 989 pages. 2007.

Vol. 4485: F. Sgallari, A. Murli, N. Paragios (Eds.), Scale Space and Variational Methods in Computer Vision. XV, 931 pages. 2007.

Vol. 4478: J. Martí, J.M. Benedí, A.M. Mendonça, J. Serrat (Eds.), Pattern Recognition and Image Analysis, Part II. XXVII, 657 pages. 2007.

Vol. 4477: J. Martí, J.M. Benedí, A.M. Mendonça, J. Serrat (Eds.), Pattern Recognition and Image Analysis, Part I. XXVII, 625 pages. 2007.

Vol. 4472: M. Haindl, J. Kittler, F. Roli (Eds.), Multiple Classifier Systems. XI, 524 pages. 2007.

Vol. 4466: F.B. Sachse, G. Seemann (Eds.), Functional Imaging and Modeling of the Heart. XV, 486 pages. 2007.

Vol. 4418: A. Gagalowicz, W. Philips (Eds.), Computer Vision/Computer Graphics Collaboration Techniques. XV, 620 pages. 2007.

Vol. 4417: A. Kerren, A. Ebert, J. Meyer (Eds.), Human-Centered Visualization Environments. XIX, 403 pages. 2007.

Vol. 4391: Y. Stylianou, M. Faundez-Zanuy, A. Esposito (Eds.), Progress in Nonlinear Speech Processing. XII, 269 pages. 2007.

Vol. 4370: P.P. Lévy, B. Le Grand, F. Poulet, M. Soto, L. Darago, L. Toubiana, J.-F. Vibert (Eds.), Pixelization Paradigm. XV, 279 pages. 2007.

Vol. 4358: R. Vidal, A. Heyden, Y. Ma (Eds.), Dynamical Vision. IX, 329 pages. 2007.

Vol. 4338: P.K. Kalra, S. Peleg (Eds.), Computer Vision, Graphics and Image Processing. XV, 965 pages. 2006.

Vol. 4319: L.-W. Chang, W.-N. Lie (Eds.), Advances in Image and Video Technology. XXVI, 1347 pages. 2006.

Vol. 4292: G. Bebis, R. Boyle, B. Parvin, D. Koracin, P. Remagnino, A. Nefian, G. Meenakshisundaram, V. Pascucci, J. Zara, J. Molineros, H. Theisel, T. Malzbender (Eds.), Advances in Visual Computing, Part II. XXXII, 906 pages. 2006.

Vol. 4291: G. Bebis, R. Boyle, B. Parvin, D. Koracin, P. Remagnino, A. Nefian, G. Meenakshisundaram, V. Pascucci, J. Zara, J. Molineros, H. Theisel, T. Malzbender (Eds.), Advances in Visual Computing, Part I. XXXI, 916 pages. 2006.

Vol. 4245: A. Kuba, L.G. Nyúl, K. Palágyi (Eds.), Discrete Geometry for Computer Imagery. XIII, 688 pages. 2006.

Vol. 4241: R.R. Beichel, M. Sonka (Eds.), Computer Vision Approaches to Medical Image Analysis. XI, 262 pages. 2006.

Vol. 4225: J.F. Martínez-Trinidad, J.A. Carrasco Ochoa, J. Kittler (Eds.), Progress in Pattern Recognition, Image Analysis and Applications. XIX, 995 pages. 2006.

Vol. 4191: R. Larsen, M. Nielsen, J. Sporring (Eds.), Medical Image Computing and Computer-Assisted Intervention – MICCAI 2006, Part II. XXXVIII, 981 pages. 2006.

Vol. 4190: R. Larsen, M. Nielsen, J. Sporring (Eds.), Medical Image Computing and Computer-Assisted Intervention – MICCAI 2006, Part I. XXXVVIII, 949 pages. 2006.

Vol. 4179: J. Blanc-Talon, W. Philips, D. Popescu, P. Scheunders (Eds.), Advanced Concepts for Intelligent Vision Systems. XXIV, 1224 pages. 2006.

Vol. 4174: K. Franke, K.-R. Müller, B. Nickolay, R. Schäfer (Eds.), Pattern Recognition. XX, 773 pages. 2006.

Vol. 4170: J. Ponce, M. Hebert, C. Schmid, A. Zisserman (Eds.), Toward Category-Level Object Recognition. XI, 618 pages. 2006.

Vol. 4153: N. Zheng, X. Jiang, X. Lan (Eds.), Advances in Machine Vision, Image Processing, and Pattern Analysis. XIII, 506 pages. 2006.

Vol. 4142: A. Campilho, M. Kamel (Eds.), Image Analysis and Recognition, Part II. XXVII, 923 pages. 2006.

Vol. 4141: A. Campilho, M. Kamel (Eds.), Image Analysis and Recognition, Part I. XXVIII, 939 pages. 2006.

Vol. 4122: R. Stiefelhagen, J.S. Garofolo (Eds.), Multimodal Technologies for Perception of Humans. XII, 360 pages. 2007.

Vol. 4109: D.-Y. Yeung, J.T. Kwok, A. Fred, F. Roli, D. de Ridder (Eds.), Structural, Syntactic, and Statistical Pattern Recognition. XXI, 939 pages. 2006.

Vol. 4091: G.-Z. Yang, T. Jiang, D. Shen, L. Gu, J. Yang (Eds.), Medical Imaging and Augmented Reality. XIII, 399 pages. 2006.

Vol. 4073: A. Butz, B. Fisher, A. Krüger, P. Olivier (Eds.), Smart Graphics. XI, 263 pages. 2006.

Vol. 4069: F.J. Perales, R.B. Fisher (Eds.), Articulated Motion and Deformable Objects. XV, 526 pages. 2006.

Vol. 4057: J.P.W. Pluim, B. Likar, F.A. Gerritsen (Eds.), Biomedical Image Registration. XII, 324 pages. 2006.

Vol. 4046: S.M. Astley, M. Brady, C. Rose, R. Zwiggelaar (Eds.), Digital Mammography. XVI, 654 pages. 2006.

Vol. 4040: R. Reulke, U. Eckardt, B. Flach, U. Knauer, K. Polthier (Eds.), Combinatorial Image Analysis. XII, 482 pages. 2006.

Vol. 4035: T. Nishita, Q. Peng, H.-P. Seidel (Eds.), Advances in Computer Graphics. XX, 771 pages. 2006.

Vol. 3979: T.S. Huang, N. Sebe, M. Lew, V. Pavlović, M. Kölsch, A. Galata, B. Kisačanin (Eds.), Computer Vision in Human-Computer Interaction. XII, 121 pages. 2006.

Vol. 3954: A. Leonardis, H. Bischof, A. Pinz (Eds.), Computer Vision – ECCV 2006, Part IV. XVII, 613 pages. 2006.

Vol. 3953: A. Leonardis, H. Bischof, A. Pinz (Eds.), Computer Vision – ECCV 2006, Part III. XVII, 649 pages. 2006.

Vol. 3952: A. Leonardis, H. Bischof, A. Pinz (Eds.), Computer Vision – ECCV 2006, Part II. XVII, 661 pages. 2006.

Vol. 3951: A. Leonardis, H. Bischof, A. Pinz (Eds.), Computer Vision – ECCV 2006, Part I. XXXV, 639 pages. 2006.

Vol. 3948: H.I. Christensen, H.-H. Nagel (Eds.), Cognitive Vision Systems. VIII, 367 pages. 2006.

Vol. 3926: W. Liu, J. Lladós (Eds.), Graphics Recognition. XII, 428 pages. 2006.

Vol. 3872: H. Bunke, A.L. Spitz (Eds.), Document Analysis Systems VII. XIII, 630 pages. 2006.

Vol. 3852: P.J. Narayanan, S.K. Nayar, H.-Y. Shum (Eds.), Computer Vision – ACCV 2006, Part II. XXXI, 977 pages. 2006.

Vol. 3851: P.J. Narayanan, S.K. Nayar, H.-Y. Shum (Eds.), Computer Vision – ACCV 2006, Part I. XXXI, 973 pages. 2006.

Vol. 3832: D. Zhang, A.K. Jain (Eds.), Advances in Biometrics. XX, 796 pages. 2005.

Vol. 3736: S. Bres, R. Laurini (Eds.), Visual Information and Information Systems. XI, 291 pages. 2006.

Vol. 3667: W.J. MacLean (Ed.), Spatial Coherence for Visual Motion Analysis. IX, 141 pages. 2006.

Vol. 3417: B. Jähne, R. Mester, E. Barth, H. Scharr (Eds.), Complex Motion. X, 235 pages. 2007.

Vol. 2396: T.M. Caelli, A. Amin, R.P.W. Duin, M.S. Kamel, D. de Ridder (Eds.), Structural, Syntactic, and Statistical Pattern Recognition. XVI, 863 pages. 2002.

Vol. 1679: C. Taylor, A. Colchester (Eds.), Medical Image Computing and Computer-Assisted Intervention – MICCAI'99. XXI, 1240 pages. 1999.